Lipids in Photosynthesis:
Structure, Function and Genetics

Advances in Photosynthesis

VOLUME 6

Series Editor:

GOVINDJEE
University of Illinois, Urbana, Illinois, U.S.A.

Consulting Editors:

Jan AMESZ, *Leiden, The Netherlands*
Eva-Mari ARO, *Turku, Finland*
James BARBER, *London, United Kingdom*
Robert E. BLANKENSHIP, *Tempe, Arizona, U.S.A.*
Norio MURATA, *Okazaki, Japan*
Donald R. ORT, *Urbana, Illinois, U.S.A.*

Advances in Photosynthesis is an ambitious book series seeking to provide a comprehensive and state-of-the-art account of photosynthesis research. Photosynthesis is the process by which higher plants, algae and certain species of bacteria transform and store solar energy in the form of energy-rich organic molecules. These compounds are in turn used as the energy source for all growth and reproduction in these organisms. As such, virtually all life on the planet ultimately depends on photosynthetic energy conversion. This series of multiauthored books spans topics from physics to agronomy, from femtosecond reactions to season long production, from the photophysics of reaction centers to the physiology of whole organisms, and from X-ray crystallography of proteins to the morphology of intact plants. The intent of this series of publications is to offer beginning researchers, graduate students, and even research specialists a comprehensive current picture of the remarkable advances across the full scope of photosynthesis research.

The titles to be published in this series are listed on the backcover of this volume.

Lipids in Photosynthesis: Structure, Function and Genetics

Edited by

Paul-André Siegenthaler
Laboratoire de Physiologie Végétale,
Université de Neuchâtel,
Neuchâtel, Switzerland

and

Norio Murata
National Institute for Basic Biology,
Myodaiji, Okazaki, Japan

KLUWER ACADEMIC PUBLISHERS
DORDRECHT / BOSTON / LONDON

A C.I.P. Catalogue record for this book is available from the Library of Congress

ISBN 0-7923-5173-8

Published by Kluwer Academic Publishers,
P.O. Box 17, 3300 AA Dordrecht, The Netherlands.

Sold and distributed in North, Central and South America
by Kluwer Academic Publishers,
101 Philip Drive, Norwell, MA 02061, U.S.A.

In all other countries, sold and distributed
by Kluwer Academic Publishers,
P.O. Box 322, 3300 AH Dordrecht, The Netherlands.

The camera ready text was prepared
by Lawrence A. Orr, Center for the Study of Early Events in Photosynthesis
Arizona State University, Tempe, Arizona 85287-1604, U.S.A.

Printed on acid-free paper

Contents

Preface

Lipids in Photosynthesis: Structure, Function and Genetics is the sixth volume in the series *Advances in Photosynthesis,* published by Kluwer Academic Publishers of the Netherlands, with Govindjee as its Series Editor. This series provides an up-to-date account of all aspects of the process of photosynthesis. The earlier volumes have covered topics such as the Molecular Biology of Cyanobacteria (Vol. 1), Anoxygenic Photosynthetic Bacteria (Vol. 2), Biophysical Techniques in Photosynthesis (Vol. 3), Oxygenic Photosynthesis: The Light Reactions (Vol. 4), and Photosynthesis and the Environment (Vol. 5).

This volume brings together topics that are generally scattered throughout various text books devoted to either lipids or photosynthesis. Photosynthesis, which is considered to be one of the major processes on earth, has been intensively studied by biochemists, biophysicists and physiologists, as well as by agronomists. On the other hand, lipids have often been neglected in the field of photosynthesis research. It is of interest that galactosyldiacylglycerols in chloroplast membranes were discovered only in the 1950s, although it is surely one of the the most abundant natural substances in the biosphere. Since their discovery, investigations have proceeded at an exponentially increasing rate, and much information is now available on the chemical structure and physico-chemical properties of glycerolipids as well as their occurrence and distribution among many genera of bacteria, algae and higher plants. At the same time, considerable work has gone into elucidating the biosynthetic pathways of glycerolipids and fatty acids. More recently, interest in glycerolipids has shifted towards understanding of their roles in the structure and the function of the thylakoid and chloroplast envelope membranes. In addition, the contribution of the methods and concepts of molecular biology has allowed us to answer many questions that had been shelved, in particular, responses of the glycerolipids of thylakoid membranes to environmental stresses.

The introductory chapter (Chapter 1) provides a brief overview of the structure, function and genetics of chloroplast lipids. The reader is guided to appropriate chapters wherein specific topics are covered in greater depth. The editors have also tried to provide cross-references between chapters to help the reader access all the materials on a particular topic.

The other chapters are organized along eight major themes: (1) The structure, composition and distribution of membrane glycerolipids in higher plant chloroplasts (Chapters 2 and 7), in algae (Chapter 3), in cyanobacteria (Chapter 4), and in anoxygenic photosynthetic bacteria (Chapter 5); (2) the physical properties of thylakoid membrane lipids and their relation to photosynthesis (Chapters 6 and 11); (3) the molecular organization and the roles of glycerolipids in photosynthetic membranes of higher plants (Chapters 7–9), (4) the lipid-protein interactions in the import of proteins into chloroplasts (Chapter 10); (5) the development of thylakoid membranes with respect to lipids (Chapter 11); (6) the biosynthetic pathways of lipids in higher-plant chloroplasts (Chapters 2 and 12), in cyanobacteria (Chapter 4) and anoxygenic photosynthetic bacteria (Chapter 5); (7) genetic engineering of the unsaturation of triacylglycerides (Chapter 12) and of membrane glycerolipids and its effects on the ability of the photosynthetic machinery to tolerate temperature stress (Chapter 13), as well as a genetic approach with mutants in investigating the structure and function of membrane lipids (Chapter 14); and finally (8) the involvement of chloroplast lipids in the reaction of plants submitted to stress (Chapter 15).

This book is intended for a wide audience, but is specifically designed for advanced undergraduate and graduate students and for researchers active in the field, as well as those scientists whose specialization include the biochemistry, physiology, molecular biology, biophysics and biotechnology of membrane glycerolipids and triacylglycerides.

The publication of this volume has involved the efforts of a number of people to whom we would like to express our gratitude. Firstly, we thank the authors for their contributions and comprehension in dealing with editorial changes, and also for their patience in waiting for the publication of this book. Secondly, Mr. Larry Orr deserves special thanks for all the advice and editorial information he provided patiently and with good humor to the editors; without his exceptional talents, his open-minded spirit and hard work, this volume could never have been realized. Thirdly, we thank Dr. Govindjee, the Editor-in-Chief of *Advances in Photosynthesis*, for inviting us to edit this book and for placing his trust in our efforts. One of us (PAS) would like to thank his wife, Madeleine, for her patience and encouragement during the preparation of this volume.

Paul-André Siegenthaler
Norio Murata

Chapter 1

Lipids in Photosynthesis: An Overview

Norio Murata
Department of Regulation Biology, National Institute for Basic Biology,
Myodaiji, Okazaki 444, Japan

Paul-André Siegenthaler
Laboratoire de Physiologie Végétale, Université de Neuchâtel,
Rue Emile-Argand 13, CH-2007 Neuchâtel, Switzerland

Summary

The intent of this book is to provide readers with a comprehensive overview of lipid structure, function and genetics of the chloroplast, with special emphasis on the photosynthetic apparatus in chloroplast thylakoid membranes. In this introductory chapter, we provide a brief review of chloroplast lipid structure and function in which we guide the reader to appropriate chapters wherein specific topics are covered in greater depth. The structures of glycerolipids and fatty acids in photosynthetic membranes are presented and details of their biosynthesis in chloroplasts and cyanobacterial cells are described. The enzymes involved in fatty acid and glycerolipid synthesis in chloroplasts and cyanobacteria for which genes or cDNAs have been cloned are also discussed. Finally, we provide a critical review of the various experimental approaches that have been proposed in the literature with the goal of defining the roles of glycerolipids and fatty acids in photosynthesis.

P.-A. Siegenthaler and N. Murata (eds): Lipids in Photosynthesis: Structure, Function and Genetics, pp. 1–20.
© *1998 Kluwer Academic Publishers. Printed in The Netherlands.*

I. Lipids and Fatty Acids in Photosynthetic Membranes

A. Photosynthetic Membranes

The early processes of photosynthesis, such as the absorption of light, photochemical reactions, electron transfer, and the synthesis of ATP, all occur in membranes, namely, the thylakoid membranes in eukaryotic chloroplasts and cyanobacterial cells, and the intracytoplasmic membranes in anoxygenic photosynthetic bacteria. Therefore, the characteristics of these membranes are obviously important for the operation of photosynthesis in photosynthetic organisms.

The main components of photosynthetic membranes are proteins and lipids. The proteins include the Photosystem I complex, the Photosystem II complex, the cytochrome *bf* complex and ATP synthase, as well as a number of peripheral proteins. The major lipids are polar glycerolipids, such as glycolipids and phospholipids. Sterols and sphingolipids, which are abundant in the plasma membrane, the tonoplast and the endoplasmic reticulum, are absent or are only minor components of photosynthetic membranes. The glycerolipids have been studied intensively in terms of their biochemistry, biosynthesis and function, but not much attention has been paid to sterols and sphingolipids and their relationship to photosynthesis.

B. Polar Glycerolipids

The thylakoid membranes of the chloroplasts of higher plants contain three glycolipids, namely, monogalactosyl diacylglycerol (MGDG), digalactosyl diacylglycerol (DGDG) and sulfoquinovosyl diacylglycerol (SQDG), and one phospholipid, phosphatidylglycerol (PG) (see Chapter 2, Joyard et al.). The chemical structures of these components are shown in Fig. 1. It seems likely that phosphatidylcholine (PC), which was initially reported as

---

Abbreviations: ACP – acyl-carrier protein; DGDG - digalactosyl diacylglycerol; GlcDG - monoglucosyl diacylglycerol; MGDG - monogalactosyl diacylglycerol; PC - phosphatidylcholine; PG - phosphatidylglycerol; SQDG - sulfoquinovosyl diacylglycerol. Each fatty acid is represented by $X:Y(Z_1,Z_2,—)$, which indicates that it contains X carbon atoms with Y double bonds in the *cis* configuration (unless otherwise mentioned) at positions $Z_1,Z_2,—$, counted from the carboxyl terminus; *sn* stands for the structural number of the glycerol backbone

a component of thylakoid membranes, is not present in the thylakoid membrane (see Chapters 2, Joyard et al. and 7, Siegenthaler). Among the four glycerolipids, MGDG and DGDG account for about 50% and 30%, respectively, of the total lipids, while SQDG and PG each account for 5–12%.

The inner membrane of the chloroplast envelope contains the same four glycerolipids at about the same relative levels as those in the thylakoid membrane (see Chapter 2, Joyard et al.). However, the glycerolipids in the outer membrane of the chloroplast envelope differ from those in the inner membrane and the thylakoid membrane. The outer membrane contains PC and phosphatidylinositol in addition to MGDG, DGDG and PG. The specific lipids of the chloroplast envelope membranes are involved in the import of cytosolically synthesized proteins (see Chapter 10, de Kruijff et al.).

The thylakoid membranes of lower plants and eukaryotic algae contain betaine lipids, such as diacylglyceryl trimethylhomoserine and/or diacylglyceryltrimethyl-β-alanine (Fig. 1), in addition to MGDG, DGDG, SQDG and PG (see Chapter 3, Harwood). Each molecule of betaine lipid has a positive and a negative charge, resembling PC in this respect.

The thylakoid membranes of cyanobacteria contain three glycolipids, namely, MGDG, DGDG and SQDG, and one phospholipid, PG, as major components (see Chapter 4, Wada and Murata), as do the thylakoid membranes of higher-plant chloroplasts. MGDG accounts for about 50% of the total lipid and the other three lipids each account for 5–20% of the total. The composition of glycerolipids of the plasma membrane is similar to that of the thylakoid membrane (see Chapter 4, Wada and Murata). In addition to these four glycerolipids, monoglucosyl diacylglycerol (GlcDG) is present as a very minor constituent (Sato and Murata, 1982).

The composition of the membrane glycerolipids in anoxygenic photosynthetic bacteria varies among species (see Chapter 5, Benning). However, the major components are phospholipids, such as PC, PE and PG, in clear contrast to the lipids in chloroplasts and cyanobacterial cells. Glycolipids, such as SQDG, and ornithine lipid are present as minor components in some species.

C. Fatty Acids

A glycerolipid molecule is esterified with fatty acids

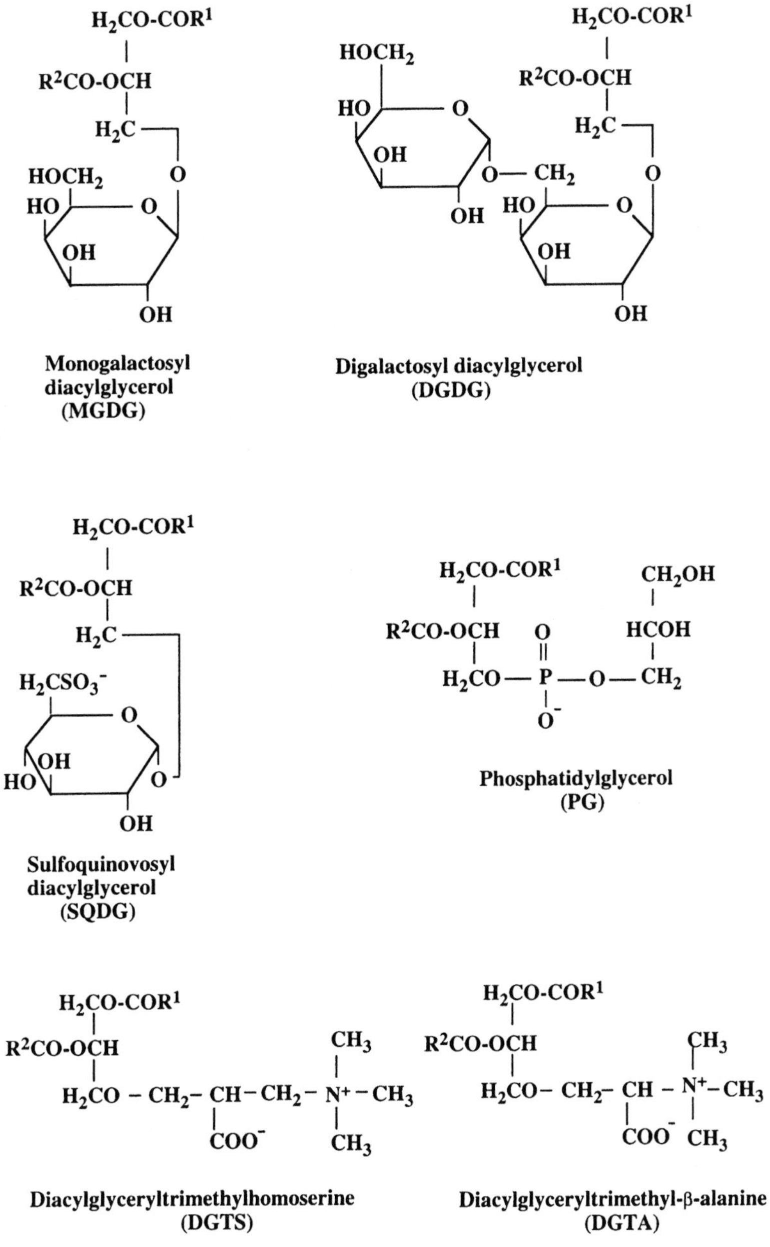

Fig. 1. Structures of the major glycerolipids in thylakoid membranes. R^1 and R^2 are fatty acids that are esterified, respectively, to the *sn-1* and *sn-2* positions of the glycerol backbone. MGDG, DGDG, SQDG and PG are present in thylakoid membranes from plants and cyanobacteria, and betaine lipids (DGTS and DGTA) are additionally present in thylakoid membranes from lower plants and eukaryotic algae.

at the *sn-1* and *sn-2* positions of the glycerol backbone (see Fig. 1). Structures of major fatty acids from plants are shown in Fig. 2. The fatty acids of the glycerolipids of higher-plant chloroplasts are highly unsaturated, and the major fatty acids are α-linolenic acid (α-18:3) in MGDG, and α-18:3 and palmitic acid (16:0) in DGDG, SQDG and PG. In addition,

PG contains Δ3-*trans*-hexadecenoic acid, that is unique in the plant kingdom and found in PG of chloroplast membranes. Certain plants also contain hexadecatrienoic acid [16:3(7,10,13)] in MGDG (see Chapters 2, Joyard et al. and 7, Siegenthaler).

The chloroplast glycerolipids of freshwater algae contain α-18:3 and 16:0 as major fatty acids,

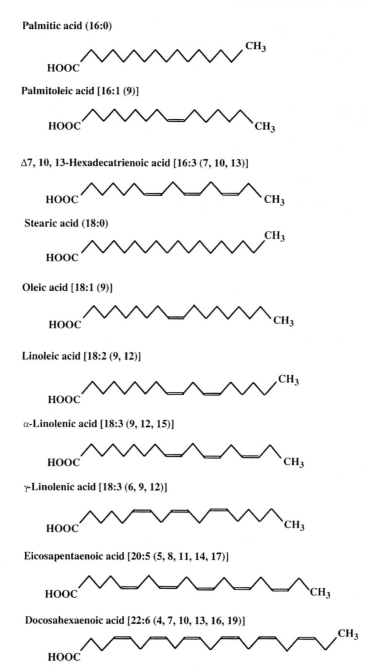

Palmitic acid (16:0)

Palmitoleic acid [16:1 (9)]

Δ7, 10, 13-Hexadecatrienoic acid [16:3 (7, 10, 13)]

Stearic acid (18:0)

Oleic acid [18:1 (9)]

Linoleic acid [18:2 (9, 12)]

α-Linolenic acid [18:3 (9, 12, 15)]

γ-Linolenic acid [18:3 (6, 9, 12)]

Eicosapentaenoic acid [20:5 (5, 8, 11, 14, 17)]

Docosahexaenoic acid [22:6 (4, 7, 10, 13, 16, 19)]

Fig. 2. Structures of major fatty acids esterified to glycerolipids in thylakoid membranes from the chloroplasts of higher plants and algae and from cyanobacterial cells.

as do those of higher plants. By contrast, the chloroplast glycerolipids of marine algae contain, in addition to α-18:3 and 16:0, eicosapentaenoic acid [20:5(5,8,11,14,17)] and docosahexaenoic acid [22:6(4,7,10,13,16,19)] (see Chapter 3, Harwood). The functions of these very-long-chain poly-unsaturated fatty acids in marine algae are not known.

The fatty acid composition of membrane lipids in cyanobacteria exhibits considerable diversity (see Chapter 4, Wada and Murata). Some strains contain a tetraunsaturated fatty acid, whereas some others contain monounsaturated but not polyunsaturated fatty acids. By contrast, anoxygenic photosynthetic bacteria contain saturated and monounsaturated fatty acids (see Chapter 5, Benning).

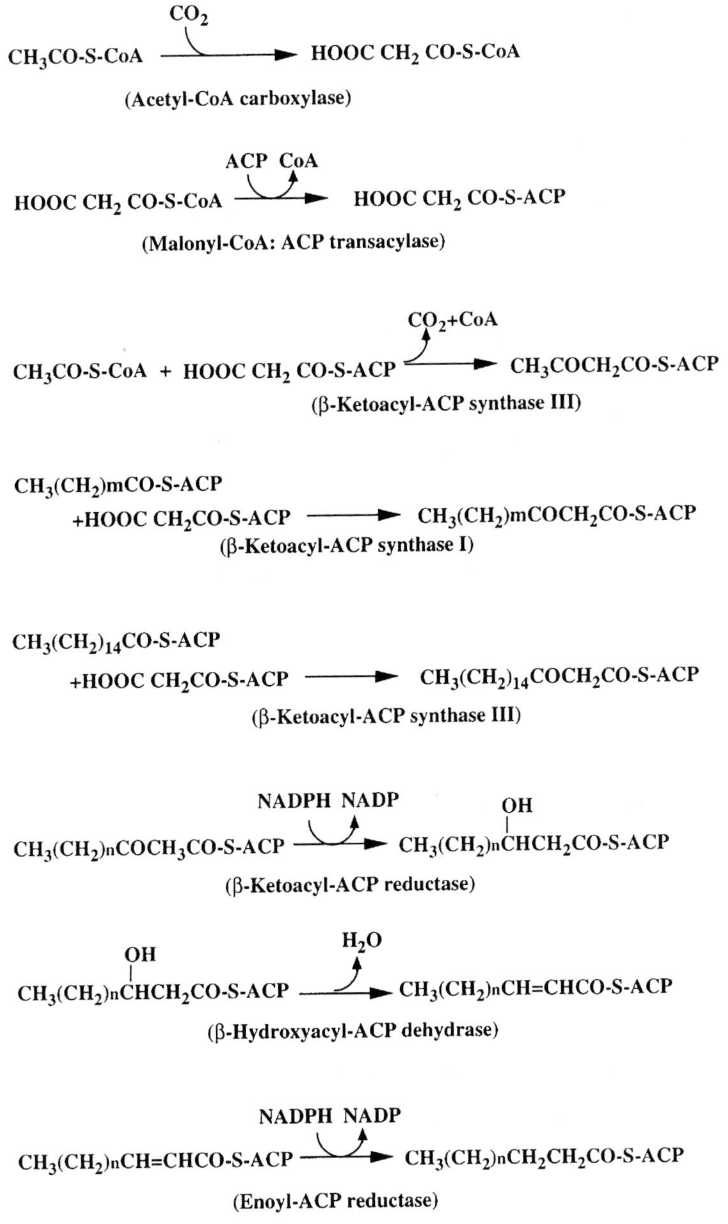

$$CH_3CO\text{-}S\text{-}CoA \xrightarrow{ CO_2 } HOOC\ CH_2\ CO\text{-}S\text{-}CoA$$

(Acetyl-CoA carboxylase)

$$HOOC\ CH_2\ CO\text{-}S\text{-}CoA \xrightarrow{ ACP\ \ CoA } HOOC\ CH_2\ CO\text{-}S\text{-}ACP$$

(Malonyl-CoA: ACP transacylase)

$$CH_3CO\text{-}S\text{-}CoA\ +\ HOOC\ CH_2\ CO\text{-}S\text{-}ACP \xrightarrow{ CO_2+CoA } CH_3COCH_2CO\text{-}S\text{-}ACP$$

(β-Ketoacyl-ACP synthase III)

$$CH_3(CH_2)mCO\text{-}S\text{-}ACP$$
$$+HOOC\ CH_2CO\text{-}S\text{-}ACP \longrightarrow CH_3(CH_2)mCOCH_2CO\text{-}S\text{-}ACP$$

(β-Ketoacyl-ACP synthase I)

$$CH_3(CH_2)_{14}CO\text{-}S\text{-}ACP$$
$$+HOOC\ CH_2\ CO\text{-}S\text{-}ACP \longrightarrow CH_3(CH_2)_{14}COCH_2CO\text{-}S\text{-}ACP$$

(β-Ketoacyl-ACP synthase III)

$$CH_3(CH_2)nCOCH_3CO\text{-}S\text{-}ACP \xrightarrow{ NADPH\ NADP } CH_3(CH_2)n\overset{\displaystyle OH}{C}HCH_2CO\text{-}S\text{-}ACP$$

(β-Ketoacyl-ACP reductase)

$$CH_3(CH_2)n\overset{\displaystyle OH}{C}HCH_2CO\text{-}S\text{-}ACP \xrightarrow{ H_2O } CH_3(CH_2)nCH=CHCO\text{-}S\text{-}ACP$$

(β-Hydroxyacyl-ACP dehydrase)

$$CH_3(CH_2)nCH=CHCO\text{-}S\text{-}ACP \xrightarrow{ NADPH\ NADP } CH_3(CH_2)nCH_2CH_2CO\text{-}S\text{-}ACP$$

(Enoyl-ACP reductase)

Fig. 3. Pathway for synthesis of fatty acids in chloroplasts. ACP, acyl-carrier protein. β-Ketoacyl-ACP synthases I, II and III catalyze the elongation of the carbon chain from C(m+2) to C(m+4), from C16 to C18, and from C2 to C4, respectively. At each step of the elongation, one type of enzyme, such as β-ketoacyl-ACP reductase, β-hydroxyacyl-ACP dehydrase and enoyl-ACP reductase, catalyzes the respective reactions. m=2,4,6,8,10 or 12; n=0,2,4,6,8,10,12 or 14.

II. Biosynthesis of Fatty Acids and Glycero-lipids in Chloroplasts and Cyanobacterial Cells

In plants and eukaryotic algae, fatty acids are synthesized in chloroplasts. The pathway for synthesis of fatty acids in higher-plant chloroplasts has been intensively studied (Fig. 3), and most of the enzymes involved in the pathway have been well characterized. The first step is the synthesis of malonyl-CoA by acetyl-CoA carboxylase. Malonyl-CoA is converted to malonyl-ACP. Elongation of fatty acids is catalyzed by four enzymes: β-ketoacyl-ACP synthase, β-ketoacyl-ACP reductase, β-hydroxyacyl-ACP dehydrase and enoyl-ACP reductase. There are three types of β-ketoacyl-ACP synthase, types I, II and III.

Table 1. Enzymes involved in fatty acid synthesis in chloroplasts and cyanobacteria for which genes or cDNAs have been cloned

Enzyme	Plant or cyanobacterium	Reference
Fatty acid synthesis		
Acetyl-CoA carboxylase		
Carboxyltransferase		
	Pisum sativum	Hirsch and Soll (1995)
	Porphyra purpurea	Reith and Munholland (1995)
	Synechocystis sp. PCC 6803	Kaneko et al. (1996)
Biotin carboxyl carrier protein		
	Anabaena sp. PCC 7120	Gornicki et al. (1993)
	Porphyra purpurea	Reith and Munholland (1995)
	Arabidopsis thaliana	Choi et al. (1995)
	Brassica napus	Elborough et al. (1996)
	Synechocystis sp. PCC 6803	Kaneko et al. (1996)
Biotin carboxylase		
	Anabaena sp. PCC 7120	Gornicki et al. (1993)
	Nicotiana tabacum	Shorrosh et al. (1995)
	Synechocystis sp. PCC 6803	Kaneko et al. (1996)
Transcarboxylase		
	Glycine max	Reverdatto et al. (1995)
	Porphyra purpurea	Reith and Munholland (1995)
	Synechocystis sp. PCC 6803	Kaneko et al. (1996)
	Brassica napus	Elborough et al. (1996)
Malonyl-CoA:ACP malonyltransferase		
	Zea mays	Verwoert et al. (1995)
Fatty acid synthase		
β-Ketoacyl-ACP synthase I		
	Hordeum vulgare	Siggand-Andersen et al. (1991)
	Cuphea lanceolata	Topfer and Martini (1994)
β-Ketoacyl-ACP synthase II		
	Hordeum vulgare	Wissenback et al. (1992)
β-Ketoacyl-ACP synthase III		
	Spinacia oleracea	Tai et al. (1993)
	Arabidopsis thaliana	Tai et al. (1994)
	Cuphea wrightii	Slabaugh et al. (1995)
	Synechocystis sp. PCC 6803	Kaneko et al. (1996)
β-Ketoacyl-ACP reductase		
	Synechocystis sp. PCC 6803	Kaneko et al. (1996)
β-Hydroxyacyl-ACP dehydrase		(not yet cloned)
Enoyl-ACP reductase		
	Brassica napus	Kater et al. (1991)
	Ricinus communis	van de Loo et al. (1995)

The type I and III enzymes specifically catalyze the elongation from C2 to C4 and from C16 to C18, respectively. The type II enzyme catalyzes the elongation from C4 to C16. At each of the elongation steps, the reductive reactions are catalyzed by the reductases and dehydrase mentioned above.

Table 1 lists enzymes involved in the synthesis of fatty acids for which genes have been cloned. Apart from the gene for β-hydroxyacyl-ACP dehydrase, genes for all the other enzymes involved in the

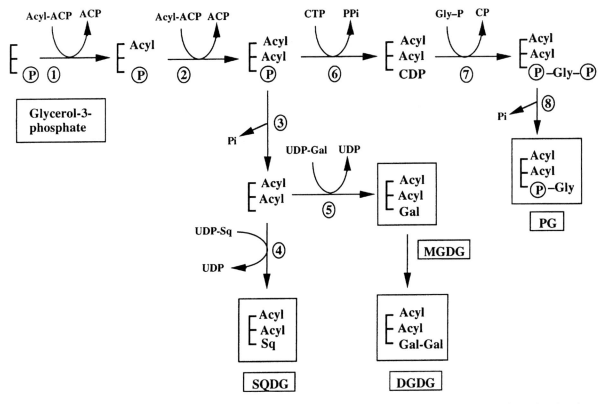

Fig. 4. Pathway for synthesis of glycerolipids in chloroplasts. Enzymes are labeled as follows: ① Acyl-ACP: glycerol-3-phosphate acyltransferase; ② acyl-ACP: *sn*-1-acylglycerol-3-phosphate acyltransferase; ③ phosphatidate phosphatase; ④ UDP-sulfoquinovose: diacylglycerol sulfoquinovosyltransferase (SQDG synthase); ⑤ UDP-galactose: diacylglycerol galactosyltransferase (MGDG synthase); ⑥ phosphatidate cytidyltransferase; ⑦ CDP-diacylglycerol: glycerol-3-phosphate phosphatidyltransferase (phosphatidylglycerophosphate synthase); ⑧ phosphatidylglycerophosphate phosphatase.

synthesis of fatty acids have been cloned (see Chapter 12, Facciotti and Knauf). The gene for trans-carboxylase, which is a subunit of acetyl-CoA carboxylase, is encoded by the chloroplast genome. By contrast, the genes for all the other cloned enzymes are localized within the nuclear genome. However, the chloroplast genome of a red alga includes the genes for three of the four subunits of acetyl-CoA carboxylase (Reith and Munholland, 1995). Enzymes involved in the synthesis of C20 and C22 fatty acids in chloroplasts of lower plants and algae have not yet been characterized.

Glycerolipids are also synthesized in plant chloroplasts (see Chapter 2, Joyard et al.). Figure 4 shows the pathway for synthesis of MGDG, DGDG, SQDG and PG in higher-plant chloroplasts. With the exception of the enzyme that catalyzes the last step in the synthesis of DGDG, all the enzymes have been well characterized. In contrast to the pathway for synthesis of MGDG from diacylglycerol in higher-plant chloroplasts, the synthesis of MGDG involves

two steps in cyanobacteria (see Chapter 4, Wada and Murata). The first step is catalyzed by UDP-glucose:diacylglycerol glucosyltransferase and results in synthesis of GlcDG, which is then converted to MGDG by an unidentified isomerase. Table 2 lists the enzymes involved in the synthesis of glycerolipids in chloroplasts. While the gene for the enzyme that catalyzes the first transfer of an acyl moiety (glycerol-3-phosphate: acyl-ACP acyltransferase) has been cloned from several plants, the gene for the second acyltransferase (*sn*-1-acylglycerol-3-phosphate:acyl-ACP acyltransferase) has not been cloned. Genes for two phosphatases also remain to be cloned.

Saturated fatty acids bound to acyl-carrier protein (ACP) are synthesized in reactions catalyzed by fatty acid synthases (see Chapter 12, Faciotti and Knauf). In plant chloroplasts, stearic acid (18:0) is desaturated, by acyl-ACP desaturase, to oleic acid [18:1(9)], which is then incorporated into glycerolipids. Further desaturation is catalyzed by Δ12 and ω3 (Δ15) acyl-lipid desaturases. Figure 5A shows the pathway for

Table 2. Enzymes involved in the synthesis of glycerolipids in chloroplasts and cyanobacteria for which genes or cDNAs have been cloned

Enzyme Plant or cyanobacterium	Reference
Transfer of acyl groups	
Acyl-ACP:glycerol-3-phosphate acyltransferase	
Cucurbita moschata	Ishizaki et al. (1988)
Pisum sativum	Weber et al. (1991)
Cucumis sativus	Johnson et al. (1992)
Arabidopsis thaliana	Nishida et al. (1993)
Carthamus tinctorius	Bhella et al. (1994)
Spinacia oleracea	Ishizaki et al. (1995)
Phaseolus vulgaris	Fritz et al. (1995)
Acyl-ACP:1-acylglycerol-3-phosphate acyltransferase	
	(not yet cloned)
Synthesis of phosphatidylglycerol	
Phosphatidate cytidyltransferase	
Synechocystis sp. PCC 6803	Kaneko et al. (1996)
CDP-Diacylglycerol:glycerol-3-phosphate phosphatidyltransferase (phosphatidylglycerophosphate synthase)	
Synechocystis sp. PCC 6803	Kaneko et al. (1996)
Phosphatidylglycerolphosphate phosphatase	
	(not yet cloned)
Synthesis of galactolipids	
Phosphatidic acid phosphatase	(not yet cloned)
UDP-Galactose:diacylglycerol galactocyltransferase (MGDG synthase)	
Cucumis sativus	Shimojima et al. (1997)
UDP-Sulfoquinovose:diacylglycerol sulfoquinovosyl transferase	
(*Rhodobacter sphaeroides*)	Rossak et al. (1995)

desaturation of fatty acids in higher-plant chloroplasts. As shown in Table 3, genes for all these desaturases have been cloned from several plant species.

No acyl-ACP desaturase has been found in cyanobacteria (Lem and Stumpf, 1984a, b), and all desaturation reactions are catalyzed by acyl-lipid desaturases after saturated fatty acids have been bound to the glycerol backbone. Figure 5B shows the pathway for desaturation of fatty acids in *Synechocystis* sp. PCC 6803, and Table 3 lists the cloned desaturases.

The enzymes involved in the biosynthesis of fatty acids and glycerolipids in chloroplasts and cyanobacteria have been well characterized in terms of their biochemistry and molecular biology. We can assume that the synthesis of fatty acids and glycerolipids in anoxygenic photosynthetic bacteria occurs in the same way as it does in chloroplasts and cyanobacterial cells (see Chapter 5, Benning).

In the cells of higher plants, glycerolipids, such as PC and phosphatidylethanolamine (PE), are synthesized in the cytoplasm. A part of PC is transported to chloroplasts and is converted to MGDG (see Chapter 2, Joyard et al.). MGDG, synthesized via PC, is characterized by a high level of 18:3 but no 16:3. Plants that contain 18:3, but no 16:3, in their MGDG and that synthesize MGDG via PC are known as '18:3 plants'. By contrast, '16:3 plants' contain 16:3 and 18:3 in their MGDG. In these plants, some of the MGDG, with 18:3 and 16:3 esterified at the *sn*-1 and *sn*-2 positions, respectively, is synthesized in chloroplasts, and the rest of the MGDG, esterified with 18:3 at both the *sn*-1 and *sn*-2 positions, is synthesized via PC (Roughan and Slack 1982; see Chapter 2, Joyard et al.). In algae, it seems likely that MGDG is synthesized in chloroplasts exclusively (see Chapter 3, Harwood).

In some of the literature about plant lipids, the glycerolipids that are synthesized in chloroplasts are

A. Plant chloroplasts

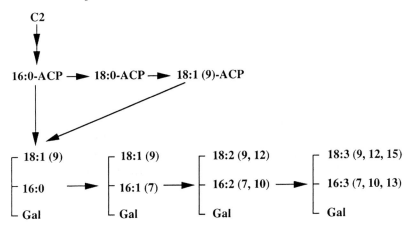

B. Cyanobacterial cells (*Synechocystis* sp. PCC 6803)

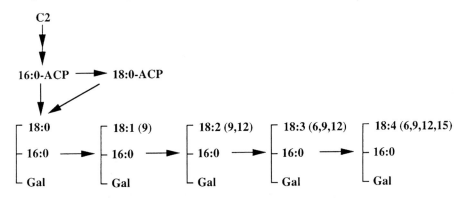

Fig. 5. Pathway for desaturation of fatty acids in MGDG. (A) Desaturation in chloroplasts of higher plants, such as spinach. (B) Desaturation in cyanobacterial cells, such as *Synechocystis* sp. PCC 6803.

designated 'prokaryotic lipids' since the distribution of fatty acids at the *sn*-positions is similar to that in glycerolipids in some cyanobacteria. On the contrary, lipids synthesized in or via the cytoplasm are designated 'eukaryotic lipids' because the distribution of fatty acids resembles that in lipids in animals. The corresponding biosynthetic pathways are termed the 'prokaryotic pathway' and the 'eukaryotic pathway', respectively. However, this terminology is confusing because the so-called 'prokaryotic pathway' is operative in eukaryotic plants, but not in prokaryotes. The pathway for synthesis of MGDG in cyanobacteria (prokaryotes) differs from that in chloroplasts (the prokaryotic pathway). The present authors strongly recommend use of the terms 'chloroplast pathway'

and 'cytoplasmic pathway' for the synthesis of glycerolipids in chloroplasts and in or via the cytoplasm, respectively.

The site of lipid synthesis in chloroplasts is controversial. Since enzymatic activities associated with some reactions in the synthesis of lipids have been identified in preparations of envelope membranes, R. Douce and J. Joyard have proposed that the biosynthetic reactions for synthesis of lipids in chloroplasts, from the acyltranfer reaction to the desaturation of fatty acids, are localized exclusively in the chloroplast envelope (see Chapter 2, Joyard et al.). However, evidence against this hypothesis is emerging. It supports the proposal that lipids are synthesized in both the envelope membrane and the

Table 3. Fatty acid desaturases in chloroplasts and cyanobacteria for which genes or cDNAs have been cloned

Enzyme	
Plant or cyanobacterium	Reference
Fatty acid desaturases	
Stearoyl-ACP desaturase	
Cucumis sativus	Shanklin et al. (1991)
Ricinus communis	Shanklin et al. (1991)
Spinacia oleracea	Nishida et al. (1992)
Brassica rapa	Knutzon et al. (1992)
Brassica napus	Slocombe et al. (1992)
Glycine max	Chen et al. (1994)
Linum usitatissimum	Singh et al. (1994)
Oryza sativa	Akagi et al. (1995)
Sesamum indicum	Yukawa et al. (1996)
$\Delta 9$ Acyl-lipid desaturase	
Synechocystis sp. PCC 6803	Sakamoto et al. (1994a)
Anbaena variabilis	Sakamoto et al. (1994a)
Synechococcus sp. PCC 7942	Ishizaki et al. (1995)
Synechococcus sp. PCC 7002	Sakamoto et al. (1997)
$\Delta 12$ Acyl-lipid desaturase	
Synechocystis sp. PCC 6803	Wada et al. (1990)
Spinacia oleracea	Schmidt et al. (1993)
Synechocystis sp. PCC 6714	Sakamoto et al. (1994b)
Arabidopsis thaliana	Falcone et al. (1994)
Glycine max	Hitz et al. (1994)
Brassica napus	Hitz et al. (1994)
Synechococcus sp. PCC 7002	Sakamoto et al. (1997)
$\omega 3$ ($\Delta 15$) Acyl-lipid desaturase	
Arabidopsis thaliana	Iba et al. (1993)
	Yadav et al. (1993)
	Gibson et al. (1994)
Brassica napus	Yadav et al. (1993)
Glycine soja	Yadav et al. (1993)
Synechocystis sp. PCC 6803	Sakamoto et al. (1994c)
$\Delta 6$ Acyl-lipid desaturase	
Synechocystis sp. PCC 6803	Reddy et al. (1993)
Spirulina platensis	Murata et al. (1996)

thylakoid membrane, with the thylakoid membrane being the major site. In cyanobacterial cells, in particular, immunogold labeling has demonstrated that fatty acid desaturases are localized mainly in the thylakoid membrane (Mustárdy et al., 1996). Essentially the same results have been obtained in chloroplasts of *Arabidopsis* by immunogold labeling of desaturases and acyl-ACP:glycerol-3-phosphate acyltransferase (Murata et al., 1996).

III. Roles of Glycerolipids in Photosynthesis

The lipids of chloroplast membranes contain poly-unsaturated fatty acids at unusually high relative levels. In most plant species, α-linolenic acid (18:3) or a combination of this acid and hexadecatrienoic acid (16:3) accounts for more than 80% of the total fatty acids of chloroplasts. Furthermore, the four lipid classes (MGDG, DGDG, SQDG and PG) found in the thylakoid membranes of higher plants consist of about 40 molecular species (Chapter 7, Siegenthaler). The physiological significance of the great diversity and varying degrees of polyunsaturation of chloroplast lipids has frequently been questioned over the past twenty years (Chapter 8, Siegenthaler and Trémolières).

In spite of considerable efforts, the precise relationship between the structure and function of acyl lipids in photosynthetic membranes remains controversial and elusive. Several reasons can be proposed to explain our failure to define the roles of glycerolipids in photosynthesis: (1) In contrast to proteins, lipids themselves have no recognizable catalytic properties. Thus, it is impossible to determine directly the function of lipids in vitro. (2) Evidence suggests that the lipids in thylakoid membranes consist of 'bulk lipids' and 'specific lipid molecules.' The bulk lipids, which predominate in the membranes, fill the spaces between the various membrane proteins and, therefore, they can be considered to be 'structural lipids.' For example, they play a role in maintaining the appropriate fluidity of the membrane, which is a consequence of the high degree of unsaturation of those lipids. Furthermore, MGDG can form, under certain conditions, non-bilayer configurations (Chapters 6, Williams, and 11, Selstam) and, therefore, can influence photosynthetic functions. By contrast, the specific lipids are bound to proteins and tend to be less unsaturated than bulk lipids. These molecules, which can also be considered as 'strategic or functional lipids' are involved in specific interactions with proteins that ensure an adequate maintenance of the conformation and/or orientation of the proteins in the membrane. Only under appropriate conditions are certain proteins able to perform their photosynthetic functions. The failure to recognize these two lipid categories has often led to misinterpretation of experimental results (Chapter 8, Siegenthaler and Trémolières). (3) The capacity of plant organs, tissues, cells, organelles and membranes to acclimate to extreme environmental conditions is often overlooked during the interpretation of experimental results. This problem is particularly apparent in studies of plant or algal mutants which are defective in certain lipids or fatty acids and of genetically-manipulated plants (Chapters 13, Gombos and Murata and 14, Vijayan et al.).

Several approaches have been proposed in the literature to overcome the above-described difficulties which are due to the particular properties of lipids and their heterogeneous distribution in thylakoid membranes (Chapter 8, Siegenthaler and Trémolières). The main experimental approaches that have been examined to date are listed in Table 4, with some relevant references.

A. Partial Degradation and Removal of Glycerolipids

Membranes can be depleted of lipids by extraction with organic solvents, by digestion of lipids with lipolytic enzymes, by removal with cyclodextrins or by treatment with detergents (Chapter 8, Siegenthaler and Trémolières). Extraction of lipids with organic solvents can provide information about the structural organization of lipids in thylakoid membranes but it is not appropriate for functional analysis (Table 4).

Hydrolysis of lipids by various lipolytic enzymes has been extensively employed in studies of the effects of changes in glycerolipid composition and of depletion of lipids on the photosynthetic functions of thylakoid membranes (Chapters 7, Siegenthaler and 8, Siegenthaler and Trémolières). Various enzymes, such as lipase from *Rhizopus arrhizus* (LRa) and phospholipases A_2, C and D (PLA$_2$, PLC and PLD), have been chosen for the hydrolysis of specific lipids. Under controlled conditions, lipids in the outer monolayer of the thylakoid membrane, are hydrolyzed first and removed, followed by those in the inner monolayer (Chapter 7, Siegenthaler). The enzymatic approach allows us to relate alterations in photosynthetic functions to topologically defined glycerolipids in the thylakoid membrane. Several functions have been studied in this way (Table 4).

Cyclodextrins are cyclic oligosaccharides that consist of 6, 7 or 8 glucopyranose units linked by α (1–4) bonds. They adopt a torus shape and are able to bind a range of small guest molecules (e.g., lipids) that have limited solubility in water within a hydrophobic cavity to form water-soluble guest-cyclodextrin inclusion complexes (Chapter 7, Siegenthaler). The cyclodextrin-induced removal of lipids from thylakoid

membranes results in important alterations of their functions (Table 4).

The composition in terms of acyl lipids and proteins, of several subthylakoid particles has been determined and has been tentatively related to specific lipid-protein associations (Chapters 6, Williams, and 7, Siegenthaler) and to particular functions (see Table 4). The purpose of such an approach is to determine the minimal specific lipid-protein association required for maintenance of the structural integrity of thylakoid membranes and retention of certain functions. However, the tight association of proteins with certain lipids in subthylakoid particles have not been demonstrated to be required for the functional expression of the particular proteins.

B. Catalytic Hydrogenation in Situ of Lipids and Studies of Mutants

A high degree of unsaturation of thylakoid membrane lipids is required for chloroplast biogenesis, for maintenance of the ultrastructural integrity and thermal stability of the photosynthetic apparatus, and for the acclimation of a plant to low temperatures (Chapters 6, Williams; 13, Gombos and Murata; 14, Vijayan et al.). Catalytic hydrogenation has been used successfully to modulate in situ and, exclusively, the extent of unsaturation of thylakoid membrane glycerolipids. The catalytic hydrogenation of the double bonds of fatty acids alters several photochemical reactions in thylakoids (Table 4). The impairment of photosynthetic activities has been interpreted in terms of a decrease in membrane fluidity when the ratio of unsaturated to saturated fatty acids is diminished upon the catalytic hydrogenation of thylakoid membranes. However, the results of studies of mutants suggest that a high level of trienoic fatty acids might not be essential, within certain limits, for the support of normal levels of photochemical activity (Chapter 14, Vijayan et al.). A number of mutants of *Arabidopsis thaliana* and *Chlamydomonas reinhardtii* have been characterized as being deficient in desaturases or glycerol-3-phosphate acyl transferases (Table 4). These mutants provide another tool for studies of the role of lipids and, in particular, the extent of lipid unsaturation in the control of thylakoid membrane functions (Chapters 8, Siegenthaler and Trémolières and 14, Vijayan et al.). The conclusions of these studies can be summarized as follows: (1) The polyunsaturation of membrane lipids is required for the biogenesis of

Table 4. Main methodological approaches that have been exploited in attempts to define the roles of lipids in the functions of thylakoid membranes

	Methodological approach	Function(s) studied	Some relevant references
1.	**Extraction of lipids**		
	Extraction of lipids by organic solvents	Structural role	Costes et al. (1972; 1978)
2.	**Enzymatic approach**		
	Digestion of lipids by LAH[1]	Photosynthetic electron flow	Krupa and Baszinski (1975); Hirayama and Matsui (1976); Rawyler and Siegenthaler (1980)
	Digestion and removal of PG by PLA$_2$[1]	Photosynthetic electron flow in outer and inner thylakoid monolayers	Siegenthaler et al. (1987a,b; 1989); Duchêne et al. (1995)
		Dark phosphorylation	Siegenthaler and Vallino (1995)
	Digestion of PG by PLC[1] and PLD[1]	Photosynthetic electron flow	Siegenthaler et al. (1984)
		Photosynthetic electron flow, fluorescence induction, thermoluminescence, structural role	Droppa et al. (1995)
	Digestion and removal of MGDG by LRa	Photosynthetic electron flow, relationship between the hydrolysis rate of MGDG, lateral packing pressure of MGDG and photophosphorylation	Rawyler and Siegenthaler (1989)
		Dark phosphorylation	Siegenthaler and Vallino (1995)
3.	**Cyclodextrin approach**		
	Partial depletion of lipids by cyclodextrins	Photosynthetic electron flow, fluorescence spectrum at 77 K, $F_{732}/F_{682\text{-}689}$ emission ratio, cation-induced fluorescence	Rawyler and Siegenthaler (1996)
4.	**Subthylakoid particles**		
	PS II[2] reaction center	Photosynthetic electron flow	Gounaris and Barber (1985)
	ATP synthase complex	ATPase	Pick et al. (1985)
	Cytochrome $b_6 f$ complex	Plastoquinol-cytochrome c reductase activity	Doyle and Yu (1985)
	PS II membrane, PS II core complex and PS II reaction center complex	Structural organization, photochemical charge separation, O_2 evolution	Murata et al. (1990)
	LHCII[2]	Trimer formation and crystallization	Nussberger et al. (1993)
5.	**Catalytic hydrogenation in situ of lipids**		
		Photosynthetic electron flow	Restall et al. (1979); Vigh et al. (1985)
		Reduction of flash-oxidized cytochrome f	Gombos et al. (1988)
		Fluorescence kinetics, PS II electron flow and thermoluminescence	Hideg et al. (1986)
		PS II organization and function	Horváth et al. (1987)

Table 4. (Continued)

Methodological approach	Function(s) studied	Some relevant references
	Photophosphorylation and electron flow	Horváth et al. (1986)
	Thermal stability of pigment-protein complex and PS I-mediated electron flow	Thomas et al. (1986); Vigh et al. (1989)
	Perception of changes in temperature	Vigh et al. (1993)
6. Studies of Mutants[3]		
fad 7 mutant of *A. thaliana*	CO_2 fixation, photosynthetic electron flow, PS II and PS I activities, F_{685}/F_{734} ratio at 77 K, energy transfer between PS II and PS I	McCourt et al. (1987)
fad 5 mutant of *A. thaliana*	CO_2 fixation, photosynthetic electron flow, thermal inactivation of electron transport	Kunst et al. (1989b)
fad 6 mutant of *A. thaliana*	Ultrastructure and thermal stability of thylakoids, electron transport, fluorescence spectra at 77 K	Hugly et al. (1989)
	Chloroplast biogenesis, adaptability to low temperature	Hugly and Somerville (1992)
fad 3, *fad* 7 and *fad* 8 mutants of *A. thaliana*	Relative fluorescence, quantum yield of PS II electron transfer	Routaboul and Browse (1995)
fad 4 mutant of *A. thaliana*	Formation of LHCII oligomers, association of LHCII and the PS II reaction center, electron transport, PS II photochemical efficiency	McCourt et al. (1985); Browse et al. (1985)
fab 1 mutant of *A. thaliana*	Response to low temperature, fluorescence Fv/Fm, quantum yield, photosynthetic capacity	Wu and Browse (1995); Wu et al. (1997)
act 1 mutant of *A. thaliana*	Distribution of excitation energy transfer, fluorescence characteristics, thermal stability of the photosynthetic apparatus	Kunst et al. (1989a)
dgd 1 mutant of *A. thaliana*	Photosynthetic quantum yield, ultrastructural features	Dörmann et al. (1995)
mf1/mf2 mutant of *Chlamydomonas reinhardtii*	Biogenesis of LHCII, oligomerization of LHCII monomers	Dubertret et al. (1994)
7. Immunological studies		
Antibodies against galactolipids	Photosynthetic electron flow	Radunz et al. (1984a)
Antibodies against PG	Photosynthetic electron flow	Radunz (1984)
Antibodies against SQDG	Photosynthetic electron flow	Radunz et al. (1984b)
Antibodies against PG	Binding of PG to D1 protein	Kruse et al. (1994); Kruse and Schmid (1995)

Table 4. (Continued)

Methodological approach	Function(s) studied	Some relevant references
8. Reconstitution studies with lipids		
Subthylakoid PS II particles	O_2 evolution	Gounaris et al. (1983); Nénonéné and Fragata (1990); Fragata et al. (1991)
Triton X-100 solubilized thylakoids	Fluorescence emission spectrum at 77 K, energy transfer	Siefermann-Harms et al. (1982)
PS II core reaction center + LHCII	Electron flow, O_2 evolution, quantum efficiency	Murphy et al. (1984)
PS II core reaction center	Variable-yield fluorescence, O_2 evolution, conversion of low-potential form of cytochrome b_{559} to high-potential form	Matsuda and Butler (1983)
Cytochrome $b_6 f$ complex	Plastoquinol-plastocyanin oxidoreductase, proton translocation, redox reaction of cytochrome b_6	Hurt et al. (1982); Chain (1985); Willms et al. (1987)
P_{700}-chlorophyll *a* protein complex + chlorophyll *a*	Photooxidation of P_{700}, transfer of electrons, fluorescence yield	Ikegami (1983); Ishikawa et al. (1984)
PS II reaction center complex + P_{700}–chlorophyll a protein complex + LHCII	Fluorescence emission spectrum at 77 K, fluorescence, energy transfer	Larkum and Anderson (1982)
LHCII	Formation of LHCII oligomers, stabilization of LHCII	Rémy et al. (1982; 1984); Krupa et al. (1992)
LHCII	Cation-dependent stacking of thylakoid membranes	Ryrie et al. (1980); McDonnel and Staehelin (1980); Ryrie and Fuad (1982)
LHCII	Low-temperature (polarized) light spectroscopy, LHCII trimerization	Peterman et al. (1996)
	Trimerization of LHCII monomers	Hobe et al. (1994); Kuttkat et al. (1995)
ATP synthetase complex	ATP hydrolysis, ATP-Pi exchange, proton conductance	Gounaris et al. (1984); Pick et al. (1984; 1987)

[1] LAH – lipolytic acyl hydrolase; LRA – lipase from *Rhizopus arrhizus*; PLA$_2$, PLC and PLD – phospholipase A$_2$, C and D; [2] PS II (PS I) – Photosystem II(I); LHCII – light-harvesting chlorophyll *a/b* protein of PS II; [3] The lipid characteristics of the mutants can be found in Ohlrogge and Browse (1995) and Chapter 14, Vijayan et al.

the chloroplast and for assembly of the oligomeric components of the photosynthetic apparatus. (2) Although a high level of trienoic acids in the membrane lipids is not critical for photosynthetic activities, it does contribute to the low-temperature fitness of the plant and has a direct effect on the thermal stability of the photosynthetic apparatus (Chapter 14, Vijayan et al.). The analysis of mutants does, however, have its own limitations. It is always possible that, in addition to the lipid mutation itself, some other mutations might occur that might result in a change in the phenotype of the organism.

C. Immunological Studies

Antibodies directed against MGDG, DGDG, SQDG and PG were used initially to determine the localization of these lipids at the surface of the thylakoid membrane, as well as in various preparations of Photosystem II (PS II) and Photosystem I (PS I) complexes and ATP synthase. Immunological studies confirmed and substantiated the great heterogeneity of lipids in the thylakoid membrane, that had been revealed by other methods (Chapter 7, Siegenthaler). Interactions between lipid-specific antibodies and the lipids in the thylakoid membrane result in the impairment of several photosynthetic activities, as listed in Table 4. It is assumed that the antibodies, upon binding to lipids, induce a conformational change in the adjacent proteins which are involved in sustaining photosynthetic activities (Chapter 8, Siegenthaler and Trémolières).

D. Reconstitution Studies

Glycerolipids have often been used in the preparation of liposomes in which one or several components of the thylakoid membrane can be inserted. After the reconstitution in vitro, the activity of the component(s) is tested and the results are interpreted by reference to the type and proportion of lipids in the liposomes. As shown in Table 4, a great variety of components of thylakoid membranes have been studied, alone or in combination, such as PS II and PS I reaction center complexes, the light-harvesting complex II (LHCII), cytochrome $b_6 f$ and the ATP synthase complex. It appears that, in some cases, specific lipids are required for reactivation of certain complexes. For example, LHCII reconstituted with a polypeptide that had been overexpressed in *E. coli*, purified pigments, and membrane lipids yielded trimers that had the same biochemical and spectroscopic properties as LHCII isolated from thylakoid membranes. PG appears to be intimately involved in formation of the LHCII trimer and it is likely to be located at the subunit interface. DGDG is essential for two- and three-dimensional crystallization, and it is likely that it binds at the periphery of the trimer (Kühlbrandt, 1994; Chapter 9, Trémolières and Siegenthaler). A main limitation to studies of lipid-protein interactions by experiments in vitro lies in the difficulty to ascertain that the observed interactions actually occur in situ, namely, in the membrane of the living organism. Therefore, approaches in vivo to lipid-protein interactions have been developed with targeting of lipids into photosynthetic membranes of different *Chlamydomonas reinhardtii* mutants that lack PG-16:1(3t) (Chapter 9, Trémolières and Siegenthaler).

E. Genetic Analysis

The functions of specific lipids and fatty acids in photosynthesis can be studied by genetic manipulation, namely, by mutagenesis and the overexpression of specific genes.

1. Insertional Inactivation of Genes

Specific genes can be mutated ('knocked out') by insertion of suitable antibiotic-resistance genes. This technique has been most successfully applied to the manipulation of the unsaturation of fatty acids of membrane lipids, with step-wise inactivation of individual desaturases in *Synechocystis* sp. PCC 6803 (Tasaka et al., 1996). This strain contains four acyl-lipid desaturases, which introduce, respectively, a double bond at the $\Delta 9, \Delta 12, \omega 3$ ($\Delta 15$) and $\Delta 6$ positions of fatty acids to produce monounsaturated (18:1) to tetraunsaturated (18:4) fatty acids. The insertional inactivation of the gene for the $\Delta 6$ desaturase resulted in the absence of 18:4, and the resultant strain contained 18:3 as the major fatty acid. The inactivation of genes for both the $\Delta 6$ and the $\omega 3$ ($\Delta 15$) desaturases eliminated 18:3(9,12,15), and the resultant strain contained 18:2(9,12) as the major fatty acid. The inactivation of genes for both the $\Delta 6$ and the $\Delta 12$ desaturases eliminated 18:2(9,12), and the resultant strain contained 18:1(9) as the major fatty acid. From a comparison of the photosynthetic activities of the various strains, we can address the issue of the specific functions of individual fatty acids. Such research has revealed that the polyunsaturated fatty acids are important for the stability of the photosynthetic machinery (in particular the Photosystem II complex) under high-intensity light at low temperature (see Chapter 13, Gombos and Murata).

2. Expression of Novel Genes

The introduction of novel genes to transform plants and cyanobacteria is another method that allows us to examine the functions of specific genes or enzymes in photosynthesis. For example, the research group of one of us (N.M.) overexpressed a cDNA for the acyl-ACP:glycerol-3-phosphate acyltransferase of squash in tobacco (Murata et al., 1992). The transformation increased the level of saturated molecular species of phosphatidylglycerol (PG) compared to the same level as in squash and rendered the tobacco plants sensitive to cold (see Chapter 13,

Gombos and Murata). This kind of study demonstrated that the level of unsaturation or saturation of fatty acids of PG is important in the tolerance of plants to cold stress.

3. Chemically Induced Mutagenesis

In higher plants, the insertional inactivation of genes has proved difficult and often impracticable. Therefore, chemically induced mutagenesis has been popular in studies of the functions of the unsaturated fatty acids of membrane lipids. In particular, J. Browse and his co-workers studied the results of mutagenesis of *Arabidopsis thaliana* with ethylmethanesulfonate (see Chapter 14, Vijayan et al.). They isolated nine different kinds of mutant which had altered relative levels of fatty acids as compared with wild-type plants. Furthermore, mutants of multiple genes with changes in fatty acid composition were produced by crossing some of the mutants of single genes. Distinct changes were observed in fatty acid composition. For example triple mutation of the three $\omega 3$ desaturases (one in the cytoplasm and two in the chloroplast) completely eliminated triunsaturated fatty acids such as 18:3 and 16:3. A double mutant of two $\Delta 12$ desaturases (one in the cytoplasm and one in the chloroplast) was completely devoid of polyunsaturated fatty acids. Comparisons of these multiple mutated strains with the wild-type strain facilitate studies of the specific functions of the unsaturation of fatty acids in membrane lipids in photosynthesis (see Chapters 8, Siegenthaler and Trémolières and 14, Vijayan et al.).

By comparing triple-mutant plants, which were devoid of 18:3 and 16:3, with wild-type plants, in which trienoic fatty acids accounted for more than 60% of the total, J. Browse and co-workers found that the trienoic fatty acids are important in protecting the photosynthetic machinery against photoinhibition at low temperatures. This result is consistent with the results obtained in cyanobacteria with respect to the importance of unsaturation of membrane lipids. However, there is clearly a difference between the two types of photosynthetic organism; in *Arabidopsis,* the presence of the third double bond in the fatty acids is important in the protection against the photoinhibition (see Chapter 14, Vijayan et al.); in cyanobacteria, the second double bond (and, thus, the presence of diunsaturated fatty acids), but not the third double bond, provides the protective effect (see Chapter 13, Gombos and Murata).

A double mutant of *Arabidopsis* that was completely devoid of polyunsaturated fatty acids was unable to grow autotrophically, suggesting that the second double bond might be essential for effective photosynthesis (see Chapter 14, Vijayan et al.). Nonetheless, both studies by insertional mutagenesis in cyanobacteria and by chemically induced mutagenesis in *Arabidopsis* clearly indicate the importance of the unsaturated fatty acids of glycerolipids in photosynthetic membranes.

Acknowledgment

The authors are grateful to Dr. Yasushi Tasaka, Research Institute for Biological Sciences, Okayama, for his help in the preparation of tables and figures.

References

Akagi H, Baba T, Shimada H and Fujimura T (1995) Nucleotide sequence of a stearoyl-acyl carrier protein desaturase cDNA from developing seeds of rice. Plant Physiol 108: 845–846

Bhella RS and MacKenzie SL (1994) Nucleotide sequence of a cDNA from *Carthamus tinctorius* encoding a glycerol-3-phosphate acyl transferase. Plant Physiol 106: 1713–1714

Browse J, McCourt P and Somerville C (1985) A mutant of *Arabidopsis* lacking a chloroplast-specific lipid. Science 227: 763–765

Chain RK (1985) Involvement of plastoquinone and lipids in electron transport reactions mediated by the cytochrome b_6-f complex isolated from spinach. FEBS Lett 180: 321–325

Chen L, Moon Y, Shanklin J, Nikolau BJ and Atherly AG (1994) Cloning and sequence of a cDNA encoding stearoyl-acyl carrier protein desaturase from *Glycine max*. Plant Physiol 109: 1498

Choi J-K, Yu F, Wurtele ES and Nikolau BJ (1995) Molecular cloning and characterization of the cDNA coding for the biotin-containing subunit of the chloroplastic acetyl-coenzyme A carboxylase. Plant Physiol 109: 619–625

Costes C, Bazier R and Lechevallier D (1972) Rôle structural des lipides dans les membranes des chloroplastes de Blé. Physiol Vég 10: 291–317

Costes C, Bazier R, Baltscheffsky H and Hallberg C (1978) Mild extraction of lipids and pigments from *Rhodospirillum rubrum* chromatophores. Plant Sci Lett 12: 241–249

Dörmann P, Hoffmann-Benning S, Balbo I and Benning C (1995) Isolation and characterization of an *Arabidopsis* mutant deficient in the thylakoid lipid digalactosyl diacylglycerol. Plant Cell 7: 1801–1818

Doyle MF and Yu C-A (1985) Preparation and reconstitution of a phospholipid deficient cytochrome b_6-f complex from spinach chloroplasts. Biochem Biophys Res Commun 131: 700–706

Droppa M, Horváth G, Hideg E and Farkas T (1995) The role of phospholipids in regulating photosynthetic electron transport activities: Treatment of thylakoids with phospholipase C.

Photosynth Res 46: 287–293

Dubertret G, Mirshahi A, Mirshahi M, Gérard-Hirne C and Trémolières A (1994) Evidence from in vivo manipulations of lipid composition in mutants that the Δ^3-*trans*-hexadecenoic acid-containing phosphatidylglycerol is involved in the biogenesis of the light-harvesting chlorophyll *a/b*-protein complex of *Chlamydomonas reinhardtii* Eur J Biochem 226: 473–482

Duchêne S, Smutny J and Siegenthaler PA (1995) Transmembrane distribution of phospholipids in spinach thylakoid inside-out vesicles and involvement of outer and inner monolayer phospholipids in the photosynthetic electron flow activity. In: Lopez-Pérez MJ, Delgado C and Cebrian-Pérez JA (eds) 9th International Conference on Partitioning in Aqueous Two-Phase Systems. Advances in the Uses of Polymers in Cell Biology, Biotechnology and Environmental Sciences, Abstract Nr, P23, University of Zaragoza, Spain

Elborough KM, Winz R, Deka RK, Markham JE, White AJ, Rawsthorne S and Slabas AR (1996) Biotin carboxyl carrier protein and carboxyltransferase subunits of the multi-subunit form of acetyl-CoA carboxylase from *Brassica napus*: Cloning and analysis of expression during oilseed rape embryogenesis. Biochem J 315: 103–112

Falcone DL, Gibson S, Lemieux B and Somerville C (1994) Identification of a gene that complements an *Arabidopsis* mutant deficient in chloroplast ω6 desaturase activity. Plant Physiol 106: 1453–1459

Fragata M, Strzalka K and Nénonéné EK (1991) $MgCl_2$-induced reversal of oxygen evolution decay in Photosystem II particles incubated with phosphatidylglycerol vesicles at high lipid/Photosystem II ratio. J Photochem Photobiol 11: 329–342

Fritz M, Heinz E and Wolter FP (1995) Cloning and sequencing of a full-length cDNA coding for *sn*-glycerol-3-phosphate acyltransferase from *Phaseolous vulgaris*. Plant Physiol 107: 1039–1040

Gibson S, Arondel V, Iba K and Somerville C (1994) Cloning of a temperature-regulated gene encoding a chloroplast ω-3 desaturase from *Arabidopsis thaliana*. Plant Physiol 106: 1615–1621

Gombos Z, Barabas K, Joó F and Vigh L (1988) Lipid saturation induced microviscosity increase has no effect on the reducibility of flash-oxidized cytochrome *f* in pea thylakoids. Plant Physiol 86: 335–337

Gornicki P, Scappino LA and Haselkorn R (1993) Genes for two subunits of acetyl coenzyme A carboxylase of *Anabaena* sp. strain PCC 7120: Biotin carboxylase and biotin carboxyl carrier protein. J Bacteriol 175: 5268–5272

Gounaris K and Barber J (1985) Isolation and characterisation of a Photosystem II reaction centre lipoprotein complex. FEBS Lett 188: 68–72

Gounaris K, Whitford D and Barber J (1983) The effect of thylakoid lipids on an oxygen-evolving Photosystem II preparation. FEBS Lett 163: 230–234

Gounaris K, Pick U and Barber J (1984) The effect of thylakoid lipids on enzymatic activity and ultrastructure of membrane protein complexes. In: Siegenthaler PA and Eichenberger W (eds) Structure, Function and Metabolism of Plant Lipids, pp 451–455. Elsevier Publication Press, Amsterdam

Hideg E, Rozsa Z, Vaas l, Vigh L and Horváth G (1986) Effect of homogeneous catalytic hydrogenation of membrane lipids on luminescence characteristics of the Photosystem II electron transport. Photobiochem Photobiophys 12: 221–230

Hirayama O and Matsui T (1976) Effects of lipolytic enzymes on the photochemical activities of spinach chloroplasts. Biochim Biophys Acta 423: 540–547

Hirsch S and Soll J (1995) Import of a new chloroplast inner envelope protein is greatly stimulated by potassium phosphate. Plant Mol Biol 27: 1173–1181

Hitz WD, Carlson TJ, Booth JR Jr, Kinney AJ, Stecca KL and Yadav NS (1994) Cloning of a higher-plant plastid ω-6 fatty acid desaturase cDNA and its expression in a cyanobacterium. Plant Physiol 105: 635–641

Hobe S, Prytulla S, Kühlbrandt W and Paulsen H (1994) Trimerization and crystallization of reconstituted light-harvesting chlorophyll *a/b* complex. EMBO J 13: 3423–3429

Horváth G, Droppa M, Szito T, Mustárdy LA, Horváth Ll and Vigh L (1986) Homogeneous catalytic hydrogenation of lipids in the photosynthetic membrane: Effects on membrane structure and photosynthetic activity. Biochim Biophys Acta 849: 325–336

Horváth G, Melis A, Hideg E, Droppa M and Vigh L (1987) Role of lipids in the organization and function of Photosystem II studied by homogeneous catalytic hydrogenation of thylakoid membranes in situ. Biochim Biophys Acta 891: 68–74

Hugly S and Somerville C (1992) A role for membrane lipid polyunsaturation in chloroplast biogenesis at low temperature. Plant Physiol 99: 197–202

Hugly S, Kunst L, Browse J and Somerville C (1989) Enhanced thermal tolerance of photosynthesis and altered chloroplast ultrastructure in a mutant of *Arabidopsis* deficient in lipid desaturation. Plant Physiol 90: 1134–1142

Hurt EC, Hauska G and Shahak Y (1982) Electrogenic proton translocation by the chloroplast cytochrome $b_6 f$ complex reconstituted into phospholipid vesicles. FEBS Lett 149: 211–216

Iba K, Gibson S, Nishiuch T, Fuse T, Nishimura M, Arondel V, Hugly S and Somerville C (1993) A gene encoding a chloroplast ω-3 fatty acid desaturase complements alterations in fatty acid desaturation and chloroplast copy number of the *fad7* mutant of *Arabidopsis thaliana*. J Biol Chem 268: 24099–24105

Ikegami I (1983) Reconstitution of antenna in P-700-enriched particles from spinach chloroplasts. Biochim Biophys Acta 722: 492–497

Ishikawa H, Hirano M-a and Takabe T (1984) Reconstitution of highly purified P700-chlorophyll a protein complexes into galactosyldiacylglycerol liposomes. Agric Biol Chem 48: 3011–3018

Ishizaki O, Nishida I, Agata K, Eguchi G and Murata N (1988) Cloning and nucleotide sequence of cDNA for the plastid glycerol-3-phosphate acyltransferase from squash. FEBS Lett 238: 424–430

Ishizaki-Nishizawa O, Azuma M, Ohtani T, Murata N and Toguri T (1995) Nucleotide sequence of cDNA from *Spinacia oleracea* encoding plastid glycerol-3-phosphate acyltransferase. Plant Physiol 108: 1342

Johnson TC, Schneider JC and Somerville C (1992) Nucleotide sequence of acyl-acyl carrier protein: Glycerol-3-phosphate acyltransferase from cucumber. Plant Physiol 99: 771–772

Kaneko T, Sato S, Kotani H, Tanaka A, Asamizu E, Nakamura Y, Miyajima N, Hirosawa M, Sugiura M, Sasamoto S, Kimura T, Hosouchi T, Matsuno A, Muraki A, Nakazaki N, Naruo K, Okumura S, Shimpo S, Takeuchi C, Wada T, Watanabe A,

Yamada M, Yasuda M and Tabata S (1996) Sequence analysis of the genome of the unicellular cyanobacterium *Synechocystis* sp. strain PCC6803. II. Sequence determination of the entire genome and assignment of potential protein-coding regions. DNA Res 3: 109–136

Kater MM, Koningstein GM, Nijkamp HJJ and Stuitje AR (1991) cDNA cloning and expression of *Brassica napus* enoyl-acyl carrier protein reductase in *Escherichia coli*. Plant Mol Biol 17: 895–909

Knutzon DS, Thompson GA, Radke SE, Johnson WB, Knauf VC and Kridl JC (1992) Modification of *Brassica* seed oil by antisense expression of a stearoyl-acyl carrier protein desaturase gene. Proc Natl Acad Sci USA 89: 2624–2628

Krupa Z and Baszynski T (1975) Requirement of galactolipids for Photosystem I activity in lyophilized spinach chloroplasts. Biochim Biophys Acta 408: 26–34

Krupa Z, Williams JP, Khan MU and Huner NPA (1992) The role of acyl lipids in reconstitution of lipid-depleted light-harvesting complex II from cold-hardened and nonhardened rye. Plant Physiol 100: 931–938

Kruse O and Schmid GH (1995) The role of phosphatidylglycerol as a functional effector and membrane anchor of the D1-core peptide from Photosystem II-particles of the cyanobacterium *Oscillatoria chalybea*. Z Naturforsch 50c: 380–390

Kruse O, Radunz A and Schmid GH (1994) Phosphatidylglycerol and β-carotene bound onto the D1-core peptide of Photosystem II in the filamentous cyanobacterium *Oscillatoria chalybea*. Z Naturforsch 49c: 115–124

Kühlbrandt W (1994) Structure and function of the plant light-harvesting complex, LHC-II. Current Opinion in Structural Biology 4: 519–528

Kunst L, Browse J and Somewille C (1989a) Altered chloroplast structure and function in a mutant of *Arabidopsis* deficient in plastid glycerol-3-phosphate acyltransferase activity. Plant Physiol 90: 846–853

Kunst L, Browse J and Somewille C (1989b) Enhanced thermal tolerance in a mutant of *Arabidopsis* deficient in palmitic acid unsaturation. Plant Physiol 91: 401–408

Kuttkat A, Grimm R and Paulsen H (1995) Light-harvesting chlorophyll *a/b*-binding protein inserted into isolated thylakoids binds pigments and is assembled into trimeric light-harvesting complex. Plant Physiol 109: 1267–1276

Larkum AWD and Anderson JM (1982) The reconstitution of a Photosystem II protein complex, P-700-chlorophyll *a*-protein complex and light-harvesting chlorophyll *a/b*-protein. Biochim Biophys Acta 679: 410–421

Lem NW and Stumpf PK (1984a) In vitro fatty acid synthesis and complex lipid metabolism in the cyanobacterium *Anabaena variabilis*. I. Some characteristics of fatty acid synthesis. Plant Physiol 74: 134–138

Lem NW and Stumpf PK (1984b) In vitro fatty acid synthesis and complex lipid metabolism in the cyanobacterium, *Anabaena variabilis*. II. Acyl transfer and complex lipid formation. Plant Physiol 75: 700–704

Matsuda H and Butler WL (1983) Restoration of high-potential cytochrome *b*-559 in Photosystem II particles in liposomes. Biochim Biophys Acta 725: 320–324

McCourt P, Browse J, Watson J, Arntzen CJ and Somerville CR (1985) Analysis of photosynthetic antenna function in a mutant of *Arabidopsis thaliana* (L.) lacking *trans*-hexadecenoic acid. Plant Physiol 78: 853–858

McCourt P, Kunst L, Browse J and Somerville CR (1987) The effects of reduced amounts of lipid unsaturation on chloroplast ultrastructure and photosynthesis in a mutant of *Arabidopsis*. Plant Physiol 84: 353–360

McDonnel A and Staehelin LA (1980) Adhesion between liposomes mediated by the chlorophyll *a/b* light-harvesting complex isolated from chloroplast membranes. J Cell Biol 84: 40–56

Murata N, Higashi S-I and Fujjmura Y (1990) Glycerolipids in various preparations of Photosystem II from spinach chloroplasts. Biochim Biophys Acta 1019: 261–268

Murata N, Ishizaki-Nishizawa O, Higashi S, Hayashi H, Tasaka Y and Nishida I (1992) Genetically engineered alteration in the chilling sensitivity of plants. Nature 356: 710–713

Murata N, Deshnium P and Tasaka Y (1996) Biosynthesis of γ-linolenic acid in the cyanobacterium *Spirulina platensis*. In: Huang Y-S and Mills DE (eds) γ-Linolenic Acid, pp 22–32. AOCS Press, Champaign, IL

Murphy DJ, Crowlher D and Woodrow IE (1984) Reconstitution of light-harvesting chlorophyll-protein complexes with Photosystem II complexes in soybean phosphatidylcholine liposomes. Enhancement of quantum efficiency at sub-saturating light intensities in the reconstituted liposomes. FEBS Lett 165: 151–155

Mustárdy L, Los DA, Gombos Z and Murata N (1996) Immunocytochemical localization of acyl-lipid desaturases in cyanobacterial cells: Evidence that both thylakoid membranes and cytoplasmic membranes are sites of lipid desaturation. Proc Natl Acad Sci USA, 93: 10524–10527.

Nénonéné EK and Fragata M (1990) Effects of pH and freeze-thaw on photosynthetic oxygen evolution of Photosystem II particles incorporated into phosphatidylglycerol bilayers. J Plant Physiol 136: 615–620

Nishida I, Beppu T, Matsuo T and Murata N (1992) Nucleotide sequence of a cDNA clone encoding a precursor to stearoyl-(acyl-carrier-protein) desaturase from spinach, *Spinacia oleracea*. Plant Mol Biol 19: 711–713

Nishida I, Tasaka Y, Shiraishi H and Murata N (1993) The gene and the RNA for the precursor to the plastid-located glycerol-3-phosphate acyltransferase of *Arabidopsis thaliana*. Plant Mol Biol 21: 267–277

Nussberger S, Dörr K, Wang DN and Kühlbrandt W (1993) Lipid-protein interactions in crystals of plant light-harvesting complex. J Mol Biol 234: 347–356

Ohlrogge J and Browse J (1995) Lipid biosynthesis. Plant Cell 7: 957–970

Peterman EJG, Hobe S, Calkoen F, van Grondelle R, Paulsen H and van Amerongen H (1996) Low-temperature spectroscopy of monomeric and trimeric forms of reconstituted light-harvesting chlorophyll *a/b* complex. Biochim Biophys Acta 1273: 171–174

Pick U, Gounaris K, Admon A and Barber J (1984) Activation of the CF_o-CF_1, ATP synthase from spinach chloroplasts by chloroplast lipids. Biochim Biophys Acta 765: 12–20

Pick U, Gounaris K, Weiss M and Barber J (1985) Tightly bound sulpholipids in chloroplast CF_o-CF_1. Biochim Biophys Acta 808: 415–420

Pick U, Weiss M, Gounaris K and Barber J (1987) The role of different thylakoid glycolipids in the function of reconstituted chloroplast ATP synthase. Biochim Biophys Acta 891: 28–39

Radunz A (1984) Serological investigations on the function of phospholipids in the thylakoid membrane. In: Sybesma C (ed), Advances in Photosynthesis Research, Vol III, pp 151–154. Martinus Nijhoff/Dr W Junk Publishers, The Hague

Radunz A, Bader KP and Schmid GH (1984a) Serological investigations of the function of galactolipids in the thylakoid membrane. Z Pflanzenphysiol 114: 227–231

Radunz A, Bader KP and Schmid GH (1984b) Influence of antisera to sulfoquinovosyl diglyceride and to β-sitosterol on the photosynthetic electron transport in chloroplasts from higher plants. In: Siegenthaler PA and Eichenberger W (eds), Structure, Function and Metabolism of Plant Lipids, pp. 479–484. Elsevier Science Publishers BV, Amsterdam

Rawyler A and Siegenthaler PA (1980) Role of lipids in functions of photosynthetic membranes revealed by treatment with lipolytic acyl hydrolase. Eur J Biochem 110: 179–187

Rawyler A and Siegenthaler PA (1989) Change in the molecular organization of monogalactosyldiacylglycerol between resting and functioning thylakoid membranes. Involvement of the CF_o-CF_1-ATP synthetase. Biochim Biophys Acta 975: 283–292

Rawyler A and Siegenthaler PA (1996) Cyclodextrins: A new tool for the controlled lipid depletion of thylakoid membranes. Biochim Biophys Acta 1287: 89–97

Reddy AS, Nuccio ML, Gross LM and Thomas TL (1993) Isolation of Δ6-desaturase gene from the cyanobacterium Synechocystis sp. strain PCC 6803 by gain-of-function expression in Anabaena sp. strain PCC 7120. Plant Mol Biol 22: 293–300

Reith ME and Munholland J (1995) Complete nucleotide sequence of the Porphyra purpurea chloroplast genome. Plant Mol Biol Rep 13: 333–335

Rémy R, Trémolières A, Duval JC, Ambard-Bretteville F and Dubacq JP (1982) Study of the supramolecular organization of light-harvesting chlorophyll protein (LHCP). FEBS Lett 137: 271–275

Rémy R, Trémolières A and Ambard-Bretteville F (1984) Formation of oligomeric light-harvesting chlorophyll a/b protein by interaction between its monomeric form and liposomes. Photobiochem Photobiophys 7: 267–276

Restall CJ, Williams WP, Percival MP, Quinn PJ and Chapman D (1979) The modulation of membrane fluidity by hydrogenation processes. III. The hydrogenation of biomembranes of spinach chloroplasts and a study of the effect of this on photosynthetic electron transport. Biochim Biophys Acta 555: 119–130

Reverdatto SV, Beilinson V and Nielsen NC (1995) The rps16, accD, psaI, ORF203, ORF151, ORF103, ORF229, and petA gene cluster in the chloroplast genome of soybean. Plant Physiol 109: 338

Rossak M, Tietje C, Heinz E and Benning C (1995) Accumulation of UDP-sulfoquinovose in a sulfolipid-deficient mutant of Rhodobacter sphaeroides. J Biol Chem 270: 25792–25797

Roughan PG and Slack CR (1982) Cellular organization of glycerolipid metabolism. Annu. Plant Physiol 33: 97–132

Routaboul JM and Browse J (1995) A role for membrane lipid trienoic fatty acids in photosynthesis at low temperatures. In: Matthis P (ed) Photosynthesis: From Light to Biosphere, Vol IV, pp 861–864. Kluwer Academic Publishers, Dordrecht

Ryrie IJ and Fuad N (1982) Membrane adhesion in reconstituted proteoliposomes containing the light-harvesting chlorophyll a/b-protein complex: The role of charged surface groups. Arch Biochem Biophys 214: 475–488

Ryrie IJ, Anderson JM and Goodchild DJ (1980) The role of the light-harvesting chlorophyll a/b-protein complex in chloroplast membrane stacking. Cation-induced aggregation of reconstituted proteoliposomes. Eur J Biochem 107: 345–354

Sakamoto T and Bryant DA (1997) Temperature-regulated mRNA accumulation and stabilization for fatty acid desaturase genes in the cyanobacterium Synechococcus sp. strain PCC 7002. Mol Microbiol 23: 1281–1292

Sakamoto T, Wada H, Nishida I, Ohmori M and Murata N (1994a) Δ9 acyl-lipid desaturase of cyanobacteria: Molecular cloning and substrate specificities in terms of fatty acids, sn-positions, and polar head groups. J Biol Chem 269: 25576–25580

Sakamoto T, Wada H, Nishida I, Ohmori M and Murata N (1994b) Identification of conserved domains in the Δ12 desaturase of cyanobacteria. Plant Mol Biol 24: 643–650

Sakamoto T, Los DA, Higashi S, Wada H, Nishida I, Ohmori M and Murata N. (1994c) Cloning of ω3 desaturase from cyanobacteria and its use in altering the degree of membrane-lipid unsaturation. Plant Mol Biol 26: 249–263

Sato N and Murata N (1982) Lipid biosynthesis in the blue-green alga, Anabaena variabilis. I. Lipid classes. Biochim Biophys Acta 710: 271–278

Schmidt H, Dresselhaus T, Buck F, Heinz E (1994) Purification and PCR-based cDNA cloning of a plastidial n-6 desaturase. Plant Mol Biol 26: 631–642

Shanklin J and Somerville C (1991) Stearoyl-acyl-carrier-protein desaturase from higher plants is structurally unrelated to the animal and fungal homologs. Proc Natl Acad Sci USA 88: 2510–2514

Shanklin J, Mullins C and Somerville C (1991) Sequence of a complementary DNA from Cucumis sativus L. encoding the stearoyl-acyl-carrier protein desaturase. Plant Physiol 97: 467–468

Shimojima M, Ohta H, Iwamatsu A, Masuda T, Shioi Y and Takamiya K (1997) Cloning of the gene for monogalactosyldiacylglycerol synthase and its evolutionary origin. Proc Natl Acad Sci USA 94: 333–337

Shorrosh BS, Roesler KR, Shintani D, van de Loo FJ and Ohlrogge JB (1995) Structural analysis, plastid localization, and expression of the biotin carboxylase subunit of acetyl-coenzyme A carboxylase from tobacco. Plant Physiol 108: 805–812

Siefermann-Harms D, Ross JW, Kaneshiro KH and Yamamoto HY (1982) Reconstitution by monogalactosyldiacylglycerol of energy transfer from light-harvesting chlorophyll a/b-protein complex to the photosystems in Triton X-100-solubilized thylakoids. FEBS Lett 149: 191–196

Siegenthaler PA and Vallino J (1995) Effect of a selective depletion of acyl lipids on the light and dark phosphorylation in spinach thylakoid membranes. In: Mathis P (ed) Photosynthesis: from Light to Biosphere, Vol III, pp 225–228. Kluwer Academic Publishers, Dordrecht

Siegenthaler PA, Smutny J and Rawyler A (1984) Involvement of hydrophilic and hydrophobic portions of phospholipid molecules in photosynthetic electron flow activities. In: Siegenthaler PA and Eichenberger W (eds) Structure, Function and Metabolism of Plant Lipids, pp 475–478. Elsevier Science Publishers BV, Amsterdam

Siegenthaler PA, Smutny J and Rawyler A (1987a) Involvement

of distinct populations of phosphatidylglycerol and phosphatidylcholine molecules in photosynthetic electron-flow activities. Biochim Biophys Acta 891:85–93

Siegenthaler PA, Rawyler A and Giroud C (1987b) Spatial organization and functional roles of acyl lipids in thylakoid membranes. In: Stumpf PK, Mudd JB and Nes WD (eds) The Metabolism, Structure and Function of Plant Lipids, pp 161–168. Plenum Press, New York

Siegenthaler, PA, Rawyler, A. and Smutny, J. (1989) The phospholipid population which sustains the uncoupled non-cyclic electron flow activity is localized in the inner monolayer of the thylakoid membrane. Biochim Biophys Acta 975: 104–111

Siggaard-Andersen M, Kauppinen S and von Wettstein-Knowles P (1991) Primary structure of a cerulenin-binding β-ketoacyl-[acyl carrier protein] synthase from barley chloroplasts. Proc Natl Acad Sci USA 88: 4114–4118

Singh S, McKinney S and Green A (1994) Sequence of a cDNA from *Linum usitatissimum* encoding the stearoyl-acyl carrier protein desaturase. Plant Physiol 104: 1075

Slabaugh MB, Tai H, Jaworski JG and Knapp SJ (1995) cDNA clones encoding β-ketoacyl-acyl carrier protein synthase III from *Cuphea wrightii*. Plant Physiol 108: 443–444

Slocombe SP, Cummins I, Jarvis RP and Murphy DJ (1992) Nucleotide sequence and temporal regulation of a seed-specific *Brassica napus* cDNA encoding a stearoyl-acyl carrier protein (ACP) desaturase. Plant Mol Biol 20: 151–155

Tai H and Jaworski JG (1993) 3-Ketoacyl-acyl carrier protein synthase III from spinach (*Spinacia oleracea*) is not similar to other condensing enzymes of fatty acid synthase. Plant Physiol 103: 1361–1367

Tai H, Post-Beittenmiller D and Jaworski JG (1994) Cloning of cDNA encoding 3-ketoacyl-acyl carrier protein synthase III from *Arabidopsis*. Plant Physiol 106: 801–802

Tasaka Y, Gombos Z, Nishiyama Y, Mohanty P, Ohba T, Ohki K and Murata N (1996) Targeted mutagenesis of acyl-lipid desaturases in *Synechocystis*: Evidence for the important roles of polyunsaturated membrane lipids in growth, respiration and photosynthesis. EMBO J 15: 6416–6425

Thomas PG, Dominy PJ, Vigh L, Mansourian AR, Quinn PJ and Williams WP (1986) Increased thermal stability of pigment-protein complexes of pea thylakoids following catalytic hydrogenation of membrane lipids. Biochim Biophys Acta 849: 131–140

Töpfer R and Martini M (1994) Molecular cloning of cDNAs or genes encoding proteins involved in de novo fatty acid biosynthesis in plants. J Plant Physiol 143: 416–425

van de Loo FJ, Turner S and Somerville C (1995) Expressed sequence tags from developing castor seeds. Plant Physiol 108: 1141–1150

Verwoert IIGS, Brown A, Slabas AR and Stuitje AR (1995) A *Zea mays* GTP-binding protein of the ARF family complements an *Escherichia coli* mutant with a temperature-sensitive malonyl-coenzyme A:acyl-carrier protein transacylase. Plant Mol Biol 27: 629–633

Vigh L, Joó F, Droppa M, Horváth LI and Horváth G (1985) Modulation of chloroplast membrane lipids by homogeneous catalytic hydrogenation. Eur J Biochem 147: 477–481

Vigh L, Gombos Z, Horváth I and Joó F (1989) Saturation of membrane lipids by hydrogenation induces thermal stability in chloroplast inhibiting the heat-dependent stimulation of Photosystem I-mediated electron transport. Biochim Biophys Acta 979: 361–364

Vigh L, Los DA, Horváth I and Murata N (1993) The primary signal in the biological perception of temperature: Pd-catalyzed hydrogenation of membrane lipids stimulated the expression of the *desA* gene in *Synechocystis* PCC 6803. Proc Natl Acad Sci USA 90: 9090–9094

Wada H, Gombos Z and Murata N (1990) Enhancement of chilling tolerance of a cyanobacterium by genetic manipulation of fatty acid desaturation. Nature 347: 200–203

Weber S, Wolter F-P, Buck F, Frentzen M and Heinz E (1991) Purification and cDNA sequencing of an oleate-selective acyl-ACP:sn-glycerol-3-phosphate acyltransferase from pea chloroplasts. Plant Mol Biol 17: 1067–1076

Willms I, Malkin R and Chain RK (1987) Oxidation-reduction reactions of cytochrome b_6-f complex. Arch Biochem Biophys 258: 248–258

Wissenbach M, Siggaard-Andersen M, Kauppinen S and von Wettstein-Knowles P (1992) Condensing enzymes of barley. In: Cherif A, Miled-Daund DB, Marzouk B, Smaoui A and Zarrouk M (eds) Metabolism, Structure and Utilization of Plant Lipids, pp 393–396. Centre National Pédagogique, Tunisie

Wu J and Browse J (1995) Elevated levels of high-melting-point phosphatidylglycerols do not induce chilling sensitivity in an *Arabidopsis* mutant. Plant Cell 7: 17–27

Wu J, Lightner J, Warwick N and Browse J (1997) Low-temperature damage and subsequent recovery of *fab*1 mutant *Arabidopsis* exposed to 2 °C. Plant Physiol 113: 347–356

Yadav NS, Wierzbicki A, Aegerter M, Caster CS, Pérez-Grau L, Kinney AJ, Hitz WD, Booth JR Jr, Schweiger B, Stecca KL, Allen SM, Blackwell M, Reiter RS, Carlson TJ, Russell SH, Feldmann KA, Pierce J and Browse J. (1993) Cloning of higher plant ω-3 fatty acid desaturases. Plant Physiol 103: 467–476

Yukawa Y, Takaiwa F, Shoji K, Masuda K and Yamada K (1996) Structure and expression of two seed-specific cDNA clones encoding stearoyl-acyl carrier protein desaturase from sesame, *Sesamum indicum* L. Plant Cell Physiol 37: 201–205

Chapter 2

Structure, Distribution and Biosynthesis of Glycerolipids from Higher Plant Chloroplasts

Jacques Joyard, Eric Maréchal, Christine Miège, Maryse A. Block,
Albert-Jean Dorne and Roland Douce

*Laboratoire de Physiologie Cellulaire Végétale, URA CNRS n°576 (CEA/CNRS/Université
Joseph Fourier, Grenoble), Département de Biologie Moléculaire et Structurale, CEA-Grenoble,
17 rue des Martyrs, F-38054, Grenoble-cédex, France*

P.-A. Siegenthaler and N. Murata (eds): Lipids in Photosynthesis: Structure, Function and Genetics, pp. 21–52.

Summary

Galactolipids (MGDG and DGDG), sulfolipid and phosphatidylglycerol are the main constituents of plastid membranes. Glycerolipid biosynthesis requires first the assembly of glycerol and esterification by fatty acids at the sn-1 and sn-2 positions of the glycerol backbone. Then, the sn-3 position of phosphatidic acid or diacylglycerol is modified, for instance by addition of a third fatty acid for triacylglycerol, of a galactose for galactolipids, of a sulfoquinovose for sulfolipid, and phosphorylglycerol for phosphatidylglycerol. Directly or indirectly, the compounds used for the biosynthesis of glycerolipids derive from photosynthesis, i.e. from endogenous CO_2 fixation by chloroplasts or from photosynthates produced in leaves. The two main MGDG molecular species found in chloroplasts have (a) 18:3 at both the sn-1 and sn-2 positions of the glycerol backbone, and (b) 18:3 and 16:3 respectively at the sn-1 and sn-2 positions of the glycerol backbone. The occurrence of such structures within plastid membranes reflects the existence of different pathways for the biosynthesis of these two types of molecules. Sulfolipid and phosphatidylglycerol molecular species also contain the typical structure of prokaryotic lipids with C16 fatty acids at the sn-2 position of glycerol. In contrast, a wide variety of diacylglycerol molecular species (i.e. with different acyl chain length and saturation levels at both sn positions) can be found in extremely variable amounts in envelope membranes where the synthesis of all typical plastid lipids takes place. Several enzymes, such as the inner envelope phosphatidate phosphatase and the outer envelope galactolipid:galactolipid galactosyltransferase, are involved in diacylglycerol formation, others, like the MGDG synthase or the sulfolipid synthase, use diacylglycerol within the inner envelope membrane as a substrate for the biosynthesis of membrane glycerolipids. A puzzling question is how such enzymes could be involved in the formation of the characteristic structural features of chloroplast glycerolipids and their final distribution within membranes. Although little molecular data are presently available on enzymes such as the phosphatidate phosphatase, the galactolipid:galactolipid galactosyltransferase or the MGDG synthase, detailed analysis of the biochemical properties of these key enzymes in galactolipid biosynthesis recently provided some clues to the problem. These observations suggest that the biochemical properties of the envelope MGDG synthase are highly responsible for the final MGDG molecular species found in plastid membranes.

I. Introduction

Most chloroplast membrane glycerolipids are glycolipids (galactolipids and sulfolipid), in contrast with other plant cell membranes in which phospholipids are the major constituent. Therefore, the total amount of galactolipids present in a plant tissue reflects the expansion of chloroplast membranes whereas the total amount of some phospholipids such as phosphatidylethanolamine reflects the expansion of extraplastidial membranes. For instance, in green photosynthetic tissues that contain fully developed thylakoid membranes, the amount of galactolipids exceeds the amount of phospholipids, whereas the reverse is true in non-photosynthetic tissues, such as potato tubers or cauliflower buds,

which contain only few thylakoid membranes. Indeed, thylakoids represent the main plastid membrane system in the cell (300–1000 m^2 per m^2 of leaves, i.e. about 20 times the envelope membranes area) and therefore contain the largest amount of plastid glycerolipids. In 1 m^2 of leaves, there are about 2.5 g of chloroplast lipids (representing 20% of a chloroplast dry mass), of which more than 2 g are galactolipids, that are mostly concentrated in the $4.1 \cdot 10^{-18}$ m^3 of the corresponding thylakoids (Lawlor, 1987). Finally, because thylakoids are probably the most widely developed membrane system on earth, monogalactosyldiacylglycerol (MGDG)—the major lipid in thylakoid membrane—is probably the most abundant polar lipid on earth (Gounaris and Barber, 1983).

This article is devoted to chloroplast glycerolipid structure, distribution and biosynthesis. For a comprehensive survey of the literature (up to 1987) on such topics, the reader is referred to reviews published in the series *Biochemistry of Plants* (Douce and Joyard, 1980; Harwood, 1980a,b; Mudd, 1980;

Abbreviations: ACP – acyl carrier protein; DGDG – digalactosyldiacylglycerol; DPG – diphosphatidylglycerol; MGDG – monogalactosyldiacylglycerol; PC – phosphatidylcholine; PE – phosphatidylethanolamine; PG – phosphatidylglycerol; PI – phosphatidylinositol; SL – sulfolipid; UDP – uridine diphosphate

Joyard and Douce, 1987; Mudd and Kleppinger-Sparace, 1987; Pollard, 1987). Several other articles provide further information concerning more recent work (Douce and Joyard, 1990; Slabas and Fawcett, 1992; Frentzen, 1993; Joyard et al., 1993, 1994; Kinney, 1993; Gunstone, 1994; Williams, 1994; Ohlrogge and Browse, 1995; Thompson, 1996; Douce and Joyard, 1996). Important aspects on the organization and role of acyl lipids in photosynthetic membranes is discussed elsewhere in this volume (Siegenthaler, Chapter 7; Siegenthaler and Trémolières, Chapter 8; Selstam, Chapter 11). We still know very little on enzymes that catalyze glycerolipid biosynthesis, i.e. the assembly of the three parts of a glycerolipid molecule namely glycerol, the polar head group and fatty acids. In contrast, major progress have been made in the understanding, at a molecular level, of enzymes involved in fatty acid biosynthesis and desaturation (for reviews, see Somerville and Browse, 1991; Heinz, 1993; Slabas et al., 1994; Ohlrogge and Browse, 1995; Harwood, 1996; Somerville and Browse, 1996; and in this volume see Wada and Murata, Chapter 4; Gombos and Murata, Chapter 13; Vijayan et al., Chapter 14). One of the question which we want to address is how such enzymes could be responsible for establishing the characteristic structural features of chloroplast glycerolipids and their final distribution within membranes. Therefore, having discussed in more general terms how lipids are made in chloroplasts, we will focus on recent observations that provide some understanding on how enzymes involved in MGDG biosynthesis could be responsible for the accumulation of specific molecular species in chloroplast membranes.

II. Structure and Distribution of Chloroplast Glycerolipids

Plastid membranes are characterized by the presence of large amounts of glycolipids: sulfolipid and galactolipids. The major plastid glycolipids, galactolipids, are neutral lipids. They contain one or two galactose molecules attached to the sn-3 position of the glycerol backbone (Fig. 1), corresponding to 1,2-diacyl-3-O-(β-D-galactopyranosyl)-sn-glycerol (or monogalactosyldiacylglycerol, MGDG) and 1,2-diacyl-3-O-(α-D-galactopyranosyl-(1$\rightarrow$6)-O-β-D-galactopyranosyl)-sn-glycerol (or digalactosyldiacylglycerol, DGDG). Galactolipids represent up to 80%

of thylakoid membrane glycerolipids, from which MGDG constitute the main part (50%). A unique feature of galactolipids is their very high content of polyunsaturated fatty acids: in some species, up to 95% of the total fatty acids is linolenic acid (18:3). Therefore, the most abundant molecular species of galactolipids have 18:3 at both the sn-1 and sn-2 positions of the glycerol backbone (Fig. 1). Some plants, such as pea, which have almost exclusively 18:3 in MGDG are called '18:3 plants'. Other plants, such as spinach, which contain large amounts of 16:3 in MGDG are called '16:3 plants' (Heinz, 1977). The positional distribution of 16:3 in MGDG is highly specific: this fatty acid is present at the sn-2 position of glycerol and is almost excluded from the sn-1 position. Therefore, different galactolipid molecular species with either C18 fatty acids at both sn position or with C18 fatty acids at the sn-1 and C16 fatty acids at the sn-2 positions can be present in plastid membranes. The proportions of these two types of MGDG molecular species vary widely among plants (Heinz, 1977). The first structure is typical of 'eukaryotic' lipids (such as phosphatidylcholine) and the second one corresponds to a 'prokaryotic' structure, since it is characteristic of cyanobacterial glycerolipids (Heinz, 1977; see also Wada and Murata, Chapter 4, this volume; refer to Siegenthaler and Murata, Chapter 1 Introduction, for the terminology). In addition, this difference is also true of other glycerolipids since any membrane lipid containing C16 fatty acids at the sn-2 position of glycerol is considered as prokaryotic. The proportion of eukaryotic to prokaryotic molecular species is not identical in all glycerolipids from a given plant. For instance, although in spinach half of MGDG has the prokaryotic structure, this holds true for only 10-15% of DGDG (Bishop et al., 1985).

The most important sulfolipid found in higher plants is a 1',2'-diacyl-3'-O-(6-deoxy-6-sulfo-α-D-glucopyranosyl)-sn-glycerol (sulfoquinovosyldiacylglycerol or SQDG). The sulfonic residue at C6 of deoxyglucose (quinovose) carries a strong negative charge at physiological pH (Fig. 1). This glycolipid is also found in cyanobacteria (Heinz, 1977). Analyses of the positional distribution of fatty acids (see for instance Siebertz et al., 1979) demonstrate that a significant proportion of SQDG in higher plants has a dipalmitoyl backbone. The presence and absence of dipalmitoyl SQDG depend on plant species (Murata and Hoshi, 1984). However, the major molecular species in SQDG contain both 16:0 and

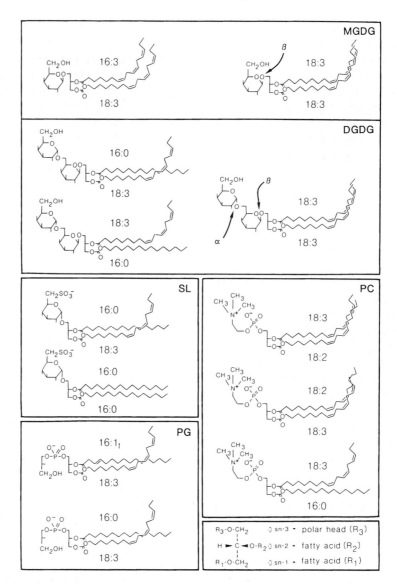

Fig. 1. Structure of chloroplast glycerolipids. Abbreviations: MGDG, monogalactosyldiacylglycerol; DGDG, digalactosyldiacylglycerol; SL, sulfolipid; PG, phosphatidylglycerol; PC, phosphatidylcholine. Reproduced with permission from Joyard and Douce (1987).

18:3 fatty acids (Siebertz et al., 1979). In fact, two distinct structures can be observed in higher plants (Fig. 1), they contain either 18:3/16:0 or 16:0/18:3 (Siebertz et al., 1979; Bishop et al., 1985). As discussed above, the structures having 16:0 at the *sn*-2 position (16:0/16:0 and 18:3/16:0), are typical of '*prokaryotic*' lipids whereas the last one (16:0/18:3), having 18:3 at the *sn*-2 position, is typical of '*eukaryotic*' lipids (see above). In a 16:3 plant, such as spinach, a higher proportion of SQDG (compared to the situation for MGDG) has a prokaryotic structure. In contrast, wheat, a 18:3 plant, contains

almost exclusively SQDG with an eukaryotic structure (Bishop et al., 1985). Interestingly, this glycolipid is present in the outer leaflet of the outer envelope membrane since it is accessible from the cytosolic face of isolated intact chloroplasts to specific antibodies, like MGDG (Billecocq et al., 1972).

Phosphatidylglycerol is a genuine constituent of all extraplastidial membranes, but it is present in low amounts (in general it represents only a few percent of the total glycerolipid content). In contrast, it is the major plastid phospholipid (Fig. 2). Phosphatidylglycerol represents (in spinach) about 7–10% of

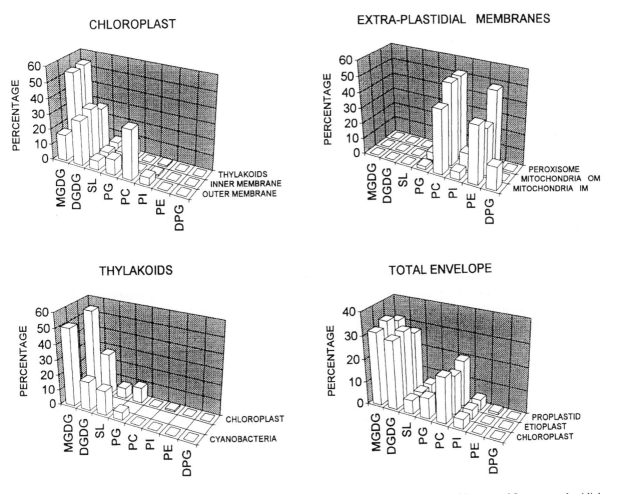

Fig. 2. Glycerolipid composition of chloroplast membranes, of envelope membranes from various plastid types and from extra-plastidial membranes. Abbreviations: IM: inner membrane; OM, outer membrane; PI, phosphatidylinositol; PE, phosphatidylethanolamine; DPG, diphosphatidylglycerol; see also Fig. 1. Reproduced with permission from Joyard et al. (1997).

the total glycerolipid of thylakoids and of the envelope (outer as well as inner) membranes (Fig. 2). However, phospholipase *c* digestion of intact chloroplasts suggest that, in contrast to phosphatidylcholine, little phosphatidylglycerol is present in the outer leaflet of the outer membrane (Dorne et al., 1985). Plastid phosphatidylglycerol has a unique structure (Dorne and Heinz, 1989) since (a) it has a typical prokaryotic structure, with C16 fatty acid esterified at the *sn*-2 position of the glycerol backbone and (b) contains a unique *trans*-hexadecenoic acid ($16:1_t$) at the *sn*-2 position of the glycerol backbone (Haverkate and Van Deenen, 1965). No eukaryotic structure of phosphatidylglycerol (with C18 fatty acids at the *sn*-2 position of the glycerol backbone) was reported in chloroplasts (Bishop et al., 1985; Dorne and Heinz,

1989). Because of this uniqueness, specific functions for chloroplast phosphatidylglycerol have been proposed in plants. Although difficult to demonstrate unambiguously, this hypothesis is supported by the following observations. The chilling sensitivity of plants can be correlated with the occurrence of high-melting 16:0/16:0 and $16:0/16:1_t$ combinations in plastid phosphatidylglycerol (Murata et al., 1982; Bishop, 1986; see also Siegenthaler and Trémolières, Chapter 8). Correlations have also been made between phosphatidylglycerol accumulation and the development of appressed membranes: for instance, mutants lacking PSII reaction centers were shown to be lacking phosphatidylglycerol (Garnier et al., 1990; Trémolières et al., 1991). Finally, Nussberger et al. (1993) have demonstrated that LHCII complex binds specifically with DGDG and phosphatidylglycerol, this phospholipid being intimately involved in trimer formation (Hobe et al., 1994) and is likely to be located at the subunit interface. These observations provide further evidence for an asymmetrical distribution of thylakoid glycerolipids. The topological studies which led to this distribution is described in this volume by Siegenthaler (Chapter 7).

Phosphatidylcholine is only a minor constituent of plastid membranes (Fig. 2). In contrast, this phospholipid represents about half of the glycerolipid content of extraplastidial membranes such as mitochondrial or peroxysomal membranes. The only plastid membrane containing significant amounts of phosphatidylcholine is the outer envelope membrane, with 30 to 35% of the glycerolipid content (Cline et al., 1981; Block et al., 1983b). The marked difference in the composition of the inner and outer membrane systems of chloroplast envelope has been established due to the development of reliable methods for the separation of these two structures (Cline et al., 1981; Block et al., 1983a). If the association between the two membrane systems is broken, the two membranes can be clearly resolved using density gradient centrifugation. The lower protein to lipid ratio in the outer membrane results in the banding of the outer membrane at a lower density (1.08 g.cm^{-3}) than the inner membrane (1.13 g.cm^{-3}). Phospholipase c digestion of intact chloroplasts provided evidence that chloroplast phosphatidylcholine is actually concentrated in the outer leaflet of the outer membrane (Dorne et al., 1985; Miquel et al., 1987). Using isolated intact chloroplasts in which envelope phosphatidylcholine was removed by phospholipase c treatment, Dorne et al. (1990) demonstrated that the presence of this phospholipid in thylakoid preparations is probably due to contaminating outer envelope membranes and therefore that thylakoids (and probably the inner envelope membrane) are devoid of phosphatidylcholine, like membranes from cyanobacteria (Chapter 4). These observations raise the problem of the apparent lack of transmembrane diffusion of phosphatidylcholine in the outer envelope membrane: although it is a major component in the outer leaflet of the outer membrane, phosphatidylcholine is not redistributed to the thylakoids (Dorne et al., 1990).

Figure 2 also shows that all plastid membranes, and most notably the outer envelope membrane, are devoid of phosphatidylethanolamine, one of the major components (together with phosphatidylcholine) of extra-plastidial membranes. This observation provides further evidence for a prokaryotic origin of envelope membranes, especially for the outer membrane.

To conclude, plastid membranes are characterized by a unique glycerolipid composition, strikingly different from other plant cell membranes, and share common features (Fig. 2). However, although the inner membrane and thylakoids are rather similar, the outer and the inner envelope membranes are not identical (Fig. 2). For instance, in addition to the presence of large amounts of phosphatidylcholine (on the cytosolic side of membrane), the outer membrane is enriched in DGDG, whereas the inner membrane (and thylakoids) are enriched in MGDG. In addition, envelope membranes from proplastids, etioplasts and chloroplasts have an almost identical lipid composition (Fig. 2). Together, these observations raise several puzzling questions and especially what controls the lipid composition of membranes, and especially plastid membranes, and how such a unique lipid composition of plastid membranes is maintained within the plant cell and among different tissues. No clear answers can be presently provided. This situation is partially relevant to the dynamic of the system which involves the massive transport of membrane lipids from their site of synthesis (the inner envelope membrane) to their site of accumulation (the thylakoids and, on a much smaller scale, the outer envelope membrane). This is especially important during plastid development (for reviews, see Douce and Joyard, 1996; and Selstam, Chapter 11).

III. Photosynthesis and Lipid Biosynthesis

Labeling studies using $^{14}CO_2$ and ^{14}C-acetate have demonstrated that lipid biosynthesis occurs in almost all plant tissues, in leaves as well as in seed cotyledons (for a review, see Roughan and Slack, 1982). Leaves mostly synthesize membrane glycerolipids (phospholipids, glycolipids, etc.) whereas seeds accumulate triacylglycerol as storage product in oil bodies (for a review, see Murphy, 1994). Glycerolipid biosynthesis requires first the assembly of glycerol and esterification by fatty acids at the sn-1 and sn-2 positions of the glycerol backbone. Then, the sn-3 position of phosphatidic acid or diacylglycerol is modified, for instance by addition of a third fatty acid for triacylglycerol or of a galactose for galactolipids.

Directly or indirectly, the compounds used for the biosynthesis of glycerolipids derive from photosynthesis, i.e. from endogenous CO_2 fixation by chloroplasts or from photosynthates produced in leaves. For instance, direct relationship between photosynthesis and triacylglycerol production has been clearly demonstrated in olive fruits (see, for instance, Sanchez, 1994, and for a review, Facciotti et al., Chapter 12).

A. The Path of Carbon in Fatty Acids

The major site of synthesis for fatty acids in the plant cell is the plastid stroma (for reviews, see Ohlrogge et al., 1993; Harwood, 1996). In this regard, the process of lipid biosynthesis in plants is different from animals and fungi, which produce fatty acids primarily in the cytosol (Ohlrogge and Browse, 1995). In addition, all fatty acid biosynthesis in plastids is acyl carrier protein (ACP)-dependent (for reviews, Ohlrogge et al., 1993; Harwood, 1996). The importance of plastid ACP isoforms in lipid biosynthesis is described in a recent and comprehensive review (Harwood, 1996). The observation that antibodies to *Neurospora* ACP cross-reacted with a protein in pea leaf mitochondria, but not with a chloroplast extract, suggested that an immunologically distinct ACP was present in pea mitochondria (Chuman and Brody, 1989). Indeed, further molecular and functional evidences for a mitochondrion-specific ACP form were provided by Shintani and Ohlrogge (1994). In addition, a cytosolic acetyl-CoA carboxylase having eukaryotic features has been characterized in leaf epidermal cells (Alban et al., 1994; and see below). Therefore, the possibility that some fatty acid synthesis could take place outside the plastidial compartment cannot be entirely ruled out, although the flux of carbon to fatty acids is essentially due to plastid metabolism.

The fatty acid biosynthetic pathway in plants is a primary metabolic pathway that is closely linked to photosynthesis: ATP and reducing equivalents used for fatty acid synthesis in the chloroplast stroma are formed by photosynthesizing thylakoids (for a review, see for instance Roughan and Slack, 1982). Fatty acid biosynthesis also takes place in the dark and in non-green plastids, but at much lower rates, and reducing equivalents are generated by oxidative processes such as glycolysis and the oxidative pentose phosphate pathway (for a review, see Browsher et al., 1996).

1. The Source of C2 Units for Fatty Acid Biosynthesis

Almost no $^{14}CO_2$ can be incorporated directly into lipids when added to isolated intact chloroplasts whereas incorporation rates as high as 1500 nmol [^{14}C]-acetate incorporated per hour per mg chlorophyll have been measured in spinach chloroplasts having high bicarbonate-dependent O_2 evolution (for a review, see Roughan and Slack, 1982). In chloroplasts, fatty acids are built up from a two-carbon molecule, acetyl-CoA. According to Ohlrogge and Browse (1995), 'our understanding of how carbon moves from photosynthesis to acetyl-CoA is clouded by an abundance of potential pathways.' Indeed, acetyl-CoA can be potentially generated by a series of biosynthetic routes and enzymes such as pyruvate decarboxylase/dehydrogenase and acetyl-CoA synthetase or from cytosolic malate and glucose-6-phosphate (for reviews, see for instance Roughan and Slack, 1982; Stymne and Stobart, 1987; Ohlrogge and Browse, 1995; Browsher et al., 1996; Harwood, 1996). For instance, chloroplasts contain a powerful acetyl-CoA synthetase that is probably the main source of C2 units for fatty acid biosynthesis in photosynthesizing organelles (for a review, see Roughan and Slack, 1982). In most non-green tissues, cytoplasmic sucrose, synthesized in leaves during photosynthesis and transported to sink tissues, is converted into hexose-phosphate, one of the major form in which carbohydrate is imported into non-green plastids, which is then converted into acetyl-CoA by the intra-plastidial glycolytic pathway (for a review, see Browsher et al., 1996). Usually, fatty acid synthesis is analyzed using [^{14}C]-acetate as a source of C2 units. However, it is clear from studies with plastids from developing castor bean endosperm and oil-seed rape embryos that other sources of C2 units, such as pyruvate or glucose-6-phosphate, can sustain much higher rates of fatty acid synthesis than acetate (for a review, see Browsher et al., 1996). Therefore, the origin of acetyl-CoA is closely linked to the plant species, the tissues analyzed and their developmental stage as well as the physiological conditions, and it is very difficult to understand how the products of photosynthesis are converted into acetyl-CoA for fatty acid synthesis.

2. Acetyl-CoA Carboxylase is a Key Step in Fatty Acid Biosynthesis

The first committed step in fatty acid biosynthesis is catalyzed by the enzyme acetyl-CoA carboxylase. Acetyl-CoA carboxylase (EC 6.4.1.2), a biotin-dependent enzyme, catalyzes the ATP-dependent formation of malonyl-CoA from acetyl-CoA and bicarbonate (Knowles, 1989; Harwood, 1996). Alban et al. (1994) demonstrated that young pea leaves contain two structurally different forms of acetyl-CoA carboxylase. A minor form (accounting for about 20% of the total activity in the whole leaf) was detected in the epidermal tissue. This enzyme is soluble and was purified to homogeneity and consist of a dimer of two identical biotinyl subunits of 220 kDa (Alban et al., 1994). This multifunctional enzyme occurs outside plastids (probably the cytosol) and is comparable to the cytosolic form found in most eukaryotes (Alban et al., 1994). Therefore, this enzyme is refereed to as the eukaryotic form of acetyl-CoA carboxylase. The function of this cytosolic form is probably to supply malonyl-CoA for a variety of pathways, including flavonoid biosynthesis and fatty acid elongation for the production of very long chain fatty acids of the cuticle (for a review, see Harwood, 1996). The other (i.e. the major) form of acetyl-CoA carboxylase is localized in plastids as a freely dissociating complex (700 kDa), the activity of which may be restored by combination of its separated constituents (Sasaki et al., 1993; Alban et al., 1994; Konishi and Sasaki, 1994). This dissociable form is able to carboxylate free D-biotin as an alternate substrate in place of the natural substrate, biotin carboxyl carrier protein (BCCP) (Alban et al., 1995). One of the constituents of the enzyme, the biotin carboxylase, free from carboxyltransferase activity, was recently purified from pea chloroplasts (Alban et al., 1995). It is composed of two firmly bound polypeptides, one of which is a biotinylated 38 kDa-polypeptide (Alban et al., 1995). The biotin carboxylase is nuclear encoded (Shorrosh et al., 1995). The gene for another polypeptide of the prokaryotic acetyl-CoA carboxylase has been identified in the plastid genome by its homology to one of the carboxyltransferase subunits of the *Escherichia coli* acetyl-CoA carboxylase (for a review, see Ohlrogge and Browse, 1995). Biochemical parameters were determined for both the eukaryotic and the prokaryotic forms from pea leaves and a different sensitivity towards the selective grass herbicide aryloxyphenoxypropionate and cyclohexanedione types was observed: only the eukaryotic form is inhibited and not the chloroplast enzyme (Alban et al., 1994; Dehaye et al., 1994). Interestingly these herbicides strongly inhibits acetyl-CoA carboxylase from monocots, such as maize or wheat, and have almost no effect on the enzyme from dicots (for a review, see Harwood, 1996). In fact, chloroplasts from monocots do not contain the prokaryotic form of acetyl-CoA carboxylase, but they contain two eukaroytic isoenzymes of acetyl-CoA carboxylase, i.e. the herbicide sensitive form (for a review, see Harwood, 1996).

To conclude from the increasing amount of data becoming available on acetyl-CoA carboxylase, this enzyme is clearly a key regulatory enzyme in fatty acid biosynthesis (for reviews, see Ohlrogge and Browse, 1995; Harwood, 1996).

3. Fatty Acid Synthesis Pathway

The organization of the enzymes involved in the formation, within plastids, of C16 and C18 fatty acids is, again, strikingly different from that observed in fungi or animals. The main difference is that the enzymes involved are not assembled as a multifunctional protein complex, like type I synthases found in eukaryotes (for a review, see Wakil, 1989), but in a somewhat loose protein complex from which all the enzymes can be easily separated (type II synthase, for review see Rock and Cronan, 1996). As described in Fig. 3, acetyl-CoA enters the fatty acid pathway both as a substrate for acetyl-CoA carboxylase and as a primer for the condensation reaction that is catalyzed by a specific 3-ketoacyl-ACP synthase (namely 3-ketoacyl-ACP synthase III; Jaworski et al., 1989) and leads to the formation of 3-ketobutyryl-ACP, a four-carbon molecule linked as a thioester to ACP. All fatty acid synthesis in plastids is ACP-dependent and malonyl-ACP is the three-carbon group acting as carbon donor for all further elongation steps that take place in the stroma (Fig. 3). Therefore, once synthesized by the acetyl-CoA carboxylase, the malonyl group from malonyl-CoA is transferred from CoA to ACP owing to a malonyl-CoA:ACP transacylase (Fig. 3). In attempt to isolate a cDNA clone corresponding to this enzyme, an *Escherichia coli* mutant (*fabD*) was complemented with a maize cDNA expression library and a *Zea mays* cDNA clone encoding a GTP-binding protein of the ARF family was isolated (Verwoert et al., 1995).

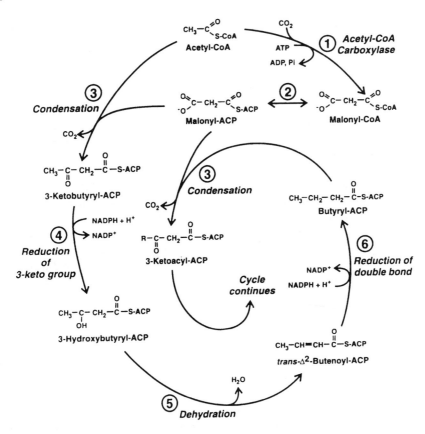

Fig. 3. Biosynthesis of saturated fatty acids in chloroplasts. The reactions involved are: (1) acetyl-CoA carboxylase, (2) malonyl-CoA:ACP transacylase, (3) 3-ketoacyl-ACP synthase, (4) 3-ketoacyl-ACP reductase, (5) a hydroacyl-ACP dehydratase and (6) enoyl-ACP reductase. Only the first condensation reaction is represented in the figure. Reproduced with persmission from Ohlrogge and Browse (1995).

Interestingly, the crystal structure of the *Escherichia coli* enzyme at 1.5 Å was obtained (Serre et al., 1996).

The biosynthesis of C16 fatty acids (Fig. 3) is characterized by a two-carbon elongation using a three-carbon building block (malonyl-ACP) with the release of one carbon atom, as CO_2. Because of the release of CO_2, the elongation reaction catalyzed by all 3-ketoacyl-ACP synthases is irreversible. The elongation of 3-ketobutyryl-ACP (C2 to C14 acyl-ACPs) to C16 fatty acids is catalyzed by a second 3-ketoacyl-ACP synthase (probably 3-ketoacyl-ACP synthase I) whereas the last elongation step, from C16 to C18 fatty acid is catalyzed by a third 3-ketoacyl-ACP synthase (3-ketoacyl-ACP synthase II). After the first condensation step, the ketobutyryl-ACP molecule formed by the condensation of acetyl-CoA and malonyl-ACP is transformed into butyryl-ACP by the successive action of three enzymes: (a) a 3-ketoacyl-ACP reductase which reduces the 3-keto

group, using NADPH as the electron donor, (b) a hydroacyl-ACP dehydratase, which remove a water molecule, and (c) an enoyl-ACP reductase, which reduces the *trans*-2 double bond (using NADH or NADPH) to synthesize a saturated molecule. The same series of reaction occurs after each elongation step, leading to the formation of 16- and 18-carbon saturated fatty acids linked to ACP (Fig. 3).

The enzymes involved in fatty acid biosynthesis have been purified first in spinach (Shimakata and Stumpf, 1982a,b,c; 1983). Since, purified enzymes are now available from various plants (avocado, rape, pea, soybean, barley, etc.) and their biochemical characteristics (cerulenin and thiolactomycin sensitivity, for instance) have been determined extensively (for a review, see Harwood, 1996). Molecular data are also available for almost all enzymes of the fatty acid biosynthetic pathway, except the hydroacyl-ACP dehydratase (for a review, see Harwood, 1996). In general, our knowledge, at both a biochemical and a molecular level, of the chloroplast enzymes involved in fatty acid biosynthesis has benefited from the numerous information now available from *Escherichia coli* (for a review, see Rock and Cronan, 1996). The different condensing enzymes or reductases involved in fatty acid elongation in *Escherichia coli* show high homologies with the corresponding chloroplast enzymes. Therefore, *Escherichia coli* enzymes were used to provide sequences data to generate probes for plant cDNAs by using PCR (polymerase chain reaction). For instance, all 3-ketoacyl-ACP synthases, from various prokaryotes (*Escherichia coli*) as well as from eukaryotes (yeast, rat, chicken, barley, spinach, etc.) share an active cysteine flanked by an upstream alanine and either an alanine or a serine upstream (Jaworski et al., 1994). The enzyme 3-ketoacyl-ACP synthase I, which is responsible for the bulk of the condensation reactions (Fig. 3), was shown (Siggaard-Andersen et al., 1991) to contain three dimeric forms ($\alpha\alpha$, $\beta\beta$ and $\alpha\beta$) that were highly homologous to the *Escherichia coli* enzymes and the gene encoding the β isoform was isolated (Kauppinen, 1992). Interestingly, the 3-ketoacyl-ACP reductase showed strong similarities with the NodG-gene product from *Rhizobium meliloti* (Slabas et al., 1992). Overexpression of the enoyl-ACP reductase gene from *Brassica* in *Escherichia coli* has allowed the preparation of large amounts of the recombinant protein for structural analyses of the crystallized protein at 1.9 Å (Rafferty et al., 1994).

4. Desaturation of Stearoyl-ACP

The biosynthetic pathway briefly described above lead to the formation of palmitoyl-ACP (16:0-ACP) and stearoyl-ACP (18:0-ACP). However, most C18 fatty acids synthesized in the chloroplasts contain a double-bond at the $\Delta 9$-position. A soluble enzyme from the plastid stroma, stearoyl-ACP desaturase (EC 1.14.99.6), catalyzes this desaturation step using reduced ferredoxin that is produced by thylakoids during photosynthesis. The critical importance of this step in fatty acid biosynthesis for the overall structure is illustrated by the *fab2* mutation of *Arabidopsis*. This mutant is characterized by a defect in plastid 18:0-ACP desaturase, possesses a severe dwarf phenotype, and many cell types in the mutant fail to expand (Lightner et al., 1994). This desaturase has been purified from a wide variety of plant species and extensively analyzed (for reviews, see Heinz,

1993; Harwood, 1996). Overexpression of the cucumber gene in *Saccharomyces cerevisae* resulted in the accumulation of functional enzyme (Shanklin and Somerville, 1991). Shanklin and Somerville (1991) found no detectable identity between cDNA corresponding to the castor bean enzyme and the sequences of the $\Delta 9$-stearoyl desaturase from rat, yeast or $\Delta 12$-oleoyl desaturase from *Synechocystis*. Crystallographic data have also been obtained from the castor bean enzyme (Schneider et al., 1992). The data were consistent with previous biochemical studies showing that the enzyme belongs to a class of O_2-activating proteins containing diiron-oxo clusters (Fox et al., 1993). The desaturation of stearoyl-ACP may proceed via enzymic generation of iron-oxo species derived from the diiron cluster (Fox et al., 1993).

B. The Fate of De Novo Synthesized Fatty Acids

The pathway described above show that de novo fatty acid biosynthesis takes place in plastids. These newly synthesized fatty acids have two basic fates. First, a large proportion of palmitoyl-ACP and oleoyl-ACP produced are used directly within plastids for the biosynthesis of specific plastid glycerolipids (see below). The second fate for newly synthesized fatty acids is their export outside plastids, across the two envelope membranes, as building block for the biosynthesis of extraplastidial phospholipids and triacylglycerol. The proportion of fatty acids that are exported outside plastids varies according to the plant type. For instance, in 18:3 plants most of the fatty acids are exported to the cytosol, in contrast with 16:3 plants which use a much larger proportion of the stroma-synthesized fatty acids for their own glycerolipid biosynthesis (for a review, see Joyard and Douce, 1987). For instance, Browse et al. (1986) calculated that in *Arabidopsis* (a 16:3 plant) 62% of the de novo synthesized fatty acids are exported from chloroplasts. Therefore, in 18:3 as well as in 16:3 plants, the chloroplast envelope membrane represents a dynamic structure through which a continuous flow of fatty acids (mostly 18:1 and, to a lesser extent, 16:0) occur, especially during photosynthesis and in plants accumulating carbon as lipid storage products.

The first step in the export of fatty acid outside of plastids is the hydrolysis of acyl-ACPs and the release of free fatty acids, owing to a soluble acyl-ACP thioesterase. The importance of this enzyme has been widely recognized for the termination of acyl chain elongation and it is one of the possible targets for manipulating the lipid composition of standard crops (especially rape) accumulating lipids in their seeds (for a review, see Harwood, 1996). However, in addition to the ubiquitous oleoyl-ACP thioesterase found in almost all tissues, acyl-ACP thioesterases with different substrate specificities have been purified from different plants, especially from crops. Several genes were characterized, showing a rather good identity in their sequences, despite the significant differences observed in substrate specificities of the purified or the overexpressed enzyme (for a review, see Harwood, 1996). Some acyl-ACP thioesterases are preferentially expressed in epidermal cells, other in seeds or leaves, suggesting that they could play a role in some specific aspects of lipid metabolism (such as wax formation in epidermal cells, or triacylglycerol accumulation in seeds). For instance, in some oil-storing seeds, medium-chain specific acyl-ACP thioesterases are present in addition to the normal long-chain specific enzyme. They are probably involved in premature chain termination and accumulation of medium chain fatty acids (see for instance, Davies, 1993).

It is not known how free fatty acids released by the acyl-ACP thioesterase (directly within the inner envelope membrane ?) move across the two envelope membranes, but this may occur by simple diffusion through the lipid bilayers and via transient contacts between both membranes. On the outer membrane of the chloroplast envelope, an acyl-CoA synthetase activity is responsible for the formation of acyl-CoA thioesters that are further available outside plastids for acyltransferase reactions to form glycerolipids within the endoplasmic reticulum (Joyard and Douce, 1977; Joyard and Stumpf, 1981; Block et al., 1983c). However, acyl-CoA thioesters have detergent properties, inhibits membrane-associated enzymes and have general damaging effects in membranes. Therefore, they should be rapidly removed from the vicinity of the outer envelope membrane and transported to the endoplasmic reticulum for rapid utilization. To date, the mechanism responsible for such a transfer is unknown, although either nonspecific lipid-transfer proteins (for a review, see Kader 1996) or specific acyl-CoA binding proteins (Hills et al., 1994) could be involved.

C. Origin of Polar Head Groups

The biosynthesis of galactolipid, sulfolipid and phospholipid head groups has been reviewed by Joyard et al. (1993) and Kinney (1993). Therefore, only the major aspects relevant to the biosynthesis of glycerol-3-phosphate, UDP-galactose and UDP-sulfoquinovose, will be discussed here.

1. Origin of Glycerol-3-Phosphate

Glycerol-3-phosphate has been identified in ^{31}P NMR spectrum obtained from plant cells and chloroplasts (Bligny et al., 1990). However, in spinach leaves, the signal obtained was much lower than those from other phosphomonoesters whereas in isolated intact chloroplasts, the concentration of glycerol-3-phosphate could be estimated to about 0.5 mM, an amount sufficient to sustain glycerolipid synthesis in chloroplasts. In a plant cell, possible sources of glycerol-3-phosphate are (a) a powerful glycerol kinase (EC 2.7.1.30), first characterized in plants by Hippman and Heinz (1976) and further analyzed by Aubert et al. (1994) and (b) a glycerol-3-phosphate:NAD$^+$ oxidoreductase (EC 1.1.1.8), also called dihydroxyacetone-phosphate reductase since the ratio of reductase activity to that of the reverse reaction (dehydrogenase activity) is 10 to 1 and therefore, under physiological conditions, it catalyzes the synthesis of glycerol-3-phosphate from dihydroxy-acetone-phosphate, i.e. a triose-phosphate deriving directly from photosynthesis (Gee et al., 1988a). Dihydroxyacetone-phosphate reductase activity probably plays the major role in glycerol-3-phosphate synthesis for glycerolipid biosynthesis in chloroplasts. At least two isoforms of the enzyme, one in the cytosol and one in the chloroplast stroma, have been characterized in plant cells (Gee et al., 1988a,b; 1993). Plastid dihydroxyacetone-phosphate reductase activity represents between 70 and 75% of the total cell activity, in pea as well as in spinach, and the remaining activity is present in the cytosol. Therefore, glycerol-3-phosphate is formed directly in the major compartments involved in glycerolipid biosynthesis. The substrate for dihydroxyacetone-phosphate reductase, dihydroxyacetone-phosphate, is a central compound in several essential metabolic pathways, such as photosynthesis and glycolysis, and its physiological concentrations are in the range of 0.1 mM (Stitt et al., 1984). From these data, it is clear that dihydroxyacetone-phosphate reductase activity is probably not a major regulatory step in glycerolipid biosynthesis. In addition, both glycerol-3-phosphate and dihydroxyacetone-phosphate can cross the chloroplast envelope membrane owing to the phosphate/triose-phosphate translocator (Flügge and Heldt, 1991). In fact, measurements of the pool sizes of glycerol-containing precursors of polar lipids in plants indicate that the pool size of glycerol-3-phosphate is much larger than that of diacylglycerol and phosphatidic acid.

2. Origin of UDP-Galactose

The formation of UDP-galactose is essential for both galactolipid synthesis and polysaccharides. No UDP-galactose could be detected in spinach chloroplast extracts, whereas the concentration of UDP-galactose in the cytosol, as determined by ^{31}P-NMR analyses, are in the range of about 0.2 to 0.5 mM (Bligny et al., 1990). UDP-galactose derives from UDP-glucose owing to UDP-glucose-4-epimerase (EC 5.1.3.2). Indeed, UDP-glucose is present in large amounts in the cytosol and the enzymes involved in UDP-glucose and UDP-galactose synthesis, i.e. UDP-glucose pyrophosphorylase and UDP-glucose-4-epimerase are cytosolic enzymes (Königs and Heinz, 1974). Therefore, UDP-glucose and UDP-galactose are in permanent equilibrium in the cytosol. All together, these observations provide evidence for an extra-plastidial location of UDP-galactose synthesis and accumulation. However, this raises the question of the availability of UDP-galactose for MGDG synthesis in plastids. MGDG synthase is located on the inner envelope membrane, at least in spinach (Block et al., 1983b), a membrane shown to be impermeable to UDP-galactose (Heber, 1974). Therefore, there is no need for UDP-galactose to accumulate in the plastid stroma only if MGDG synthesis occurs on the outer surface of the inner envelope membrane, but this remains to be demonstrated. If this is true, the concentrations of UDP-glucose (about 4–5 mM) and UDP-galactose (about 0.2 to 0.5 mM) found in the cytosol are high enough to sustain optimal rates of galactolipid biosynthesis under normal physiological conditions.

A cDNA clone was recently characterized and functionally expressed in the *Saccharomyces cerevisae* gal10-mutant and in *Escherichia coli* (Dörmann and Benning, 1996). The sequence shows high degree of similarity with UDP-glucose epimerase from bacteria, rat and yeast. Only one gene was characterized in *Arabidopsis* which was expressed in all tissues with highest expression in stems and roots (Dörmann and Benning, 1996). Interestingly, crystal structures of the oxidized and reduced forms of UDP-glucose epimerase from *Escherichia coli* were recently obtained (Thoden et al., 1996) that could be rather similar to the plant enzyme.

3. Origin of UDP-Sulfoquinovose

Sulfate is a suitable precursor for SQDG synthesis: when intact and purified chloroplasts were incubated in the light and in presence of ^{35}SO$_4^{2-}$, high rates of SQDG synthesis were demonstrated (Haas et al., 1980; Kleppinger-Sparace et al., 1985; Joyard et al., 1986). Adenosine 5'-phosphosulfate (APS) and 3'phosphoadenosine 5'-phosphosulfate (PAPS) are also good precursors for SQDG biosynthesis (Hoppe and Schwenn, 1981; Kleppinger-Sparace et al, 1990). The light dependency for SQDG biosynthesis is due to the light requirement for sulfate reduction and activation, i.e. for APS synthesis, since an ATP-generating system allows SQDG synthesis in the dark (Benson, 1963; Kleppinger-Sparace et al, 1990). In non-green plastids, the oxidative pentose phosphate pathway also functions to provide reductant for sulfate assimilation. In fact, little is known on the biosynthetic pathway involved in sulfate incorporation into the polar head group of SQDG, and especially on the origin of the carbon-sulfur bond occurring in the sulfoquinovose moiety of SQDG (for a review, see Joyard et al., 1993).

Benning and Somerville (1992a,b) have developed an approach combining biochemistry and molecular biology to characterize enzymes involved in the biosynthesis of SQDG precursors. They have generated a series of mutants (deficient in the genes *sqdA*, *sqdB*, *sqdC* and *sqdD*) from *Rhodobacter sphaeroides*, a photosynthetic bacterium containing SQDG, that were demonstrated to be deficient in SQDG accumulation (Benning and Sommerville, 1992a,b; Rossak et al., 1995). One of them, deficient in the *sqdD* gene, was also shown to accumulate UDP-sulfoquinovose (Rossak et al., 1995), thus demonstrating a direct relationship between inhibition of SQDG synthesis and UDP-sulfoquinovose accumulation. This result is in good agreement with previous observations by Heinz et al. (1989) who demonstrated that chemically synthesized UDP-sulfoquinovose can be used by isolated envelope membranes for SQDG synthesis. One of the *Rhodobacter sphaeroides* mutant was complemented by the *sqdB* gene which was demonstrated to encode a 46-kDa protein having slight homology with UDP-glucose epimerase from bacteria and yeast (Benning and Somerville, 1992b). Together, these observations demonstrate that the photosynthetic bacterium *Rhodobacter sphaeroides* is a very good model for enzymes involved in SQDG biosynthesis, especially those involved in the formation of the polar head group (for a review, see Benning, Chapter 5). The same is probably true for cyanobacteria, since a null mutant of *Synechococcus* sp. PCC 7942 was demonstrated to be deficient in SQDG (Guler et al., 1996).

IV. Diversity of Diacylglycerol Molecular Species Produced in Plastid Envelope Membranes

Diacylglycerol molecules are formed within chloroplast envelope membranes by several different enzymatic pathways. First, the inner envelope membrane contains the enzymes involved in phosphatidic acid biosynthesis and a very specific phosphatidate phosphatase (Joyard and Douce, 1977). Diacylglycerol molecules synthesized through this pathway (the Kornberg-Pricer pathway) contain 18:1 and 16:0 fatty acids, respectively, at the *sn*-1 and *sn*-2 position of the glycerol backbone, a structure described as prokaryotic since it is found in cyanobacterial glycerolipids (Heinz, 1977). Some minor molecular species such as diacylglycerol containing only 16:0 are also synthesized by the enzymes of the Kornberg-Pricer pathway. However, they are mostly found in sulfolipid and phosphatidylglycerol, but not in galactolipids. Diacylglycerol is also produced in envelope membranes by a second type of enzyme, a galactolipid:galactolipid galactosyltransferase (van Besouw and Wintermans, 1978), localized on the cytosolic side of the outer envelope membrane (Dorne et al., 1982). This enzyme catalyzes the transfer of a galactose from one MGDG molecule to another and leads to the formation of DGDG and diacylglycerol (van Besouw and Wintermans, 1978). The diacylglycerol molecules formed by this enzyme present the same structure as their parent galactolipids: i.e. either the typical prokaryotic structure (see above) or a eukaryotic structure, with C18 fatty acids at both *sn* positions, like in typical eukaryotic

glycerolipids such as phosphatidylcholine. The origin of the diacylglycerol used for the biosynthesis of MGDG molecules with this so-called eukaryotic structure is still poorly understood, but it is assumed that it could derive from phosphatidylcholine (see below). In vivo labeling studies of spinach leaves with [^{14}C]-acetate (Joyard et al., 1980) have demonstrated that radioactive envelope MGDG contained a whole set of more or less saturated molecular species. The possibility that at least some of them could derive from newly synthesized diacylglycerol molecules cannot be entirely ruled out. Together, these observations suggest that diacylglycerol molecules having various fatty acids at their *sn*-1 and *sn*-2 positions could be available for further metabolism, and especially galactolipid biosynthesis, in envelope membranes.

A. Diacylglycerol Formation Through the Envelope Kornberg-Pricer Pathway

1. The Envelope Kornberg-Pricer Pathway

For many years, a central role in plant glycerolipid biosynthesis was assumed for the endoplasmic reticulum. It is now clear that in chloroplasts, the envelope is the site for acylation of *sn*-glycerol-3-phosphate for further metabolism into plastid glycerolipids. This biosynthetic pathway involves first a soluble glycerol-3-phosphate acyltransferase (Joyard and Douce, 1977), closely associated with the inner envelope membrane, that catalyzes the transfer of oleic acid (18:1), from 18:1-ACP to the *sn*-1 position of glycerol (Frentzen et al., 1983; Frentzen, 1993) producing mostly 1-oleoyl-*sn*-glycerol-3-phosphate (lysophosphatidic acid). In fact this is strongly dependent upon the plant species. For instance, when a choice of substrate is given to the soluble acyltransferase (i.e. mixtures of 16:0 and 18:1, or 18:0 and 18:1), the purified enzyme shows a marked preference for 18:1. However, in some plants some isoforms of the soluble acyltransferase are rather unselective (Frentzen, 1993). In the case of *Amaranthus*, for instance, 16:0 is incorporated into lysophosphatidic acid with almost the same efficiency as 18:1 (Frentzen, 1993). Lysophosphatidic acid is further acylated to form 1,2-diacyl-*sn*-glycerol-3-phosphate (phosphatidic acid) by the action of an envelope-bound 1-acylglycerol-3-phosphate acyltransferase (Joyard and Douce, 1977). Since lysophosphatidic acid used for this reaction is esterified at the *sn*-1 position, the enzyme will direct fatty acids, almost exclusively 16:0, to the available *sn*-2 position (Frentzen et al., 1983; Frentzen, 1993). In contrast with the soluble acyltransferase, the membrane bound enzyme displays a strict selectivity: when both 16:0 and 18:1 are offered together as substrates to the membrane-bound enzyme, more than 90% of the fatty acids incorporated at the *sn*-2 position of the glycerol backbone are 16:0 (Frentzen, 1993). Therefore, the two plastid acyltransferases have distinct specificities and selectivities for acylation of *sn*-glycerol-3-phosphate (Fig. 4). Together, they lead to the formation, within the inner envelope membrane, of phosphatidic acid with 18:1 fatty acid at the *sn*-1 and 16:0 fatty acid at the *sn*-2

positions of the glycerol backbone. This structure is typical of the so-called prokaryotic glycerolipids. A minor proportion of phosphatidic acid with palmitic acid at both *sn* positions can also be synthesized in chloroplasts. In contrast, extraplastidial acyl-transferases have distinct localization and properties (nature of the acyl donor, specificities, selectivities, etc.) as discussed by Frentzen (1986, 1993). Phosphatidic acid, synthesized in envelope membranes, is further metabolized into either diacyl-glycerol or phosphatidylglycerol (Figs. 4 and 5). Deriving from phosphatidic acid, chloroplast phosphatidylglycerol has the same structure and therefore contains a C16 fatty acid (i.e. the unique 16:1, fatty acid) at the *sn*-2 position of the glycerol

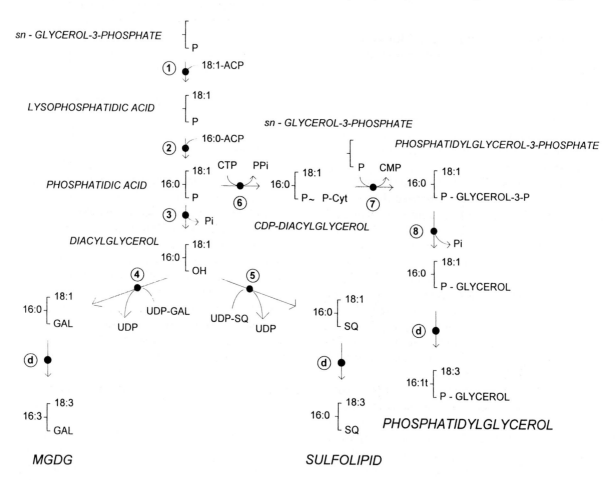

Fig. 4. Biosynthesis of plastid glycerolipids by plastid envelope membranes. The enzymes involved are: (1) glycerol-3-phosphate acyltransferase, (2) 1-acylglycerol-3-phosphate acyltransferase, (3) phosphatidate phosphohydrolase, (4) MGDG synthase, (5) sulfolipid synthase, (6) phosphatidate cytidyltransferase, (7) CDP-diacylglycerol-glycerol-3-phosphate 3-phosphatidyltransferase, (8) phosphatidylglycerophosphatase, (d) desaturases. This biosynthetic pathway leads to the synthesis of MGDG, sulfolipid and phosphatidylglycerol with unsaturated C 18 and C 16 fatty acids at the *sn*-1 and *sn*-2 positions of glycerol respectively. Reproduced with permission from Joyard et al. (1991).

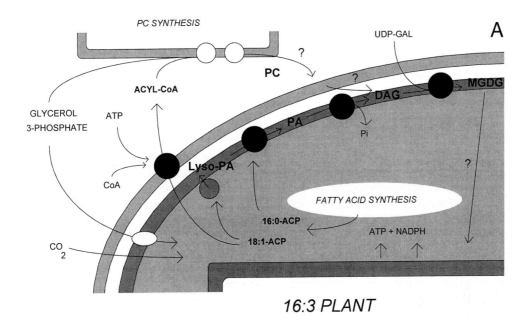

16:3 PLANT

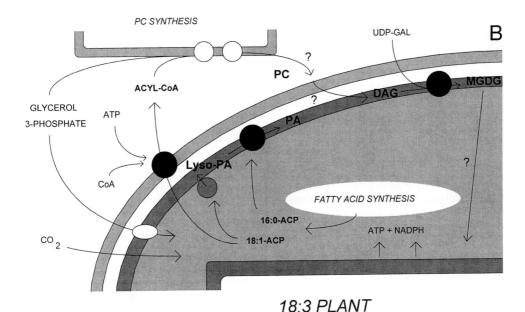

18:3 PLANT

Fig. 5. Biosynthesis of MGDG in 16:3- and 18:3-plants. The formation of MGDG with the 18:1/16:0 diacylglycerol backbone (prokaryotic structure) involves the Kornberg-Pricer pathway of the chloroplast envelope. The formation of C18/C18 (eukaryotic structure) is not clearly established. In 16:3-plants (A), both pathways are generally active, whereas, in 18:3-plants (B), only the latter pathway is operating. The lack of phosphatidate phosphatase in 18:3 plants prevents the formation (in the inner envelope membrane) of diacylglycerol, the substrate for MGDG synthase. The C18/C18 diacylglycerol used for MGDG synthesis in envelope membranes probably derives from phosphatidylcholine synthesized outside the chloroplast, in the endoplasmic reticulum, but this has not been conclusively demonstrated yet. These MGDG molecular species formed in envelope membranes are then desaturated in envelope membranes (for a review, see for instance Heinz, 1993) and transported to the thylakoids (for a review, see for instance Joyard and Douce, 1994). Reproduced from Douce and Joyard (1996).

backbone (Fig. 4). Phosphatidylglycerol synthesis is localized in the inner envelope membrane (Andrews and Mudd, 1985).

2. Properties of the Envelope Phosphatidate Phosphatase

Phosphatidate phosphatase (3-*sn*-phosphatidate phosphohydrolase; E.C. 3.1.3.4) catalyzes the dephosphorylation of phosphatidic acid (Fig. 6). In chloroplasts, it is a membrane-bound enzyme (Joyard and Douce, 1977; 1979), exclusively located on the inner envelope membrane (Block et al., 1983b; Andrews et al., 1985). In contrast to chloroplasts from 16:3 plants, those from 18:3 plants have a rather low phosphatidate phosphatase activity (Heinz and Roughan, 1983) and cannot deliver diacylglycerol fast enough to sustain the full rate of glycolipid synthesis. This observation may explain why 18:3 plants contain only small amounts of galactolipids and sulfolipid with C16 fatty acids at the *sn*-2 position, but contain phosphatidylglycerol with such a structure. It is not yet known whether the reduced level of phosphatidate phosphatase activity is due to lower expression (species-specific) of the gene coding for the enzyme or to the presence of regulatory molecules which control the activity of the enzyme.

Analyses of the biochemical properties of the envelope phosphatidate phosphatase are difficult to perform with either isolated envelope membranes or intact chloroplasts (Joyard and Douce, 1979; Gardiner and Roughan, 1983; Gardiner et al., 1984; Malherbe et al., 1992). For instance, in isolated envelope membranes, the major problem is due to the presence of the galactolipid:galactolipid galactosyltransferase (see below) which is responsible for the accumulation of diacylglycerol during the course of envelope preparation. This problem can be prevented by the incubation of isolated intact chloroplasts with the protease thermolysin. Dorne et al. (1982) observed that such a mild proteolytic treatment destroys the galactolipid:galactolipid galactosyltransferase and no further accumulation of diacylglycerol is therefore possible during envelope purification. Using envelope membranes from thermolysin-treated chloroplasts, Malherbe et al. (1992, 1995) have set up an in vitro assay to measure the enzyme activity and to analyze its properties after solubilization.

The envelope phosphatidate phosphatase exhibits biochemical properties clearly different from similar enzymes described in the various cell fractions from

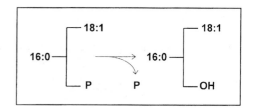

Fig. 6. Reaction catalyzed by phosphatidate phosphatase.

animals or yeast (for reviews, see Bishop and Bell, 1988; Carman and Henry, 1989). The enzyme described in extraplastidial compartments from plant tissues strongly resemble their animal counterpart and are very different from the envelope enzyme (Stymne and Stobart, 1987; Joyard and Douce, 1987). First, the envelope enzyme is tightly membrane-bound, whereas in yeast or animal cells the activity is recovered in both cytosolic and microsomal fractions. Furthermore, the pH optimum for the envelope enzyme is alkaline (9.0) and cations, such as Mg^{2+}, are powerful inhibitors of the enzyme. After solubilization, phosphatidate phosphatase remains sensitive to Mg^{2+} and to a wide range of other cations Mn^{2+}, Cu^{2+} and Zn^{2+} were found to be the most potent inhibitors of the solubilized enzyme (Malherbe et al., 1995). In marked contrast, the yeast or animal phosphatidate phosphatases are active in the pH range of 5.5 to 7.5 and their activity is strongly dependent upon the presence of Mg^{2+} (for a review, see Joyard and Douce, 1987).

Diacylglycerol is a competitive inhibitor of the envelope phosphatidate phosphatase (Malherbe et al., 1992). Using isolated intact chloroplasts, Malherbe et al. (1992) demonstrated that phosphatidate phosphatase activity (a) reached maximal values only when diacylglycerol levels in chloroplasts were low (i.e. in presence of UDP-galactose or at the early stages of acetate incorporation) and (b) decreased when diacylglycerol levels increased. In vivo, the steady state activity of phosphatidate phosphatase is therefore sensitive to the diacyl-glycerol/phosphatidic acid molar ratio. Feedback inhibition of phosphatidate phosphatase (and consequently of galactolipid and sulfolipid synthesis) by diacylglycerol might lead to accumulation of phosphatidic acid and therefore will favor phosphatidylglycerol synthesis (see Fig. 4). Therefore, the rate of diacylglycerol formation is tightly related to the rate of its utilization by the envelope enzymes involved in galactolipids and sulfolipid biosynthesis.

B. Possible Origin of Dioleoylglycerol

As discussed above, the specificities of the envelope acyltransferases do not allow the formation of phosphatidic acid and diacylglycerol exclusively with C18 fatty acids. Cyanobacteria are also unable to synthesize such structures. In vivo kinetics of acetate incorporation into chloroplast lipids suggest that phosphatidylcholine could provide the diacylglycerol backbone for eukaryotic plastid glycerolipids. The reader is referred to reviews by Heinz (1977), Douce and Joyard (1980), Roughan and Slack (1982) and Joyard and Douce (1987) for detailed presentations of the arguments in favor of this hypothesis. The outer envelope phosphatidylcholine is apparently of extraplastidial origin because (a) envelope membranes are unable to synthesize phosphatidylcholine (Joyard and Douce, 1976a) and (b) plastid and extraplastidial phosphatidylcholine have the same diacylglycerol structure (Heinz, 1977; Siebertz et al., 1979). Therefore, mechanisms should exist for the transfer of phosphatidylcholine molecules from their site of synthesis (probably the endoplasmic reticulum) to the outer envelope membrane. To date, no clear evidence for a precise mechanism is available. For instance, phospholipid transfer proteins have been demonstrated to mediate net in vitro transfer of phosphatidylcholine between plant cell membranes, but it is not yet clear whether such proteins are active *in vivo* since they have been demonstrated to have features of secreted proteins (Kader, 1996). Another mechanism that was recently proposed involves lysophosphatidylcholine: by preincubating microsomal membranes with labeled lysophosphatidylcholine, Bessoule et al. (1995) have shown that lysophosphatidylcholine could be transferred by a partition process to chloroplasts where it is acylated to yield phosphatidylcholine. Once, in the outer envelope membrane, phoshatidylcholine should be transformed into diacylglycerol and the demonstration that this indeed operates in plants has yet to be provided. In addition, nothing is known on the acyltransferase, especially its specificity and selectivity, that could be involved in the acylation of lysophosphatidic acid. Furthermore, envelope membranes apparently lack a phospholipase c activity which would produce diacylglycerol (Joyard and Douce, 1976b; Roughan and Slack, 1977; Siebertz et al., 1979). Of course, this can be done in vitro by addition of phospholipase c to isolated intact chloroplasts. For instance, Dorne et al. (1982) have

demonstrated that all the outer envelope phosphatidylcholine could be converted into diacylglycerol by addition of very small amounts of phospholipase c for a very brief period of time. However, although almost all envelope phosphatidylcholine is present in the outer leaflet of the outer envelope membrane, this experiment do not prove that the action of soluble phospholipase c on the envelope phosphatidylcholine has any physiological significance.

C. Diacylglycerol Formation by the Galactolipid: Galactolipid Galactosyltransferase

1. Properties of the Galactolipid:Galactolipid Galactosyltransferase

The galactolipid:galactolipid galactosyltransferase catalyzes an interlipid exchange of galactose between two MGDG molecules and led to the formation of DGDG and diacylglycerol (Fig. 7), as described by van Besouw and Wintermans (1978). Galactolipids having up to 3 (tri-GDG) to 4 (tetra-GDG) galactose molecules can be formed in vitro by this enzyme (Joyard and Douce, 1976b). Therefore, diacylglycerol molecules deriving from the various galactolipid molecular species present in the chloroplast envelope can be formed within the membranes during the course of their purification (van Besouw and Wintermans, 1978; Dorne et al., 1982). This enzyme was demonstrated to be localized on the outer envelope membrane (Dorne et al., 1982; Heemskerk et al., 1986) and was demonstrated to be present also in non-green plastids (Alban et al., 1989). During envelope purification, the two envelope membranes probably fuse together, making the substrate (MGDG and DGDG) present in the inner envelope membrane available to the outer envelope enzyme. In addition, the enzyme has a broad pH optimum (around 6.5), is

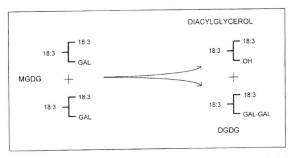

Fig. 7. Reaction catalyzed by galactolipid:galactolipid galactosyltransferase.

still active at 4 °C and is stimulated by 5 mM Mg^{2+}. Since magnesium is present in all the medium used for envelope purification (Douce et al., 1973), the enzyme is active during the course of envelope preparation leading to massive accumulation of diacylglycerol (up to 15% of total glycerolipids) in isolated envelope membranes (Joyard and Douce, 1976b). The presence of such large amounts of diacylglycerol in isolated envelope membranes reflects the in vitro activity of the galacto-lipid:galactolipid galactosyltransferase, and does not represent the in vivo situation where diacylglycerol is only a minor lipid component of envelope membranes. The galactolipid:galactolipid galacto-syltransferase is indeed an extremely active enzyme: in vitro, activities up to 3000 nmol MGDG converted/ mg protein/h were obtained (Heemskerk et al., 1986). These values were obtained in envelope membrane supplied with additional MGDG and deoxycholate (Heemskerk et al., 1986). Such a high in vitro activity, which is quite uncommon for enzymes involved in lipid biosynthesis, raises questions about the physiological significance of the galactolipid: galactolipid galactosyltransferase.

2. Possible Physiological Roles of the Galacto-lipid:Galactolipid Galactosyltransferase

The galactolipid:galactolipid galactosyltransferase is the only DGDG-forming enzyme clearly described in plastids to date (Van Besouw and Wintermans, 1978, Heemskerk and Wintermans, 1987; Heemskerk et al., 1990). For instance, Heemskerk et al. (1990) analyzed galactolipid synthesis in chloroplasts or chromoplasts from 8 species of 16:3 or 18:3 plants and found that digalactosyldiacylglycerol formation is never stimulated by UDP-galactose or any other nucleoside 5'-diphosphodigalactoside; in all cases, DGDG formation was reduced by thermolysin digestion of intact organelles, a treatment that also hydrolyzes the galactolipid:galactolipid galacto-syltransferase (Dorne et al., 1982). In vitro, the galactolipid:galactolipid galactosyltransferase does not show strong specificity for any MGDG molecular species. However, if this enzyme is indeed the DGDG-synthesizing enzyme, it should discriminate in vivo between the various MGDG molecular species that are available since (a) the proportion of eukaryotic molecular species is higher in DGDG than in MGDG (Heinz, 1977, Bishop et al., 1985) and (b) DGDG contains 16:0 fatty acids (up to 10-15%) at both the

sn-1 and sn-2 position and very little 16:3 (in 16:3 plants), whereas MGDG contains little 16:0, but (in 16:3 plants) 16:3 at the sn-2 position (Heinz, 1977). To understand the complex processes involved in DGDG biosynthesis, a genetic approach was recently developed by Dörmann et al. (1995) who reported the isolation of a DGDG-deficient Arabidopsis mutant. This mutant carries a recessive nuclear mutation at a single locus (dgd1) which resulted in more than a 90% decrease of the normal DGDG level. Backcrossed mutants show stunted growth, pale green leaf color, reduced photosynthetic capability, and altered thylakoid membrane ultra-structure. As discussed by Dörmann et al. (1995), if the galactolipid:galactolipid galactosyltransferase is indeed the only DGDG-forming enzyme, a mutation at this biosynthetic step should result in an accumulation of the eukaryotic MGDG (i.e. containing 18:3 fatty acids at both sn positions). However, the amount of MGDG is only slightly increased in the mutant. In addition, DGDG synthesis was still very active in chloroplasts isolated from the mutant. Furthermore, the fatty acid composition of these DGDG molecules was very different from those synthesized in normal chloroplasts, thus suggesting that the mutation affected in vivo DGDG synthesis, but not the galactolipid:galactolipid galactosyltransferase in the outer envelope membrane. Therefore, the nature of the enzyme responsible for DGDG synthesis remains an open question.

Sakaki et al. (1990a,b; 1994) proposed another physiological significance for the envelope galacto-lipid:galactolipid galactosyltransferase. Using ozone-fumigated spinach leaves (the reactions of plants to ozone is discussed in this volume by Harwood, Chapter 15), they demonstrated in vivo (Sakaki et al., 1990a,b) that MGDG was converted into diacylglycerol by the galactolipid:galactolipid galactosyltransferase (see however Carlsson et al., 1994 and Hellgren et al., 1995) and then to triacylglycerol (by acylation with 18:3-CoA), owing to a diacylglycerol acyltransferase associated with envelope membranes (Martin and Wilson, 1984). Together, these observations demonstrate that DGDG synthesis in chloroplasts is even more complex than expected previously. Analysis of the gene affected in the dgd1 mutant (Dörmann et al., 1995), as well as further characterization of the mutant should allow to gain a deeper insight in the biosynthesis of DGDG. Therefore, the role of the galactolipid:galactolipid galactosyltransferase as the DGDG-synthesizing enzyme remains to be demonstrated, as well as the proposition that DGDG synthesis takes only place in the outer envelope membrane (Van Besouw and Wintermans, 1978, Heemskerk and Wintermans, 1987; Heemskerk et al., 1990; Williams and Khan, 1996). Because DGDG formation and diacylglycerol production are closely linked due to the reaction catalyzed by the galactolipid:galactolipid galactosyltransferase, it is not yet clear whether the synthesis of diacylglycerol (and DGDG) in the outer envelope membrane is a process specifically regulated, for instance by environmental conditions, or whether a continuous production of diacylglycerol (and DGDG) occurs.

IV. MGDG Synthesis and Diacylglycerol Metabolism in Envelope Membranes

Diacylglycerol molecules formed de novo in the inner envelope membrane constitute a pool of substrate shared by both MGDG synthase and SQDG synthase (Joyard et al., 1986). SQDG is synthesized by a 1,2-diacylglycerol 3-β-sulfoquinovosyltransferase. However, although very little is known of the plant enzyme involved in SQDG synthesis, the use of *Rhodobacter sphaeroides* mutants that are deficient in SQDG accumulation (Benning and Sommerville, 1992a,b; Rossak et al., 1995) is likely to rapidly provide some key information on SQDG synthase in prokaryotes, and therefore in higher plants.

More biochemical data are presently available on MGDG synthesis which is catalyzed by a UDP-galactose:1,2-diacylglycerol 3-β-D-galactosyltransferase (E.C. 2.4.1.46), or MGDG synthase. This enzyme transfers a galactose from a water-soluble donor, UDP-galactose, to a hydrophobic acceptor molecule, diacylglycerol (Fig. 8). MGDG synthase is localized in envelope membranes (Douce, 1974) and—at least in spinach—in the inner envelope membrane (Block et al., 1983a,b). A minor, but interesting, difference exists for MGDG synthesis between cyanobacteria and plastids. In cyanobacteria, MGDG biosynthesis is a two-step mechanism involving the formation of a lipid intermediate, monoglucosyldiacylglycerol (MGluDG). Then, MGDG is formed from MGluDG by epimerization of glucose into galactose (for a review, see Murata and Nishida, 1987). In higher plants, UDP-galactose which is used for MGDG synthesis derives from

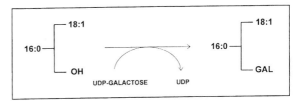

Fig. 8. Reaction catalyzed by MGDG synthase.

cytosolic UDP-glucose (see above).

A. Partial Purification of the Envelope MGDG Synthase

Progress in the understanding of the functioning of MGDG synthase has been hindered by the difficulty of handling enzyme and substrate with limited aqueous solubility. Covès et al. (1986, 1987) described solubilization of MGDG synthase from spinach chloroplast envelope membranes and the development of a specific assay for the solubilized activity which allowed a partial purification of this enzyme (Covès et al., 1986). Teucher and Heinz (1991) and Maréchal et al. (1991) independently purified several hundred fold MGDG synthase activity from spinach chloroplast envelope, but in each case, unambiguous characterization of the polypeptide associated with MGDG synthase activity was still very critical. A polypeptide with a Mr around 20 kDa (Teucher and Heinz, 1991; Maréchal et al., 1991) was proposed to be associated with the enzyme. However, the final amount of enzyme was so low (in the range of μg when starting the purification from 100 mg envelope proteins), that further analyses of the protein were almost impossible. Starting from cucumber microsomal fraction, Ohta et al. (1995) obtained an active fraction containing a 47 kDa polypeptide as a major constituent. The cDNA corresponding to this cucumber polypeptide (1.6 kbp) was subsequently isolated and expressed as a glutathione-S-transferase fusion in *Escherichia coli* (K.-I. Takamiya and H. Ohta, personal communication). The differences in Mr (Teucher and Heinz, 1991; Maréchal et al., 1991; Ohta et al., 1995) as well as sequence analyses compared with available biochemical data suggest the possible existence of two different MGDG synthases which might be correlated with the localization of the galactosyltransferase activity. Whereas in spinach, a '16:3 plant', this enzyme is localized in the inner envelope membrane (Block et al., 1983b), it has been shown to be associated with the outer envelope membrane in pea (Cline and Keegstra, 1983). As previously indicated, pea and cucumber are '18:3 plants'. One can expect that availability of cDNAs encoding MGDG synthase from various plant species and overexpression of the corresponding enzyme in an active form will help a better understanding of enzyme structure and localization. However, all the information presently available on MGDG synthase provide further evidence for the conclusion that, despite its importance for chloroplast membrane biogenesis, MGDG synthase is present only in vanishing amounts (about 0.5 μg/mg protein) in envelope membranes.

Finally, it is not yet known whether the sequence of MGDG synthase from cucumber shows any homology with *sqdA*, *sqdC* or *sqdD* genes from *Rhodobacter sphaeroides* (Benning and Somerville, 1992a,b; Rossak et al., 1995). These genes could be involved in the last step of SQDG biosynthesis and therefore, at least one of them could correspond to a SQDG synthase.

B. Reconstitution of MGDG Synthase Activity in Mixed Micelles

Van Besouw and Wintermans (1979) proposed that MGDG synthase functions according to a ping-pong mechanism: the enzyme first reacting with UDP-galactose, forming a transitory complex with galactose with the release of UDP, and then reacting with diacylglycerol to synthesize MGDG. However, such a mechanism remained to be clearly established, mostly because critical steps in studies on enzyme regulation are the availability of the hydrophobic substrate, namely diacylglycerol, and the membrane or micelle localization of the protein making, again, clear kinetic analyses very difficult. Among the different kinetic models available for such studies, the 'surface dilution' kinetic model proposed by Deems et al. (1975) and Dennis (1983) for cobra venom phospholipase A2 (i.e. for enzymes inter-acting at the interface of macromolecular aggre-gates) provides a theoretical basis for the use of standard kinetic techniques in mixed micelles, since substrate concentration can be varied according to the surface of the lipid structure. This can be easily compared to the situation in membranes since the two-dimensional surface of the structure is con-sidered, and not the whole volume of the incubation mixture, increasing tremendously the collision probability between the enzyme and its lipidic

substrate. The 'surface dilution' kinetic model was applied to studies of MGDG synthase and kinetic experiments were carried out by Maréchal et al. (1994a) in mixed micelles containing the partially purified delipidated enzyme, the substrate (diacyl-glycerol) and the detergent CHAPS.

Covès et al. (1988) have shown that addition of anionic glycerolipids (mostly phosphatidylglycerol, but also to a lesser extent sulfolipid) to the incubation medium was essential to measure in vitro MGDG synthase activity, thus suggesting that phosphatidyl-glycerol could create a favorable environment for enzyme activity. In fact, Maréchal et al. (1994a) demonstrated that diacylglycerol was solubilized at low CHAPS concentration only if phosphatidyl-glycerol was present in the mixed micelles: the apparent CMC for dioleoylglycerol solubilization by CHAPS, for example, was reduced from about 8 mM, in absence of phosphatidylglycerol, to less than 3 mM, in presence of phosphatidylglycerol (Maréchal et al., 1994a). These results explain therefore why addition of phosphatidylglycerol was so critical for an optimal activity of enzymes using diacylglycerol as a substrate, such as choline phosphotransferase (Miller and Weinhold, 1981; Mantel et al., 1993) or MGDG synthase (Covès et al., 1988). In the case of envelope MGDG synthase, since diacylglycerol is only poorly solubilized at 6 mM CHAPS in absence of phos-phatidylglycerol, the enzyme activity in such experimental conditions is very low, in contrast with experiments done in presence of phosphatidylglycerol where diacylglycerol is well solubilized. Cholate probably behaves like CHAPS and phosphatidyl-glycerol since no phosphatidylglycerol stimulation of MGDG synthase activity was observed by Teucher and Heinz (1991) in presence of this anionic detergent. In fact, cholate and phosphatidylglycerol both provide an anionic hydrophobic environment which probably favors diacylglycerol solubilization within mixed micelles.

C. MGDG Synthase Has Two Distinct Substrate-Binding Sites

As already mentioned, MGDG synthase catalyses the transfer of a β-galactose from a nucleotidic donor (UDP-galactose), to the *sn*-3 carbon of 1,2-*sn*-diacylglycerol. Therefore, understanding of how both UDP-galactose and diacylglycerol interact with the enzyme is essential to understand regulation of MGDG synthase activity.

MGDG synthase inhibition by UDP is strictly competitive relatively to UDP-galactose (van Besouw and Wintermans, 1979; Covès et al., 1988; Maréchal et al., 1994a) and noncompetitive relatively to 1,2-dioleoylglycerol (Maréchal et al., 1994a). UDP, an obvious structural analog of UDP-galactose, binds to UDP-galactose binding site whereas it does not interfere with 1,2-diacylglycerol binding. This demonstrates that UDP-galactose binds to MGDG synthase at least by part of its nucleotidic side, on a site topologically distinct from that of 1,2-dioleoylglycerol binding. MGDG synthase affinity for 1,2-diacylglycerol is dependent upon acyl chain length (16 or 18 carbons), their position on the glycerol backbone (*sn*-1 or *sn*-2), and the number of double bonds they have (0 to 3 double bonds on each fatty acid) (see below). Therefore, 1,2-diacylglycerol binding on MGDG synthase involves, at least partially, its acyl part. In addition, Maréchal et al.

(1995) demonstrated that inactivation of MGDG synthase with *t*-butoxycarbonyl-L-methionine hydrosuccinimidyl ester, NEM or *ortho*-phenanthroline could be prevented by 1,2-diacylglycerol, whereas UDP-galactose had no effect. Together, these results indicate that MGDG synthase active site contains 3 distinct parts: two independent substrate-binding sites (for UDP-galactose and 1,2-diacylglycerol) and the catalytic site *sensu stricto* (where galactose transfer occurs). Relative topology of the substrate-binding sites is schematically presented in Fig. 9: either binding sites are completely separated and catalysis occurs after a deep conformationnal change (Fig. 9A), or binding sites partly overlap (Fig. 9B). Therefore the hypothesis of a ping-pong mechanism proposed for MGDG synthase by van Besouw and Wintermans (1979) is not valid. In contrast, the data obtained by Maréchal et al. (1994a) is characteristic of sequential catalytic mechanisms

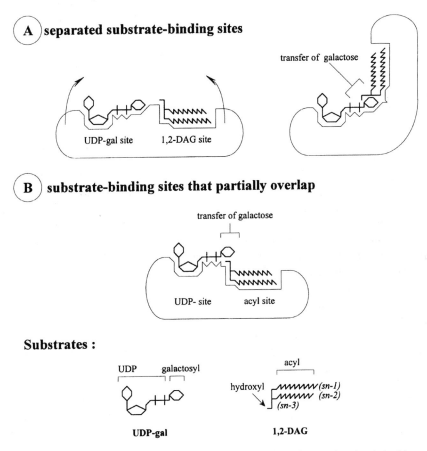

Fig. 9. UDP-galactose and 1,2-diacylglycerol (1,2-DAG) binding sites in MGDG synthase active site. A, in this case substrate binding sites are totally separated, catalysis implies a deep conformational change to allow galactose transfer. B, in this case substrate binding sites partly intersect, and domains which binds the nucleotidic part of UDP-galactose (UDP-) and the acyl groups of 1,2-diacylglycerol are topologically distinct (Maréchal et al., 1995).

for MGDG synthase which are compatible with either a random or an ordered bireactant system (Segel, 1975). Since MGDG synthase inactivation by citraconic anhydride is prevented after preincubation with either substrate, substrate binding is therefore likely to be random (Maréchal et al., 1995).

1. Characterization of the UDP-Galactose-Binding Site

MGDG synthase recognizes the nucleotidic (UDP) part of UDP-galactose since UDP is a competitive inhibitor of the enzyme (van Besouw and Wintermans, 1979; Covès et al., 1988; Maréchal et al., 1994a). Moreover, the protection of MGDG synthase by UDP-galactose against citraconic anhydride (Maréchal et al., 1995) indicates that the enzyme has lysine residues close to its UDP-galactose binding site. Both results indicates that MGDG synthase has some similarities with other galactosyltransferases (Beyer et al., 1981; Doering et al., 1989; Yadav and Brew, 1990, 1991).

DTT (a reduced cystein protecting agent) is essential to prevent loss of galactosylation activity after solubilization of envelope proteins by CHAPS (Covès et al., 1986, 1987). Since high concentrations of DTT did not alter MGDG synthase, it seems very likely that reduced cysteins from MGDG synthase are protected against oxidation, and that no disulfide bonds between two other cysteins are accessible. However, no protection of MGDG synthase against NEM (a reduced cystein blocking agent) was observed when MGDG synthase was preincubated with 1 mM UDP-galactose (Maréchal et al., 1995).

MGDG synthase is inhibited by zinc cation at pH 7.8 (Maréchal et al., 1991), whereas it is not affected by 1 mM $ZnCl_2$ at pH 6, thus suggesting that MGDG synthase could possess at least one imidazole group deprotonated at pH 7.8, which confers sensitivity to metals. At pH 6, protonated histidine residues do not bind bivalent cations any more. Since EDTA, a hydrophilic chelating agent, is sufficient to prevent inhibition of MGDG synthase by Zn^{2+} or Cu^{2+} (Maréchal et al., 1991), it is likely that this imidazole group should be exposed to the hydrophilic portion of the enzyme.

2. Characterization of Diacylglycerol-Binding Site

Comparison of MGDG synthase with 1,2-diacylglycerol-binding proteins provides some puzzling similarities. Enzymes from the protein kinase C family (PKCs), possessing a 1,2-diacylglycerol regulatory domain, and eukaryotic 1,2-diacylglycerol kinases (DGKs) contain 1,2-diacylglycerol-binding domains which homology is very high (Sakane et al., 1990, Schaap et al., 1990). Kinetic analyses of DGKs showed a functional independence of cosubstrates (1,2-diacylglycerol and ATP) binding (Kanoh and Ohno, 1981, Bishop et al., 1986, Wissing and Wagner, 1992). Amino acid sequence analyses of porcine DGK (Sakane et al., 1990) and human DGK (Schaap et al., 1990) also suggest that 1,2-diacylglycerol binds to a domain distinct than that of ATP. Moreover, like MGDG synthase (Maréchal et al., 1994a,b), PKCs (Hannun et al., 1991) and DGKs (Bishop et al., 1986) have a specificity for 1,2-diacylglycerol which differs according to their fatty acid composition.

MGDG synthase inactivation by NEM (see above) was demonstrated to strongly depend on the enzyme micro-environment (Maréchal et al., 1995): when the enzyme is embedded in its native lipidic envelope membrane, high concentrations of NEM are required to inactivate the enzyme, while concentrations of NEM in the micromolar range are sufficient to inactivate solubilized MGDG synthase, and to a lesser extent partially purified fraction. This dependence upon the hydrophobic environment (envelope lipid bilayer, envelope lipids included in CHAPS micelles, or delipidated CHAPS micelles) suggests that the MGDG synthase cysteins, sensitive to NEM, are located in a hydrophobic portion of the protein. Indeed, when MGDG synthase was preincubated in the presence of 1,2-dioleoylglycerol, its lipidic substrate, the enzyme was protected against NEM, whereas UDP-galactose did not affect the inactivation by NEM (Maréchal et al., 1995). This result demonstrates that reduced cysteins lie in the vicinity of the 1,2-diacylglycerol binding site.

MGDG synthase (apo-MGDG synthase) can be inactivated by ortho-phenanthroline, a hydrophobic chelating agent, whereas incubation of MGDG synthase with hydrophilic chelating agents (EDTA, 8-hydroxyquinoline), for a long period (16 h), never led to any inactivation (Maréchal et al., 1995). This result demonstrate that a metal could be associated with MGDG synthase and could be located in the hydrophobic core of the enzyme, which accessibility is limited to hydrophobic reagents such as ortho-phenanthroline. Maréchal et al. (1995) also demonstrated that 1,2-diacylglycerol protected

MGDG synthase against *ortho*-phenanthroline, while the hydrophilic substrate (UDP-galactose) did not affect MGDG synthase inactivation by *ortho*-phenanthroline, indicating that MGDG synthase associated metal is localized in the vicinity of the 1,2-diacylglycerol binding site. Analysis of the kinetic of MGDG inactivation by *ortho*-phenanthroline suggest that apo-MGDG synthase might be associated with one or more zinc cations, maybe associated with the presence of magnesium (Maréchal et al., 1995). However, the cDNA corresponding to MGDG synthase from cucumber does not encode for any consensus Cys-His cluster or any other known metal-binding consensus sequences (K.-I Takamiya and H. Ohta, personal communication), although this does not exclude that cucumber MGDG synthase contains a new class of metal-binding site. The different metal-binding properties of MGDG synthase suggest the possible existence of two different proteins involved in MGDG synthesis in spinach and cucumber (see above).

To conclude, some biochemical relationship probably exists between MGDG synthase and enzymes from the PKC family and from the eukaryotic 1,2-diacylglycerol kinase family.

D. MGDG Synthase Discriminates Between Diacylglycerol Molecular Species

Using acetone powder from spinach leaves as a source of enzyme, Mudd et al. (1969) were the first to investigate the specificity of the enzyme involved in MGDG synthesis and concluded that the requirement for galactolipid synthesis was satisfied best by diacylglycerol with the highest unsaturation. In contrast, using envelope membranes incubated with phosphatidylcholine and phospholipase *c* to generate endogenously diacylglycerol within the vesicles, Heemskerk et al. (1987) found that the highest rates of galactolipid biosynthesis were obtained with saturated species of diacylglycerol. In fact, these early studies were hampered by several major limitations: (a) the enzyme preparations used were rather crude (acetone powder extract in one case, isolated envelope membranes in the other) and contained the galactolipid:galactolipid galactosyltransferase since they were able to synthesize galactolipids with more than one galactose, (b) it was not obvious whether all diacylglycerol molecular species were indeed available to the enzyme, (c) specificities of the enzyme for the different

diacylglycerol molecules were analyzed by comparison of the initial velocity of the reaction for only one concentration of substrate. The use of the surface dilution kinetic model, allowed Maréchal et al. (1994a,b) to perform a direct comparison of the affinity of the enzyme for various diacylglycerol molecular species. They demonstrated that the envelope MGDG synthase does not have the same affinity for all diacylglycerol molecular species analyzed since the corresponding Km values varied from 1 to 8: values ranging between 0.0089-mol fraction to 0.066-mol fraction were obtained, but the differences observed were not really related to the unsaturation of the molecule. The enzyme had the lowest affinity for saturated diacylglycerol molecules like 18:0/18:0 and 16:0/16:0, in agreement with the observations by Mudd et al. (1969), but 18:2/18:2 was a much better substrate than 18:3/18:3. The diacylglycerol molecular species synthesized within chloroplasts, i.e., 18:1/16:0, was a rather good substrate, better than 16:0/18:1. However, Maréchal et al. (1994a,b) demonstrated that MGDG synthase has a much higher affinity (Km = 0.0089-mol fraction) for dilinoleoylglycerol (18:2/18:2) than for any other diacylglycerol molecular species analyzed. From these data, one can suggest that the low affinity of the envelope MGDG synthase towards dipalmitoylglycerol is probably responsible for its exclusion from MGDG. In contrast, the enzyme shows a better affinity for 1-oleoyl-2-palmitoylglycerol which, in addition, is produced in more significant amounts by the envelope phosphatidate phosphatase. This could explain why this typical prokaryotic structure is readily incorporated into MGDG.

Analysis of envelope lipids have demonstrated that MGDG does not contain dipalmitoyl species, in contrast to sulfolipid (Siebertz et al., 1979) whereas MGDG and sulfolipid synthase both compete for the same pool of diacylglycerol molecules (Joyard et al., 1986; Williams and Khan, 1996). Dipalmitoylglycerol is only synthesized in small amounts within envelope membranes by the enzymes of the Kornberg-Pricer pathway whereas 1-oleoyl-2-palmitoylglycerol is the major product (Frentzen et al., 1983). In addition, the affinity of the envelope MGDG synthase for 1-oleoyl-2-palmitoylglycerol is higher than for dipalmitoylglycerol (Maréchal et al., 1994a,b). Furthermore, when envelope membranes loaded with 16:0/16:0 and/or 18:1/16:0 diacylglycerol were incubated in presence of UDP-galactose or UDP-sulfoquinovose, 16:0/16:0 was incorporated with a much greater efficiency into sulfolipid than into MGDG whereas 18:1/16:0 was incorporated into both MGDG and sulfolipid with almost the same efficiency (Seifert and Heinz, 1992). These observations strongly suggest that sulfolipid synthase should have a very high affinity for 16:0/16:0, but this remains to be demonstrated.

Comparison of the affinity of MGDG synthase for diacylglycerol molecules with the eukaryotic structure, C18/C18, is also interesting. Most of these molecular species, such as distearoylglycerol and dioleoylglycerol, are usually not present in envelope membranes and, since no plant glycerolipids contain this structure (Heinz, 1977), there is little chance that the corresponding MGDG molecules could be formed in vivo, whatever the affinity of the enzyme for such substrates could be. In contrast, dilinolenoylglycerol can be formed from MGDG within envelope membranes during the course of their purification by the outer envelope membrane enzyme galacto-lipid:galactolipid galactosyltransferase (see above). The envelope membranes thus obtained can contain up to 15% of polyunsaturated diacylglycerol which can be rapidly converted into MGDG by the MGDG synthase (Joyard and Douce, 1976b) despite its rather low Km for this substrate. Finally, and because MGDG synthase has such a high affinity for dilinoleoylglycerol, one should question whether this molecule could actually be present (or formed) in envelope membranes. Phospholipids (most likely phosphatidylcholine) are probably the sole source of dilinoleoylglycerol in a plant cell, but most of them are present almost exclusively in extraplastidal membranes (Douce and Joyard, 1980). In chloroplasts, phosphatidylcholine is concentrated in the outer leaflet of the outer envelope membrane (Dorne et al., 1985, 1990), and contains significant amounts of 18:2 at both sn-1 and sn-2 position of the glycerol backbone (Siebertz et al., 1979), but the enzyme which could generate diacylglycerol from phosphatidylcholine, i.e. phospholipase c, has not been found in envelope membranes. Therefore, if phosphatidylcholine is indeed the source of dilinoleoylglycerol in the outer membrane, only limited amounts would be delivered to the inner membrane where MGDG synthase is located. In addition, nothing is known on the possible mechanisms that could be involved in diacylglycerol transfer from the outer to the inner envelope membrane (spontaneous diffusion of free monomers, lateral diffusion of lipids between membranes at regions of direct intermembrane contact). Therefore, only the high affinity of the envelope MGDG synthase for dilinoleoylglycerol would explain the presence of C18 fatty acids at both sn position in MGDG.

To conclude, a given diacylglycerol should either be present in appreciable amounts within the envelope (like 1-oleoyl,2-palmitoylglycerol, deriving from the envelope Kornberg-Pricer pathway, or dilinolenoyl-glycerol, formed by the galactolipid:galactolipid galactosyltransferase in given experimental conditions), or should present a very low Km value (like dilinoleoylglycerol) to be used efficiently by the envelope MGDG synthase. However, one should keep in mind that the amount of diacylglycerol necessary for optimal MGDG synthase activity could be totally different if all the enzymes are either dispersed within the membrane or tightly coupled together. In the latter situation, which would favor channeling between the active sites of phosphatidate phosphatase and MGDG synthase, local substrate concentration at the level of MGDG synthase would be sufficient for optimal enzyme activity with most diacylglycerol molecular species. But there is no evidence that all the enzymes involved in galactolipid biosynthesis could form a sort of protein complex within the membrane. The efficiency of channeling, i.e. the extent to which the product of the first reaction is preferentially transferred to the second site is not known. The chloroplast envelope phosphatidate phosphatase can be competitively inhibited by diacylglycerol (Malherbe et al., 1992). Therefore, diacylglycerol concentrations necessary for optimal MGDG synthase activity would be high enough to cause significant phosphatidate phosphatase inhibition unless diacylglycerol is used immediately after its formation. Indeed the specific activity of MGDG synthase is much higher than that of the envelope phosphatidate phosphatase. Finally, if diacylglycerol is transported from the outer to the inner envelope membrane, the activity of diacylglycerol-forming enzymes in the outer membrane would also modulate the activity of enzymes involved in the production and utilization of diacylglycerol in the inner envelope membrane. This would provide further possibilities for diacylglycerol to play a central role in the regulation of envelope glycerolipid biosynthesis. Clearly, analyses of the mechanisms involved require purification of envelope proteins, investigations of enzyme topography as well as on the nature of membrane interactions within the envelope, but these studies are still in their infancy. This is important for understanding the molecular events involved in regulation of enzyme activity, channeling of the different precursors, synthesis and distribution of complex lipids, i.e. all events essential for plastid membrane biogenesis (for a review, see Selstam, Chapter 11).

VI. The Last Step in Glycerolipid Biosynthesis: Fatty Acid Desaturation

Once synthesized, plastid glycerolipids are further modified by the desaturation of their fatty acids, C16:0 and C18:1, into the corresponding polyunsaturated fatty acids, C16:3 and C18:3. Our understanding of chloroplast desaturases activities has benefited considerably from the characterization of *Arabidopsis* mutants, each one being deficient in a specific desaturation step (for review Ohlrogge and Browse 1995). Five loci, i.e. *fad4*, *fad5*, *fad6*, *fad7* and *fad8*, affect chloroplast lipid desaturation (Fig. 10). Two of these desaturases are highly substrate-

specific: the *fad4* gene product is responsible for inserting a Δ3-*trans* double bond into the 16:0 esterified to *sn*-2 of phosphatidylglycerol, and the *fad5* gene product is responsible for the synthesis of Δ7 16:1 on MGDG and possibly on DGDG. The 16(18):1 desaturase is encoded by the *fad6* gene, whereas two 16(18):2 isozymes are encoded by *fad7* and *fad8* (reviewed in Ohlrogge and Browse 1995). With the exception of the *fab2* mutant of *Arabidopsis* (see above), none of the single-gene desaturase mutants shows any visible phenotype at normal growth temperatures (25 °C) or following several days of exposure to low temperature (Somerville and Browse 1996). Such results provide further evidence that polyunsaturation is not essential for normal functioning of the membranes. However, plants with the *fad5* and *fad6* mutations fail to accumulate normal levels of chlorophyll, and show a up to four-fold decrease in the amount of chloroplast membranes when grown at low temperatures (Hugly and Somerville, 1992). In addition, the rate of damage and repair of the D1 protein at low temperature was

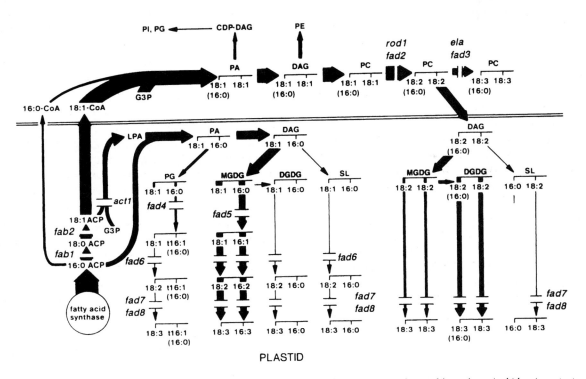

Fig. 10. Abbreviated diagram showing putative deficiencies of lipid biosynthetic enzymes observed in various *Arabidopsis* mutants. Adapted with permission from Somerville and Browse (1991). The breaks indicate the putative enzyme deficiencies, mostly in fatty acid desaturation. Abbreviations, see Figs. 1 and 2.

much slower in transgenic tobacco plants in which the proportion of saturated plastid phosphatidyl-glycerol was increased by expression of an isoform of sn-glycerol-3-phosphate acyl transferase with a preference for saturated fatty acyl substrates (Moon et al., 1995). Various aspects of the relation between lipid composition and temperature is addressed in several chapters of this volume (Wada and Murata, Chapter 4; Siegenthaler and Trémolières, Chapter 8; Gombos and Murata, Chapter 13; Vijayan et al., Chapter 14; Harwood, Chapter 15).

Investigation of membrane desaturases by traditional biochemical approaches has been limited because solubilizing and purifying them have proven very difficult (Schmidt et al., 1994). In fact, the nature of the enzymes involved in vivo in the cis-desaturation of fatty acyl chains engaged in glycerolipids is not yet clearly established. In addition, there is a large number of enzymes involved in desaturation (as shown for instance by the number of genes affecting chloroplast lipid desaturation, as shown in Fig. 10). Until recently, most of our knowledge of chloroplast desaturases was limited to the soluble components of the 18:0 to 18:1 desaturation system, i.e. ferredoxin, ferredoxin: NADP oxidoreductase, stearoyl-ACP desaturase (see above). Because intact chloroplasts are able to catalyze desaturation of MGDG-linked unsaturated or monounsaturated fatty acids to polyunsaturated fatty acids (Roughan et al., 1979), the addition of electron transport inhibitors was expected to provide information about electron donors and transport components involved in this type of desaturation (Andrews et al., 1989). Indeed, fatty acid desaturation in newly synthesized MGDG by intact chloroplasts can be blocked by compounds that interfere with light-driven electron transport in thylakoids (Andrews et al., 1989). In the case of envelope-bound desaturase, preliminary in vitro evidence for the involvement of a ferredoxin:NADPH oxidoreductase (FNR) was obtained by the use of anti-FNR-IgG (Schmidt and Heinz 1990a,b). Experiments with isolated chloroplasts (Norman et al., 1991) suggest that O_2 is the final electron acceptor whereas reduced ferredoxin $(E_0' = -0.4$ v) is the source of electrons for the reduction of O_2 to H_2O $(E_0' = -0.8$ v). Since ferredoxin delivers only one electron at a time, the desaturase has to oxidize two reduced ferredoxin, and store the first electron before the double bond is formed (Heinz 1993). This is possible only in the presence of a complex electron transfer chain that has not yet been characterized in envelope membranes. Interestingly,

some preliminary information on a possible electron transfer chain were recently provided: Jäger-Vottero et al. (1996) have shown that envelope membranes from spinach chloroplasts contain (a) semiquinone and flavosemiquinone radicals, (b) a series of iron-containing electron-transfer centers and (c) flavins (mostly FAD) loosely associated with proteins. Therefore, envelope membranes probably contain all the enzymatic machinery including various electron carriers involved in the formation and reduction of semiquinone radicals (quinol oxidase, NADPH-quinone and NADPH-semiquinone reductases) and that could be constituents of a redox chain essential for fatty acid desaturation. However, unambiguous evidence that the true electron donor for the envelope-bound fatty acid desaturase is indeed reduced ferredoxin, as proposed by Schmidt and Heinz (1990 a,b), is still lacking.

Acknowledgments

The authors are grateful to H. Ohta and K.-I. Takamiya for sharing unpublished data. We also wish to acknowledge financial support of our work as a continuing program by the Commissariat à l'Energie Atomique (CEA) and the Centre National de la Recherche Scientifique (CNRS).

References

Alban C, Joyard J and Douce R (1989) Comparison of glycerolipid biosynthesis in non-green plastids from sycamore (*Acer pseudoplatanus*) cells and cauliflower (*Brassica oleracea*) buds. Biochem J 259: 775–783

Alban C, Baldet P and Douce R (1994) Localization and characterization of two structurally different forms of acetyl-CoA carboxylase in young pea leaves, of which one is sensitive to aryl-oxyphenoxypropionate herbicides. Biochem J 300: 557–565

Alban C, Jullien J, Job D and Douce R (1995) Isolation and characterization of biotin carboxylase from pea chloroplasts. Plant Physiol 109: 927–935

Andrews J and Mudd JB (1985) Phosphatidylglycerol synthesis in pea chloroplasts. Pathways and localization. Plant Physiol 79: 259–265

Andrews J, Ohlrogge J and Keegstra K (1985) Final steps of phosphatidic acid synthesis in pea chloroplasts occurs in the inner envelope membrane. Plant Physiol 78: 459–465

Andrews J, Schmidt H and Heinz E (1989) Interference of electron transport inhibitors with desaturation of mono-galactosyl diacylglycerol in intact chloroplasts. Arch Biochem Biophys 270: 611–622

Aubert S, Gout E, Bligny R. and Douce R. (1994) Multiple effects of glycerol on plant cell metabolism. J. Biol. Chem. 269: 21420–21427

Benning C (1998) Membrane lipids in anoxygenic photosynthetic bacteria. In: Siegenthaler P-A and Murata N (eds) Lipids in Photosynthesis: Structure, Function and Genetics, pp 83–101. Kluwer Academic Publishers, Dordrecht

Benning C and Somerville CR (1982a) Isolation and genetic complementation of a sulfolipid-deficient mutant of *Rhodobacter sphaeroides*. J Bacteriol 174: 2352–2360

Benning C and Somerville CR (1982b) Identification of an operon involved in sulfolipid biosynthesis in *Rhodobacter sphaeroides*. J Bacteriol 174: 6479–6487

Benson AA (1963) The plant sulfolipid. Adv Lipid Res 1: 387–394

Bessoule JJ, Testet E and Cassagne C (1995) Synthesis of phosphatidylcholine in the chloroplast envelope after import of lysophosphatidylcholine from endoplasmic reticulum membranes. Eur J Biochem 228: 490–497

Beyer TA, Sadler JE, Rearick JI, Paulson JC and Hill R (1981) Glycosyltransferases and their use in assessing oligosaccharide structure and structure-function relationships. Adv Enzymol 52: 23–175

Billecocq A, Douce R and Faure M (1972) Structure des membranes biologiques: Localisation des galactosyldiglycérides dans les chloroplastes au moyen des anticorps spécifiques. CR Acad Sci Paris 275: 1135–1137

Bishop DG (1986) Chilling sensitivity in higher plants: The role of phosphatidylglycerol. Plant Cell Environment 9: 613–616

Bishop DG, Sparace SA and Mudd JB (1985) Biosynthesis of sulfoquinovosyldiacylglycerol in higher plants: The origin of the diacylglycerol moiety. Arch Biochem Biophys 240: 851–858

Bishop WR and Bell RM (1988) Assembly of phospholipids into cellular membranes: Biosynthesis, transmembrane movement and intracellular location. Annu Rev Cell Biol 4: 579–610

Bishop WR, Ganong BR and Bell RM (1986) Attenuation of *sn*-1,2-diacylglycerol second messengers by diacylglycerol kinase. Inhibition by diacylglycerol analogs in vitro and in human platelets. J Biol Chem 261: 6993–7000

Bligny R, Gardeström P, Roby C and Douce R (1990) ^{31}P NMR studies of spinach leaves and their chloroplasts. J Biol Chem 265: 1319–1326

Block MA, Dorne AJ, Joyard J and Douce R (1983a) Preparation and characterization of membrane fractions enriched in outer and inner envelope membranes from spinach chloroplasts. I – Electrophoretic and immunochemical analyses. J Biol Chem 258: 13273–13280

Block MA, Dorne AJ, Joyard J and Douce R (1983b) Preparation and characterization of membrane fractions enriched in outer and inner envelope membranes from spinach chloroplasts. II – Biochemical characterization. J Biol Chem 258: 13281–13286

Block MA, Dorne AJ, Joyard J and Douce R (1983c) The acyl-CoA synthetase and the acyl-CoA thioesterase are located respectively on the outer and on the inner membrane of the chloroplast envelope. FEBS Lett 153: 377-381

Browse J, Warwick N, Somerville CR and Slack CR (1986) Fluxes through the prokaryotic and eukaryotic pathways of lipid synthesis in the '16:3' plant *Arabidopsis thaliana*. Biochem J 235: 25–31

Browsher CG, Tetlow IJ, Lacey AE, Hanke GT and Emes MJ (1996) Integration of metabolism in non-photosynthetic plastids of higher plants. CR Acad Sci Paris 319: 853–860

Carlsson AS, Hellgren LI, Selden G and Sandelius AS (1994) Effects of moderately enhanced levels of ozone on the acyl lipid composition of leaf lipids of garden pea (*Pisum sativum*). Physiol Plant 91: 754–762

Carman GM and Henry SA (1989) Phospholipid biosynthesis in yeast. Annu Rev Biochem 58: 635–669

Chuman L and Brody S (1989) Acyl carrier protein is present in the mitochondria of plants and eucaryotic micro-organisms. Eur J Biochem 184: 643–649

Cline K and Keegstra K (1983) Galactosyltransferases involved in galactolipid biosynthesis are located in the outer membrane of pea chloroplast envelopes. Plant Physiol 71: 366–372

Cline K, Andrews J, Mersey B, Newcomb EH and Keegstra K (1981) Separation and characterization of inner and outer envelope membranes of pea chloroplasts. Proc Natl Acad Sci USA 78: 3595–3599

Covès J, Block MA, Joyard J and Douce R (1986) Solubilization and partial purification of UDP-galactose:diacylglycerol galactosyltransferase activity from spinach chloroplast envelope. FEBS Lett 208: 401–406

Covès J, Pineau B, Block MA, Joyard J and Douce R (1987) Solubilization and partial purification of chloroplast envelope proteins: Application to UDP-galactose:diacylglycerol galactosyltransferase. In: Leaver C and Sze H (eds.) Plant Membranes: Structure, Function, Biogenesis, pp. 103–112. Alan R. Liss, New York

Covès J, Joyard J and Douce R (1988) Lipid requirement and kinetic studies of solubilized UDP-galactose:diacylglycerol galactosyltransferase activity from spinach chloroplast envelope membranes. Proc Natl Acad Sci USA 85: 4966–4970

Davies HM (1993) Medium chain acyl-ACP hydrolysis activity of developing oilseeds. Phytochemistry 33: 1353–1356

Deems RA, Eaton BR and Dennis EA (1975) Kinetic analysis of phospholipase A2 activity towards mixed micelles and its implication for the study of lipolytic enzymes. J Biol Chem 250: 9013–9020

Dehaye L, Alban C, Job C, Douce R and Job D (1994) Kinetics of the two forms of acetyl-CoA carboxylase from *Pisum sativum*. Eur J Biochem 22: 1113–1123

Dennis EA (1983) The phospholipases. In: Boyer PD (ed) The Enzymes. Vol 16, pp. 307–353. Academic Press, London

Doering TL, Masterson WJ, Englund PT and Hart GW (1989) Biosynthesis of the glycosyl phosphatidylinositol membrane anchor of the trypanosome variant surface glycoprotein: origin of the non-acetylated glucosamine. J Biol Chem 264: 11168–11173

Dörmann P and Benning C (1996) Functional expression of uridine 5′-diphospho-glucose-4-epimerase (EC 5.1.3.2) from *Arabidopsis thaliana* in *Saccharomyces cerevisae* and *Escherichia coli*. Arch Biochem Biophys 327: 27–34

Dörmann P, Hoffmann-Benning S, Balbo I and Benning C (1995) Isolation and characterization of an *Arabidopsis* mutant deficient in the thylakoid lipid digalactosyl diacylglycerol. Plant Cell 7: 1801–1810

Dorne A-J and Heinz E (1989) Position and pairing of fatty acids in phosphatidylglycerol from pea leaf chloroplasts and mitochondria. Plant Sci 60: 39–46

Dorne A-J, Block MA, Joyard J and Douce R (1982) The galactolipid:galactolipid galactosyltransferase is located on the outer membrane of the chloroplast envelope. FEBS Lett 145: 30–34

Dorne A-J, Joyard J, Block MA and Douce R (1985) Localization of phosphatidylcholine in outer envelope membrane of spinach chloroplasts. J Cell Biol 100: 1690–1697

Dorne A-J, Joyard J and Douce, R (1990) Do thylakoids really contain phosphatidylcholine? Proc Natl Acad Sci USA 87: 71–74

Douce R (1974) Site of synthesis of galactolipids in spinach chloroplasts. Science 183: 852–853

Douce R and Joyard J (1980) Plant galactolipids. In: Stumpf PK (ed.) The Biochemistry of Plants, Lipids: Structure and Function, Vol 4, pp 321–362. Academic Press, New York

Douce R and Joyard J (1990) Biochemistry and function of the plastid envelope. Annu Rev Cell Biol 6: 173–216

Douce R and Joyard J (1996) Biosynthesis of thylakoid membrane lipids. In: Ort DR and Yocum CF (eds) Oxygenic Photosynthesis: The Light Reactions, pp 69–101. Kluwer Academic Publishers, Dordrecht

Douce R, Holtz RB and Benson AA (1973) Isolation and properties of the envelope of spinach chloroplasts. J Biol Chem 248: 7215–7222

Faccioti D, Knauf VC and Stumpf PK (1998) Triglycerides as products of photosynthesis. In: Siegenthaler P-A and Murata N (eds) Lipids in Photosynthesis: Structure, Function and Genetics, pp 225–248. Kluwer Academic Publishers, Dordrecht

Flügge UI and Benz R (1984) Pore-forming activity in the outer membrane of the chloroplast envelope. FEBS Lett 169: 85–89

Flügge UI and Heldt HW (1991) Metabolite translocators of the chloroplast envelope. Annu Rev Plant Physiol Plant Mol Biol 42: 129-144

Fox BG, Shanklin J, Somerville C and Munck E (1993) Stearoyl-acyl carrier $\Delta 9$ desaturase from *Ricinus communis* is a diiron-oxo protein. Proc Natl Acad Sci USA 90: 2486–2490

Frentzen M (1986) Biosynthesis and desaturation of the different diacylglycerol moieties in higher plants. J Plant Physiol 124: 193–209

Frentzen M (1993) Acyltransferases and triacylglycerol. In: Moore ST Jr (ed) Lipid Metabolism in Plants, pp 195–230. CRC Press, Boca Raton

Frentzen M, Heinz E, McKeon TA and Stumpf PK (1983) Specificities and selectivities of glycerol-3-phosphate acyltransferase and monoacylglycerol-3-phosphate acyltransferase from pea and spinach chloroplasts. Eur J Biochem 129: 629–636

Gardiner SE and Roughan PG (1983) Relationship between fatty-acyl composition of diacylgalactosylglycerol and turnover of chloroplast phosphatidate. Biochem J 210: 949–952

Gardiner SE, Roughan PG and Browse J (1984) Glycerolipid labeling kinetics in isolated intact chloroplasts. Biochem J 224: 637–643

Garnier J, Wu B, Maroc J, Guyon D and Trémolières A (1990) Restoration of both an oligomeric form of the light-harvesting antenna CPII and a fluorescence state II-state I transition by Δ^3-*trans*-hexadecenoic acid-containing phosphatidylglycerol, in cells of a mutant of *Chlamydomonas reinhardtii*. Biochim Biophys Acta 1020: 153–162

Gee RW, Byerrum RU, Gerber DW and Tolbert NE (1988a) Dihydroxyacetone phosphate reductase in plants. Plant Physiol 86: 98–103

Gee RW, Goyal A, Gerber DW, Byerrum RU and Tolbert NE (1988b) Isolation of dihydroxyacetone phosphate reductase from *Dunaliella* chloroplasts and comparison with isozymes from spinach leaves. Plant Physiol 88: 896–903

Gee RW, Goyal A, Gerber DW, Byerrum RU and Tolbert NE (1993) Two isoforms of dihydroxyacetone phosphate reductase

from the chloroplasts of *Dunaliella tertiolecta*. Plant Physiol 103: 243–249

Gombos Z and Murata (1998) Genetically engineered modulation of the unsaturation of glycerolipids and its consequences in tolerance of photosynthesis to temperature stresses. In: Siegenthaler P-A and Murata N (eds) Lipids in Photosynthesis: Structure, Function and Genetics, pp 249–262. Kluwer Academic Publishers, Dordrecht

Gounaris K and Barber J (1983) Monogalactosyldiacylglycerol: The most abundant polar lipid in Nature. Trends Biochem Sci 9: 378–381

Guler S, Seeliger A, Hartel H, Renger G and Benning C (1996) A null mutant of *Synechococcus* sp. PCC 7942 deficient in the sulfolipid sulfoquinovosyl diacylglycerol. J Biol Chem 271: 7501–7507

Gunstone F.D. (1994) High resolution ^{13}C NMR. A technique for the study of lipid structure and composition. Prog Lipid Res 33: 19–28

Haas R, Siebertz HP, Wrage K and Heinz E (1980) Localization of sulfolipid labeling within cells and chloroplasts. Planta 148: 238–244

Hannun YA, Loomis CR and Bell RM (1991) Activation of protein kinase C by triton-X100 mixed micelles containing diacylglycerol and phosphatidylserine. J Biol Chem 260: 10039–10043

Harwood JL (1980a) Plant acyl lipids: Structure, distribution, and analysis. In: Stumpf PK (ed) The Biochemistry of Plants, Vol 4, Lipids: Structure and Function, pp 2–55. Academic Press, New York

Harwood JL (1980b) Sulfolipids. In: Stumpf PK (ed) The Biochemistry of Plants, Vol 4, Lipids: Structure and Function, pp 301–320. Academic Press, New York

Harwood JL (1996) Recent advances in the biosynthesis of plant fatty acids. Biochim Biophys Acta 1301: 7–56

Harwood JL (1998) Involvement of chloroplast lipids in the reaction of plants submitted to stress. In: Siegenthaler P-A and Murata N (eds) Lipids in Photosynthesis: Structure, Function and Genetics, pp 287–302. Kluwer Academic Publishers, Dordrecht

Haverkate F and Van Deenen LLM (1965) Isolation and chemical characterization of phosphatidylglycerol from spinach leaves. Biochim Biophys Acta 106: 78–92

Heber U (1974) Metabolite exchange between chloroplasts and cytoplasm. Annu Rev Plant Physiol 25: 393–321

Heemskerk JWM and Wintermans JFGM (1987) The role of the chloroplast in the leaf acyl-lipid synthesis. Physiol Plant 70: 558–568

Heemskerk JWM, Wintermans JFGM, Joyard J, Block MA, Dorne A-J and Douce R (1986) Localization of galacto-lipid:galactolipid galactosyltransferase and acyltransferase in outer envelope membrane of spinach chloroplasts. Biochim Biophys Acta 877: 281–289

Heemskerk JHW, Jacobs FHH, Scheijen MAM, Heslper JPFG and Wintermans JFGM (1987) Characterization of galacto-syltransferases in spinach chloroplast envelope. Biochim Biophys Acta 918: 189–203

Heemskerk JHW, Storz T, Schmidt RR and Heinz E (1990) Biosynthesis of digalactolsyldiacylglycerol in plastids from 16:3 and 18:3 plants. Plant Physiol 93: 1286–1294

Heinz E (1977) Enzymatic reactions in galactolipid biosynthesis. In: Tevini M and Licthenthaler HK (eds) Lipids and Lipid

Polymers, pp 102–120. Springer Verlag, Berlin

Heinz E (1993) Biosynthesis of polyunsaturated fatty acids. In: Moore ST Jr (ed) Lipid Metabolism in Plants, pp 34–89. CRC Press, Boca Raton

Heinz E and Roughan PG (1983) Similarities and differences in lipid metabolism of chloroplasts isolated from 18:3 and 16:3 plants. Plant Physiol 72: 273–279

Heinz E, Schmidt H, Hoch M, Jung K-H, Binder H and Schmidt RR (1989) Synthesis of different nucleoside 5'-diphospho-sulfoquinovoses and their use for studies on sulfolipid biosynthesis in chloroplasts. Eur J Biochem 184: 445–453

Hellgren LI, Carlsson AS, Sellden G and Sandelius AS (1995) In situ leaf lipid metabolism in garden pea (*Pisum sativum* L.) exposed to moderately enhanced levels of ozone. J Exp Bot 4: 221–230

Hills MJ, Dann R, Lydiate D and Sharpe A (1994) Molecular cloning of a cDNA from *Brassica napus* L. for a homologue of acyl-CoA-binding protein. Plant Mol Biol 25: 917–920

Hippman H and Heinz E (1976) Glycerol kinase in leaves. Z Pflanzenphysiol 79: 408–418

Hobe S, Prytulla S, Külbrandt W and Paulsen H (1994) Trimerization and crystallization of reconstituted light-harvesting chlorophyll *a/b* complex. EMBO J 13: 3423–3429

Hoppe W and Schwenn JD (1981) In vitro biosynthesis of the plant sulfolipid: on the origin of the sulfonate group. Z Naturforsch 36: 820–826

Hugly S and Somerville C (1992) A role for membrane lipid polyunsaturation in chloroplast biogenesis at low temperature. Plant Physiol 99: 197–202

Jäger-Vottero P, Dorne AJ, Jordanov J, Douce R and Joyard J (1997) Redox chains in chloroplast envelope membranes: Spectroscopic evidences for the presence of electron carriers, including iron-sulfur centers. Proc Natl Acad Sci USA 94: 1597–1602

Jaworski JG, Clough RC and Barnum SR (1989) A cerulenin insensitive short chain 3-ketoacyl-acyl carier protein synthase in *Spinacia oleracea* leaves. Plant Physiol 90: 41–44

Jaworski JG, Tai H, Ohlrogge JB and Post-Beittenmiller D (1994) The initial reactions of fatty acid biosynthesis in plants. Prog Lipid Res 33: 47-54

Joyard J and Douce R (1976a) L'enveloppe des chloroplastes est-elle capable de synthétiser la phosphatidylcholine? CR Acad Sci Paris 282: 1515–1518

Joyard J and Douce R (1976b) Mise en évidence et rôle des diacylglycérols dans l'enveloppe des chloroplastes d'épinard. Biochim Biophys Acta 424: 126–131

Joyard J and Douce R (1977) Site of synthesis of phosphatidic acid and diacylglycerol in spinach chloroplasts. Biochim Biophys Acta 486: 273–285

Joyard J and Douce R (1979) Characterization of phosphatidate phosphohydrolase activity associated with chloroplast envelope membranes. FEBS Lett 102: 147–150

Joyard J and Douce R (1987) Galactolipid biosynthesis. In: Stumpf PK (ed.) The Biochemistry of Plants, Vol 9, Lipids: Structure and Function, pp 215–274. Academic Press, New York

Joyard J and Stumpf PK (1981) Synthesis of long-chain acyl-CoA in chloroplast envelope membranes. Plant Physiol 67: 250–256

Joyard J, Douce R, Siebertz HP and Heinz E (1980) Distribution of radioactive lipids between envelopes and thylakoids from in

vivo labelled chloroplasts. Eur J Biochem 108: 171–176

Joyard J, Blée E and Douce R (1986) Sulfolipid synthesis from $^{35}SO_4^{2-}$ and [1-^{14}C]-acetate in isolated intact spinach chloroplasts. Biochim Biophys Acta 879: 78–87

Joyard J, Block MA, Malherbe A, Maréchal E and Douce R (1993) Origin and synthesis of galactolipid and sulfolipid head groups. In: Moore ST Jr (ed) Lipid Metabolism in Plants, pp 231–258. CRC Press, Boca Raton, Florida

Joyard J, Maréchal E, Malherbe A, Block MA, and Douce R (1994) Importance of diacylglycerol in glycerolipid biosynthesis by spinach chloroplast envelope membranes. Prog Lipid Res 33: 105–118

Kader JC (1996) Lipid transfer proteins. Annu Rev Plant Physiol Plant Mol Biol 47: 627–654

Kanoh H and Ohno K (1981) Partial purification and properties of diacylglycerol kinase from rat liver cytosol. Arch Biochem Biophys 209: 266–275

Kauppinen S (1992) Structure and expression of the *KAS12* gene encoding a β-ketoacyl-acyl carrier protein synthase I isozyme from Barley. J Biol Chem 267: 23999–24006

Kinney AJ (1993) Phospholipid head groups. In: Moore ST Jr (ed) Lipid Metabolism in Plants, pp 259–284. CRC Press, Boca Raton

Kleppinger-Sparace KF, Mudd JB and Sparace SA (1990) Biosynthesis of plant sulfolipids. In: Rennenberg H, Brunold Ch, Dekok LJ and Stulen I (eds) Sulfur Nutrition and Sulfur Assimilation in Higher Plants, pp 77–88. SPB Academic Publishing, The Hague

Knowles JR (1989) The mechanism of biotin-dependent enzymes. Annu Rev Biochem 58: 195–221

Königs B and Heinz E (1974) Investigation of some enzymatic activities contributing to the biosynthesis of galactolipid precursors in *Vicia faba*. Planta 118: 159–169

Konishi T and Sasaki Y (1994) Compartmentalization of two forms of acetyl-CoA carboxylase in plants and the origin of their tolerance toward herbicides. Proc Natl Acad Sci USA 91: 3598–3601

Lawlor DW (1987) Photosynthesis: Metabolism, Control and Physiology. Longman Scientific and Technical, Harlow, UK

Lightner J, James DW Jr, Dooner HK and Browse J (1994) Altered body morphology is caused by increased stearate levels in a mutant of *Arabidopsis*. Plant J 6: 401–412

Malherbe A, Block MA, Joyard J and Douce R (1992) Feedback inhibition of phosphatidate phosphatase from spinach chloroplast envelope membranes by diacylglycerol. J Biol Chem 267: 23546–23553

Malherbe A, Block MA, Douce R and Joyard J (1995) Solubilization and biochemical properties of phosphatidate phosphatase from spinach chloroplast envelope membranes. Plant Physiol Biochem 33: 149–161

Mantel CR, Schulz AR, Miyazawa K and Broxmeyer HE (1993) Kinetic selectivity of cholinephosphotransferase in mouse liver: the Km for CDP-choline depends on diacylglycerol structure. Biochem J 289: 815–820

Maréchal E, Block MA, Joyard J and Douce R (1991) Purification de l'UDP-galactose:1,2-diacylglycérol galactosyltransferase de l'enveloppe des chloroplastes d'épinard. CR Acad Sci Paris 313: 521–528

Maréchal E, Block MA, Joyard J and Douce R (1994a) Kinetic properties of monogalactosyldiacylglycerol synthase from spinach chloroplast envelope membranes. J Biol

Chem 269: 5788–5798

Maréchal E, Block MA, Joyard J and Douce R (1994b) Comparison of the kinetic properties of MGDG synthase in mixed micelles and in envelope membranes from spinach chloroplast. FEBS Lett 352:307–310

Maréchal E, Miège C, Block MA, Douce R and Joyard J (1995) The catalytic site of monogalactosyldiacylglycerol synthase from spinach chloroplast envelope membranes. Biochemical analysis of the structure and of the metal content. J Biol Chem 270: 5714–5722

Martin BA and Wilson RF (1984) Subcellular localization of triacylglycerol biosynthesis in spinach leaves. Lipids 19: 117–121

Miller JC and Weinhold PA (1981) Cholinephosphotransferase in rat lung. J Biol Chem 256: 12662–12665

Miquel M, Block MA, Joyard J, Dorne A-J, Dubacq J-P, Kader J-C and Douce R (1987) Protein-mediated transfer of phosphatidylcholine from liposomes to spinach chloroplast envelope membranes. Biochim Biophys Acta 937: 219–228

Moon BY, Higashi S, Gombos Z and Murata N (1995) Unsaturation of the membrane lipids of chloroplasts stabilizes the photosynthetic machinery against low-temperature photoinhibition in transgenic tobacco plants. Proc Natl Acad Sci USA 92: 6219–6223

Mudd JB (1980) Phospholipid biosynthesis. In: PK Stumpf (ed.) The Biochemistry of Plants, Vol 4, Lipids: Structure and Function, pp 250–282. Academic Press, New York

Mudd JB and Kleppinger-Sparace KF (1987) Sulfolipids. In: Stumpf PK (ed) The Biochemistry of Plants, Vol. 9, Lipids: Structure and Function, pp 275–289. Academic Press, New York

Mudd JB, Van Vliet HHDM and Van Deenen LLM (1969) Biosynthesis of galactolipids by enzyme preparations from spinach leaves. J Lipid Res 10: 623–630

Murata N and Hoshi H (1984) Suloquinovosyl diacylglycerols in chilling-sensitive and chilling-resistant plants. Plant Cell Physiol 25: 1241–1245

Murata N and Nishida I (1987) Lipids of blue-green algae (cyanobacteria). In: Stumpf PK (ed) The Biochemistry of Plants, Vol 9, Lipids: Structure and Function, pp 315–347. Academic Press, New York

Murata N, Sato N, Takahashi N and Hamazaki Y (1982) Compositions and positional distribution of fatty acids in phospholipids from leaves of chilling-sensitive and chilling-resistant plants. Plant Cell Physiol 23: 1071–1079

Murphy DJ (1994) Biogenesis, function and biotechnology of plant storage lipids. Prog Lipid Res 33: 71–85

Norman HA, Pillai P and St John JB (1991) In vitro desaturation of monogalactosyldiacylglycerol and phosphatidylcholine molecular species by chloroplast homogenates. Phytochemistry 30: 2217–2222

Nussberger S, Dörr K, Wang DN and Kühlbrandt W (1993) Lipid-protein interactions in crystals of plant light-harvesting complex. J Mol Biol 234: 347–356

Ohlrogge J and Browse J (1995) Lipid biosynthesis. Plant Cell 7: 957–970

Ohta H, Shimojima M, Arai T, Masuda T, Shioi Y and Takamiya KI (1995) UDP-galactose:diacylglycerol galacto-syltransferase in cucumber seedlings: Purification of the enzyme and its activation by phosphatidic acid. In: Kader JC

and Mazliak P (eds) Plant Lipid Metabolism, pp 152–155. Kluwer Academic Publishers, Dordrecht

Pollard MR (1987) Analysis and structure determination of acyl lipids. In: Stumpf PK (ed) The Biochemistry of Plants, Vol 9, Lipids: Structure and Function, pp 1–30. Academic Press, New York

Rafferty JB, Simon JW, Stuije AR, Slabas AR, Fawcett T and Rice DW (1994) Crystallization of the NADH-specific enoyl acyl carrier protein reductase from Brassica napus. J Mol Biol 237: 240–242

Rock CO and Cronon JE (1996) Escherichia coli as a model for the regulation of dissociable (type II) fatty acid biosynthesis. Biochim Biophys Acta 1302: 1–16

Rossak M, Tietje C, Heinz E and Benning C (1995) Accumulation of UDP-sulfoquinovose in a sulfolipid-deficient mutant of Rhodobacter sphaeroides. J Biol Chem 270: 25792–25797

Roughan PG and Slack CR (1977) Long-chain acyl-coenzyme A synthetase activity of spinach chloroplasts is concentrated in the envelope. Biochem J 162: 457–459

Roughan PG and Slack CR (1982) Cellular organization of glycerolipid metabolism. Annu Rev Plant Physiol 33: 97–132

Roughan PG, Mudd JB, McManus TT and Slack CR (1979) Linoleate and α-linolenate synthesis by isolated spinach (Spinacia oleracea) chloroplasts. Biochem J 184: 571–574

Sakaki T, Saito K, Kawaguchi A, Kondo N and Yamada M (1990a) Conversion of monogalactosyldiacylglycerols to triacylglycerol in ozone-fumigated spinach leaves. Plant Physiol 94: 766–772

Sakaki T, Kondo N and Yamada M (1990b) Pathway for the synthesis of triacylglycerol from monogalactosyldiacylglycerols in ozone-fumigated spinach leaves. Plant Physiol 94: 773–780

Sakaki T, Tanaka K and Yamada M (1994) General metabolic changes in leaf lipids in response to ozone. Plant Cell Physiol 35: 53–62

Sakane F, Yamada K, Kanoh H, Yokoyama C and Tanabe T (1990) Porcine diacylglycerol kinase sequence has a zinc finger and a E-F hand motifs. Nature 344: 345–348

Sanchez J (1994) Lipid photosynthesis in olive fruit. Prog Lipid Res 33: 97–104

Sasaki Y, Hakamada K, Suama Y, Nagano Y, Furusawa I and Matsuno R (1993) Chloroplast-encoded protein as a subunit of acetyl-CoA carboxylase in pea plant. J Biol Chem 268: 25118–25123

Schaap D, De Widt J, Van der Wal J, Vanderkerkhove J, Van Damme J, Gussow D, Ploegh HL, Van Blitterswijk J and Van der Bend RL (1990) Purification, cDNA-cloning and expression of human diacylglycerol kinase. FEBS Lett 275: 151–158

Schmidt H and Heinz E (1990a) Involvement of ferredoxin in desaturation of lipid-bound oleate in chloroplasts. Plant Physiol 94: 214–220

Schmidt H and Heinz E (1990b) Desaturation of oleoyl groups in envelope membranes from spinach chloroplasts. Proc Natl Acad Sci USA 87: 9477–9480

Schmidt H, Dresselhaus T, Buck F and Heinz E (1994) Purification and PCR-based cDNA cloning of a plastidial n-6 desaturase. Plant Mol Biol 26: 631–642

Schneider G, Lindqvist Y, Shanklin J and Somerville CR (1992) Preliminary crystallographic data for stearoyl-acyl carrier protein desaturase from castor seed. J Mol Biol 225: 561–564

Segel IH (1975) Enzyme Kinetics, John Wiley and Sons, New York

Seifert U and Heinz E (1992) Enzymatic characteristics of UDP-sulfoquinovose: diacylglycerol sulfoquinovosyltransferase from chloroplast envelopes. Bot Acta 105: 197–205

Selstam E (1998) Development of thylakoid membranes with respect to lipids. In: Siegenthaler P-A and Murata N (eds) Lipids in Photosynthesis: Structure, Function and Genetics, pp 209–204. Kluwer Academic Publishers, Dordrecht

Serre L, Verbree EC, Dauter Z, Stuije AR and Derewenda ZS (1995) The *Escherichia coli* malonyl-CoA:acyl carrier protein transacylase at 1.5 Å resolution. Crystal structure of a fatty acid synthase component. J Biol Chem 270: 12961–12964

Shanklin J and Somerville C (1991) Stearoyl-acyl-carrier-protein desaturase from higher plants is structurally unrelated to the animal and fungal homologs. Proc Natl Acad Sci USA 88: 2510-2514

Shimakaya T and Stumpf PK (1982a) Isolation and function of spinach leaf β-ketoacyl-[acyl carrier protein] synthases. Proc Natl Acad Sci USA 79: 5808–5812

Shimakaya T and Stumpf PK (1982b) The purification and function of acetyl coenzyme A:acyl carrier protein transacylase. J Biol Chem 258: 3592–3598

Shimakaya T and Stumpf PK (1982c) Purification and characteristics of β-ketoacyl-[acyl carrier protein] reductase, β-hydroxyacyl-[acyl carrier protein] dehydrase and enoyl-[acyl carrier protein] reductase from *Spinacia oleracea* leaves. Arch Biochem Biophys 218: 77–91

Shimakaya T and Stumpf PK (1983) Purification and characteristics of β-ketoacyl-[acyl carrier protein] synthase I from *Spinacia oleracea* leaves. Arch Biochem Biophys 220: 39–45

Shintani DK and Ohlrogge JB (1994) The characterization of a mitochondrial acyl carrier isform isolated from *Arabidopsis thaliana*. Plant Physiol 104: 1221-1229

Shorrosh BS, Roesler KR, Shintani D, van de Loo FJ and Ohlrogge JB (1995) Structural analysis, plastid localisation, and expression of the biotine carboxylase subunit of acetyl-CoA carboxylase from tobacco. Plant Physiol 108: 805–812

Siebertz HP, Heinz E, Linscheid M Joyard J and Douce R (1979) Characterization of lipids from chloroplast envelopes. Eur J Biochem 101: 429–438

Siegenthaler PA (1998) Molecular organization of acyl lipids in photosynthetic membranes of higher plants. In: Siegenthaler P-A and Murata N (eds) Lipids in Photosynthesis: Structure, Function and Genetics, pp 119–144. Kluwer Academic Publishers, Dordrecht

Siegenthaler PA and Trémolières A (1998) Role of acyl lipids in the function of photosynthetic membranes in higher plants. In: Siegenthaler P-A and Murata N (eds) Lipids in Photosynthesis: Structure, Function and Genetics, pp 145–173. Kluwer Academic Publishers, Dordrecht

Siggaard-Andersen M, Kauppinen S and von Wettstein-Knowles P (1991) Primary structure of a cerulenin-binding β-ketoacyl-[acyl carrier protein] synthase from barley chloroplasts. Proc Natl Acad Sci USA 88: 4114–4118

Slabas AR and Fawcett T (1992) The biochemistry and molecular biology of plant lipid biosynthesis. Plant Mol Biol 19: 169–191

Slabas AR, Chase D, Nishida I, Murata N, Sidebottom C, Safford R, Sheldon PS, Kekwick RG, Hardie DG and Mackintosh RW (1992) Molecular cloning of higher-plant-3-oxoacyl-(acyl carrier protein) reductase. Sequence identities with the nodG-gene of the nitrogen-fixing soil bacterium *Rhizobium meliloti*. Biochem J 283: 321-326

Slabas AR, Brown A, Sinden BS, Swinhoe R, Simon JW, Ashton AR, Whitfeld PR and Elborough KM (1994) Pivotal reactions in fatty acid synthesis. Prog Lipid Res 33: 39–46

Somerville C and Browse J (1991) Plant lipids: Metabolism, mutants, and membranes. Science 252: 80–87

Somerville C and Browse J (1996) Dissecting desaturation: Plants prove advantageous. Trends Cell Biol 6: 148–153

Stitt M, Kürzel B and Heldt HW (1984) Control of photosynthetic sucrose synthesis by fructose 2,6-bisphosphate. Plant Physiol 75: 544–560

Stymne S and Stobart AK (1987) Triacylglycerol biosynthesis. In: Stumpf PK (ed) The Biochemistry of Plants, Vol 9, Lipids: Structure and Function, pp 175–214. Academic Press, New York

Teucher T and Heinz E (1991) Purification of UDP-galactose:diacylglycerol from chloroplast envelopes of spinach (*Spinacia oleracea* L.). Planta 184: 319–326

Thoden JB, Frey PA and Holden HM (1996) Crystal structures of the oxidized and reduced forms of UDP-galactose-4-epimerase isolated from *Escherichia coli*. Biochemistry 35: 2557–2567

Thompson GA (1996) Lipids and membrane function in green algae. Biochim Biophys Acta 1302: 17–45

Trémolières A, Roche O, Dubertret G, Maroc J, Guyon D and Garnier J (1991) Restoration of thylakoid appression by Δ^3-*trans*-hexadecenoic acid-containing phosphatidylglycerol, in a mutant of *Chlamydomonas reinhardtii*. Relationship with the regulation of excitation energy distribution. Biochim Biophys Acta 1059: 286–292

Van Besouw A and Wintermans JFGM (1978) Galactolipid formation in chloroplast envelopes. I. Evidence for two mechanisms in galactosylation. Biochim Biophys Acta 529: 44–53

Van Besouw A and Wintermans JFGM (1979) The synthesis of galactolipids by chloroplast envelopes. FEBS Lett 102: 33–37

Verwoert II, Brown A, Slabas AR and Stuije AR (1995) A *Zea mays* GTP-binding protein of the ARF family complements an *Escherichia coli* mutant with a temperature-sensitive malonyl-CoA:acyl carrier protein transacylase. Plant Mol Biol 27: 629–633

Vijayan P, Routaboul JM and Browse J (1998) A genetic approach to investigating membrane lipid structure and photosynthetic function. In: Siegenthaler P-A and Murata N (eds) Lipids in Photosynthesis: Structure, Function and Genetics, pp 263–285. Kluwer Academic Publishers, Dordrecht

Wada H and Murata N (1998) Membrane lipids in cyanobacteria. In: Siegenthaler P-A and Murata N (eds) Lipids in Photosynthesis: Structure, Function and Genetics, pp 65–81. Kluwer Academic Publishers, Dordrecht

Wakil SJ (1989) Fatty acid synthase, a proficient multifunctional enzyme. Biochemistry 28: 4523–4530

Williams WP (1994) The role of lipids in the structure and function of photosynthetic membranes. Prog Lipid Res 33: 119–127

Williams JP and Khan MU (1996) Lipid metabolism in leaves of an 18:4-plant, *Echium plantagineum*: A model of galactolipid biosynthesis in 18:3- and 18:4-plants. Plant Physiol Biochem 34:93–100

Wissing JB and Wagner KG (1992) Diacylglycerol kinase from suspension cultured plant cells. Characterization and subcellular localization. Plant Physiol 98: 1148–1153

Yadav SP and Brew K (1990) Identification of a region of UDP-galactose:N-acetylglucosamine β-4-galactosyltransferase involved in UDP-galactose binding by differential labeling. J Biol Chem 265: 14163–14169

Yadav SP and Brew K (1991) Structure and function in galactosyltransferase. Sequence locations of α-lactalbumin binding site, thiol group, and disulfide bond. J Biol Chem 266: 698–703

Chapter 3

Membrane Lipids in Algae

John L. Harwood

*School of Molecular and Medical Biosciences, University of Wales College of Cardiff,
P.O. Box 911, Cardiff, CF1 3US, U.K.*

Summary

Eukaryotic algae are a diverse group of organisms. Their lipid compositions have been less studied than those of higher plants but, nevertheless, we now have sufficient data to be able to make some broad generalizations. Algae contain many of the major lipids of plants, such as the glycosylglycerides and the usual phosphoglycerides. In addition, more unusual compounds such as the betaine lipids, chlorosulfolipids or various other sulfolipids may be major components of some species or orders.

Where information is available specifically about chloroplast membranes, it seems that the three glycosylglycerides, monogalactosyldiacylglycerol, digalactosyldiacylglycerol and sulfoquinovosyl-diacylglycerol, and phosphatidylglycerol are the main acyl lipids, as in plants. These four lipids have characteristic fatty acid compositions and are often highly enriched in polyunsaturated fatty acids which may contain as many as six double bonds. The positional distribution of fatty acids on the thylakoid lipids of *Chlamydomonas* indicates that the former are made exclusively within the chloroplast.

P.-A. Siegenthaler and N. Murata (eds): Lipids in Photosynthesis: Structure, Function and Genetics, pp. 53–64.
© *1998 Kluwer Academic Publishers. Printed in The Netherlands.*

I. Introductory Remarks

In general terms the eukaryotic algae have been rather poorly studied when compared to higher plants. This statement applies as much to lipid structure, occurrence and metabolism as to other aspects of biochemistry. Only a few individual species, such as *Chlorella* or *Chlamydomonas*, have had more than a handful of papers published about them. Therefore, it is imperative to realize that remarks made in this chapter are made with relatively little evidence which prevents them being used as anything more than broad generalizations. Indeed, with the discovery of a major new membrane lipid, the β-alanine ether lipid (see below), in the last few years it seems likely that there will be many more exciting new observations to be made in the foreseeable future.

Algae have been traditionally divided into groups, based on their predominant pigment contents. Within this classification the blue-green algae are now termed cyanobacteria (to reflect their prokaryotic nature) and are discussed in the next chapter. I will use these classifications in my discussion of algal membrane lipids and their metabolism. But, first, a few remarks about the nature of algal lipids.

II. The Nature of Algal Lipids

Most of the eukaryotic algae contain a range of glycerol-based lipids, many of which are found in higher plants (see Chapter 2). Thus, the glycosylglycerides (monogalactosyldiacylglycerol, digalactosyldiacylglycerol, sulfoquinovosyldiacylglycerol, Fig. 1) are major constituents while phosphoglycerides (such as phosphatidylcholine, phosphatidylethanolamine and phosphatidylglycerol, Fig. 2) are usually significant (see Pohl and Zurheide, 1979a,b). *Chlorella* has been reported to contain trigalactosyl-

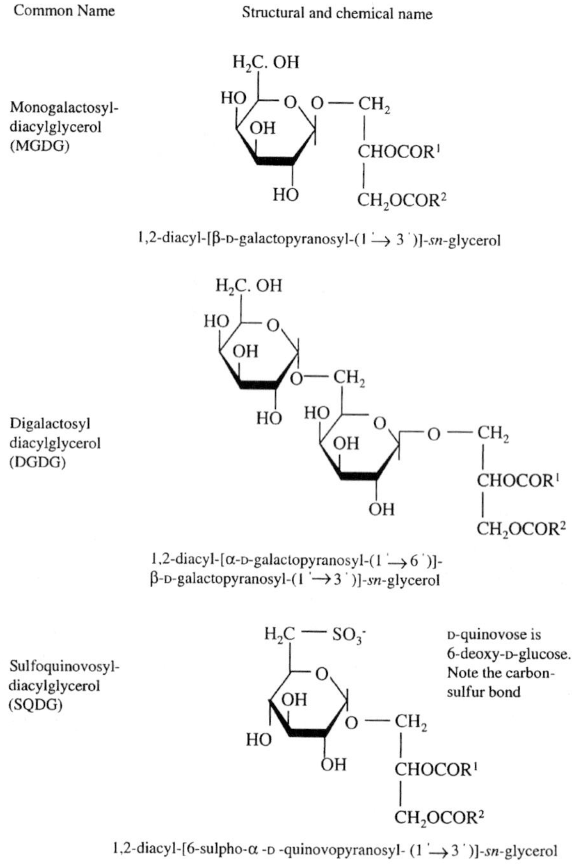

Abbreviations: DGDG – digalactosyldiacylglycerol; DGTA – diacylglycerylhydroxymethyltrimethylalanine; DGTS – diacylglyceryltrimethylhomoserine; DPG – diphosphatidylglycerol (cardiolipin); MGDG – monogalactosyldiacylglycerol; PC – phosphatidylcholine; PE – phosphatidylethanolamine; PG – phosphatidylglycerol; PI – phosphatidylinositol; SQDG – sulfoquinovosyldiacylglycerol.

Fatty acids are abbreviated with two numbers separated by a colon (e.g. 18:1). The figure before the colon represents the number of carbon atoms while the number afterwards shows the number of double bonds. Where the position of the double bonds is known then these can be shown numbering from the carboxyl group (Δ) or from the methyl end of the chain (n-).

Fig. 1. The main glycosylglycerides of algae.

diacylglycerol (Benson et al., 1958) and sugars other than galactose may be found. For example, in the red algae, *Polysiphonia lanosa* and *Chondrus crispus,* mannose and rhamnose were also detected in their glycolipids (T. R. Pettitt, A. L. Jones and J. L. Harwood, unpublished). Moreover, Pham-Quang and Laur (1976a,b,c) reported a range of novel glycolipids (and phospho- and sulfolipids) in three brown algae, *Fucus serratus, F. vesiculosus* and *Pelvetia canaliculata.*

Because the thylakoids of chloroplasts are the main intra-cellular membranes of algae, their lipid composition dominates extracts from total cells. Four glycerolipids are present in large amounts—monogalactosyldiacylglycerol (MGDG), digalactosyldiacylglycerol (DGDG), sulfoquinovosyldiacylglycerol (SQDG) and phosphatidylglycerol (PG). In fact, although minor amounts of other lipids have been found in isolated chloroplasts (see Section III.B and Table 4), it seems by analogy with cyanobacterial lipids (Chapter 4) that these minor components are

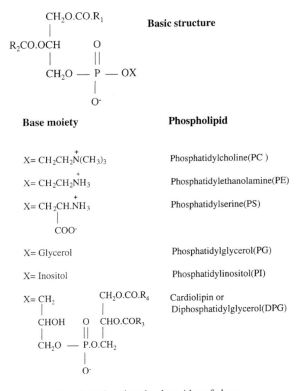

Fig. 2. Major phosphoglycerides of algae.

present in the chloroplast envelopes and/or are due to contamination of the purified chloroplasts with other organellar membranes. Thus, algal chloroplasts have MGDG as their main lipid, with smaller amounts of DGDG and the two negatively charged lipids, SQDG and PG. Because it is exceptionally difficult to isolate purified chloroplasts from most algae, we have a dearth of knowledge about the lipid composition of photosynthetic membranes from these organisms. However, those analyses that have been made support the idea that algal chloroplast thylakoids generally have a similar percentage lipid composition to those from higher plants (see Chapter 2) or to the thylakoids of cyanobacteria (see Chapter 4). Thus, MGDG will probably be present at 40–55%, DGDG at 15–35%, SQDG at 10–20% and PG at 10–20%. Whole tissue analyses support this proposal although, of course, the evidence is indirect and, in brown marine algae, interpretation of whole tissue analysis is complicated by the presence of SQDG apparently in non-chloroplast membranes (Smith and Harwood, 1984; Harwood and Jones, 1989).

The use of radiolabeled sulfate has enabled the detection of a number of unusual sulfur-containing lipids in algae. One of these, the sulfonium analogue

of phosphatidylcholine, phosphatidylsulfocholine, has been found in *Euglena*, red algae and diatoms (see Harwood and Jones, 1989) (Fig. 3). In addition, Kates and coworkers have found several more interesting structures in *Nitzschia alba* and other diatoms (Kates, 1987). Overall, sulfolipids are present in very high concentrations in diatoms. For example, in the non-photosynthetic diatom *N. alba* they account for 74% of the total polar lipids while in the photosynthetic *N. pelliculose* they represent over 30%.

Another class of lipid are the chlorosulfolipids (Fig. 4). These compounds have been detected in some algae and are major components of *Ochromonas danica* where they represent almost half of the total membrane lipids. A whole family of compounds are found with two sulfate ester functions and from one to six chlorines (Haines, 1973). Some other structures (e.g. trichloro derivatives) have been reported but not fully characterized (Haines, 1973).

Benson has discussed the presence of arsenic-containing lipids in marine organisms, including algae. A major compound is arsenoribosylphos-phatidylglycerol, which is commonly distributed in aquatic plants and algae (Benson, 1987).

Betaine lipids (Fig. 5), that are often major lipids of lower plants, are found in many algae. The homoserine ether lipid, 1(3),2-diacylglyceryl-(3)-0-4′-(*N,N,N*-trimethyl)homoserine (DGTS) is important in green algae where it has been suggested to act as a substitute for phosphatidylcholine. The β-alanine analogue, DGTA [1(3),2-diacylglycerol-(3)-0-hydroxymethyl (*N,N,N*-trimethyl)β-alanine], is a major component of brown algae, *Crytomonas* and *Ochromonas,* are the only organisms known to produce both of these betaine lipids (Eichenberger and Hofmann, 1992). In a survey concerning the distribution of betaine lipids in plants, it was concluded that betaine lipids are limited to cryptogamic plants and heterotrophs. Thus, DGTS is a typical component of ferns, mosses, green algae and certain chromophytes. DGTA was found in 52 out of 88 species of brown algae (Phaeophyceae) (Eichenberger and Hofmann, 1992). Many of the betaine lipid-containing algae do not contain phosphatidylcholine thus suggesting their functional substitution for zwitterionic membrane phospho-glycerides, as remarked above. A recent, compre-hensive, survey of the betaine lipids has been made (Dembitsky, 1996).

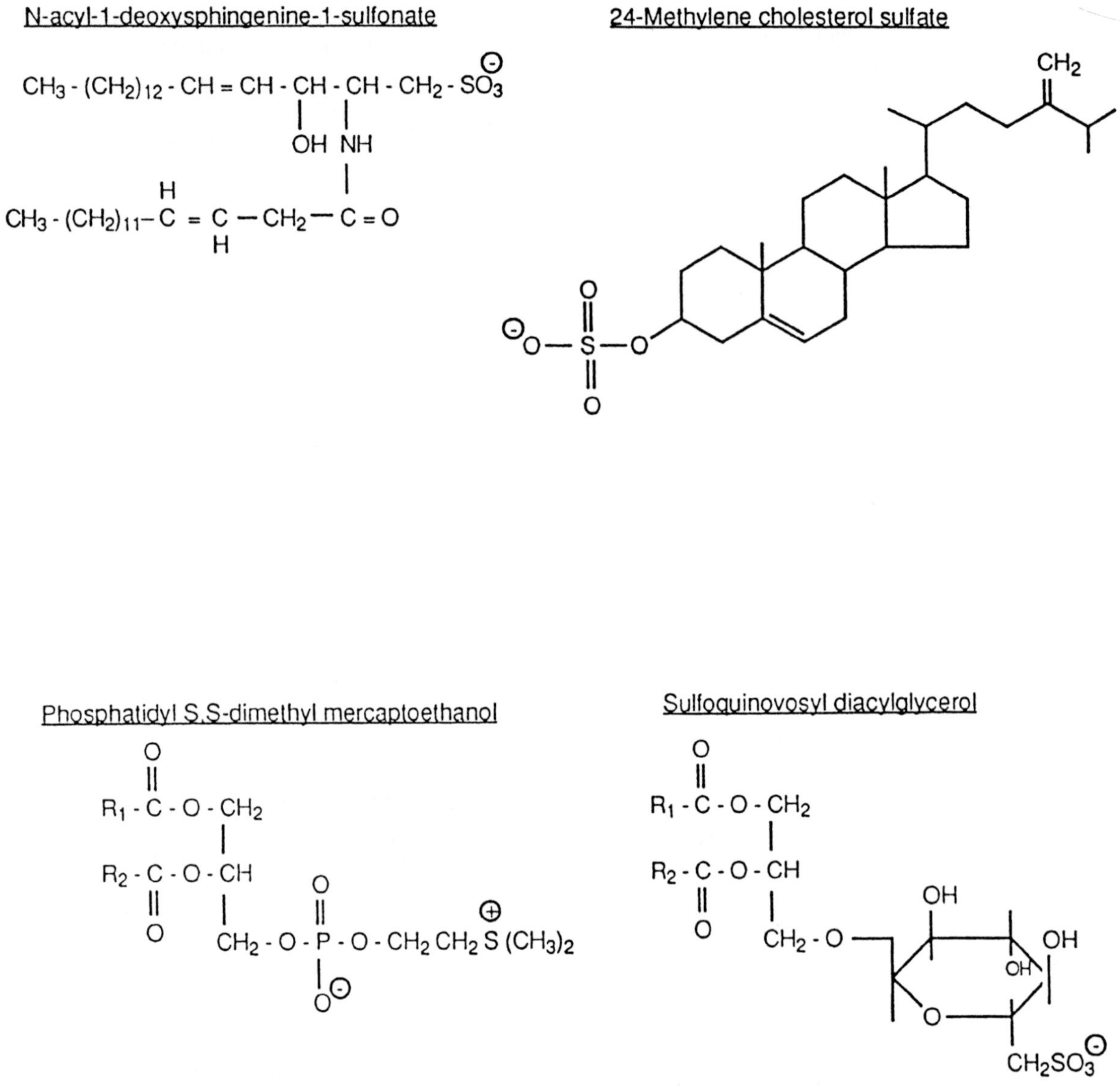
N-acyl-1-deoxysphingenine-1-sulfonate
24-Methylene cholesterol sulfate
Phosphatidyl S,S-dimethyl mercaptoethanol
Sulfoquinovosyl diacylglycerol

Fig. 3. Some sulfur-containing lipids detected in algae. SQDG, the major sulphur-containing lipid, was shown in Fig. 1.

III. Lipid Composition of Algae

A. Fatty Acids

Most of the analyses of different algae for their fatty acid content have used total lipid extracts rather than examining individual lipid classes. Nevertheless, more data is now becoming available with regard to the latter, including (in some cases) the separation of molecular species.

Typically, freshwater algae contain similar fatty acids to higher plants, though in rather different proportions. Even chain fatty acids in the range C_{14}–C_{22} usually account for over 98% of the total. Freshwater algae usually have significantly higher levels of C_{16} acids and lower proportions of C_{18} acids (especially α-linolenate) when compared to higher plant leaves (see Chapter 2 and Harwood et al., 1989).

There are several problems with the interpretation of data for total fatty acid contents of whole lipid extracts as pointed out by Harwood and Jones (1989). Thus, it is well known that storage lipids, such as triacylglycerols, often have a much lower proportion

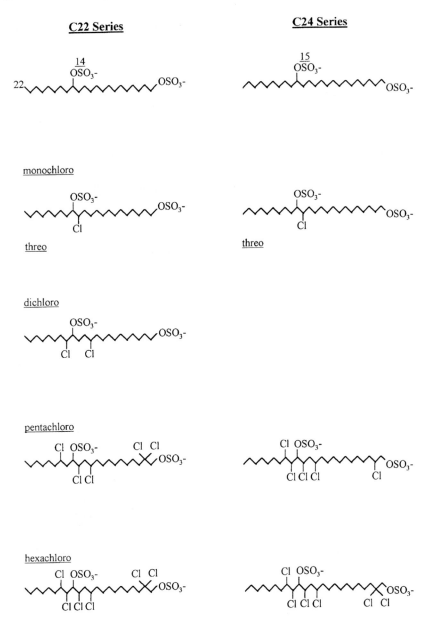

Fig. 4. Some chlorosulfolipids from *Ochromonas* spp. Other structures have also been characterized (see Haines, 1973). Re-drawn from Harwood (1980) with permission.

of polyunsaturated fatty acids than membrane lipids. When these storage lipids accumulate, their contribution to the total algal composition may give misleading results. In addition, where isomers of individual acids are present, they may have a markedly different distribution within lipid classes. Thus, Eichenberger et al. (1986) analyzed two species of *Chlamydomonas reinhardtii* and noted that, whereas the glycosylglycerides, phosphatidylglycerol and phosphatidylinositol contained mainly or exclusively

the n-3 isomer of linolenic acid (α-linolenate), phosphatidylethanolamine and DGTS contained the n-6 isomer (γ-linolenate). Perhaps, most importantly, environmental factors can have a very significant effect on fatty acid (and lipid) composition. This is seldom taken into consideration when collecting algae from natural sites.

Some fatty acid compositions for freshwater algae and salt-tolerant spp. are shown in Table 1. In the examples shown C_{20} fatty acids were confined to the

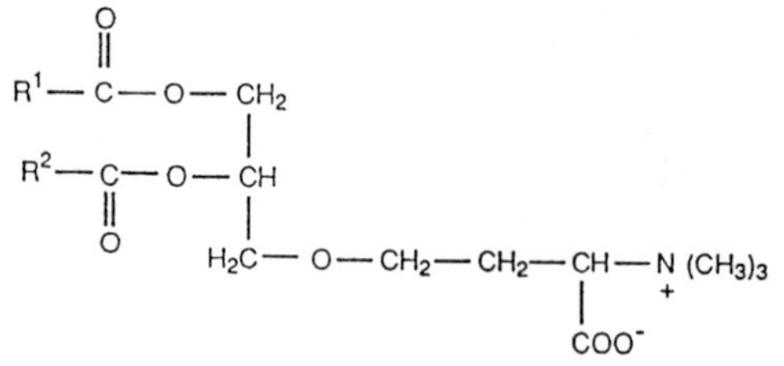

R^1—C—O—CH$_2$

R^2—C—O—CH

H$_2$C—O—CH$_2$—CH$_2$—CH—N (CH$_3$)$_3$

COO$^-$

Diacylglyceryl-O-(N, N, N - trimethyl) homoserine (DGTS)

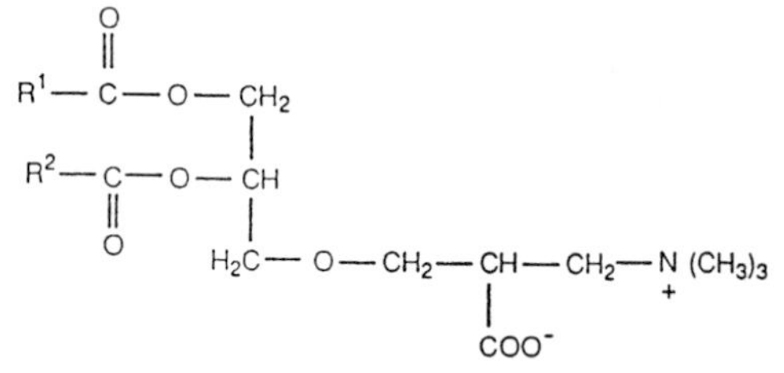

R^1—C—O—CH$_2$

R^2—C—O—CH

H$_2$C—O—CH$_2$—CH—CH$_2$—N (CH$_3$)$_3$

COO$^-$

Diacylglyceryl-O- (hydroxymethyl)-(N, N, N - trimethyl)β-alanine (DGTA)

Fig. 5. Betaine membrane lipids.

salt-tolerant species and were highly unsaturated. When compared with higher plants, the existence of tetraenoic acids of C$_{16}$ and C$_{18}$ is notable in algae and, as mentioned above, γ-linolenate is present whereas only the α-isomer is normally found in higher plant membrane lipids.

Within marine algae there tends to be a much more variable selection of fatty acid components than in freshwater species (see Harwood and Jones, 1989; Pohl and Zurheide, 1979a,b). Examples for different classes are given in Table 2. Of the saturated acids, myristate can be a major constituent in some cases

but palmitate is the main example. As noted above for freshwater algae, the C$_{16}$ acids represent a higher proportion of the total than in higher plants. However, in contrast to many freshwater species, hexadeca-tetraenoate is not usually a significant component. By contrast, octadecatetraenoate can be present at up to 30% of the total acids. All the species analyzed contained high amounts of C$_{20}$ polyunsaturated fatty acids. For marine macroalgae, arachidonate is frequently a main component. This is specifically true for those Phaeophyceae and Rhodophyceae which have been analyzed. Whereas phytoplankton may have significant quantities of docosahexaenoate, this fatty acid is rarely found in more than trace amounts in macroalgae.

Examination of the double bond position of unsaturated fatty acids in marine algae (Pohl and Zurheide, 1979a) reveals that there are clearly a number of desaturase systems present in such organisms. Apart from the commonly occurring $\Delta 6$, $\Delta 9$, $\Delta 12$ and $\Delta 15$ double bonds, the longer chain acids 20:5 ($\Delta 5, 8, 11, 14, 17$) and 22:6 ($\Delta 4, 7, 10, 13, 16, 19$) must have been formed with the action of $\Delta 4$ and $\Delta 5$ desaturases. On the other hand, fatty acids such as 16:1 ($\Delta 7$) may have been formed by an ω-9 desaturase.

Odd chain-length fatty acids (e.g. 15:0, 17:0, $\Delta 9$, 12-17:2) may occur as very minor components. However, a distinguishing feature of algae (especially marine species) are the high amounts of long chain polyunsaturated fatty acids which are separate from the α-linolenate characteristic of higher plants. Only the Chlorophyceae have large amounts of the latter acid. By contrast, the marine Chrysophyceae and

Table 1. Major fatty acids of some algae

	Fatty acid composition (% total)										
	16:0	16:1	16:2	16:3	16:4	18:1	18:2	18:3	18:4	20:5	22:6
Freshwater spp.											
Scenedesmus obliquus	35	2	tr.	tr.	15	9	6	30	2	-	-
Chlorella vulgaris	26	8	7	2	-	2	34	20	-	-	-
Chamydomonas reinhardtii	20	4	1	4	22	7	6	30	3	-	-
Salt-tolerant spp.											
Ankistrodesmus spp.	13	3	1	1	14	25	2	29	2	1	-
Isochrysis spp.	12	6	-	-	15	4	6	17	-	2	13
Nannochloris spp.	9	20	7	9	-	4	1	1	-	27	-

Taken from Harwood and Jones (1989) with permission. Fatty acids are abbreviated with the figure before the colon indicating the number of carbon atoms, and the figure after the colon indicating the number of double bonds contained. Dashes mean not detected.

Table 2. The fatty acid composition of some marine algae.

	Fatty acid composition (% total)												
	14:0	16:0	16:1	16:2	16:3	16:4	18:1	18:2	18:3	18:4	20:4	20:5	22:6
Phytoplankton													
Monochrysis lutheri (Chrysophyceae)	10	13	22	5	7	1	3	1	tr.	2	1	18	7
Olisthodiscus spp. (Xanthophyceae)	8	14	10	2	2	1	4	4	6	18	2	19	2
Lauderia borealis (Bacillariophyceae)	7	12	21	3	12	1	2	1	tr.	-	1	3	-
Amphidinium carterae (Dinophyceae)	3	24	1	1	tr.	-	5	1	2	15	-	14	25
Dunaliella salina (Chlorophyceae)	tr.	41	15	tr.	-	-	11	8	19	-	-	-	-
Hemiselmis brunescens (Cryptophyceae)	1	13	3	3	tr.	tr.	2	tr.	9	30	tr.	14	-
Macroalgae													
Fucus vesiculosus (Phaeophyceae)	-	21	2	tr.	-	tr.	26	10	7	4	15	8	-
Chondrus crispus (Rhodophyceae)	-	34	6	tr.	-	-	9	1	1	4	18	22	-
Ulva lactuca (Chlorophyceae)	1	18	2	tr.	1	18	9	2	17	24	1	2	tr.

For fatty acid abbreviations see Table 1. Taken from Harwood and Jones (1989) with permission.

Dinophyceae have appreciable quantities of 18:4 ($\Delta 6, 9, 12, 15$), 20:5 ($\Delta 5, 8, 11, 14, 17$) and 22:6 ($\Delta 4, 7, 10, 13, 16, 19$) while the marine Phaeophyceae and Rhodophyceae contain large amounts of 20:4 ($\Delta 5, 8, 11, 14$) and 20:5 ($\Delta 5, 8, 11, 14, 17$) (see Pohl and Zurheide, 1979a).

As discussed above, overall fatty acid percentages may hide characteristic patterns for individual lipid classes. Thus, a common theme for marine algae, with the exception of the Chlorophyceae, is that C_{20} polyenoic acids are major components. These acids may be particularly enriched in certain lipids. For example, arachidonate accounted for 67% of the total fatty acids of phosphatidylcholine in *F. serratus* (Smith and Harwood, 1984) while in *Chondrus crispus* eicosapentaenoate represented more than 25% of the total acids of the glycosylglycerides (Pettitt et al., 1989). The unicellular alga, *Chattonella antiqua* (Raphidophyceae), was also found to contain rather high amounts of eicosapentaenoate in all lipid classes (Sato et al., 1987).

The presence of the unusual *trans*-Δ-Hexadecenoate at the *sn*-2 position of thylakoid membrane phosphatidylglycerol was mentioned in Chapter 2 and its possible function in Chapter 9. In view of suggestions that it can function in higher plant granal stacking (Chapter 9) it is of considerable interest to see whether it is present in marine algae which have a quite different chloroplast morphology. *trans*-$\Delta 3$-Hexadecenoate is found exclusively in phosphatidylglycerol of a wide variety of algae including brown species (Smith and Harwood, 1983), red algae (Pettitt and Harwood, 1986), Raphidophyceae (Sato et al., 1987) and diatoms (Kawaguchi et al., 1987). The wide variety of chloroplast structures in these organisms argues against a common function for this fatty acid in gross plastid morphology.

B. Acyl Lipids

There are a wide variety of acyl lipids found in eukaryotic algae. In all types, the glycosylglycerides (MGDG, DGDG, SQDG; Fig. 1) are major components. So far as the phosphoglycerides (Fig. 2) are concerned, all the main classes can usually be detected. At present, it is difficult to make generalizations about the distribution of lipids within algal divisions but a few representative examples are shown in Table 3. For more references to particular analyses the reader is referred to Dembitsky (1996), Harwood and Jones (1989) and Pohl and Zurheide (1979a,b).

Typically, green algae such as *Chlorella* or *Chlamydomonas* contain DGTS and PC in inter-changeable amounts. This is, presumably, because these two lipids have equivalent membrane functions. For green macroalgae, the freshwater species usually have mainly PC and little DGTS while marine species have much more DGTS (Dembitsky, 1996). This may reflect the low availability of phosphate in the marine environment. Phosphatidylglycerol is usually the main phosphoglyceride detected in green algae as befits its role as the only important phospholipid in chloroplast thylakoids. Indeed, this can be easily seen with the data for *Chlamydomonas* shown in Table 4. PE, on the other hand, a typical extra-thylakoid lipid, is deficient in isolated chloroplast lamellae (Mendiola-Morgenthaler et al., 1995).

MGDG is enriched in chloroplast thylakoids compared to either whole cells or envelope membranes in *Chlamydomonas* (Table 4). In addition, the ratio of DGDG/SQDG is also enriched. DGDG is found in especially high amounts in envelope membranes when compared to the other glycosyl-glycerides. These facts are reminiscent of the situation in higher plant chloroplasts (Chapter 2 and Harwood, 1980). DGTS appears to be concentrated in the extra-chloroplastic membranes (Table 4) which is as expected if it substitutes for PC, which is the major extra-chloroplastic membrane lipid of higher plants (Harwood, 1980).

In general the green algae have glycosylglyceride compositions rather similar to the leaves of higher plants (Chapter 2). However, the ratio of MGDG to DGDG varies greatly for algal species. Thus,

Chlorella, *Chlamydomonas* and *Dunaliella* have been reported to have ratios of 0.7, 4.2 and 1.6, respectively. Also, the level of SQDG is usually higher than in plants and is comparable to that of DGDG (Kates, 1990).

Red algae such as *Chondrus crispus* or *Poly-siphonia lanosa* tend to have generally higher amounts of phosphoglycerides compared to green algae (Pettitt et al., 1989a). Whether this reflects merely a shift in the balance of intracellular membranes or denotes chloroplasts of unusual lipid composition is not known, in the absence of satisfactory methods for the subcellular fractionation of red macroalgae.

Brown algae are rather diverse and are subdivided into a number of orders such as the Laminariales, Chordariales or Fucales. All have high amounts of glycosylglycerides, which usually represent over 60% of their total acyl lipids (see e.g. Araki et al., 1991). The brown algae tend to have rather large levels of SQDG which represents 30% of the total lipids of *Hizikia fusiformis* (Fucales). For certain brown algal species, which lack PC, a novel phospholipid has been detected. This has been identified as *N*-(1-carboxy-3-amino-propyl-3)-1,2-diacyl-*sn*-glycero-3-phosphorylethanolamine or *N*-CAPE for short (Fig. 6 and Schmid et al., 1994). The original identification was for *Ectocarpus siliculosus* but the lipid has been detected in various species of the orders Ectocarpales, Fucales and Sphacelaries.

A major lipid found in a majority of brown algae is DGTA. This betaine lipid appears to take the place of DGTS which is found in other algal types (see above). Thus, DGTS is typical of green algae and certain chromophytes but seems to be absent from Prasinophyceae and Chlorococcales. DGTA is characteristic of the majority of Phaeophyceae (52 out of 88 species examined) and both betaine lipids

Table 3. Percentage acyl lipid compositions of some algae

	PC	PE	PI	PG	DPG	MGDG	DGDG	SQDG	DGTS	DGTA
Chattonella antiqua	5	3	1	3	tr.	29	18	29	6	n.d.
Dunaliella tertiolecta	4	2	3	8	tr.	22	21	10	8	n.d.
Acetabalaria mediterranea	tr.	1	1	3	tr.	37	20	20	20	n.d.
Chlamydomonas reinhardtii	n.d.	5	2	10	tr.	47	16	7	16	n.d.
Chondrus crispus	30	2	tr.	8	2	17	15	16	n.d.	n.d.
Fucus vesiculosus	4	6	3	2	5	15	11	22	n.d.	24
Eisenia bicyclis	12	12	6	13	tr.	26	4	16	n.d.	n.d.

For lipid abbreviations see text. n.d. = not detected, tr. = trace. Data taken from Harwood and Jones (1989), Evans et al. (1982) for *Dunaliella* and Araki et al. (1991) for *Eisenia*.

Table 4. Lipid composition of whole cells and chloroplast envelopes and thylakoids from the green alga, *Chlamydomonas reinhardtii*

Lipid	% total acyl lipids		
	Cells	Thylakoids	Envelopes
MGDG	42	55	27
DGDG	12	20	31
SQDG	13	13	8
DGTS	18	5	21
PE	5	1	5
PG	4	6	4
PI	4	3	4
DPG	1	0	0

Data taken from Mendiola-Morgenthaler et al. (1985).

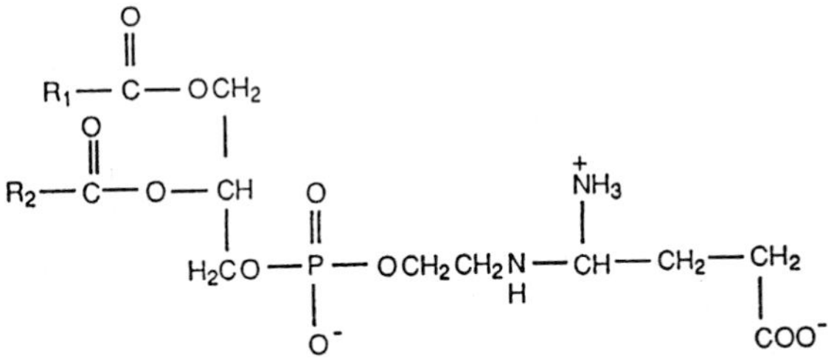

Fig. 6. A new glycerophospholipid, *N*-CAPE, as detected in *Ectocarpus siliculosus* and other brown algae.

have been observed in Chrysophyceae and Cryptophyceae (Eichenberger and Hofmann, 1992). For a detailed discussion of the distribution and levels of betaine lipids in different algae see Dembitsky (1996).

Many freshwater algae contain chlorosulfolipids (Mercer and Davies, 1979). Work with *Ochromonas danica* first revealed significant amounts of these lipids (Haines and Block, 1962). The compounds were soon identified as sulfur esters that were present in amounts greater than for most glycosylglycerides or phosphoglycerides in the cell. Many of these sulfate esters also contain chlorine (Fig. 4) and, hence, are known as chlorosulfolipids. In *O. danica* they represent about 15% of the total lipids and they were detected in all the freshwater species but not in any of the marine species examined by Mercer and Davies (1979). For further information about these unusual lipids see Haines (1973a,b,c).

Certain diatoms have been found to contain other novel sulfur-containing lipids (Kates, 1987). Characteristically, all species examined contain three sulfolipids (Fig. 1) in addition to SQDG. They were

identified using *Nitzchia alba* (Anderson et al., 1978a,b; Kates et al., 1978). Overall, sulfolipids are major components of diatoms. For example, they represent over 30% of the total polar lipids in *N. pelliculosa* and over 74% in *N. alba*. Phosphatidylsulfocholine has also been found in small amounts in marine red algae such as *Chondrus crispus* and *Polysiphonia lanosa* (Pettitt et al., 1989b).

In general, the diatoms (even non-photosynthetic species like *N. alba*) contain high amounts of the four typical chloroplast lipids (MGDG, DGDG, SQDG, PG). PC is often present in comparable amounts to PG and the phosphatidylsulfocholine is also usually a major component (Kates, 1987).

C. Fatty Acid Composition of Individual Lipid Classes

As pointed out earlier, detailed analysis of lipid classes often reveals some important features. For example, in Table 5, the composition and positional distribution of major fatty acids in the principal chloroplast lipids is shown for the green alga,

Table 5. Composition and positional distribution of major fatty acids in chloroplast lipids of *Chlamydomonas reinhardtii* (Giroud et al. 1988)

Lipid	Position	16:0	16:1*	16:3	16:4	18:1 (n-9)	18:1 (n-7)	18:2	18:3 (n-3)
MGDG	1	2	1	tr.	tr.	27	1	15	52
	2	2	14	10	64	tr.	-	tr.	tr.
DGDG	1	4	tr.	-	tr.	55	2	16	19
	2	76	14	2	2	tr.	-	tr.	tr.
SQDG	1	68	-	-	-	6	9	5	10
	2	99	tr.	tr.	-	-	-	tr.	1
PG	1	6	2	-	-	11	11	47	21
	2	27	51	11	-	-	-	-	tr.

* *trans*-Δ3-Hexadecenoate in the case of PG. Dashes mean none detected.

Chlamydomonas reinhardtii. So far as composition is concerned, the four lipids show distinct differences. For example, MGDG contains almost all of the hexadecatetraenoate and is generally enriched in polyunsaturated acids. DGDG on the other hand has higher amounts of palmitate and oleate while SQDG is exceptionally enriched in palmitate. PG, as in other photosynthetic eukaryotes, contains the unique *trans*-Δ3-hexadecenoate. When the positional distribution of these acids is examined, then it will be seen that many components are (virtually) exclusively located at one position only. Thus, the C_{16} unsaturated acids are found at the *sn*-2 position while the C_{18} unsaturated fatty acids are found at the *sn*-1 position. These distributions most probably relate to the mechanism of their synthesis and, by analogy with work using higher plants (see Browse and Somerville, 1991), suggest strongly that the fatty acids of these lipids are formed exclusively in the chloroplast.

The features referred to above are often reciprocated in other algal classes. Some representative examples are shown in Table 6. Thus, MGDG tends to have high contents of polyunsaturated fatty acids while PG and SQDG are generally more saturated. In most algae, SQDG has a high level of palmitate. Octadecatetraenoate is a major component of the chloroplast lipids of many eukaryotic algae. In contrast, *trans*-Δ3-hexadecenoate, when it is found, is exclusively located on PG. For further details of the detailed fatty acid composition of the major acyl lipids of algae, the reader is referred to Pohl and Zurheide (1979a,b) and to Dembitsky (1996).

Some general comments on the distribution of fatty acids in individual classes from different types of algae have been made by Kates (1990). Thus, green algae such as *Chlorella* and the Euglenophyta have MGDG and, to a lesser extent DGDG, with high amounts of 16:4 as well as 16:3. In contrast, although *Dunaliella* spp. have 16:4 in their galactosylglycerides, they lack 16:3. As noted above the SQDG of green algae is much more saturated and contains palmitate as a major component. Therefore, the green algae appear to resemble the so-called 16:3 plants, presumably because of their use of the 'prokaryotic' pathway for fatty acid synthesis (Browse and Somerville, 1991).

In contrast to green algae, red algae contain much less C_{16} or C_{18} polyunsaturated fatty acids, which are replaced with C_{20} acids, especially 20:5. Both DGDG and SQDG are more saturated, with significant levels of palmitate.

Photosynthetic diatoms also have fairly high amounts of 20:5 and low amounts of C_{18} polyunsaturated fatty acids in their galactosylglycerides.

Table 6. Some examples of fatty acid compositions of chloroplast lipids from different algal classes

		16:1	16:1*	18:1	18:2	18:3	18:4	20:4
Chrysophyta								
Ochromonas danica	MGDG	1	tr.	2	30	34	23	1
(Vogel and Eichenberger	DGDG	1	1	3	25	41	15	1
1992)	SQDG	8	-	10	33	22	3	2
	PG	6	-	tr.	44	25	6	8
Xanthophyta								
Nephrochloria salina	MGDG	24	3	5	1	1	10	47
(Dembitsky 1996)	DGDG	20	10	10	1	12	6	26
	SQDG	29	10	7	2	8	19	10
	PG	15	2	1	1	4	21	38
Chromophyta								
Cryptomonas CR-1	MGDG	5	1	1	4	31	45	tr.
(Sato 1991)	DGDG	4	1	1	5	53	32	tr.
	SQDG	47	2	1	4	32	7	tr.
	PG	12	39	2	1	5	-	tr.
Phaeophyta								
Fucus vesiculosus	MGDG	6	tr.	7	4	10	36	5
(Jones and Harwood, 1992)	DGDG	9	2	11	4	12	22	5
	SQDG	43	1	26	3	13	tr.	5
	PG	19	18	7	4	18	22	tr.

*All isomers but mainly *trans*-Δ3-hexadecenoate in PG. Dashes mean none detected.

High amounts of C_{16} unsaturated fatty acids have also been noted (Kates, 1990).

IV. Metabolism of Algal Lipids

A discussion of the metabolism of algal lipids is beyond the scope of this chapter and the reader is referred to summaries that have been made (Harwood and Jones, 1989; Harwood et al., 1989). Some significant progress has been made recently concerning the synthesis of betaine lipids, of which DGTA had not been identified in 1989. From studies with *Ochromonas* (Vogel and Eichenberger, 1990) and brown algae (Eichenberger and Hofmann, 1992), it appears that DGTA is formed from DGTS. These results agree with those of Sato (1988, 1991a,b) who showed that methionine acts as the precursor of the polar group in DGTS for *Chlamydomonas* and also of DGTA in *Cryptomonas*. Other studies on the formation of the betaine lipids are summarized by Dembitsky (1996).

Effects of the environment on algal lipid metabolism have been discussed by Harwood and Jones (1989). These include the actions of light, temperature and heavy metals. Temperature effects have been studied particularly in *Dunaliella salina*, which can be fractionated conveniently to allow the study of individual subcellular membranes (Cho and Thompson, 1987). The experiments with *Dunaliella* have included attempts to elucidate the possible involvement of inositol lipids in the control of cellular functions as a result of environmental stress (e.g. Einspahr et al., 1990).

Acknowledgments

Work in the author's lab on algal lipid biochemistry has been supported by the Natural Environment Research Council and the Biotechnology and Biological Sciences Research Council.

References

Anderson R, Livermore BP, Kates M and Volcani BE (1978a) The lipid composition of the non-photosynthetic diatom *Nitzschia alba*. Biochim Biophys Acta 528: 77–88

Anderson R, Kates M and Volcani BE (1978b) Identification of the sulfolipids in the non-photosynthetic diatom, *Nitzschia alba*. Biochim Biophys Acta 528: 89–106

Araki S, Eichenberger W, Sakurai T and Sato N (1991) Distribution of diacylglycerylhydroxymethyl-beta-alanine (DGTA) and phosphatidylcholine in brown algae. Plant Cell Physiol 32: 623–628

Benson AA (1987) Little known facts of plant lipid metabolism. In: Stumpf PK, Mudd JB and Nes WD (eds) The Metabolism Structure and Function of Plant Lipids, pp 599–601. Plenum, New York

Benson AA, Wiser R, Ferrari RA and Miller JA (1958) Photosynthesis of galactolipids. J Am Chem Soc 80: 4740

Browse J and Somerville C (1991) Glycerolipid synthesis: Biochemistry and regulation. Ann Rev Plant Physiol Plant Mol Biol 42: 467–506

Cho SH and Thompson GA (1987) Metabolism of galactolipids in *Dunaliella salina*. In: Stumpf PK, Mudd JB and Nes WD (eds) The Metabolism Structure and Function of Plant Lipids, pp 623–629. Plenum, New York

Dembitsky VM (1996) Betaine ether-linked glycerolipids: chemistry and biology. Prog Lipid Res 35: 1–51

Eichenberger W and Hofmann M (1992) Metabolism and distribution of betaine lipids in algae. In: Cherif A, Miled-Daoud D, Marzouk B, Smaoui A and Zarrouk M (eds) Metabolism Structure and Utilization of Plant Lipids. pp 18–21, Centre National Pedagogique, Tunis

Eichenberger W, Boschetti A and Michel HP (1986) Lipid and pigment composition of a chlorophyll *b*-deficient mutant of *Chlamydomonas reinhardii*. Physiol Plant 66: 589–594

Einspahr KJ, Rodriguez Rosales MP, Ha KS, Herrin DL and Thompson GA (1990) Transmembrane signalling in the green alga *Dunaliella salina* under salt stress. In: Quinn PJ and Harwood JL (eds) Plant Lipid Biochemistry Structure and Utilization, pp 351–356. Portland Press, London

Evans RW, Kates M and Ginzburg BZ (1982) Lipid composition of halotolerant algae, *Dunaliella parva* and *Dunaliella tertiolecta*. Biochim Biophys Acta 712: 186–195

Giroud C, Gerber A and Eichenberger W (1988) Lipids of *Chlamydomonas reinhardtii*. Analysis of molecular species and intracellular site(s) of biosynthesis. Plant Cell Physiol 29: 587–595

Haines TH (1973a) Halogen- and sulfur-containing lipids of *Ochromonas*. Ann Rev Microbiol 27: 403–411

Haines TH (1973b) The halogenated sulphatides. In: Goodwin TW (ed) The Biochemistry of Lipids, pp 271–286. Butterworths, London

Haines TH (1973c) Sulfolipids and halosulfolipids. In: Erwin JA (ed) Lipids and Biomembranes of Eukaryotic Microorganisms, pp 197–232. Academic Press, New York.

Haines TH and Block RJ (1962) Sulfur metabolism in algae. J. Protozool. 9: 33–38

Harwood JL (1980) Plant acyl lipids: Structure, distribution and analysis. In: Stumpf, PK and Conn EE (eds) The Biochemistry of Plants, Vol 4, pp 1–55. Academic Press, New York.

Harwood JL and Jones AL (1989) Lipid metabolism in algae. Adv Bot Res 16: 1–53

Harwood JL, Pettitt TP and Jones AL (1989) Lipid metabolism. In: Gallon JR and Rogers LJ (eds) Biochemistry of Algae and Cyanobacteria, pp 49–67. Oxford University Press, Oxford

Jones AL and Harwood JL (1992) Lipid composition of the brown algae *Fucus vesiculosus* and *Ascophyllum nodosum*. Phytochemistry 31: 3397–3403

Kates M (1987) Lipids of diatoms and of halophilic *Dunaliella* species. In: Stumpf, PK, Mudd JB and Nes WD (eds) The

Metabolism Structure and Function of Plant Lipids, pp 613–621. Plenum, New York

Kates M (1990) Glycolipids of higher plants, algae, yeasts and fungi. In: Kates M (ed) Handbook of Lipid Research, Vol 6, pp 235–320. Plenum, New York

Kates M, Tremblay P, Anderson R and Volcani BE (1978) Identification of the free and conjugated sterol in the non-photosynthetic diatom *Nitzschia alba* as 24-methylene cholesterol. Lipids 13: 34–41

Kawaguchi Y, Arao T and Yamada Y (1987) Composition and positional distribution of fatty acids in lipids from the diatom *Phaeodactylum tricornutum*. In: Stumpf PK, Mudd JB and Nes WD (eds) The Metabolism Structure and Function of Plant Lipids, pp 653–656. Plenum, New York

Mendida-Morgenthaler L, Eichenberger W and Boschetti A (1985) Isolation of chloroplast envelopes from *Chlamydomonas*. Lipid and polypeptide composition. Plant Sci 41: 97–104

Mercer EI and Davies CL (1979) Distribution of chlorosulfolipids in algae. Phytochemistry 18: 457–462

Pettitt TR and Harwood JL (1986) Lipid characterisation and metabolism in two red marine algae. Biochem Soc Trans 14: 148–149

Pettitt TR, Jones AL and Harwood JL (1989a) Lipids of the marine red algae *Chondrus crispus* and *Polysiphonia lanosa*. Phytochemistry 28: 399–405

Pettitt TR, Jones AL and Harwood JL (1989b) Lipid metabolism in the red marine algae *Chondrus crispus* and *Polysiphonia lanosa* as modified by temperature. Phytochemistry 28: 2053–2058

Pham Quang L and Laur M-H (1976a) Structures, levels and compositions of sulfuric, sulfonic and phospheric esters of glycosyldiglycerides from three Fucaceae. Biochimie 58: 1367–1380

Pham Quang L and Laur M-H (1976b) The sulfated, aliphatic alcohols: new polar lipids isolated from various Fucacea. Biochimie 58: 1381–1396

Pham Quang L and Laur M-H (1976c) Levels, composition and cytologic distribution of sulfur and phosphorus polar lipids of *Pelvetia canaliculata*, *Fucus vesiculosus* and *F. serratus*. Phycologia 15: 367–376

Pohl P and Zurheide F (1979a) Fatty acids and lipids of marine algae and the control of their biosynthesis by environmental factors. In: Hoppe HA, Levring T and Tanaka Y (eds) Marine Algae in Pharmaceutical Science, pp 473–523. Walter de Gruyter, Berlin

Pohl P and Zurheide F (1979b) Control of fatty acid and lipid formation in Baltic marine algae by environmental factors. In: Appelqvist L-A and Liljenberg C (eds) Advances in the Biochemistry and Physiology of Plant Lipids, pp 427–432. Elsevier, Amsterdam

Sato N (1988) Dual role of methionine in the biosynthesis of diacylglyceryltrimethylhomoserine in *Chlamydomonas reinhardtii*. Plant Physiol 86: 931–934

Sato N (1991a) Lipids in *Cryptomonas* CR-1. 1. Occurrence of betaine lipids. Plant Cell Physiol 32: 819–825

Sato N (1991b) Lipids in *Crytomonas* CR-1. 2. Biosynthesis of betaine lipids and galactolipids. Plant Cell Physiol 32: 845–851

Sato N, Nemoto Y and Furuya M (1987) Lipids of *Chattonella antiqua* (Raphidophyceae). In: Stumpf PK, Mudd JB and Nes WD (eds) The Metabolism Structure and Function of Plant Lipids, pp 661–664. Plenum, New York

Schmid CE, Muller DG and Eichenberger W (1994) Isolation and characterisation of a new phospholipid from brown algae-intracellular-localisation and site of biosynthesis. J Plant Physiol 143: 570–574

Smith KL and Harwood JL (1983) The effect of trace metals on lipid metabolism in the brown alga *Fucus serratus*. Biochem Soc Trans 11: 394–395

Smith KL and Harwood JL (1984) Lipids and lipid metabolism in the brown alga, *Fucus serratus*. Phytochemistry 23: 2469–2473

Vogel G and Eichenberger W (1992) Betaine lipids in lower plants—biosynthesis of DGTS and DGTA in *Ochromonas danica* (Chrysophyceae) and the possible role of DGTS in lipid metabolism. Plant Cell Physiol 33: 427–436

Chapter 4

Membrane Lipids in Cyanobacteria

Hajime Wada[1,2] and Norio Murata[2]

[1]*Department of Biology, Faculty of Science, Kyushu University, Ropponmatsu, Fukuoka 810, Japan; and [2]Department of Regulation Biology, National Institute for Basic Biology, Myodaiji, Okazaki 444, Japan*

Summary

Cyanobacteria, a large group of microorganisms in the prokaryotic kingdom, perform oxygenic photosynthesis using two photosystems that resemble those in the chloroplasts of eukaryotic plants. Cyanobacteria contain three glycolipids, monogalactosyldiacylglycerol, digalactosyldiacylglycerol and sulfoquinovosyldiacylglycerol, and a phospholipid, phosphatidylglycerol, as major glycerolipids. The lipid composition of most cyanobacteria is similar to that of the inner envelope membranes and thylakoid membranes of the chloroplasts of higher plants, and it is different from that of the membranes of most bacteria, which contain phospholipids as major glycerolipids. Cyanobacteria can be classified into four groups with respect to the composition of the fatty acids of their glycerolipids. Since some strains, such as *Synechocystis* sp. PCC 6803 and *Synechococcus* sp. PCC 7942, take up exogenous DNA autonomously, they are naturally transformable. Thus, molecular biological techniques, for example, transformation and gene targeting, can easily be applied to these strains and they provide useful systems for studying the molecular aspects of the biosynthesis of lipids and fatty acids, as well as of the functions of membrane lipids in oxygenic photosynthesis. Furthermore, since cyanobacteria respond to changes in a variety of environmental conditions by altering their membrane lipids, they are also useful systems for studying the acclimation of the photosynthetic machinery to environmental factors such as temperature.

P.-A. Siegenthaler and N. Murata (eds): Lipids in Photosynthesis: Structure, Function and Genetics, pp. 65–81.
© *1998 Kluwer Academic Publishers. Printed in The Netherlands.*

I. Introduction

Cyanobacteria are Gram-negative bacteria that form a large group in the prokaryotic kingdom (Stanier and Cohen-Bazire, 1977). They can be divided in terms of morphology into two groups, namely, unicellular and filamentous strains. The membranes of cyanobacteria are similar in structure to those of chloroplasts in higher plants and algae, which have outer and inner envelope membranes and thylakoid membranes. The envelope of each cyanobacterial cell is composed of an outer membrane and an inner (cytoplasmic or plasma) membrane, which are separated by a peptidoglycan layer (Stanier, 1988). In addition to the envelope membranes, the cells contain intracellular membranes, namely, thylakoid membranes. Figure 1 shows a picture of electron micrograph of a unicellular cyanobacterium *Synechocystis* sp. PCC 6803.

Cyanobacteria perform oxygenic photosynthesis exploiting two types of photochemical reaction, as do the chloroplasts of eukaryotic plants. Some strains, such as *Synechocystis* sp. PCC 6803 and *Synechococcus* sp. PCC 7942, are naturally transformable so that these strains can easily be modified by transformation and gene targeting (Golden et al., 1987). The photosynthetic transport of electrons occurs in the thylakoid membranes. It is likely that the respiratory transport of electrons also occurs in the same membranes (Omata and Murata, 1984b, 1985). The thylakoid membranes contain protein-bound chlorophyll *a* (Nichols, 1973), β-carotene (Omata and Murata, 1983, 1984a) and phycobilins (Sidler, 1994) as major photosynthetic pigments. Phycobiliproteins form phycobilisomes, which function as the light-harvesting complexes of Photosystem II and are attached to the outer surface of the thylakoid membranes (Gantt and Conti, 1969). The cytoplasmic membranes, by contrast, are involved in the transport of anions such as bicarbonate (Ogawa, 1991), nitrate (Omata et al., 1989) and sulfate (Green and Grossman, 1988), and contain xanthophylls and

---

Abbreviations: ACP – acyl carrier protein; CoA – coenzyme A; DGDG – digalactosyldiacylglycerol; GlcDG – monoglucosyldiacylglycerol; MGDG – monogalactosyldiacylglycerol; PG – phosphatidylglycerol; SQDG – sulfoquinovosyldiacylglycerol; X:Y – fatty acids are abbreviated as X:Y, where X represents the number of carbon atoms and Y represents the number of double bonds, or as $X:Y(Z_1,Z_2,\cdots)$ where, Z_1, Z_2, $\cdots$ represent the positions of double bonds in the *cis* configuration, counted from the carboxyl terminus

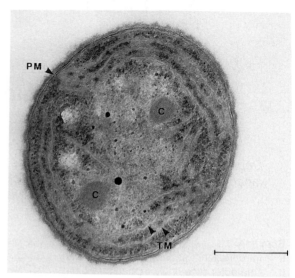

Fig. 1. Electron micrograph of a thin section of the unicellular cyanobacterium *Synechocystis* sp. PCC 6803. PM, plasma membrane; TM, thylakoid membrane; C, carboxysome. This picture was kindly provided by Dr. Laszlo Mustardy. Scale = $0.5\mu m$.

a precursor of chlorophyll *a* but no chlorophyll *a* itself (Murata et al., 1981; Omata and Murata, 1983, 1984a; Resch and Gibson, 1983; Jürgens and Weckesser, 1985; Peschek et al., 1989).

Cyanobacterial membranes are composed of glycerolipids and proteins, as other biological membranes. As described below, the thylakoid and cytoplasmic membranes contain MGDG, DGDG, SQDG and PG as major glycerolipids, as do the inner envelope membranes and thylakoid membranes of chloroplasts in higher plants (see Chapter 2 by Joyard and Douce). The outer membranes contain lipopolysaccharides and hydrocarbons in addition to the glycerolipids that are found in the cytoplasmic membranes. GlcDG is found as a minor glycerolipid in the outer, inner and thylakoid membranes. This lipid is synthesized from diacylglycerol and UDP-glucose and is immediately converted to MGDG by epimerization of the constituent glucose to galactose (Sato and Murata, 1982c). By contrast, the MGDG in eukaryotic chloroplasts is synthesized directly from diacylglycerol and UDP-galactose (Teucher and Heinz, 1991).

In this chapter, we shall describe recent developments in research on the membrane lipids of cyanobacteria, including details of the characteristics and biosynthesis of fatty acids and lipids, as well as changes in membrane lipids in response to changes in environmental factors, such as temperature. Several reviews of earlier researches related to these topics include those by Nichols (1973), Quinn and Williams (1983) and Murata and Nishida (1987).

II. Characteristics of Cyanobacterial Lipids

A. Lipid Classes

The lipid composition of cyanobacteria differs from that of bacteria, in which phospholipids are the major glycerolipids (Murata and Nishida, 1987; Lechevalier and Lechevalier, 1988; see Chapter 5 by Benning). Cyanobacterial cells contain four glycerolipids, namely, MGDG, DGDG, SQDG and PG, and these glycerolipids are the major glycerolipids in the inner envelope and thylakoid membranes of the chloroplasts of higher plants (Block et al., 1983, see Chapter 2 by Joyard and Douce). The relative level of MGDG in cyanobacterial cells is slightly higher than 50% of the total glycerolipids, and relative levels of DGDG, SQDG and PG range from 5% to 25%. Cyanobacteria also contain a minor glycerolipid, GlcDG, which is absent from chloroplasts (Sato and Murata, 1982a). The level of GlcDG is relatively low as compared to the levels of other glycerolipids, and it accounts for less than 1% of the total glycerolipids in cyano-bacterial cells. However, in a recent study, Sato (1994) has showed that cells of *Synechocystis* sp. PCC 6803 accumulate GlcDG when they are grown in the presence of glucose. While the level of GlcDG in *Synechocystis* sp. PCC 6803 is 0.6% of the total glycerolipids when the cells are grown under photoautotrophic conditions, the level of GlcDG reaches as high as 12% when the cells are grown in the presence of glucose under photoheterotrophic conditions.

Phosphatidylcholine, phosphatidylinositol, phos-phatidylethanolamine, diphosphatidylglycerol (cardiolipin) and phosphatidylserine are generally not found in cyanobacteria. However, it has been reported that *Chlorogloeopsis fritschii* contains phosphatidylcholine in addition to MGDG, DGDG, SQDG and PG (Sallal et al., 1987). *Synechococcus* sp. BP-1 contains a unique sulfolipid that contains a thioic O-acid ester (Kaya et al., 1993), designated thionsulfolipid, which has not been found in any other cyanobacteria.

Table 1 shows the composition of the glycerolipids in the thylakoid membranes, cytoplasmic membranes and intact cells of *Synechocystis* sp. PCC 6803. MGDG accounts for a little more than 50% of the total glycerolipids in these membranes and intact cells, and the relative levels of DGDG, SQDG and PG range from about 5% to about 25%. There are no major differences between the two types of membranes from this cyanobacterium. By contrast, Ritter and Yopp (1993) reported that the cytoplasmic membranes of the halophilic cyanobacterium *Aphanothece halophytica* do not contain SQDG, whereas the thylakoid membranes do contain all four of the glycerolipids found in other strains. In this strain, the ratio of MGDG plus DGDG to PG decreases in cytoplasmic membranes, but not in thylakoid membranes, when the salinity of the growth medium is increased (Ritter and Yopp, 1993). It is possible that the alteration in lipid composition might be one of the major factors involved in the acclimation of this strain to increased external salinity.

In some species, such as *Anabaena cylindrica*, the cells differentiate to form heterocysts which fix nitrogen when sources of nitrogen in the growth medium are insufficient (Haselkorn, 1978). Hetero-cysts contain unique glycolipids and hydrocarbons that are not found in vegetative cells (Nichols and Wood, 1968; Murata and Nishida, 1987).

B. Fatty Acids

The fatty acid composition of the glycerolipids of

Table 1. Glycerolipid composition of *Synechocystis* sp. PCC 6803

Sample	Lipid (mol%)				
	GlcDG	MGDG	DGDG	SQDG	PG
Intact cells[a]	t[c]	59	17	16	8
Thylakoid membranes[b]	t	62	14	18	6
Cytoplasmic membranes[b]	t	59	18	13	10

[a]Wada and Murata, 1989; [b]Gombos et al., 1996; [c]t trace amount (less than 0.5%).

cyanobacteria differs from that of most bacteria, which can synthesize only saturated and mono-unsaturated fatty acids, such as 16:0, 16:1(9), 18:0 and 18:1(11), and the unsaturated fatty acids are mainly synthesized via an anaerobic pathway (see Chapter 5 by Benning). The most abundant fatty acids in cyanobacteria are hexadecanoic acid or palmitic acid (16:0), Δ9-cis-hexadecenoic acid or palmitoleic acid [16:1(9)], Δ9,12-cis-hexadeca-dienoic acid [16:2(9,12)], octadecanoic acid or stearic acid (18:0), Δ9-cis-octadecenoic acid or oleic acid [18:1(9)], Δ9,12-cis-octadecadienoic acid or linoleic acid [18:2(9,12)], Δ6,9,12-cis-octadecatrienoic acid or γ-linolenic acid [18:3(6,9,12)], Δ9,12,15-cis-octadecatrienoic acid or α-linolenic acid [18:3 (9,12,15)], and Δ6,9,12,15-cis-octadecatetraenoic acid [18:4(6,9,12,15)]. In addition to these fatty acids, Δ7-cis-hexadecenoic acid [16:1(7)], which is found in the chloroplasts of 16:3 plants as a precursor of 16:3(7,10,13) esterified to the sn-2 position of MGDG, has been detected in Oscillatoria limnetica (Jahnke et al., 1989). Some species, such as Anacystis nidulans and Synechococcus lividus, contain Δ11-cis-octadecenoic acid (cis-vaccenic acid) (Fork et al., 1979; Sato et al., 1979; Wada et al., 1994). The composition of fatty acids in cyanobacteria depends on the species and on the environmental conditions, as described in Section V. Fatty acids with 20 or more carbon atoms, which are present in marine algae, have not been found in cyanobacteria. Branched-chain fatty acids and Δ3-trans-hexadecenoic acid, which is a ubiquitous fatty acid in the PG from chloroplasts of higher plants, have also not been found in cyanobacteria to date.

Kenyon (1972) and Kenyon et al. (1972) invest-igated the fatty acid composition of several strains of cyanobacteria and divided them into four groups on the basis of the fatty acid composition of each. Strains in the first group contain only saturated and monounsaturated fatty acids, such as 16:1(9) and 18:1(9). The strains in the other groups contain polyunsaturated fatty acids that are characteristic of each group, in addition to saturated and mono-unsaturated fatty acids. The second group is characterized by α-18:3, the third group by γ-18:3, and the fourth group by 18:4(6,9,12,15). Table 2 shows the fatty acid composition of two species of cyanobacteria from each group.

Table 3 shows the fatty acid composition of individual lipids in thylakoid and cytoplasmic membranes prepared from Synechocystis sp. PCC

6803, which belongs to the fourth group. The fatty acid composition of MGDG is similar to that of DGDG, whereas the fatty acid composition of SQDG is similar to that of PG in both types of membranes. The fatty acid composition of each lipid in the thylakoid membranes is essentially the same as that of the corresponding lipid in the cytoplasmic membranes. However, the extent of unsaturation of each lipid in the cytoplasmic membranes is slightly lower than that of the corresponding lipid in the thylakoid membranes.

C. Molecular Species

The molecular species of glycerolipids are charac-terized by the combinations of acyl groups at the sn-1 and sn-2 positions of the glycerol moieties of the glycerolipids. The most abundant molecular species in all lipid classes from cyanobacteria are those that contain the sn-1-C_{18}/sn-2-C_{16}-glycerol moiety, which is represented as C_{18}/C_{16} species. In the cyanobacterial species that belong to the third and fourth groups, the sn-2 positions of all glycerolipids are exclusively esterified with 16:0, which is the only C_{16} fatty acid in these species. By contrast, in species that belong to the first and second groups, unsaturated C_{16} fatty acids are also esterified at the sn-2 position. In Anabaena variabilis, which belongs to the second group, 16:0, 16:1(9) and 16:2(9,12) are esterified at the sn-2 position of the glycerol moiety (Sato et al., 1979). In Anacystis nidulans, which belongs to the first group, 16:0 and 16:1(9) are esterified at the sn-2 position (Murata et al., 1992; Sato et al., 1979).

Quoc et al. (1993, 1994) have reported that Spirulina platensis, which belongs to the third group, can synthesize C_{18}/C_{18} species of glycerolipids when oleate or linoleate is added exogenously at a high concentration to the growth medium. This finding suggests that oleate or linoleate can act as substrates for the 1-monoacylglycerol-3-phosphate acyltrans-ferase and, as a consequence, C_{18}/C_{18} species are synthesized in the cyanobacterium under such extreme conditions.

III. Biosynthesis of Lipids

A. Saturated Fatty Acids

Acyl carrier proteins, which are essential for the synthesis of fatty acids in prokaryotic cells and

Table 2. Fatty acid composition of total lipids from various strains of cyanobacteria[a]

Strain	Growth temp. (°C)	Fatty acid (mol%)										
		14:0	14:1 (9)	16:0	16:1 (9)	16:2 (9,12)	18:0	18:1 (9)	18:2 (9,12)	18:3 (9,12,15)	18:3 (6,9,12)	18:4 (6,9,12,15)
Group 1												
Synechococcus sp. PCC 7942	34	1	0	49	36	0	4	10[b]	0	0	0	0
Mastigocladus laminosus	28	1	0	19	50	0	1	27[b]	0	0	0	0
Group 2												
Anabaena variabilis	22	0	0	29	22	3	t[c]	7	15	24	0	0
Synechococcus sp. PCC 7002	22	1	0	35	19	0	t	10	25	10	0	0
Group 3												
Spirulina platensis	32	t	t	53	3	0	1	1	13	0	29	0
Synechocystis sp. PCC 6714	22	t	t	52	3	0	1	2	19	0	23	0
Group 4												
Tolypothrix tenius	30	0	0	55	3	0	1	2	5	6	11	17
Synechocystis sp. PCC 6803	22	t	0	51	3	0	t	2	6	8	21	8

[a]Murata et al., 1992; [b]mixture of oleic acid [18:1(9)] and cis-vaccenic acid [18:1(11)]; [c]t, trace (less than 0.5%).

Table 3. Fatty acid composition of lipids from the thylakoid and cytoplasmic membranes of *Synechocystis* sp. PCC 6803 after growth at 34 °C[a]

Membrane	Lipid	Fatty acid (mol %)						
		14:0	16:0	16:1 (9)	18:0	18:1 (9)	18:2 (9,12)	18:3 (6,9,12)
TM								
	MGDG	t[b]	55	4	t	9	15	17
	DGDG	1	57	4	1	6	13	20
	SQDG	t	74	4	2	13	7	t
	PG	1	63	2	3	12	17	2
CM								
	MGDG	t	61	3	2	11	12	11
	DGDG	1	57	4	1	6	13	20
	SQDG	t	82	1	4	10	3	0
	PG	1	64	5	8	12	9	1

[a]Gombos et al., 1996; TM, thylakoid membranes; CM, cytoplasmic membranes; [b]t, trace (less than 0.5%)

chloroplasts, have been purified from both *Anabaena variabilis* and *Synechocystis* sp. PCC 6803 (Froehlich et al., 1990). Analysis of the amino acid components indicated that ACPs from both strains are highly charged acidic proteins that resemble other previously characterized ACPs from higher plants (Ohlrogge, 1987). Analysis of their sequences revealed that cyanobacterial ACPs are more closely related to ACP from *E. coli* than ACPs from higher plants. However, essentially the same metabolic products were obtained when synthesis of fatty acids was assayed with cyanobacterial ACPs and either bacterial or plant fatty acid synthases.

The enzymes involved in the synthesis of fatty acids in cyanobacteria were characterized in crude extracts of *Anabaena variabilis* (Lem and Stumpf, 1984; Stapleton and Jaworski, 1984a,b). The fatty acid synthase of *Anabaena variabilis* is composed of several separate enzymes that catalyze the individual steps in the synthesis of fatty acids (Stumpf, 1987), and it requires ACP as a cofactor (Stumpf, 1987). Such enzyme complexes have been found in most bacterial cells and in the chloroplasts of higher plants (Cronan and Rock, 1987; Harwood, 1988). The biochemical characteristics of the individual cyanobacterial enzymes are also similar to those found in most bacterial cells and chloroplasts. Elongation of fatty acids from 14:0 to 16:0 and from 16:0 to 18:0 requires ACP and NADPH, but not CoA or NADH (Lem and Stumpf, 1984; Stapleton and Jaworski, 1984b). Malonyl-CoA: ACP transacylase

was also purified from *Anabaena variabilis* and the characteristics of the enzyme were similar to those of the enzyme from spinach (Stapleton and Jaworski, 1984a).

B. Lipid Classes

As described in the previous section, cyanobacteria contain MGDG, DGDG, SQDG and PG as the major glycerolipids and GlcDG as a minor glycerolipid. The biosynthesis of glycerolipids in cyanobacteria has been studied by pulse-chase experiments. Feige et al. (1980) conducted pulse-chase experiments with $H^{14}CO_3^-$ in 30 species of cyanobacteria, and they proposed that GlcDG might be a precursor of MGDG in the biosynthesis of glycolipids in cyanobacteria. By contrast, in higher plants, MGDG is synthesized by the transfer of galactose from UDP-galactose to diacylglycerol. Sato and Murata (1982a) studied the biosynthesis of glycerolipids in *Anabaena variabilis* and demonstrated that GlcDG is converted to MGDG by a stereochemical isomerization at the C-4 atom of the glucose unit and not by replacement of glucose by galactose. They further demonstrated that DGDG is synthesized by the transfer of galactose from an unidentified carrier to MGDG, and that SQDG and PG are synthesized directly and not via GlcDG.

The activity of UDP-glucose: diacylglycerol glucosyltransferase, which catalyzes the transfer of glucose from UDP-glucose to diacylglycerol, is associated with membrane fractions from *Anabaena variabilis* (Sato and Murata, 1982c). In *Anacystis nidulans* this activity is found in both the thylakoid and the cytoplasmic membranes (Omata and Murata, 1986).

The activities of glycerol-3-phosphate acyltransferase and 1-monoacylglycerol-3-phosphate acyltransferase, which are involved in the first steps in the biosynthesis of glycerolipids, are characterized in crude extracts of *Anabaena variabilis* (Lem and Stumpf, 1984). These enzymes use acyl-ACP and not acyl-CoA as their substrate. The characteristics of these enzymes are similar to those of the corresponding acyltransferases in the chloroplasts of higher plants, but may differ from those of enzymes in the cytoplasm, which use acyl-CoA as their substrate. Chen et al. (1988) reported that a membrane fraction from *Anabaena variabilis* had acyl-ACP: lyso-MGDG acyltransferase activity that catalyzes the transfer of the acyl moiety from acyl-ACP to lyso-MGDG. This activity is also identified in thylakoid membranes from *Synechocystis* sp. PCC 6803 (Wada et al., 1993b). However, the function of this enzyme in the biosynthesis of glycerolipids in cyanobacteria is still open to question.

The specificity of acyltransferases with respect to the chain length of fatty acids is studied by feeding cells of *Synechocystis* sp. PCC 6803 with heptanoic acid, a C_7 fatty acid (Higashi and Murata, 1993). This aliphatic acid is elongated to C_{15}, C_{17} and C_{19} fatty acids, all of which are incorporated into glycerolipids in the cells. Analysis of the distribution of fatty acids at the two *sn* positions of the glycerol moiety revealed that the C_{17}, C_{18} and C_{19} fatty acids are located at the *sn*-1 position, whereas the C_{15} and C_{16} fatty acids are located at the *sn*-2 position. This result suggests that glycerol-3-phosphate acyltransferase is specific for 17:0, 18:0 and 19:0, whereas 1-monoacylglycerol-3-phosphate acyltransferase is specific for 15:0 and 16:0.

A tentative scheme, based on the available information, is presented in Fig. 2. However, the biosynthesis of glycerolipids in cyanobacteria remains to be fully clarified.

C. Desaturation of Fatty Acids

Sato and Murata (1982b) studied the desaturation of fatty acids of glycerolipids in *Anabaena variabilis* by pulse-labeling with $H^{14}CO_3^-$ and a subsequent chase. The primary products of lipid synthesis were 18:0/ 16:0-GlcDG, 18:0/16:0-PG and 18:0/16:0-SQDG.

These findings suggest that, in cyanobacteria, saturated fatty acids are the final products of fatty acid synthesis and that the desaturation of fatty acids occurs after the fatty acids have been esterified to the glycerol moiety of glycerolipids. The lipid-linked desaturation of 16:0 to 16:1(9) in MGDG was confirmed by a combination of feeding with $H^{13}CO_3^-$ and mass spectrometric analysis of the 2-acylglycerol moiety of ^{13}C-enriched MGDG (Sato et al., 1986).

The absence of desaturation of fatty acids as acyl-ACP or acyl-CoA is suggested by the in vitro experiments of Lem and Stumpf (1984) and Stapleton and Jaworski (1984b). Under optimal conditions for fatty acid synthesis in cell homogenates, only saturated fatty acids are produced from acetate, and neither exogenously added 18:0-ACP nor exogenously added 18:0-CoA is desaturated. Similar results are obtained by Al'Araji and Walton (1980).

Murata et al. (1992) analyzed the positions of double bonds in fatty acids and the distribution of fatty acids at the *sn*-1 and the *sn*-2 positions of the glycerol moiety of glycerolipids in various strains of cyanobacteria. On the basis of the results they estimated the mode of desaturation of fatty acids in cyanobacteria and reevaluated the classification of cyanobacteria proposed by Kenyon (1972) and Kenyon et al. (1972) that was described in Section II. As shown in Table 4, cyanobacterial strains can be divided into four groups. Strains in the first group can only introduce one double bond at the Δ9 position of fatty acids, either at the *sn*-1 or of the *sn*-2 position. Strains in the second group can introduce double bonds at the Δ9, Δ12 and Δ15(ω3) positions of C_{18} fatty acids at the *sn*-1 position and also at the Δ9 and Δ12 positions of C_{16} fatty acids at the *sn*-2 position. Strains in the third group can also introduce three double bonds, but these double bonds are at the Δ6, Δ9 and Δ12 positions of C_{18} fatty acids at the *sn*-1 position. Strains in the fourth group can introduce double bonds at the Δ6, Δ9, Δ12 and Δ15(ω3) positions of C_{18} fatty acids at the *sn*-1 position. The third and fourth groups cannot introduce double bonds into fatty acids that are esterified at the *sn*-2 position. The desaturation at the *sn*-2 position in the first and second groups and the Δ6 desaturation at the *sn*-1 position in the third and fourth groups are confined to MGDG, and they do not occur in SQDG and PG. It is likely that desaturation does not occur when fatty acids are bound to DGDG. Among the characteristics of the four groups, those of the second group are the most similar to those of the chloroplasts

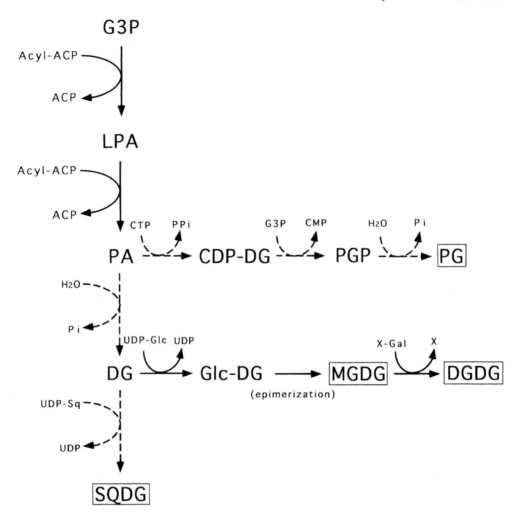

Fig. 2. A scheme for the biosynthesis of glycerolipids in cyanobacteria. Reactions indicated by solid arrows have been demonstrated biochemically, whereas those indicated by broken arrows have been demonstrated by the presence of the corresponding genes in *Synechocystis* sp. PCC 6803 (Kaneko et al., 1995, 1996). G3P, Glycerol-3-phosphate; LPA, 1-acylglycerol-3-phosphate; PA, phosphatidic acid; DG, 1,2-diacylglycerol; CDP-DG, cytidine diphosphate diacylglycerol; PGP, phosphatidylglycerolphosphate; Acyl, fatty acid; Sq, sulfoquinovose; Glc, glucose; Gal, galactose; X, unidentified galactose carrier.

of higher plants in terms of desaturation of fatty acids.

IV. Fatty Acid Desaturases

Fatty acid desaturases introduce double bonds into fatty acids. There are three types of desaturases, namely, acyl-CoA desaturase, acyl-ACP desaturase and acyl-lipid desaturase (Murata and Wada, 1995). Acyl-CoA desaturases introduce double bonds into fatty acids that are bound to CoA. These enzymes are present in animal, yeast and fungal cells. Acyl-ACP desaturases introduce double bonds into fatty acids

that are bound to ACP. These enzymes are present in plastids such as chloroplasts. Acyl-lipid desaturases introduce double bonds into fatty acids that are esterified to glycerolipids. These enzymes are present in plants and cyanobacteria; in plants, one type is located in the cytoplasm and the other is in plastids. In cyanobacterial cells, the enzymes are associated with both cytoplasmic and thylakoid membranes (Mustardy et al., 1996).

All known cyanobacterial desaturases are of the acyl-lipid, membrane-bound type. Since purification of these enzymes by conventional biochemical methods has proved to be difficult, a molecular genetic approach has been used for studies of

Table 4. The sites (Δ) of double bonds in fatty acids from cyanobacteria in groups 1 through 4[a]

Group	Sites of double bonds (Δ)			
	MGDG and DGDG		SQDG and PG	
	sn-1	sn-2	sn-1	sn-2
	(C_{18})	(C_{16})	(C_{18})	(C_{16})
1	9[b]	9	9[b]	N[c]
2	9, 12, 15(ω3)	9, 12	9, 12, 15(ω3)	N
3	6, 9, 12	N	9, 12	N
4	6, 9, 12, 15(ω3)	N	9, 12, 15(ω3)	N

[a]Murata et al., 1992
[b]Mixture of C_{18} and C_{16}
[c]N represents the absence of a double bond at the sn-2 position

cyanobacterial desaturases.

A gene (desA) for the Δ12 desaturase was isolated from Synechocystis sp. PCC 6803 (Wada et al., 1990). The isolated desA gene encodes a polypeptide of 351 amino acid residues that includes four putative membrane-spanning regions. When Synechococcus sp. PCC 7942, which has no Δ12 desaturase activity, is transformed with the desA gene, the resultant transformed cells exhibit Δ12 desaturase activity and produce 16:2(9,12) and 18:2(9,12), neither of which is found in the wild-type cells (Wada et al., 1990). Overexpression of the desA gene in E. coli, which does not contain any desaturases, results in desaturation of fatty acids at the Δ12 position in a homogenate of E. coli cells (Wada et al., 1993a). Homologs of the desA gene have also been isolated from Synechococcus sp. PCC 7002, Synechocystis sp. PCC 6714, Anabaena variabilis and Spirulina platensis (Murata et al., 1996; Sakamoto et al., 1994a). Comparisons of amino acid sequences of Δ12 desaturases revealed four conserved regions and demonstrated that certain residues, in particular histidine residues, are strongly conserved in these desaturases. Furthermore, cDNAs and genes for Δ12 acyl-lipid desaturases have also been isolated from higher plants (Hitz et al., 1994; Okuley et al., 1994; Schmidt et al., 1994), and the analogous histidine residues are also strongly conserved in the latter desaturases.

The desC genes for the Δ9 desaturase were isolated from Anabaena variabilis and Synechocystis sp. PCC 6803 (Sakamoto et al., 1994b). Since the expression of the desC gene from each strain in E. coli results in Δ9 desaturase activity in vivo in E. coli cells, it appears that each desC gene encodes a Δ9 acyl-lipid desaturase. The deduced amino acid sequences of the acyl-lipid Δ9 desaturases of Synechocystis sp. PCC 6803 and Anabaena variabilis are found to be similar to those of the Δ9 acyl-CoA desaturases from rat (Thiede et al., 1986), mouse (Ntambi et al., 1988; Kaestner et al., 1989) and yeast (Stukey et al., 1990). However, no significant homology to the Δ9 acyl-ACP desaturases from higher plants (Shanklin and Somerville, 1991; Thompson et al., 1991) has been found.

The desB gene for the ω3 desaturase was isolated from Synechocystis sp. PCC 6803 (Sakamoto et al., 1994c). The insertional disruption of the desB gene in Synechocystis sp. PCC 6803 causes the loss of ω3 desaturase activity, and the transformation of Synechococcus sp. PCC 7942 with the desB gene from Synechocystis sp. PCC 6803 results in detectable ω3 desaturase activity. The deduced amino acid sequence of the polypeptide encoded by the desB gene resembles those of the ω3 desaturases from higher plants with 45–50% homology. These findings demonstrate that the desB gene of Synechocystis sp. PCC 6803 encodes a ω3 acyl-lipid desaturase. Reddy et al. (1993) cloned the desD gene for the Δ6 acyl-lipid desaturase from the same strain. Thus the genes for all the acyl-lipid desaturases of Synechocystis sp. PCC 6803 have been cloned.

When we compared the deduced amino acid sequences of acyl-lipid desaturases from cyano-bacteria, we found that the enzymes include some well-conserved regions of amino acid sequence, in particular three histidine-cluster motifs, as shown in Table 5 (Murata and Wada, 1995; Murata et al., 1996); one HXXXH motif and two HXXHH motifs in the Δ12 and ω3 desaturases. However, in the Δ6

Table 5. Structures of three histidine clusters that are conserved in and specific to the individual acyl-lipid desaturases of cyanobacteria and higher plants

| Desaturase | Histidine cluster | | |
	1st	2nd	3rd
Δ12	HxxxH	HxxHH	HxxHH
ω3 (Δ15)	HxxxH	HxxHH	HxxHH
Δ9	HxxxxH	HxxHH	HxxHH
Δ6	HxxxH	HxxxHH	xxxHH

desaturase, two HXXHH motifs are replaced by HXXXHH and XXXHH, while in the Δ9 desaturase the HXXXH motif is replaced by HXXXXH. These histidine motifs are conserved in other acyl-lipid desaturases from higher plants and acyl-CoA desaturases from animals and yeast (Shanklin et al., 1994).

Histidine residues are potential ligands of iron atoms, which are assumed to constitute the catalytic centers of the acyl-CoA desaturases (Strittmatter et al., 1974). The importance of the conserved histidine residues in the Δ12 desaturase of *Synechocystis* sp. PCC 6803 was studied by site-directed mutagenesis (Avelange-Macherel et al., 1995). The substitution of five conserved histidine residues, namely, His-90 (at position 90 of the Δ12 desaturase encoded by the *desA* gene of *Synechocystis* sp. PCC 6803), His-109, His-129, His-287 and His-290, to arginine residues by the mutagenesis leads to loss of the activity. This result demonstrates that each of the histidine residues is essential for the enzymatic activity of the Δ12 desaturase.

The hydropathy profiles of all the desaturases reveals two clusters of hydrophobic regions, namely, putative membrane-spanning domains (Murata and Wada, 1995). It has been suggested that each domain spans the membrane twice, thus, each individual desaturase spans the membrane four times (Murata and Wada, 1995). The three histidine clusters conserved in all desaturases (Murata and Wada, 1995) are located at similar positions in all desaturases. The predicted structure of cyanobacterial desaturases is shown in Fig. 3.

Mustardy et al. (1996) have recently examined immunocytochemically the localization of all the desaturases of *Synechocystis* sp. PCC 6803 and have found that all desaturases are located in both cytoplasmic and thylakoid membranes. It seems likely that the conserved histidine clusters are located on the cytoplasmic side of the cytoplasmic and thylakoid membranes. Cyanobacterial desaturases require reduced ferredoxin for desaturation of fatty acids as an electron donor (Wada et al., 1993b). Since ferredoxin is present in the cytoplasm of cyanobacterial cells (Matsubara and Wada, 1987), the region of desaturases which interacts with ferredoxin may also be located at the cytoplasmic side of the cytoplasmic and thylakoid membranes.

V. Changes in Membrane Lipids in Response to Environmental Factors

A. Temperature

Many organisms, including both prokaryotes and eukaryotes, respond to changes in environmental temperature by altering the fatty acid composition of their membrane lipids. When the growth temperature is decreased, the extent of unsaturation increases and/or the chain length of fatty acids decreases (Russell, 1984). Similar changes in fatty acid composition with growth temperature are observed in cyanobacteria. As described below, species in group 1 change both the extent of unsaturation and the chain length of fatty acids with alterations of growth temperatures, whereas species in the other groups change only the extent of unsaturation of fatty acids. Species in group 2 change the extent of unsaturation of C_{16} and C_{18} fatty acids esterified to the *sn*-2 and *sn*-1 positions of membrane lipids, respectively. Species in groups 3 and 4 change the extent of unsaturation of C_{18} fatty acids esterified to the *sn*-1 position of membrane lipids.

An increase in the extent of unsaturation and a shortening of the chain lengths of fatty acids decrease the phase-transition temperature and increase the fluidity of membrane lipids (Chapman, 1975). The changes in fatty acid composition with alterations of growth temperature in cyanobacteria can be regarded as an acclimation to the temperature of the environment (Murata and Nishida, 1987). As described in Chapter 13 by Gombos and Murata the effects of unsaturation of fatty acids in membrane lipids on physiological characteristics, such as growth and chilling tolerance, have been studied in genetically manipulated strains of *Synechocystis* sp. PCC 6803 and *Synechococcus* sp. PCC 7942. It has been found that polyunsaturated fatty acids are important for growth and the ability to tolerate photoinhibition of photosynthesis at low temperature (Wada et al., 1990, 1992, 1994; Gombos et al., 1992, 1994; Tasaka et al., 1996; see Chapter 13 by Gombos and Murata).

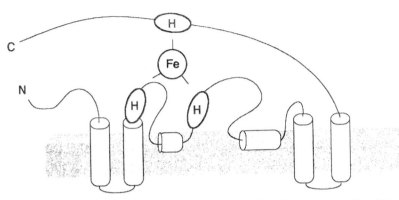

Fig. 3. The predicted structure of cyanobacterial acyl-lipid desaturases. Histidine clusters conserved in all desaturases and membrane-spanning domains are represented by circled H and cylinders, respectively.

1. Species in Group 1

Sato et al. (1979) studied the effects of growth temperature on the unsaturation of fatty acids in *Anacystis nidulans* (*Synechococcus* sp. PCC 6301), which belongs to the first group of cyanobacteria. They found that, with a decrease in growth temperature, the average chain length of mono-unsaturated fatty acids was reduced at the *sn*-1 position of all major lipid classes. Moreover, the level of 16:1 at the *sn*-2 position of MGDG and DGDG increased while that of 16:0 decreased concomitantly.

Murata et al. (1979) studied the low temperature-induced changes in fatty acid composition of *Anacystis nidulans*. In this strain, almost all of the 16:0 at the *sn*-1 position of MGDG, of SQDG and of PG is converted to 16:1 within 10 h after a shift in temperature from 38 °C to 28 °C. The chain length of monounsaturated fatty acids at the *sn*-1 positions of all major lipid classes also decreases. In addition, the slow desaturation of 16:0 to 16:1 occurs at the *sn*-2 positions of MGDG and DGDG. It seems likely that the rapid desaturation of 16:0 to yield 16:1 at the *sn*-1 position of the lipids might play an important role in the acclimation of *Anacystis nidulans* to lower temperatures. Species in group 1 change both the extent of unsaturation and the chain length of fatty acids with changes in growth temperature.

In the thermophilic cyanobacterium, *Synecho-coccus lividus*, a decrease in growth temperature induces a decrease in the levels of 16:0 and 18:0 and an increase in the level of 16:1 in MGDG and DGDG. In SQDG and PG, by contrast, the level of 18:0 decreases while levels of 16:1 and 18:1 increase

(Fork et al., 1979). A decrease in growth temperature of another thermophilic cyanobacterium *Mastigocladus laminosus* reduces levels of 16:0 and 18:0 and increases the level of 18:1 (Hirayama and Kishida, 1991).

2. Species in Group 2

Anabaena variabilis, a member of the second group that contains α-18:3 at the *sn*-1 position and 16:1 (16:2 in some species) at the *sn*-2 position as characteristic unsaturated fatty acids, responds to changes in growth temperature in a manner different from the cyanobacteria in the first group. When cells are grown under isothermal conditions, major changes occur in levels of C_{18} fatty acids but not in levels of C_{16} fatty acids. With decreases in growth temperature, levels of 18:1 and 18:2 decrease and the level of α-18:3 increases at the *sn*-1 position of all major lipid classes, whereas the levels of C_{16} fatty acids remain nearly constant with the exception of a minor decrease in the level of 16:1 and an increase in 16:2 at the *sn*-2 positions of MGDG and DGDG (Sato et al., 1979; Sato and Murata, 1980).

When the growth temperature is shifted from 38°C to 22 °C in *Anabaena variabilis*, a decrease in the level of 16:0 and a concomitant increase in that of 16:1 occur for 10 h at the *sn*-2 position of MGDG (Sato and Murata, 1980), while the de novo synthesis of fatty acids and lipids is totally blocked. Then, as the synthesis of fatty acids and lipids resumes, the ratio of levels of 16:0 to 16:1 slowly returns to the value determined prior to the shift in temperature. This type of transient desaturation, namely, the conversion of 16:0 to 16:1, is not observed in the other lipid classes. The rapid changes in the extent of unsaturation of the C_{16} fatty acids after the changes in temperature does not require the synthesis of fatty acids de novo (Sato and Murata, 1981). Mass-spectrometric analysis of ^{13}C-enriched MGDG indicated that the changes in unsaturation of 16:0 occur via the lipid-linked desaturation of 16:0, and not by replacement of 16:0 with previously synthesized 16:1 (Sato et al., 1986). Molecular oxygen, rather than light, is required for the desaturation (Sato and Murata, 1981). Moreover, the desaturation of 16:0 is suppressed by chlor-amphenicol, an inhibitor of prokaryotic protein synthesis, and by rifampicin, an inhibitor of RNA synthesis. These findings suggest that a fatty acid desaturase, specific to 16:0 at the *sn*-2 position of MGDG, might be synthesized transiently after the downward shift in temperature (Sato and Murata, 1981).

Decreases in levels of 18:1 and 18:2 and an increase in the level of α-18:3 occur in MGDG, SQDG and PG, but these changes occur more slowly than those in the C_{16} fatty acids of MGDG. DGDG responds more slowly to the shift in temperature. The desaturation of 18:1 and of 18:2 after a change in temperature is also suppressed by inhibitors of the synthesis of protein or RNA (Sato and Murata, 1981). The rapid and transient desaturation of 16:0 to 16:1, which takes place at the *sn*-2 position of MGDG, can be regarded as an 'emergency' acclimation that allows cells to compensate for the decrease in membrane fluidity caused by a decrease in temperature (Sato and Murata, 1981).

3. Species in Group 3

Murata et al. (1992) studied the effects of growth temperature on the fatty acid composition of *Synechocystis* sp. PCC 6714 which belongs to the third group. In this cyanobacterium, the extent of unsaturation of the C_{18} fatty acids esterified to the *sn*-1 position of all the lipid classes increased with a decrease in growth temperature.

4. Species in Group 4

Synechocystis sp. PCC 6803, which belongs to the fourth group, responds to a change in growth temperature by altering the extent of unsaturation of the C_{18} fatty acids at the *sn*-1 position of all major lipid classes as observed in *Synechocystis* sp. PCC

6714 in group 3. The C_{16} fatty acids are unaffected by growth temperature, with the exception of those in SQDG (Wada and Murata, 1990). A decrease in growth temperature induces an increase in levels of γ-18:3 and 18:4 in MGDG and DGDG, as well as in α-18:3 in SQDG and PG. In SQDG, the level of 16:0 is higher than that in the other lipids and it decreases with decreases in the growth temperature.

Wada and Murata (1990) studied the low temperature-induced desaturation of fatty acids of membrane lipids in *Synechocystis* sp. PCC 6803. When the growth temperature was shifted from 38 °C to 22 °C, the total amount of fatty acids did not increase for the first 10 h. However, the extent of unsaturation of the C_{18} fatty acids, but not of the C_{16} fatty acids, changed significantly after the shift in temperature. 18:1 and 18:2 in MGDG were desaturated to γ-18:3 and then to 18:4, whereas 18:1 and 18:2 in PG were desaturated to α-18:3, and 18:1 in SQDG was desaturated to 18:2 and α-18:3. The fatty acid composition of DGDG did not change significantly in response to the downward shift in temperature. The low temperature-induced desaturation of C_{18} fatty acids in *Synechocystis* sp. PCC 6803 occurred only in the light and was inhibited by 3-(3,4-dichlorophenyl)-1,1-dimethylurea, an inhibitor of the photosynthetic transport of electrons in thylakoid membranes (Wada and Murata, 1990). These findings suggest that the low temperature-induced desaturation of fatty acids requires the reactions of the photosynthetic electron transport system in the thylakoid membranes.

The $\Delta 9$ and $\Delta 12$ desaturases require the reduced form of ferredoxin as an electron donor (Wada et al., 1993b). Since ferredoxin is reduced by electrons from the photosynthetic electron transport in thylakoid membranes, light is required to reduce ferredoxin.

Chloramphenicol and rifampicin inhibit the low temperature-induced desaturation of fatty acids (Wada and Murata, 1990), as also observed in *Anabaena variabilis* (Sato and Murata, 1981). This finding suggests that *Synechocystis* sp. PCC 6803 synthesizes those desaturases that are specific to C_{18} fatty acids at the *sn*-1 position of lipids after a downward shift in temperature.

As described in Section IV, the genes for the various acyl lipid desaturases of *Synechocystis* sp. PCC 6803 have been cloned. Thus, examination of the effects of low temperature on their expression is possible. Los et al. (1993) and Los and Murata

(1994) investigated the levels of mRNAs for the desaturases of *Synechocystis* sp. PCC 6803 by northern blotting analysis. The levels of mRNAs for the Δ6, Δ12 and ω3 desaturases increase with decreases in ambient temperature. By contrast, the level of the mRNA for the Δ9 desaturase remains at a constant level regardless of the ambient temperature (Los and Murata, 1994). These results suggest that the low temperature-induced desaturation of C_{18} fatty acids at the *sn*-1 positions of lipids is caused by the low temperature-induced upregulation of the expression of genes for the Δ6, Δ12 and ω3 desaturases.

Fujii and Fulco (1977) observed that, in the bacterium *Bacillus megaterium*, the activity of fatty acid desaturation at the Δ5 position transiently increases after a downward shift in growth temperature. They proposed that the desaturation of fatty acids induced by low temperature can be explained by the low temperature-induced synthesis of the desaturase. By contrast, Skriver and Thompson (1979) inferred that the accelerated desaturation of fatty acid in *Tetrahymena pyriformis*, upon a downward shift in temperature, results in the activation of desaturases by the changes in membrane fluidity. Our findings that chloramphenicol and rifampicin inhibit the low temperature-induced desaturation of fatty acids in *Anabaena variabilis* and *Synechocystis* sp. PCC 6803 (Sato and Murata, 1981; Wada and Murata, 1990) and that the genes for Δ6, Δ12 and ω3 desaturases of *Synechocystis* sp. PCC 6803 are upregulated at low temperature (Los et al., 1993) strongly suggest that the cyanobacterial cells desaturate the fatty acids after a downward shift in temperature by the induced synthesis of desaturases, as in *B. megaterium*.

The primary signal in the biological perception of temperature was studied by catalytic hydrogenation of membrane lipids of *Synechocystis* sp. PCC 6803 (Vigh et al., 1993). A decrease in the unsaturation of cytoplasmic membranes of *Synechocystis* sp. PCC 6803 by the hydrogenation induces expression of the *desA* gene for the Δ12 desaturase, as observed after a downward shift in temperature. This finding demonstrates that a decrease in the unsaturation of glycerolipids in the cytoplasmic membrane by the hydrogenation can mimic the effect of a decrease in temperature on the regulation of the expression of the *desA* gene, and suggests that a decrease in membrane fluidity is the primary signal in the perception of temperature.

B. Other Environmental Factors

Al-Hasan et al. (1989) studied the effects of light on the fatty acid composition of five species of marine cyanobacteria. In *Phormidium jenkelianum* and *Synechocystis* sp., the fatty acid composition of cells from light- and dark-incubated cultures were similar, whereas in *Anabaena constricta*, *Phormidium corium* and *Spirulina subsalsa*, incubation in darkness caused an increase in the level of 18:1(9) in all lipid classes.

The effects of the concentration of nitrate in the medium on the fatty acid composition of four species of cyanobacteria were studied by Piorreck and Pohl (1984). The concentration of nitrate affected the fatty acid composition in *Anacystis nidulans* and *Spirulina platensis*, but not in *Microcystis aeruginosa* and *Oscillatoria rubescens*.

Gombos et al. (1987) examined the effects of nitrate starvation on the growth, lipid and fatty acid composition of cells of *Anacystis nidulans*. A remarkable decrease in the protein content and a disruption of endomembrane system in the cells were observed when the cells were transferred from a normal growth medium to a nitrate-free medium. However, the cells retain their ability even for days. An increase in the level of 18:0 with an accompanying decrease in the level of 16:1 in all lipid classes was also observed after the transfer of the cells from a normal growth medium to a nitrate-free medium. By contrast, the lipid composition did not change and no significant changes in the fluidity or in the phase-transition temperature of cytoplasmic membranes were observed during the nitrate starvation. Since an increase in relative levels of saturated fatty acids and the elongation of fatty acids reduce fluidity of membranes (Chapman, 1975, see Chapter 6 by Williams), it seems likely that the increase in the level of 18:0 accompanied by the decrease in the level of 16:1 during the nitrate deprivation is important for the maintenance of the membrane fluidity at an optimum level under such severe conditions.

The effects of salinity on the lipid and fatty acid composition of membranes were studied in *Synechococcus* sp. PCC 6311 (Huflejt et al., 1990; Khomutov et al., 1990; Molitor et al., 1990). Growth of *Synechococcus* sp. PCC 6311 in the presence of high concentration of NaCl was accompanied by significant changes in fatty acid and lipid composition. Upon transfer of cells from a growth medium that contained 15 mM NaCl to one that contained 500 mM NaCl, a rapid decrease in the level of

16:1(9) occurred with a concomitant increase in the levels of two C_{18} fatty acids, namely, 18:1(9) and 18:1(11) (Huflejt et al., 1990). Salinity also affected the lipid composition. The level of MGDG decreased, whereas those of DGDG and PG increased after transfer of cells from low to high salinity. These changes in the lipid and fatty acid composition of membranes are observed in both the thylakoid and the cytoplasmic membranes.

Ritter and Yopp (1993) investigated the effects of salinity on the lipid composition of the halophilic cyanobacterium *Aphanothece halophytica*. Growth of this strain under high salinity conditions was also accompanied by significant changes in membrane lipid composition. The levels of MGDG and DGDG in the cytoplasmic membranes decreased as the external salinity increased, whereas the level of PG increased. By contrast, the composition of glycerolipids in the thylakoid membranes was unaffected by changes in the salinity of the growth medium. These findings suggest that an increase in the level of the anionic lipid PG and a decrease in that of MGDG might be important for acclimation of the cells to high salinity.

VI. Conclusion and Future Perspectives

The basic metabolic pathways for the biosynthesis of membrane lipids in cyanobacteria are now fairly well understood, an`d most of the genes for enzymes which are involved in the biosynthesis of membrane lipids have been identified in the genomic sequence of *Synechocystis* sp. PCC 6803 (Kaneko et al., 1996, 1997). In higher plants, on the other hand, almost all of the soluble enzymes involved in the synthesis of fatty acids and lipids have been purified and the corresponding genes have been cloned (Ohlrogge and Browse, 1995). Several of the membrane-bound enzymes which are difficult to purify by a biochemical approach have now been characterized by alternative strategies, such as map-based cloning, T-DNA or transposon tagging, and the complementation of mutants (Ohlrogge and Browse, 1995). Since the genes for the enzymes which are involved in the metabolism of lipids in cyanobacteria have been identified, it is possible to manipulate lipid metabolism by molecular biological techniques and to address questions such as: (1) how are lipid biosynthesis genes regulated? (2) how is the lipid metabolism regulated by environmental factors? and (3) what is the function of glycolipids in photosynthesis? The

molecular genetic modification of the unsaturation of membrane lipids (Wada et al., 1992, 1994; Murata and Wada, 1995; see Chapter 13 by Gombos and Murata) is a successful example of such studies.

Acknowledgment

The authors thank Dr. Z. Gombos for helpful discussions and for his careful reading of the manuscript.

References

Al'Araji ZT and Walton TJ (1980) Fatty acid biosynthesis in cell free preparations of *Anabaena cylindrica*. In: Mazliak P, Benveniste P, Costes C and Douce R (eds) Biogenesis and Function of Plant Lipids, pp 259–262. Elsevier/North-Holland Biochemical Press, Amsterdam

Al-Hasan RH, Ali AM and Radwan SS (1989) Effects of light and dark incubation on the lipid and fatty acid composition of marine cyanobacteria. J Gen Microbiol 135: 865–872

Avelange-Macherel M-H, Macherel D, Wada H and Murata N (1995) Site-directed mutagenesis of histidine residues in the Δ12 acyl-lipid desaturase of *Synechocystis*. FEBS Lett 361: 111–114

Benning C (1998) Membrane lipids in anoxygenic photosynthetic bacteria. In: Siegenthaler P-A and Murata N (eds) Lipids in Photosynthesis: Structure, Function and Genetics, pp 83–101, Kluwer Academic Publishers, Dordrecht

Block MA, Dorne A-J, Joyard J and Douce R (1983) Preparation and characterization of membrane fractions enriched in outer and inner envelope membranes from spinach chloroplasts. II. Biochemical characterization. J Biol Chem 258: 13281–13286

Chapman D (1975) Phase transition and fluidity characteristics of lipids and cell membranes. Quart Rev Biophys 8: 185–235

Chen HH, Wickrema A and Jaworski JG (1988) Acyl-acyl carrier protein: Lysomonogalactosyldiacylglycerol acyltransferase from the cyanobacterium *Anabaena variabilis*. Biochim Biophys Acta 963: 493–500

Cronan Jr JE and Rock CA (1987) Biosynthesis of membrane lipids. In: Ingraham JL, Low KB, Magasanik B and Schaechter M (eds) *Escherichia coli* and *Salmonella typhimurium*, Vol 1, pp 474–479. American Society for Microbiology, Washington, DC

Feige GB, Heinz E, Wrage K, Cochems N and Ponzelar E (1980) Discovery of a new glyceroglycolipid in blue-green algae and its role in galactolipid synthesis. In: Mazliak P, Benveniste P, Costes C and Douce R (eds) Biogenesis and Function of Plant Lipids, pp 135–140. Elsevier/North-Holland Biochemical Press, Amsterdam

Fork DC, Murata N and Sato N (1979) Effect of growth temperature on the lipid and fatty acid composition, and the dependence on temperature of light-induced redox reactions of cytochrome *f* and of energy redistribution in the thermophilic blue-green alga *Synechococcus lividus*. Plant Physiol 63: 524–530

Froehlich JE, Poorman R, Reardon E, Barnum SR and Jaworski JG

(1990) Purification and characterization of acyl carrier protein from two cyanobacteria species. Eur J Biochem 193: 817–825

Fujii DK and Fulco AJ (1977) Biosynthesis of unsaturated fatty acids by Bacilli. Hyperinduction and modulation of desaturase synthesis. J Biol Chem 252: 3660–3670

Gantt E and Conti SF (1969) Ultrastructure of blue-green algae. J Bacteriol 97: 1486–1493

Golden SS, Brusslan J and Haselkorn R (1987) Genetic engineering of the cyanobacterial chromosome. Meth Enzymol 153: 215–231

Gombos Z and Murata N (1998) Genetically engineered modulation of the unsaturation of glycerolipids and its consequences in tolerance of photosynthesis to temperature stresses. In: Siegenthaler P-A and Murata N (eds) Lipids in Photosynthesis: Structure, Function and Genetics, pp 249–262. Kluwer Academic Publishers, Dordrecht

Gombos Z, Kis M, Pali T and Vigh L (1987) Nitrate starvation induces homeoviscous regulation of lipids in the cell envelope of the blue-green alga, Anacystis nidulans. Eur J Biochem 165: 461–465

Gombos Z, Wada H and Murata N (1992) Unsaturation of fatty acids in membrane lipids enhances tolerance of the cyanobacterium Synechocystis PCC6803 to low temperature photoinhibition. Proc Natl Acad Sci USA 89: 9959–9963

Gombos Z, Wada H and Murata N (1994) The recovery of photosynthesis from low-temperature photoinhibition is accelerated by the unsaturation of membrane lipids: A mechanism of chilling tolerance. Proc Natl Acad Sci USA 91: 8787–8791

Gombos Z, Wada H, Varkonyi Z, Los D and Murata N (1996) Characterization of the Fad12 mutant of Synechocystis that is defective in Δ12 acyl-lipid desaturase activity. Biochim Biophys Acta 1299: 117–123

Green LS and Grossman AR (1988) Changes in sulfate transport characteristics and protein composition of Anacystis nidulans R2 during sulfur deprivation. J Bacteriol 170: 583–587

Harwood JL (1988) Fatty acid metabolism. Annu Rev Plant Physiol Plant Mol Biol 39: 101–138

Haselkorn R (1978) Heterocysts. Annu Rev Plant Physiol 29: 319–344

Higashi S and Murata N (1993) An in vivo study of substrate specificities of acyl-lipid desaturases and acyltransferases in lipid synthesis in Synechocystis PCC6803. Plant Physiol 102: 1275–1278

Hirayama O and Kishida T (1991) Temperature-induced changes in the lipid molecular species of a thermophilic cyanobacterium, Mastigocladus laminosus. Agric Biol Chem 55: 781–785

Hitz WD, Carlson TJ, Booth Jr JR, Kinney AJ, Stecca KL and Yadav NS (1994) Cloning of a higher-plant plastid ω-3 fatty acid desaturase cDNA and its expression in a cyanobacterium. Plant Physiol 105: 635–641

Huflejt ME, Trémolières A, Pineau B, Lang JK, Hatheway J and Packer L (1990) Changes in membrane lipid composition during saline growth of the freshwater cyanobacterium Synechococcus 6311. Plant Physiol 94: 1512–1521

Jahnke LL, Lee B, Sweeney MJ and Klein HP (1989) Anaerobic biosynthesis of unsaturated fatty acids in the cyanobacterium, Oscillatoria limnetica. Arch Microbiol 152: 215–217

Joyard J and Douce R (1998) Lipids in thylakoid and envelope membrane of higher plant chloroplasts. In: Siegenthaler P-A and Murata N (eds) Lipids in Photosynthesis: Structure, Function and Genetics, pp 21–52. Kluwer Academic

Publishers, Dordrecht

Jürgens UJ and Weckesser J (1985) Carotenoid-containing outer membrane of Synechocystis sp. strain PCC6714. J Bacteriol 164: 384–389

Kaestner KH, Ntambi JM, Kelly Jr TJ and Lane MD (1989) Differentiation-induced gene expression in 3T3-L1 preadipocytes. A second differentially expressed gene encoding stearoyl-CoA desaturase. J Biol Chem 264: 14755–14761

Kaneko T, Tanaka A, Sato S, Kotani H, Sazuka T, Miyajima N, Sugiura M and Tabata S (1995) Sequence analysis of the genome of the unicellular cyanobacterium Synechocystis sp. strain PCC6803. I. Sequence features in the 1 Mb region from map positions 64% to 92% of the genome. DNA Res 2: 153–166

Kaneko T, Sato S, Kotani H, Tanaka A, Asamizu E, Nakamura Y, Miyajima N, Hirosawa M, Sugiura M, Sasamoto S, Kimura T, Hosouchi T, Matsuno A, Mukai A, Nakazaki N, Naruo K, Okumura S, Shimpo S, Takeuchi C, Wada T, Watanabe A, Yamada M, Yasuda M and Tabata S (1996) Sequence analysis of the genome of the unicellular cyanobacterium Synechocystis sp. strain PCC6803. II. Sequence determination of the entire genome and assignment of potential protein-coding regions. DNA Res 3: 109–136

Kaya K, Sano T, Watanabe MM, Shiraishi F and Ito H (1993) Thioic O-acid ester in sulfolipid isolated from freshwater picoplankton cyanobacterium, Synechococcus sp. Biochim Biophys Acta 1169: 39–45

Kenyon CN (1972) Fatty acid composition of unicellular strains of blue-green algae. J Bacteriol 109: 827–834

Kenyon CN, Rippka R and Stanier RY (1972) Fatty acid composition and physiological properties of some filamentous blue-green algae. Arch Mikrobiol 83: 216–236

Khomutov G, Fry IV, Huflejt ME and Packer L (1990) Membrane lipid composition, fluidity, and surface charge changes in response to growth of the freshwater cyanobacterium Synechococcus 6311 under high salinity. Arch Biochem Biophys 277: 263–267

Lechevalier H and Lechevalier MP (1988) Chemotaxonomic use of lipids – an overview. In: Ratledge C and Wilkinson SG (eds) Microbial Lipids, Vol 1, pp 869–902. Academic Press, San Diego

Lem NW and Stumpf PK (1984) In vitro fatty acid synthesis and complex lipid metabolism in the cyanobacterium Anabaena variabilis. Plant Physiol 71: 134–138

Los DA and Murata N (1994) Low-temperature induced accumulation of the desaturase gene transcript in Synechocystis PCC 6803 results from both acceleration of transcription and increase in mRNA stability. Russ J Plant Physiol 41: 146–151

Los DA, Horvath I, Vigh L and Murata N (1993) The temperature-dependent expression of the desaturase gene desA in Synechocystis PCC6803. FEBS Lett 318: 57–60

Matsubara H and Wada K (1987) Soluble cytochromes and ferredoxin. Meth Enzymol 167: 387–410

Molitor V, Trnka M, Erber W, Steffan I, Riviere M-E, Arrio B, Springer-Lederer H and Peschek GA (1990) Impact of salt adaptation on esterified fatty acids and cytochrome oxidase in plasma and thylakoid membranes from the cyanobacterium Anacystis nidulans. Arch Microbiol 154: 112–119

Murata N and Nishida I (1987) Lipids of blue-green algae (cyanobacteria). In: Stumpf PK (ed) The Biochemistry of Plants, Vol 9, pp 315–347. Academic Press, Orlando

Murata N and Wada H (1995) Acyl-lipid desaturases and their

importance in the tolerance and acclimatization to cold of cyanobacteria. Biochem J 308: 1–8

Murata N, Ono T and Sato N (1979) Lipid phase of membrane and chilling injury in the blue-green alga, *Anacystis nidulans*. In: Lyons JM, Graham D and Raison JK (eds) Low Temperature Stress in Crop Plants: The Role of the Membrane, pp 337–345. Academic Press, New York

Murata N, Sato N, Omata T and Kuwabara T (1981) Separation and characterization of thylakoid and cell envelope of the blue-green alga (cyanobacterium) *Anacystis nidulans*. Plant Cell Physiol 22: 855–866

Murata N, Wada H and Gombos Z (1992) Modes of fatty-acid desaturation in cyanobacteria. Plant Cell Physiol 33: 933–941

Murata N, Deshnium P and Tasaka Y (1996) Biosynthesis of γ-linolenic acid in the cyanobacterium *Spirulina platensis*. In: Huang Y.-S. and Mills DE (eds) γ-Linolenic acid, pp 22–32. AOCS Press, Champaign, Illinois

Mustardy L, Los DA, Gombos Z and Murata N (1996) Immunocytological localization of acyl-lipid desaturases in cyanobacterial cells: Evidence that both thylakoid membranes and cytoplasmic membranes are the sites of lipid desaturations. Proc Natl Acad Sci USA 93: 10524–10527

Nichols BW (1973) Lipid composition and metabolism. In: Carr NG and Whitton BA (eds) The Biology of Blue-Green Algae, pp 144–161. University of California Press, Berkeley

Nichols BW and Wood BJB (1968) New glycolipid specific to nitrogen-fixing blue-green algae. Nature 217: 767–768

Ntambi JM, Buhrow SA, Kaestner KH, Christy RJ, Sivley E, Kelly Jr TJ and Lane MD (1988) Differentiation-induced gene expression in 3T3-L1 preadipocytes. Characterization of differentially expressed gene encoding stearoyl-CoA desaturase. J Biol Chem 263: 17291–17300

Ogawa T (1991) A gene homologous to the subunit-2 gene of NADH dehydrogenase is essential to inorganic carbon transport of *Synechocystis* PCC6803. Proc Natl Acad Sci USA 88: 4275–4279

Ohlrogge J (1987) Biochemistry of plant acyl carrier proteins. In: Stumpf PK (ed) The Biochemistry of Plants, Vol 9, pp 137–157. Academic Press, New York

Ohlrogge J and Browse J (1995) Lipid biosynthesis. Plant Cell 7: 957–970

Okuley J, Lightner J, Feldmann K, Yadav N, Lark E and Browse J (1994) *Arabidopsis FAD2* gene encodes the enzyme that is essential for polyunsaturated lipid synthesis. Plant Cell 6: 147–158

Omata T and Murata N (1983) Isolation and characterization of the cytoplasmic membranes from the blue-green alga (cyanobacterium) *Anacystis nidulans*. Plant Cell Physiol 24: 1101–1112

Omata T and Murata N (1984a) Isolation and characterization of three types of membrane from the cyanobacterium (blue-green alga) *Synechocystis* PCC 6714. Arch Microbiol 139: 113–116

Omata T and Murata N (1984b) Cytochromes and prenylquinones in preparations of cytoplasmic and thylakoid membranes from the cyanobacterium (blue-green alga) *Anacystis nidulans*. Biochim Biophys Acta 766: 395–402

Omata T and Murata N (1985) Electron-transport reactions in cytoplasmic and thylakoid membranes prepared from the cyanobacteria (blue-green algae) *Anacystis nidulans* and *Synechocystis* PCC 6714. Biochim Biophys Acta 810: 354–361

Omata T and Murata N (1986) Glucolipid synthesis activities in cytoplasmic and thylakoid membranes from the cyanobacterium *Anacystis nidulans*. Plant Cell Physiol 27: 485–490

Omata T, Ohmori M, Arai N and Ogawa T (1989) Genetically engineered mutant of the cyanobacterium *Synechococcus* PCC 7942 defective in nitrate transport. Proc Natl Acad Sci USA 86: 6612–6616

Peschek GA, Hinterstoisser B, Wastyn M, Kuntner O, Pineau B, Missbichler A and Lang J (1989) Chlorophyll precursors in the plasma membrane of a cyanobacterium, *Anacystis nidulans*. J Biol Chem 264: 11827–11832

Piorreck M and Pohl P (1984) Formation of biomass, total protein, chlorophylls, lipids and fatty acids in green and blue-green algae during one growth phase. Phytochemistry 23: 217–223

Quoc KP, Dubacq J-P, Justin A-M, Demandre C and Mazliak P (1993) Biosynthesis of eukaryotic lipid molecular species by the cyanobacterium *Spirulina platensis*. Biochim Biophys Acta 1168: 94–99

Quoc KP, Dubacq J-P, Demandre C and Mazliak P (1994) Comparative effects of exogenous fatty acid supplementations on the lipids from the cyanobacterium *Spirulina platensis*. Plant Physiol Biochem 32: 501–509

Quinn PJ and Williams WP (1983) The structural role of lipids in photosynthetic membranes. Biochim Biophys Acta 737: 223–266

Reddy AS, Nuccio ML, Gross LM and Thomas TL (1993) Isolation of a Δ6-desaturase gene from the cyanobacterium *Synechocystis* sp. strain PCC6803 by gain-of-function expression in *Anabaena* sp. strain PCC7120. Plant Mol Biol 27: 293–300

Resch CM and Gibson J (1983) Isolation of the carotenoid-containing cell wall of three unicellular cyanobacteria. J. Bacteriol 155: 345–350

Ritter D and Yopp JH (1993) Plasma membrane lipid composition of the halophilic cyanobacterium *Aphanothece halophytica*. Arch Microbiol 159: 435–439

Russell NJ (1984) Mechanisms of thermal adaptation in bacteria: blueprints for survival. Trends Biochem Sci 9: 108–112

Sakamoto T, Wada H, Nishida I, Ohmori M and Murata N (1994a) Identification of conserved domains in the Δ12 desaturases of cyanobacteria. Plant Mol Biol 24: 643–650

Sakamoto T, Wada H, Nishida I, Ohmori M, Murata N (1994b) Δ9 acyl-lipid desaturases of cyanobacteria. J Biol Chem 269: 25576–25580

Sakamoto T, Los DA, Higashi S, Wada H, Nishida I, Ohmori M and Murata N (1994c) Cloning of ω3 desaturase from cyanobacteria and its use in altering the degree of membrane-lipid unsaturation. Plant Mol Biol 26: 249–263

Sallal A-K, Ghannoum MA, Al-Hasan RH, Nimer NA and Radwan SS (1987) Lanosterol and diacylglycerophosphocholines in lipids from whole cells and thylakoids of the cyanobacterium *Chlorogloeopsis fritschii*. Arch Microbiol 148: 1–7

Sato N (1994) Effect of exogenous glucose on the accumulation of monoglucosyl diacylglycerol in the cyanobacterium *Synechocystis* PCC 6803. Plant Physiol Biochem 32: 121–126

Sato N and Murata N (1980) Temperature shift-induced responses in lipids in the blue-green alga, *Anabaena variabilis*. The central role of diacylmonogalactosylglycerol in thermoadaptation. Biochim Biophys Acta 619: 353–366

Sato N and Murata N (1981) Studies on the temperature shift-induced desaturation of fatty acids in monogalactosyl diacylglycerol in the blue-green alga (cyanobacterium), *Anabaena variabilis*. Plant Cell Physiol 22: 1043–1050

Sato N and Murata N (1982a) Lipid biosynthesis in the blue-green alga, *Anabaena variabilis*. I. Lipid classes. Biochim Biophys Acta 710: 271–278

Sato N and Murata N (1982b) Lipid biosynthesis in the blue-green alga, *Anabaena variabilis*. II. Fatty acids and lipid molecular species. Biochim Biophys Acta 710: 279–289

Sato N and Murata N (1982c) Lipid biosynthesis in the blue-green alga (cyanobacterium), *Anabaena variabilis*. III. UDP-glucose: diacylglycerol glucosyltransferase activity in vitro. Plant Cell Physiol 23: 1115–1120

Sato N, Murata N, Miura Y and Ueta N (1979) Effect of growth temperature on lipid and fatty acid compositions in the blue-green algae, *Anabaena variabilis* and *Anacystis nidulans*. Biochim Biophys Acta 572: 19–28

Sato N, Seyama Y and Murata N (1986) Lipid-linked desaturation of palmitic acid in monogalactosyl diacylglycerol in the blue-green alga (cyanobacterium) *Anabaena variabilis* studied in vivo. Plant Cell Physiol 27: 819–835

Schmidt H, Dresselhaus T, Buck F and Heinz E (1994) Purification and PCR-based cDNA cloning of a plastidial n-6 desaturase. Plant Mol Biol 26: 631–642

Shanklin J and Somerville C (1991) Stearoyl-acyl-carrier-protein desaturase from higher plants is structurally unrelated to the animal and fungal homologs. Proc Natl Acad Sci USA 88: 2510–2514

Shanklin J, Whittle E, and Fox BG (1994) Eight histidine residues are catalytically essential in a membrane-associated iron enzyme, stearoyl-CoA desaturase, and are conserved in alkane hydroxylase and xylene monooxygenase. Biochemistry 33: 12787–12794

Sidler WA (1994) Phycobilisome and phycobiliprotein structures. In: Bryant DA (ed) The Molecular Biology of Cyanobacteria, pp 139–216. Kluwer Academic Publishers, Dordrecht

Skriver L and Thompson GA Jr (1979) Temperature-induced changes in fatty acid unsaturation of *Tetrahymena* membranes do not require induced fatty acid desaturase synthesis. Biochim Biophys Acta 572: 376–381

Stanier (Cohen-Bazire) G (1988) Fine structure of cyanobacteria. Meth Enzymol 167: 157–172

Stanier RY and Cohen-Bazire G (1977) Phototrophic prokaryotes: the cyanobacteria. Annu Rev Microbiol 31: 225–274

Stapleton SR and Jaworski JG (1984a) Characterization and purification of malonyl-coenzyme A: [acyl-carrier-protein] transacylase from spinach and *Anabaena variabilis*. Biochim Biophys Acta 794: 240–248

Stapleton SR and Jaworski JG (1984b) Characterization of fatty acid biosynthesis in the cyanobacterium *Anabaena variabilis*. Biochim Biophys Acta 794: 249–255

Strittmatter P, Spatz L, Corcoran D, Rogers MJ, Setlow B and Redline R (1974) Purification and properties of rat liver microsomal stearyl coenzyme A desaturase. Proc Natl Acad Sci USA 71: 4565–4569

Stukey JE, McDonough VM and Martin CE (1990) The *OLE1* gene of *Saccharomyces cerevisiae* encodes the Δ9 fatty acid desaturase and can be functionally replaced by the rat stearoyl-CoA desaturase gene. J Biol Chem 265: 20144–20149

Stumpf PK (1987) The biosynthesis of saturated fatty acids. In: Stumpf PK (ed) The Biochemistry of Plants, Vol 9, pp 121–136. Academic Press, Orlando

Tasaka Y, Gombos Z, Nishiyama Y, Mohanty P, Ohba T, Ohki K and Murata N (1996) Targeted mutagenesis of acyl-lipid desaturases in *Synechocystis*: Evidence for the important roles of polyunsaturated membrane lipids in growth, respiration and photosynthesis. EMBO J 15: 6416–6425

Teucher T and Heinz E (1991) Purification of UDP-galactose: diacylglycerol galactosyltransferase from chloroplast envelopes of spinach (*Spinacia oleracea* L.). Planta 184: 319–326

Thiede MA, Ozols J and Strittmatter P (1986) Construction and sequence of cDNA for rat liver stearyl coenzyme A desaturase. J Biol Chem 261: 13230–13235

Thompson GA, Scherer DE, Aken SF-V, Kenny JW, Young HL, Shintani DK, Kridl JC and Knauf VC (1991) Primary structure of the precursor and mature forms of stearoyl-acyl carrier protein desaturase from safflower embryos and requirement of ferredoxin for enzyme activity. Proc Natl Acad Sci USA 88: 2578–2582

Vigh L, Los DA, Horváth I and Murata N (1993) The primary signal in the biological perception of temperature: Pd-catalyzed hydrogenation of membrane lipids stimulated the expression of the *desA* gene in *Synechocystis* PCC6803. Proc Natl Acad Sci USA 90: 9090–9094

Wada H and Murata N (1989) *Synechocystis* PCC6803 mutants defective in desaturation of fatty acids. Plant Cell Physiol 30: 971–978

Wada H and Murata N (1990) Temperature-induced changes in the fatty acid composition of the cyanobacterium, *Synechocystis* PCC6803. Plant Physiol 92: 1062–1069

Wada H, Gombos Z and Murata N (1990) Enhancement of chilling tolerance of a cyanobacterium by genetic manipulation of fatty acid desaturation. Nature 347: 200–203

Wada H, Gombos Z, Sakamoto T and Murata N (1992) Genetic manipulation of the extent of desaturation of fatty acids in membrane lipids in the cyanobacterium *Synechocystis* PCC6803. Plant Cell Physiol 33: 535–540

Wada H, Avelange-Macherel M-H and Murata N (1993a) The *desA* gene of the cyanobacterium *Synechocystis* sp. strain PCC6803 is the structural gene for Δ12 desaturase. J Bacteriol 175: 6056–6058

Wada H, Schmidt H, Heinz E and Murata N (1993b) In vitro ferredoxin-dependent desaturation of fatty acids in cyano-bacterial thylakoid membranes. J Bacteriol 175: 544–547

Wada H, Gombos Z and Murata N (1994) Contribution of membrane lipids to the ability of the photosynthetic machinery to tolerate temperature stress. Proc Natl Acad Sci USA 91: 4273–4277

Williams WP (1997) The physical properties of thylakoid membrane lipids and their relation to photosynthesis. In: Siegenthaler P-A and Murata N (eds) Lipids in Photosynthesis: Structure, Function and Genetics, pp 103–118. Kluwer Academic Publishers, Dordrecht

Chapter 5

Membrane Lipids in Anoxygenic Photosynthetic Bacteria

Christoph Benning
Institut für Genbiologische Forschung Berlin GmbH, Ihnestrasse 63, 14195 Berlin, Germany

Summary

Photosynthetic bacteria represent a diverse group of microorganisms that convert light into metabolic energy. The anoxygenic photosynthetic bacteria cannot utilize water as an electron donor for photosynthetic electron transport and contain only one photosystem. Not only have these bacteria been intensely studied to elucidate the molecular events involved in the primary reactions of photosynthesis, but they also provide a rich source for the analysis of membrane lipids. Structural data has become available and novel lipids have been discovered, such as a betaine lipid and a glycolipid of the purple bacterium *Rhodobacter sphaeroides,* or a glycosphingolipid of the green sulfur bacterium *Chlorobium limicola.* As new data accumulate, some of the postulated structures for lipids thought to occur in certain species need to be reevaluated. Thus it is currently debated, whether the sulfolipid sulfoquinovosyl diacylglycerol or rather a structurally related compound may be present in green

P.-A. Siegenthaler and N. Murata (eds): Lipids in Photosynthesis: Structure, Function and Genetics, pp. 83–101.
© *1998 Kluwer Academic Publishers. Printed in The Netherlands.*

sulfur bacteria. Increasing consideration is given to specific growth conditions, such as phosphate-limitation, that strongly affect the membrane lipid composition of *Rb. sphaeroides*. Surprisingly, it appears that even a xenobiotic compound such as Tris-buffer is incorporated by certain strains of *Rb. sphaeroides* into the novel lipid phosphatidyl-Tris. This considerable tolerance with regard to membrane lipid composition, our detailed knowledge about the photosynthetic apparatus, and the genetic versatility of anoxygenic photosynthetic bacteria make them ideal model organisms to study the biosynthesis and function of lipids associated with photosynthetic membranes. As a first step in this direction, mutants of *Rb. sphaeroides* deficient in the biosynthesis of sulfoquinovosyl diacylglycerol or phosphatidylcholine have been isolated. This has led to the characterization of the respective genes and provides the basis for a detailed analysis of the biosynthesis and function of these lipids.

I. Introduction

The conversion of sunlight into metabolic energy, called photosynthesis, is dependent on highly organized pigment protein complexes which are embedded in the polar lipid matrix of the photosynthetic membrane. Although much is known about the function of photosynthetic pigment protein complexes, the analysis of the polar lipids found in photosynthetic membranes is to a large extent descriptive. Photosynthetic organisms have evolved in the three main branches of the phylogenetic tree that were newly defined as *Archaea*, *Bacteria*, and *Eucarya* (Woese et al., 1990). Photosynthetic bacteria of the branch *Bacteria* can be divided into two groups, based on the presence of chlorophyll or bacteriochlorophyll. The overwhelming majority of chlorophyll containing bacteria are cyanobacteria, a phylogenetic coherent group (Woese, 1987). Cyanobacteria are able to utilize water as the electron donor for photosynthetic electron transport and release oxygen during photosynthesis. They contain two photosystems with reaction centers resembling those of higher plants. However, anoxygenic photosynthetic bacteria, the focus of this chapter, are characterized by the presence of bacteriochlorophyll and a single photosystem. These bacteria do not release photosynthetic oxygen and can be found interspersed within the *Bacteria* (Woese, 1987). For a more detailed overview of the biology of anoxygenic photosynthetic bacteria, the interested reader is referred to a previous publication in this series (Blankenship et al., 1995).

Polar lipids are characterized by the presence of a hydrophilic headgroup and a lipophilic tail. Due to their amphiphilic properties, these molecules aggregate in an aqueous environment to form vesicular or bilayer structures. They provide the lipid matrix for biological membranes. The analysis of

polar lipids and fatty acids has been used extensively for the classification of anoxygenic photosynthetic bacteria (e.g., Imhoff et al., 1982; Imhoff, 1988; Imhoff, 1991) and detailed reviews are available (Kenyon, 1978; Smith, 1988; Imhoff and Bias-Imhoff, 1995). Here, a condensed overview will be presented focusing on those lipids of anoxygenic photosynthetic bacteria, for which detailed structural information is available. Characteristic polar lipids of the outer membrane of gram-negative bacteria, the lipopolysaccharides, and nonpolar lipids are not discussed in this chapter. Special emphasis will be given to the use of bacterial lipid mutants for the study of biosynthesis and function of polar lipids found in photosynthetic membranes. Recent progress towards the elucidation of the biochemical reactions involved in lipid headgroup biosynthesis will be presented. Furthermore, the effects of growth conditions on the polar lipid composition of anoxygenic photosynthetic bacteria will be discussed.

A. Taxonomy

Anoxygenic photosynthetic bacteria are found interspersed with nonphotosynthetic bacteria along the *Bacteria* branch of the phylogenetic tree (Woese, 1987). Classically, the anoxygenic photosynthetic bacteria have been divided into the purple nonsulfur, the purple sulfur, the green nonsulfur and the green sulfur bacteria (Tindall and Grant, 1986). The 'purple bacteria and related bacteria' were combined into the taxon *Proteobacteria* (Stackebrandt et al., 1988), which can be divided into four subgroups based on rRNA sequence comparison (Woese, 1987). Within the *Proteobacteria* one can also find a strictly aerobe bacteriochlorophyll containing bacterium, *Erythrobacter longus* (Shiba and Harashima, 1986; Shimada, 1995). Some stem nodulating *Rhizobia* exist, that were shown to be photosynthetically active (Eaglesham et al., 1990; Ladha et al., 1990). In addition to gram-negative photosynthetic bacteria, there is a small group of gram-positive photosynthetic bacteria, the *Heliobacteria* (Brockmann and Lipinski, 1983; Gest and Favinger, 1983; Beer-Romero and Gest, 1987; Beer-Romero et al., 1988). Table 1 gives an overview of the anoxygenic photosynthetic bacteria and their classification based on rRNA sequence comparison. To obtain a measure for the research activity, the different genera of photosynthetic bacteria were searched for entries in the GenBank database. Species for which a particularly high number of sequences has been deposited are shown. According to this measure, by far the most is known about the purple nonsulfur bacteria *Rb. capsulatus* and *Rb. sphaeroides*, and much of this chapter will concern these bacteria.

B. Bacterial Photosynthesis

Much of what is known today about the primary reactions of photosynthesis and the photosynthetic electron transport chain is derived from experiments with anoxygenic photosynthetic bacteria (for reviews see, Prince, 1990; Deisenhofer and Norris, 1993; Blankenship, 1994; Blankenship et al., 1995, and references therein). After all, the first high resolution X-ray structure of a photosynthetic reaction center to be solved was that of *Rhodopseudomonas viridis* (Deisenhofer et al., 1985; Deisenhofer and Michel, 1989).

Generally, anoxygenic photosynthetic bacteria have a single photosystem containing different types of bacteriochlorophyll, depending on the individual species. Photosynthetic electron transport proceeds

Table 1. Number of entries in GenBank for different genera and selected species of photosynthetic bacteria

Classification[a]	Genus	Species[b]	Entries in GenBank[c]
Proteobacteria α-Subgroup	*Rhodobacter*		268
		Rb. capsulatus	124
		Rb. sphaeroides	122
	Rhodospirillum		56
		Rs. rubrum	37
	Rhodopseudomonas		52
		Rp. palustris	21
		Rp. viridis	9
	Erythrobacter		9
Proteobacteria β-Subgroup	*Rhodocyclus*		6
Proteobacteria γ-Subgroup	*Chromatium*		20
		Cm. vinosum	18
	Ectothiorhodospira		7
	Thiocystis		4
	Thiocapsa		2
Green Nonsulfur Bacteria	*Chloroflexus*		19
		Cf. aurantiacus	15
Green Sulfur Bacteria	*Chlorobium*		26
		Cb. limicola	11
Gram[+] Bacteria	*Heliobacterium*		8

[a] Classification according to Woese (1987) and Stackebrandt et al. (1988).
[b] Only selected species are shown, for which a large number of sequences were deposited. The entries listed for individual species represent a subset of those obtained for the entire genus.
[c] GenBank release 95.0, 6/15/96, and Updates 95.0+, 08/19/96 at the National Center for Biological Information were searched.

in a cyclic manner and no oxygen is released during the process. With the exception of strictly aerobic bacteria such as *Erythrobacter longus*, anoxygenic photosynthetic bacteria perform photosynthesis under anaerobic conditions. The oxygen concentration in the medium is one of the factors controlling the expression of photosystem genes in *Rb. sphaeroides* and *Rb. capsulatus* (for a review, see Bauer et al., 1993). As the oxygen concentration decreases, an internal membrane system carrying the photosynthetic complexes develops by invagination of the plasma membrane. This internal membrane system is characteristic for purple bacteria. However, cells of green bacteria contain distinct subcellular structures, the chlorosomes, that underlie the plasma membrane (Staehelin et al., 1980).

Comparison of the molecular architecture of the different photosystems provides an insight into the evolution of photosynthesis. The photosystems of purple bacteria and of the green nonsulfur bacterium *Chloroflexus aurantiacus* resemble each other and are very similar to photosystem II of higher plants (e.g., Blankenship, 1994). Although green sulfur bacteria have chlorosomes comparable to those of *Cf. aurantiacus*, the structure of their photosystem is nevertheless different, resembling that of the higher plant photosystem I (Goldbeck, 1993). A slightly less complex photosystem is found in *Heliobacteria* (Blankenship, 1994).

Compared to our extensive knowledge about bacterial photosystems, little information is available on the lipids found in bacterial photosynthetic membranes. One of the questions addressed in this chapter will be whether anoxygenic photosynthetic bacteria can serve as model organisms to study the biosynthesis and function of polar lipids associated with photosynthetic complexes.

II. Polar Lipids of Anoxygenic Photosynthetic Bacteria

The bulk of the polar lipids of anoxygenic photosynthetic bacteria belong to one of two classes, phospholipids or nonphosphorous glycolipids. In addition, ornithine containing lipids as well as betaine lipids can be present in substantial amounts. The relative abundance of polar lipids is highly variable among different species of anoxygenic photosynthetic bacteria and the growth conditions can drastically affect the lipid composition of a given bacterial strain. To provide an impression of the degree of variability, the membrane lipid composition of closely related bacterial species of the *Proteobacteria* α-subgroup is shown in Table 2. In the following an overview of the structure and occurrence of different polar lipids of anoxygenic photosynthetic bacteria will be presented.

Table 2. Polar lipid composition of selected species of the *Proteobacteria* α-subgroup

Species	Phospholipids[a] (%)[b]				Glycolipids[c] (%)[b]				Others[d] (%)[b]	
	PC	PE	PG	CL	SQD	GAD	MHD	DHD	OL	BL
Rhodobacter sphaeroides[e]	27.7	39.9	22.8	+	2.2	-	-	tr	5.5	tr
Rhodobacter sphaeroides[f]	2.9	6.8	12.2	+	16.6	-	+	31.1	11.2	31.1
Rhodobacter capsulatus[g]	7.0	18.7	62.5	8.1	0.7	-	-	-	0.9	-
Rhodospirillum rubrum[g]	5.1	37.8	37.5	11.3	2.6	-	-	-	1.4	-
Rhodopseudomonas viridis[h]	37.5	6.2	8.6	5.9	-	11.7	4.0	13.7	11.7	-

[a] PC, phosphatidylcholine; PE, phosphatidylethanolamine; PG, phosphatidylglycerol; CL, cardiolipin
[b] Values are given in mol % or weight % (*Rb. capsulatus* and *Rs. rubrum*); -, not detected; +, present, but not quantified; tr, traces
[c] SQD, sulfoquinovosyl diacylglycerol; GAD, glucuronosyl diacylglycerol; MHD, monohexosyl diacylglycerol (monogalactosyl diacylglycerol for *Rp. viridis*); DHD, dihexosyl diacylglycerol (glucosylgalactosyl diacylglycerol for *Rb. sphaeroides* or digalactosyl diacylglycerol for *Rp. viridis*)
[d] OL, ornithine lipid; BL, betaine lipid
[e] Benning et al. (1993); cells grown in complete medium
[f] Benning et al. (1993); cells grown in phosphate depleted medium
[g] Russel and Harwood (1979)
[h] E. Heinz, personal communication

A. Phospholipids

The phospholipids found in substantial amounts in anoxygenic photosynthetic bacteria are shown in Fig. 1. Their structure is characterized by a 1,2-diacyl-3-phospho-*sn*-glycerol, also known as phosphatidyl moiety, and a variable headgroup attached to the phosphate. The only phospholipid present in the membranes of all anoxygenic photosynthetic bacteria is phosphatidylglycerol. This comes as no surprise, because phosphatidylglycerol is found in the membranes of virtually all photosynthetic organisms. For example, in cyanobacteria, it is the only phospholipid present (see Chapter 4, this volume). With few exceptions, membranes of anoxygenic photosynthetic bacteria contain substantial amounts of phosphatidylethanolamine and cardiolipin. Phosphatidylethanolamine is thought to be absent from certain species of green sulfur bacteria (Kenyon and Gray, 1974; Imhoff, 1988), the green nonsulfur bacterium *Cf. aurantiacus* (Kenyon and Gray, 1974; Knudsen et al., 1982; Imhoff, 1988) and few species of purple nonsulfur bacteria (Imhoff, 1991). Cardiolipin was not detected in some species of *Rhodobacter* (Imhoff, 1991).

Phosphatidylcholine is rarely found in membranes of bacteria. However, all strains of the genus *Ectothiorhodospira* contain this lipid (Asselineau and Trüper, 1982; Imhoff et al., 1982), whereas the closely related *Chromatiaceae* of the *Proteobacteria* *γ*-subgroup, e.g. *Chromatium*, are devoid of it (Imhoff et al., 1982). Furthermore, phosphatidylcholine is present in many species of purple nonsulfur bacteria including *Rb. sphaeroides* and *Rb. capsulatus* (Kenyon, 1978; Imhoff, 1991). There appears to be only one anoxygenic photosynthetic bacterium, *Cf. aurantiacus*, for which the presence of phosphatidylinositol has been confirmed (Kenyon and Gray, 1974; Knudsen et al., 1982). However, the question remains, whether cells of *Cf. aurantiacus* grown in their natural environment contain this lipid, because yeast extract added to the medium could be the source of phosphatidylinositol (Kenyon, 1978). The membranes of *Heliobacteria* contain the phospholipids phosphatidylglycerol, phosphatidylethanolamine, and cardiolipin, but are apparently devoid of nonphosphorus polar lipids (Imhoff, 1988). The list of phospholipids presented above is by far not complete. Unidentified phospholipids have been described for several species (Kenyon, 1978; Imhoff and Bias-Imhoff, 1995). Precursors for the biosynthesis of the major phospholipids, e.g. phosphatidic acid, phosphatidylserine, and *N*-methylated derivatives of phosphatidylethanolamine should be present in small amounts in photosynthetic bacteria, but have been rarely identified.

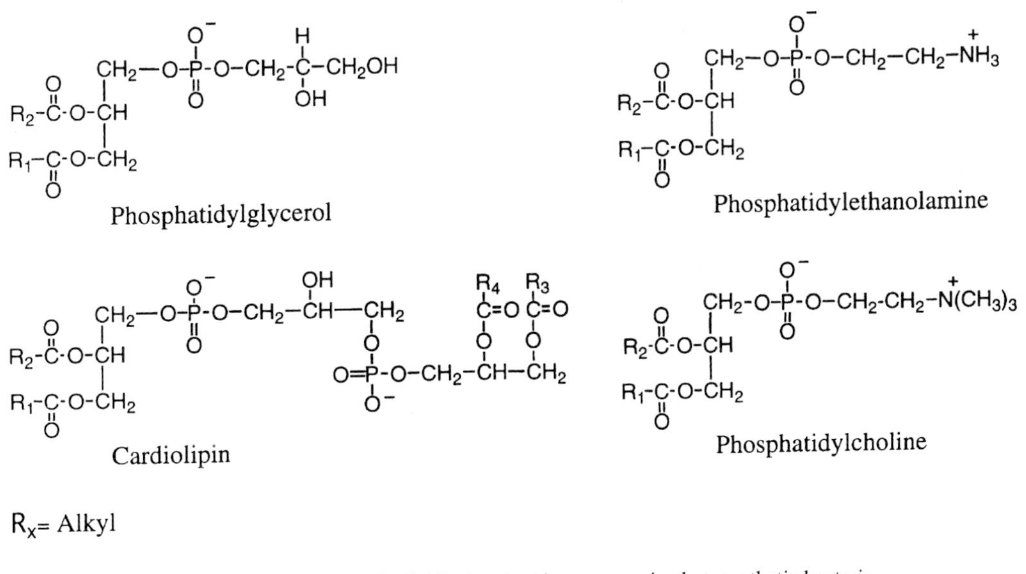

Fig. 1. Structure of phospholipids abundant in anoxygenic photosynthetic bacteria.

B. Glycolipids

The membranes of many anoxygenic photosynthetic bacteria contain glycolipids. Genera reported to be devoid of glycolipids are *Ectothiorhodospira* and *Heliobacterium* (Asselineau and Trüper, 1982; Imhoff, 1988). In most cases the identification of glycolipids has not been pursued beyond the stage of co-chromatography with lipid standards and sugar-specific staining on thin layer chromatography plates. Typical glycolipids found in photosynthetic membranes do not contain phosphorus and are characterized by a 1,2-diacyl-*sn*-glycerol moiety to which a mono- or a disaccharide is attached at the *sn*-3 position. The glycolipid structures of anoxygenic photosynthetic bacteria, that were confirmed by

rigorous structural elucidation including different spectroscopic methods, are shown in Fig. 2. The sulfolipid α- D-sulfoquinovosyl diacylglycerol was first discovered in *Rs. rubrum* (Benson et al., 1959). Based on IR and mass spectroscopy, the structure of the sulfolipid present in *Rb. sphaeroides* is identical to that of the sulfolipid present in chloroplasts of higher plants, with the exception of the fatty acid substituents (Radunz, 1969; Gage et al., 1992). The occurrence of this particular lipid in different photosynthetic organisms has been widely investigated, because for a long time it was thought to be important for photosynthesis (Barber and Gounaris, 1986). This idea was to a large extent based on the erroneous assumption that all photosynthetic organisms contain sulfoquinovosyl diacylglycerol,

Glucuronosyl diacylglycerol

Sulfoquinovosyl diacylglycerol

Monogalactosyl diacylglycerol

β–Digalactosyl diacylglycerol

Glucosylgalactosyl diacylglycerol

R_x= Alkyl

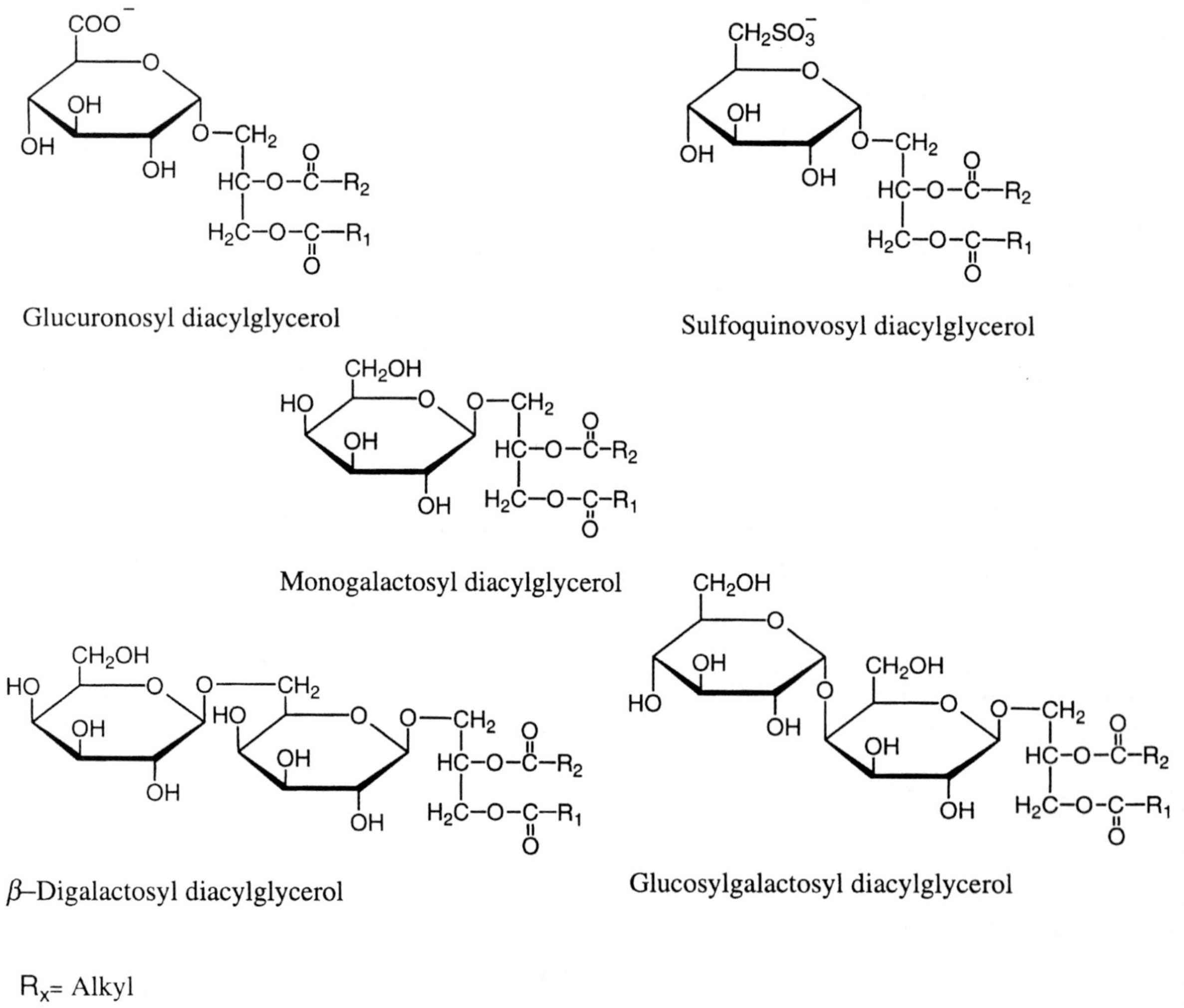

Fig. 2. Glycolipids of anoxygenic photosynthetic bacteria for which structural information is available.

but that it is absent in nonphotosynthetic organisms. Closely scrutinizing the literature, the correlation does not hold, because the nonphotosynthetic bacterium *Rhizobium meliloti* was found to contain a substantial amount of this lipid (Cedergreen and Hollingsworth, 1994) and the presence of sulfo-quinovosyl diacylglycerol in many anoxygenic photosynthetic bacteria remains uncertain. Although this lipid represents between 2% and 16% of polar lipids in *Rb. sphaeroides* depending on the growth conditions (Benning et al., 1993), several species, including *Rb. capsulatus*, seem to contain only traces of it. Small amounts can be detected only after sulfur-labeling of the cells (Imhoff, 1984). In *Rhodopseudomonas viridis* sulfolipid has not been identified, but an anionic lipid with similar biophysical properties, glucuronosyl diacylglycerol was dis-covered (Övermöhle et al., 1995). Currently, there are doubts whether the sulfolipid sulfoquinovosyl diacylglycerol or rather a structurally related sulfur-labeled compound is present in bacteria of the *Proteobacteria* γ-subgroup (*Chromatiaceae*) and the green sulfur bacteria (Imhoff and Bias-Imhoff, 1995). Structural elucidation of these sulfolipids will be required to clarify this point.

Several glycolipids containing galactose have been reported for anoxygenic photosynthetic bacteria. The galactolipids mono- and β-digalactosyl diacylglycerol have been identified in *Rp. viridis* (Övermöhle et al., 1995). Based on NMR spectroscopy, the structure of the monogalactosyl diacylglycerol (β-D-galactosyl diacylglycerol) is identical to that of the major galactolipid of chloroplast membranes in higher plants. However, the two galactose moieties of the digalactosyl diacylglycerol of *Rp. viridis* are β (1→6) linked not α (1→6) as described for the equivalent lipid of higher plants. In *Rb. sphaeroides* the related lipid α-D-glucosyl-(1→4)-O-β-D-galactosyl diacyl-glycerol (Fig. 2) and its presumed precursor monogalactosyl diacylglycerol accumulate under phosphate-limiting growth conditions (Benning et al., 1995). A glycolipid containing galactose and glucose has been identified in the green nonsulfur bacterium *Cf. aurantiacus* (Knudsen et al., 1982). It is interesting to note that also in cyanobacteria a glucose containing lipid, monoglucosyl diacyl-glycerol, is present (Sato and Murata, 1982). Green sulfur bacteria contain monogalactosyl diacylglycerol (Cruden and Stanier, 1970) and a number of glycolipids which still await further structural elucidation (Imhoff and Bias-Imhoff, 1995).

Galactolipids appear to be absent from membranes of *Cm. vinosum*, a bacterium of the *Proteobacteria* γ-subgroup (*Chromatiaceae*), but different glycolipids tentatively identified as monoglucosyl-, mannosyl-glucosyl- and dimannosyl diacylglycerol are present (Steiner et al., 1969). In membranes of the closely related bacterium *Thiocapsa floridana*, glucose and rhamnose containing glycolipids were discovered (Takacs and Holt, 1971). The *Chromatiaceae* as well as the green sulfur and nonsulfur bacteria appear to be rich in different glycolipids (Imhoff et al., 1982; Imhoff, 1988) that await further structural elucidation.

C. Other Polar Lipids

In the following, a number of unusual lipids of anoxygenic bacteria are described, which cannot be classified as phospholipids or glycolipids, or which are only conditionally synthesized.

1. Ornithine Lipids

Phosphorus-free ornithine-containing membrane lipids have been identified in different species of gram-negative bacteria (Ratledge and Wilkinson, 1988). They generally appear to be present in purple nonsulfur bacteria (*Proteobacteria* α-subgroup) but are absent in other photosynthetic bacteria (Imhoff et al., 1982). Detailed structural analysis is available for the ornithine lipid of *Rb. sphaeroides* (Gorchein, 1968; Gorchein, 1973) and that of *Rs. rubrum* (Brooks and Benson, 1972). The structure originally proposed for the ornithine lipid of *Rb. sphaeroides* has later been corrected (Gorchein, 1973) and was shown to be identical to that proposed for the ornithine lipid of *Rs. rubrum*. The characteristic feature of this lipid is the long chain 3-hydroxy fatty acid attached via an amide bond to the α-amino group of ornithine. A second long chain fatty acid is esterified to the 3-hydroxy group of the first fatty acid (Fig. 3). Although the major ornithine lipids of purple nonsulfur bacteria seem to be identical, the existence of additional ornithine containing lipids in purple nonsulfur bacteria has been reported (e.g., Brooks and Benson, 1972).

2. Betaine Lipids

Recently, the nonphosphorus betaine lipid diacyl-glycerol-*N*-trimethylhomoserine (Fig. 3) was discovered in cells of *Rb. sphaeroides* grown under

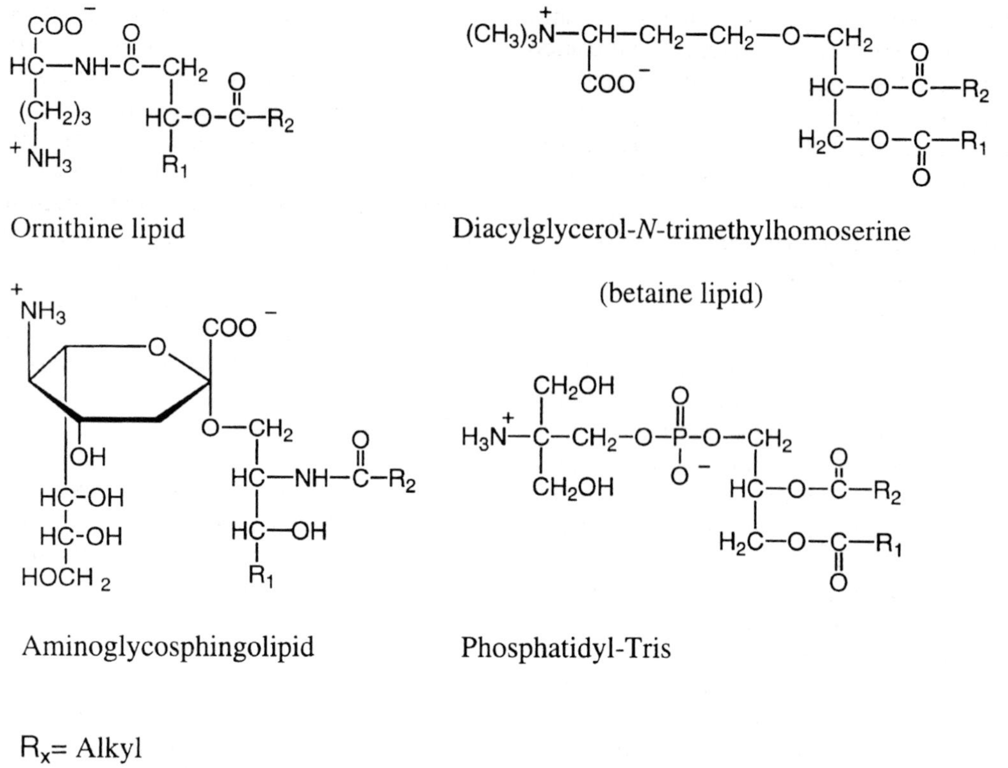

Fig. 3. Unusual polar lipids of anoxygenic photosynthetic bacteria.

phosphate limitation (Benning et al., 1995). Detailed structural elucidation suggests that the headgroup of this lipid is identical to that of the betaine lipid first discovered in the alga *Ochromonas danica* (Brown and Elovson, 1974). Previously, it has been assumed that the occurrence of this lipid is restricted to algae and lower plants (Sato, 1992). However, it is now certain that during the course of evolution the ability to synthesize betaine lipids was not a newly acquired trait by eucaryotic algae, but had already evolved in photosynthetic bacteria. It remains to be seen whether other photosynthetic bacteria contain betaine lipids and further studies of the distribution of these lipids may provide clues to the evolution of photosynthetic organisms.

3. Sphingolipids

Aminolipids devoid of phosphorus have been repeatedly described to be present in green sulfur bacteria (e.g., Kenyon and Gray, 1974; Knudsen et al., 1982). A preliminary investigation showed that this lipid is a secondary or tertiary amide containing

one mole of myristic acid (Olson et al., 1983). Detailed structural elucidation suggested that the aminolipid of *Cb. limicola* is an aminoglycosphingolipid (Jensen et al., 1991). The headgroup consists of the carbohydrate neuraminic acid which is attached to the sphingosine backbone (Fig. 3). Myristic acid is linked to the sphingosine by an amide bond. This lipid is absent from the green nonsulfur bacterium *Cf. aurantiacus* (Kenyon and Gray, 1974; Knudsen et al., 1982; Imhoff, 1988; Jensen et al., 1991), but sphingolipids similar in structure have been reported to be present in a selected number of nonphotosynthetic bacteria (see references in Jensen et al., 1991).

4. Xenobiotic Lipids

It has been reported that the addition of the buffer substance tris(hydroxymethyl)aminomethane (Tris) induces the accumulation of the unusual phospholipid *N*-acylphosphatidylserine in certain strains of *Rb. sphaeroides* (Donohue et al., 1982a,b; Cain et al., 1982). The structural elucidation of this compound was partially based on IR and proton NMR spectra (Donohue et al., 1982a). However, comparing proton NMR spectra for chemically synthesized N-acylphosphatidylserine and phosphatidyl-Tris, it became apparent that the spectrum presented by Donohue et al. (1982a) is identical to that of phosphatidyl-Tris (Schmid et al., 1991). Based on this indirect evidence, it seems plausible that in the presence of the buffer substance Tris, certain strains of *Rb. sphaeroides* synthesize the xenobiotic lipid phosphatidyl-Tris (Fig. 3). This would be another example for the remarkable metabolic flexibility of anoxygenic photosynthetic bacteria and their ability to tolerate drastic changes in membrane lipid composition. The capability to synthesize phosphatidyl-Tris may not be restricted to *Rb. sphaeroides,* because *Ectothiorhodospira halophila* accumulates a similar lipid when grown in the presence of Tris (Raymond and Sistrom, 1969). However, additional experiments will be required to confirm the structure of the compound directly, following the isolation from the respective organism, and to reconcile conflicting observations. For example, a small amount of 'N-acylphosphatidylserine' was found in cells of *Rb. sphaeroides* grown in the absence of Tris (Donohue et al., 1982b). Furthermore, the presence of a large amount of an unidentified phospholipid with similar chromatographic properties to phosphatidyl-Tris was reported for a strain of *Rb. sphaeroides*, that was grown in a medium supposedly lacking Tris (Onishi and Niederman, 1982).

III. Fatty Acid Constituents of Polar Lipids

The fatty acids of bacterial cells can be easily identified employing gas chromatography of the respective methylester derivatives, and the fatty acid composition of most of the described species of anoxygenic photosynthetic bacteria is known. To a certain extent, the fatty acid composition of different classes of photosynthetic bacteria is of taxonomic value. However, it has to be taken into account that the fatty acid composition of photosynthetic bacteria may be drastically affected by the growth conditions (for a recent review see Imhoff and Bias-Imhoff, 1995). Fatty acids of the polar lipids found in anoxygenic photosynthetic bacteria are typically saturated or monounsaturated with a chain length of 16 or 18 carbon atoms. For example, the predominant fatty acid found in species of *Rhodobacter* and many

other *Proteobacteria* is cis-Δ^{11} octadecenoic acid (vaccenic acid). However, green sulfur bacteria contain predominantly myristic (14:0) and palmitic acid (16:0). Polyunsaturated fatty acids are generally absent in anoxygenic photosynthetic bacteria. These bacteria lack the fatty acid desaturases characteristic for some species of cyanobacteria (see Chapter 4). Small amounts of fatty acids with an odd carbon number or cyclopropane fatty acids are present in anoxygenic photosynthetic bacteria. Branched chain fatty acids are characteristic for species of *Heliobacterium* (Beck et al., 1990). Apparently, lipids of gram-negative bacteria lack this class of fatty acids.

For each polar membrane lipid a number of molecular species can be distinguished on the basis of their fatty acid constituents. Not all fatty acids present in a bacterial cell may be evenly distributed among the different lipid classes. Specific fatty acids can be more abundant in one class than in another. For example, the sulfolipid sulfoquinovosyl diacylglycerol of *Rb. sphaeroides* and of other photosynthetic organisms contains a substantial amount of palmitic acid compared to other lipids (Radunz, 1969; Gage et al., 1992). Myristic acid is found preferentially linked via an amide bond to the sphingosine backbone of the aminoglycosphingolipid in *Cb. limicola* (Jensen et al., 1991). Furthermore, the 3-hydroxy fatty acid of purple nonsulfur bacteria is a key constituent of the ornithine lipid, but is rarely present in other membrane lipids (Brooks and Benson, 1972; Gorchein, 1973). The presence of different fatty acids in each polar lipid class and the resulting large number of molecular species add to the complexity of lipid metabolism. A large number of genes are expected to encode the respective enzymes, but only a few have been isolated to date. In the following section, a brief overview of our current knowledge on lipid biosynthesis in anoxygenic bacteria will be provided.

IV. Biosynthesis of Polar Lipids

A. Phospholipid Biosynthesis

Much of what is known about phospholipid biosynthesis in photosynthetic bacteria has been derived from experiments with the purple nonsulfur bacterium *Rb. sphaeroides*. In this bacterium, the formation of the intracytoplasmic membrane system carrying the photosynthetic apparatus can be easily induced by lowering the oxygen concentration of the medium. This feature has been widely used to study the assembly of the intracytoplasmic membrane, the biosynthesis of lipids, as well as the insertion of proteins.

1. Phosphatidylglycerol and Phosphatidylethanolamine

The biosynthesis of phospholipids is commonly thought to proceed as was proposed for other gram-negative bacteria (Cronan and Rock, 1987). The crucial intermediate phosphatidic acid is formed by transfer of fatty acids from acyl-carrier protein onto sn-glycerol-3-phosphate. A sn-glycerol-3-phosphate acyltransferase from Rb. sphaeroides has been characterized (Luecking and Goldfine, 1975; Cooper and Luecking, 1984), and an acyl-carrier protein from Rb. sphaeroides has been purified (Cooper et al., 1987). Furthermore, rapid pulse chase labeling experiments with Rb. sphaeroides confirmed that phosphatidic acid is the precursor for the biosynthesis of phosphatidylglycerol and phosphatidylethanolamine (e.g., Cain et al., 1983). Phosphatidic acid and CTP are converted to CDP-diacylglycerol. In the case of phosphatidylglycerol biosynthesis, the CMP moiety is replaced by sn-glycerol-3-phosphate, giving rise to phosphatidylglycerolphosphate, which is subsequently dephosphorylated. On the contrary, for the biosynthesis of phosphatidylethanolamine, CMP is replaced by L-serine giving rise to phosphatidylserine, which is decarboxylated. Pulse chase labeling of the Tris-induced phospholipid 'N-acylphosphatidylserine' suggested that phosphatidylserine is not the direct precursor (Cain et al., 1983). At the time the experiment was done, this result was difficult to explain. However, given that the structure of this lipid has to be reinterpreted as phosphatidyl-Tris (see above), it appears plausible that this lipid is formed from CDP-diacylglycerol via replacement of CMP by Tris. Therefore, the pulse chase labeling experiments provide independent evidence that the lipid accumulating in some strains of Rb. sphaeroides is not N-acylphosphatidylserine, but phosphatidyl-Tris as proposed by Schmid et al. (1991).

The enzymes involved in the biosynthesis of phosphatidylglycerol and phosphatidylethanolamine have been assayed in crude extracts of Rb. sphaeroides (Cain et al., 1984). The phosphatidylglycerol-phosphate and phosphatidylserine synthases were partially purified and characterized in greater detail (Radcliff et al., 1989). Recently, the gene (pgsA) encoding the phosphatidylglycerolphosphate synthase of Rb. sphaeroides has been isolated and characterized. The CDP-diacylglycerol synthase and the phosphatidylglycerolphosphate synthase were localized to the cytoplasmic membrane, while the phosphatidylserine synthase and carboxylase were either described as closely membrane bound or associated with distinct regions of the cytoplasmic membrane (Cain et al., 1984; Radcliff et al., 1985, 1989). Within limits, this result is consistent with the idea that phospholipids are synthesized in the cytoplasmic membrane and subsequently transferred to the intracytoplasmic membrane system (e.g., Cain et al., 1981).

2. Phosphatidylcholine

Phosphatidylcholine is rarely present in bacteria. However, Rb. sphaeroides contains a substantial amount of phosphatidylcholine and has provided an ideal opportunity to study the biosynthesis of this lipid. Pulse chase labeling experiments indicated a product-precursor relationship between phosphatidylethanolamine and phosphatidylcholine (e.g., Cain et al., 1983). It is now certain that phosphatidylcholine is formed via a sequential methylation of phosphatidylethanolamine in Rb. sphaeroides. Recently, a genetic mutant of Rb. sphaeroides deficient in the biosynthesis of phosphatidylcholine has been isolated and the corresponding gene (pmtA) was cloned by complementation of this mutant (Arondel et al., 1993). The gene encodes a 22.9 kDa protein, that is soluble based on hydropathy analysis. This result is in good agreement with localization data for phosphatidylethanolamine methyltransferase activity, which was found to be predominantly associated with the soluble fraction (Cain et al., 1984). Functional expression of the pmtA gene in E. coli resulted in the accumulation of phosphatidylcholine (Arondel et al., 1993). Because methylated derivatives of phosphatidylethanolamine are absent in E. coli, this result clearly demonstrates that the pmtA gene encodes a phosphatidylethanolamine methyltranferase, which catalyzes the sequential transfer of three methyl groups onto phosphatidylethanolamine. The methyl group donor is S- adenosylmethionine (Arondel et al, 1993).

3. Temporal Patterns and Regulation of Phospholipid Biosynthesis

As a result of experiments with synchronous cell populations of *Rb. sphaeroides*, a discontinuous increase in the net accumulation of lipids (Luecking et al., 1978), but a continuous accumulation of membrane proteins (Fraley et al., 1978) in the intracytoplasmic membrane was observed during the cell cycle. Labeling experiments suggested a dramatic increase in the biosynthesis of phospholipids just prior to cell division (Fraley et al., 1979a; Knacker et al., 1985), although cell cycle dependent transfer of lipids from the cytoplasmic membrane to the intracytoplasmic membrane system mediated by a phospholipid transfer protein is discussed as well (Cain et al., 1981; Tai and Kaplan, 1984, 1985). Regardless of the mechanistic basis for the discontinuous phospholipid accumulation in intracytoplasmic membranes, corresponding changes in the fluidity (Fraley et al., 1979b) and the ultrastructure of the intracytoplasmic membrane (Yen et al., 1984) were observed. Furthermore, the changes in lipid/protein ratio during the cell cycle were shown to influence the activity of membrane bound enzymes (e.g., Hoger et al., 1987).

Light appears to be an important factor regulating phospholipid biosynthesis. After a high-to-low light transition of high light adapted *Rb. sphaeroides* cells, a rapid inhibition of total phospholipid biosynthesis was observed (Campbell and Luecking, 1983). This inhibition coincided with an accumulation of guanosine-5′-diphosphate-3′-diphosphate. In *E. coli* the accumulation of this intracellular signal compound is part of the stringent response following starvation. This compound acts as a negative effector of bacterial phospholipid biosynthesis (for a review see Cashel and Rudd, 1987). The apparent involvement of the stringent response system in *Rb. sphaeroides* during high to low light transition experiments suggests that phospholipid biosynthesis is only indirectly affected by light under these conditions. Thus far, a direct regulation of phospholipid biosynthesis by light has not been demonstrated.

It should also be noted that amino acid deprivation of *Rb. sphaeroides* cells leads to an inhibition of phospholipid biosynthesis without the concomitant accumulation of guanosine-5′-diphosphate-3′-diphosphate (Acosta and Luecking, 1987). Apparently, different stress factors trigger the stringent response system in different ways in *Rb. sphaeroides*.

B. Diacylglycerol-N-Trimethylhomoserine

The discovery of the betaine lipid diacylglycerol-*N*-trimethylhomoserine in *Rb. sphaeroides* (Benning et al., 1995) provides a novel opportunity to study the biosynthesis of this lipid. Contrary to *Rb. sphaeroides*, other organisms which contain this lipid, such as algae and lower plants, are not very well suited for biochemical and genetic analysis. For this reason, Hofmann and Eichenberger (1995) began to use *Rb. sphaeroides* as a model system to identify the precursors and enzymes involved in the biosynthesis of diacylglycerol-*N*-trimethylhomoserine. They were able to show that methionine provides the carbon skeleton, and is the donor of the *N*-methylgroups for this lipid in *Rb. sphaeroides*. This result suggests that the biosynthesis proceeds in a similar fashion as postulated for algae and lower plants (Sato, 1992). As a result of pulse chase experiments, Hofmann and Eichenberger (1995) were able to identify putative intermediates, presumably partially *N*-methylated precursors of diacylglycerol-*N*-trimethylhomoserine. This result points towards lipid-linked *N*-methylation as the final step as is known for the biosynthesis of phosphatidylcholine (see above). However, a different methylase must be involved in the biosynthesis of diacylglycerol-*N*-trimethylhomoserine, because phosphatidylcholine-deficient strains of *Rb. sphaeroides* lacking phosphatidylethanolamine methyltransferase are still able to synthesize diacylglycerol-*N*-trimethylhomoserine (M. Hofmann and W. Eichenberger, personal communication).

C. Sulfoquinovosyl Diacylglycerol

The biosynthesis of the sulfolipid sulfoquinovosyl diacylglycerol in higher plants has been investigated for over thirty years, but little is known about the reactions leading to the synthesis of the sulfoquinovose headgroup. Recently, the purple nonsulfur bacterium *Rb. sphaeroides* has been employed to develop a bacterial model system to solve the pathway for sulfolipid biosynthesis. Sulfolipid-deficient mutants of this bacterium and four genes complementing these mutants were isolated (Benning and Somerville, 1992a,b). The genes of *Rb. sphaeroides* involved in sulfolipid biosynthesis were designated *sqdA*, *sqdB*, *sqdC*, and *sqdD*. Three of these genes

A

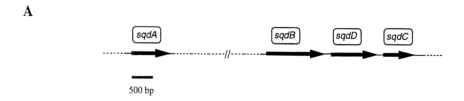

B

Fig. 4. Sulfolipid biosynthesis in *Rb. sphaeroides*. A. Organization of genes involved in sulfolipid biosynthesis. Arrows indicate the direction of transcription. B. Hypothesis for the biosynthesis of sulfolipid. The possible involvement of the *sqdB* and *sqdD* gene products is indicated. DAG, diacylglycerol; R-SO$_3^-$, hypothetical sulfur donor; SQDG, sulfoquinovosyl diacylglycerol; UDP-Glc, UDP-glucose; UDP-SQ, UDP-sulfoquinovose.

are organized in a transcriptional unit (Fig. 4A). Null-mutants were constructed and tested for the accumulation of precursors of sulfolipid biosynthesis. Insertional inactivation of *sqdD* resulted in the accumulation of UDP-sulfoquinovose (Rossak et al., 1995) suggesting that the respective gene product is involved in catalyzing the last step of sulfolipid biosynthesis, the transfer of sulfoquinovose from UDP-sulfoquinovose onto diacylglycerol (Fig. 4B). This conclusion was further supported by the sequence similarity between the predicted *sqdD*-gene product and glycogenin, a UDP-glucose dependent glycosyltransferase (Benning and Somerville, 1992a). Another mutant lacking the *sqdC* gene product accumulates a sulfur-labeled, water-soluble compound of unknown structure (M. Rossak and C. Benning, unpublished). The predicted *sqdC* gene product shows no sequence similarity to known proteins and its role in the biosynthesis of sulfolipid remains unclear. Insertional inactivation of *sqdA* and *sqdB* did not result in the accumulation of sulfur-labeled compounds suggesting that the respective gene products catalyze a reaction prior to, or involved in the incorporation of sulfur into a sulfolipid precursor (M. Rossak and C. Benning, unpublished). The predicted *sqdB* gene product shows sequence

similarity to UDP-glucose epimerases and sugar nucleotide dehydratases (Benning and Somerville, 1992b). Based on this observation, it seems plausible that the *sqdB* gene product is involved in the biosynthesis of UDP-sulfoquinovose as proposed in Fig. 4B. This hypothesis is inspired by the known reaction mechanism for UDP-glucose epimerases and sugar nucleotide dehydratases (for a review see Gabriel, 1987). An enzyme-bound UDP-5,6-glucose is proposed to be the critical intermediate that can be converted to UDP-sulfoquinovose by transfer of the respective sulfur group from a postulated donor. This hypothesis represents a strong contrast to the originally proposed 'glycolytic pathway' for sulfolipid biosynthesis (Benson, 1963). At present, experiments are conducted in my laboratory to test this hypothesis and to elucidate the specific function of the different *sqd* gene products in *Rb. sphaeroides*.

V. Factors Affecting Membrane Lipid Composition

A. Subcellular Location

In most photosynthetic bacteria, one can distinguish

three subcellular membrane systems, the outer membrane, common to all gram-negative bacteria, the cytoplasmic membrane, and the intracytoplasmic membrane harboring the photosynthetic apparatus. Considering the different protein composition and function of the three membrane systems, as well as the observation that polar lipids are synthesized by enzymes that are preferentially associated with the cytoplasmic membrane (see above), differences in the polar lipid composition of the three membrane systems would be expected. Very few attempts have been made to address this point, but some data are available for purple nonsulfur bacteria which suggest that there is no major difference in polar lipid composition of the cytoplasmic membrane and the intracytoplasmic membrane (for a discussion, see Imhoff and Bias-Imhoff, 1995). Similarly, careful analysis of subcellular membranes of *Ectothio-rhodospira mobilis*, a bacterium of the *Proteobacteria* γ-subgroup, revealed only subtle differences in polar lipid composition of the outer membrane and the cytoplasmic membrane and no differences between the cytoplasmic membrane and the intracytoplasmic membrane (Ditandy and Imhoff, 1993). The presence of lyso-phosphatidylethanolamine and an increased content of saturated and short-chain fatty acids were found to be characteristic for the outer membrane. On the contrary, the protein profiles of the three membrane systems of *Ec. mobilis* were clearly distinguishable.

In green photosynthetic bacteria, a specialized subcellular structure, the chlorosome, harbors the photosynthetic apparatus. These structures can be separated from other cell membranes and were found to be enriched in the glycolipid monogalactosyl diacylglycerol (Cruden and Stanier, 1970). Based on these data and their own structural observations, Staehelin et al. (1980) proposed a model for the chlorosome ultrastructure, that featured a monolayer of monogalactosyl diacylglycerol covering the cytoplasmic boundary of the chlorosome. However, in a subsequent study, no correlation was detected between the cellular content of galactolipids and the bacteriochlorophyll content or the total amount of chlorosome material in cells of *Chlorobium* and *Chloroflexus* (Holo et al., 1985). Therefore, a specific enrichment of galactolipids in chlorosomes remains questionable.

Olson et al. (1983) found that an aminolipid, later identified as aminoglycosphingolipid (see Fig. 3; Jensen at al., 1991), was the major polar lipid component of membrane fractions of the green sulfur bacterium *Cb. limicola*, that were enriched in photosynthetic reaction centers. Since this lipid is found in the cytoplasmic membrane but not in the chlorosome (Olsen et al., 1984), it was proposed to serve as a membrane anchor for the bacterio-chlorophyll α-protein (Jensen et al., 1991). In *Cb. limicola*, this protein is thought to form a crystal plate between the cytoplasmic membrane and the chlorosome (Staehelin et al. 1980). However, further experiments will be required to test this idea.

B. Growth Conditions

1. Oxygen Concentration and Light

In purple nonsulfur bacteria, phototrophic growth is induced by low oxygen concentration in the medium and depends on the quality and photon flux density of the incident light. Several studies describing the effects of oxygen and light, or phototrophic versus chemotrophic growth on membrane lipid composition in purple nonsulfur bacteria have been conducted, but the results are inconsistent. An early study suggested that the relative amounts of phosphatidylethanolamine and phosphatidylglycerol increase with increasing light intensity in *Rb. sphaeroides* (Steiner et al., 1970). Russel and Harwood (1979) compared the lipid composition of chemo- and phototrophically grown cells of different purple nonsulfur bacteria. In their hands the relative amount of phosphatidylglycerol was increased at the cost of phosphatidylethanolamine. These conflicting results may be explained in part by the fact that the light intensity is affecting the polar lipid composition during photosynthetic growth. Thus, Onishi and Niederman (1982) observed no change in aerobically versus phototrophically grown cells of *Rb. sphaer-oides* under high light conditions. But cells grown under low light contained increased relative amounts of phosphatidylethanolamine at the cost of an uniden-tified lipid, presumably phosphatidyl-Tris (Onishi and Niederman, 1982). Overall, there appears to be no dramatic difference in lipid composition of chemo-versus phototrophically grown cells of *Rb. sphaer-oides*. This conclusion is supported by the general observation that there is no substantial difference in lipid composition between the cytoplasmic membrane and the intracytoplasmic membrane, the predominant membrane system of phototrophically grown cells (see above).

Numerous reports are available that describe differences in the fatty acid composition in chemo- versus phototrophically grown cells of purple nonsulfur bacteria. For a recent discussion, the reader is referred to the review by Imhoff and Bias-Imhoff (1995).

2. Phosphate Limitation

The factors discussed above had only moderate effects on the membrane lipid composition of *Rb. sphaeroides*. On the contrary, phosphate-limiting growth conditions result in the accumulation of novel lipids in *Rb. sphaeroides* (see Table 2; Benning et al., 1993; Benning et al., 1995). Under these conditions, the relative amounts of phospholipids decrease, but those of sulfoquinovosyl diacylglycerol and the ornithine lipid increase. In addition, novel glycolipids, monogalactosyl diacylglycerol and glucosyl- galactosyl diacylglycerol, as well as the betaine lipid diacylglycerol-*N*-trimethylhomoserine accumulate (Benning et al., 1995). The observed changes in lipid composition can be viewed as a requirement of the cell to maintain certain ratios of negatively-to- positively charged, and bilayer-to-non-bilayer forming lipids in order to provide a functional membrane (Webb and Green, 1991). Under phosphate-limiting growth conditions the role of negatively charged lipids, such as phosphatidylglycerol is taken over by the sulfolipid sulfoquinovosyl diacylglycerol (Benning et al., 1993) and the zwitter-ionic phosphatidylcholine is replaced by the betaine lipid diacylglycerol-*N*-trimethylhomoserine, as in many marine algae (Sato, 1992). The amine lipid phosphatidylethanolamine is substituted by the ornithine lipid, and glucosylgalactosyl diacylglycerol is synthesized to provide the bulk of the membrane lipids. This phenomenon is not restricted to *Rb. sphaeroides* or photosynthetic bacteria. The replacement of phospholipids by nonphosphorus acidic glycolipids and ornithine lipids in membranes of *Pseudomonas* sp. is well known (Minnikin and Abdolrahimzadeh, 1974; Minnikin et al., 1974). Whether bacterial cells can measure the biophysical parameters of the membrane or respond directly to decreasing phosphate levels in the medium remains unknown. A first step toward the elucidation of the regulatory mechanism would be the isolation of genes involved in the biosynthesis of specific lipids. Studying the expression of these genes may provide clues about the process by which bacteria alter their membrane composition in response to environmental changes.

3. Additions to the Medium and Salt Concentration

Some anoxygenic photosynthetic bacteria have a remarkable ability to utilize compounds from the medium as precursors for membrane lipid biosyn- thesis. One example is the presumed incorporation of the xenobiotic buffer substance Tris into phosphatidyl-Tris by some strains of *Rb. sphaeroides* (Schmid et al., 1991) and presumably *Ectothio- rhodospira* (Raymond and Sistrom, 1969). This phenomenon has to be kept in mind, when comparing the polar lipid composition of different bacteria. Strains, which require a rich but undefined medium for optimal growth in the laboratory, may reveal a different lipid composition than the same strains grown in the natural environment. One example presumably is the highly unusual presence of phosphatidylinositol in the green nonsulfur bacterium *Cf. aurantiacus*, which is grown in a medium containing yeast extract (see above; Kenyon, 1978).

Halotolerant anoxygenic photosynthetic bacteria of the genus *Ectothiorhodospira* show a strong increase in the relative amount of phospha- tidylglycerol and a concomitant decrease in the amounts of phosphatidylethanolamine and cardiolipin with increasing salinity of the medium (Thiemann and Imhoff, 1991). A corresponding increase in excess negative charges carried by the phospholipids and in the ratio of bilayer-to-non-bilayer forming lipids can be calculated. This apparent salinity-dependent phosphatidylglycerol / phosphatidylethanolamine antagonism is not an adaptive mechanism specific to photosynthetic bacteria, but has been described for a large number of non-photosynthetic bacteria (for a review, see Kates, 1986). In general, the observed adaptations are interpreted as an attempt by the halotolerant bacterial cell to maintain membrane bilayer stability.

VI. The Function of Membrane Lipids in Anoxygenic Photosynthesis

There is no doubt that polar lipids are crucial building blocks of photosynthetic membranes. They provide the lipid matrix surrounding the pigment protein complexes. Thus, membrane bound enzyme complexes of purple nonsulfur bacteria lose their activity upon removal of phospholipids, and regain activity following the subsequent reconstitution by phospholipids (e.g. Klemme et al., 1971). Furthermore, an investigation of the interaction between the boundary lipids and pigment protein complexes of the photosynthetic membrane of *Rb. sphaeroides* led to the conclusion that these complexes preferentially associate with negatively charged lipids (Birrell et al., 1978). The most abundant anionic lipid in membranes of *Rb. sphaeroides* is phosphatidylglycerol. The importance of this lipid for anoxygenic photosynthesis has been suggested by Russell and Harwood (1979), but direct experimental evidence has not been provided.

Another polar lipid, that has been repeatedly suggested to play an important role in photosynthetic membranes, is the sulfolipid sulfoquinovosyl diacylglycerol (Barber and Gounaris, 1986). However, a sulfolipid-deficient null-mutant of *Rb. sphaeroides* showed the same growth rates under optimal phototrophic growth conditions (Benning et al., 1993). The ratios of the various components of the photosynthetic electron transport chain, as well as the electron transfer rates during cyclic electron transport were not altered in the mutant. No aspect of photosynthesis analyzed was affected by the absence of sulfolipid in the mutant, making it unlikely that sulfolipid has an intrinsic, unique, direct or indirect function in the photosynthetic apparatus of *Rb. sphaeroides*. Nevertheless, sulfolipid appears to be important under conditions of phosphate limitation, which cause the sulfolipid-deficient mutant of *Rb. sphaeroides* to cease growth much earlier than the wild type (Benning et al., 1993). Under these conditions, wild type cells show reduced amounts of phospholipids including phosphatidylglycerol, and an increased amount of sulfolipid. It appears that under phosphate limitation the anionic phospholipid phosphatidylglycerol is replaced as a boundary lipid by the anionic glycolipid sulfoquinovosyl diacylglycerol in the wild type, an adaptation not possible for the sulfolipid-deficient mutant. This hypothesis provides a possible answer to the question, why sulfolipid has evolved preferentially in photosynthetic organisms. Frequently, these organisms live in an environment depleted of phosphate, and sulfolipid provides a mechanism to conserve phosphate bound in membrane lipids.

Similar experiments, employing molecular genetics to elucidate the function of specific lipids have not been performed for lipids other than sulfolipid in anoxygenic photosynthetic bacteria. However, as more genes become available, this approach can be applied to other lipids. Already, the role of phosphatidylcholine and phosphatidylglycerol could be tested in the same way, because the genes for the respective biosynthetic enzymes of *Rb. sphaeroides* are available (Arondel et al., 1993; Dryden and Dowhan, 1996).

VII. Perspectives

Although anoxygenic photosynthetic bacteria include a large number of different microorganisms, only a selected number of species has been thoroughly investigated (see Table 1). With regard to the analysis of function and biosynthesis of membrane lipids, the purple nonsulfur bacterium *Rb. sphaeroides* is now used as a model organism. This bacterium was the first photosynthetic organism, of which genetic mutants were isolated lacking a complete class of membrane lipids (Benning and Somerville, 1992a,b; Arondel et al., 1993). These mutants provided the basis for the isolation of *sqd* and *pmt* genes coding for enzymes involved in the biosynthesis of sulfoquinovosyl diacylglycerol and phosphatidylcholine, respectively. The analysis of the *sqd* genes of *Rb. sphaeroides* is already well underway (e.g. Rossak et al., 1995), and it is expected that they provide the key to a deeper understanding of sulfolipid biosynthesis in *Rb. sphaeroides* and other photosynthetic organisms. In addition, the results of the physiological analysis of sulfolipid-deficient null mutants of *Rb. sphaeroides* has led to a new hypothesis with regard to the role of sulfolipid in photosynthetic membranes (Benning et al., 1993). Furthermore, employing the *sqd* genes of *Rb. sphaeroides,* it has been possible to isolate the corresponding genes of the cyanobacterium *Synechococcus* sp. PCC7942 (Güler et al., 1996) and of the higher plant *Arabidopsis thaliana* (S. Güler and C. Benning, unpublished). The sequence similarity between the homologous *sqd* genes of the purple bacterium, the cyanobacterium and the higher plant suggests that conclusions drawn based on experiments with *Rb. sphaeroides* may be applicable to most photosynthetic organisms, a prerequisite for a model organism.

Purple nonsulfur bacteria provide a range of opportunities to study the function and biosynthesis of membrane lipids with widespread occurrence in photosynthetic organisms. For example, the presence of monogalactosyl diacylglycerol in membranes of *Rp. viridis* (Övermöhle et al., 1995) or in *Rb. sphaeroides* (Benning et al., 1995) may allow us to characterize a bacterial monogalactosyl diacylglycerol synthase, an enzyme that is very difficult to study in higher plants (Joyard et al., 1993). Furthermore, initial experiments have been performed suggesting that *Rb. sphaeroides* can be used as a bacterial model organism to study the biosynthesis of the betaine lipid diacylglycerol-*N*-trimethylhomoserine (Hofmann and Eichenberger, 1995), an abundant membrane constituent of algae and lower plants.

Interesting research opportunities have arisen, that directly concern the biology of purple nonsulfur bacteria. Apparently, *Rb. sphaeroides* is remarkably flexible with regard to its membrane lipid composition in response to environmental factors. The mechanism by which the biosynthesis of novel lipids during phosphate-limited growth is induced is completely unknown. Furthermore, it would be interesting to know, how certain strains of *Rb. sphaeroides* discriminate against the incorporation of Tris into the xenobiotic lipid phosphatidyl-Tris.

In summary, anoxygenic bacteria can provide a wealth of information that is applicable to other photosynthetic organisms. Particularly, the purple nonsulfur bacterium *Rb. sphaeroides* is being developed to become a model system for the analysis of function and biosynthesis of membrane lipids present in photosynthetic organisms. The genes isolated from *Rb. sphaeroides* can provide probes for the isolation of lipid biosynthetic genes of higher plants. The continued application of genetic, biochemical and molecular approaches to study membrane biogenesis in *Rb. sphaeroides* will eventually allow us to understand the biosynthesis and function of thylakoid lipids in photosynthetic organisms.

Acknowledgments

I would like to thank N. Focks, P. Dörmann and B. Essigmann for their help during preparation of the manuscript. I gratefully acknowledge W. Eichenberger and E. Heinz for communicating unpublished results. Financial support was provided in part by the Genzentrum Berlin (BMBF 0316301 A, Project 2).

References

Acosta R and Lueking DR (1987) Stringency in the absence of ppGpp accumulation in *Rhodobacter sphaeroides*. J Bacteriol 169: 908–912

Arondel V, Benning C and Somerville CR (1993) Isolation and functional expression in *Escherichia coli* of a gene encoding phosphatidylethanolamine methyltransferase (EC 2.1.1.17) from *Rhodobacter sphaeroides*. J Biol Chem 268: 16002–16008

Asselineau J and Trüper HG (1982) Lipid composition of six species of the phototrophic bacterial genus *Ectothiorhodospira*. Biochim Biophys Acta 712: 111–116

Barber J and Gounaris K (1986) What role does sulpholipid play within the thylakoid membrane? Photosynth Res 9: 239–249

Bauer C, Buggy J and Mosley C (1993) Control of photosystem genes in *Rhodobacter capsulatus*. Trends Genetics 9: 56–60

Beck H, Hegeman GD and White D (1990) Fatty acids and lipopolysaccharide analyses of three *Heliobacterium* ssp. FEMS Microbiol Lett 69: 229–232

Beer-Romero P and Gest H (1987) *Heliobacillus mobilis*, a peritrichously flagellated anoxyphototroph containing bacteriochlorophyll *g*. FEMS Lett 41: 109–114

Beer-Romero P, Favinger JL and Gest H (1988) Distinctive properties of bacilliform photosynthetic heliobacteria. FEMS Lett 49: 451–454

Benning C and Somerville CR (1992a) Isolation and genetic complementation of a sulfolipid-deficient mutant of *Rhodobacter sphaeroides*. J Bacteriol 174: 2352–2360

Benning C and Somerville CR (1992b) Identification of an operon involved in sulfolipid biosynthesis in *Rhodobacter sphaeroides*. J Bacteriol 174: 6479–6487

Benning C, Beatty JT, Prince RC and Somerville CR (1993) The sulfolipid sulfoquinovosyldiacylglycerol is not required for photosynthetic electron transport in *Rhodobacter sphaeroides* but enhances growth under phosphate limitation. Proc Natl Acad Sci USA 90: 1561–1565

Benning C, Huang ZH and Gage DA (1995) Accumulation of a novel glycolipid and a betaine lipid in cells of *Rhodobacter sphaeroides* grown under phosphate limitation. Arch Biochem Biophys 317: 103–111

Benson AA (1963) The plant sulfolipid. Adv Lipid Res 1: 387–394

Benson AA, Daniel H and Wiser R (1959) A sulfolipid in plants. Proc Natl Acad Sci USA 45: 1582–1587

Birrell GB, Sistrom WR and Griffith OH (1978) Lipid-protein associations in chromatophores from the photosynthetic bacterium *Rhodopseudomonas sphaeroides*. Biochemistry 17: 3768–3773

Blankenship RE (1994) Protein structure, electron transfer and evolution of prokaryotic photosynthetic reaction centers. Ant van Leeuwenhoek 65: 311–329

Blankenship RE, Madigan MT and Bauer CE (eds) (1995) Anoxygenic Photosynthetic Bacteria. Kluwer Academic Publishers, Dordrecht

Brockman H Jr. and Lipinski A (1983) Bacteriochlorophyll *g*. A new bacteriochlorophyll from *Heliobacterium chlorum*. Arch Microbiol 136: 17–19

Brooks JL and Benson AA (1972) Studies on the structure of an ornithine-containing lipid from *Rhodospirillum rubrum*. Arch Biochem Biophys 152: 347–355

Brown AE and Elovson J (1974) Isolation and characterization of a novel lipid, 1(3), 2-diacylglyceryl-(3)-*O*-4′-(*N, N, N*-trimethyl)homoserine from *Ochromonas danica*. Biochemistry 13: 3476–3482

Cain BD, Deal CD, Fraley RT and Kaplan S (1981) In vivo intermembrane transfer of phospholipids in the photosynthetic bacterium *Rhodopseudomonas sphaeroides*. J Bacteriol 145: 1154–1166

Cain BD, Donohue TJ and Kaplan S (1982) Kinetic analysis of *N*-acylphosphatidylserine accumulation and implications for membrane assembly in *Rhodopseudomonas sphaeroides*. J Bacteriol 152: 607–615

Cain BD, Singer M, Donohue TJ and Kaplan S (1983) In vivo metabolic intermediates of phospholipid biosynthesis in *Rhodopseudomonas sphaeroides*. J Bacteriol 156: 375–385

Cain BD, Donohue TJ, Shepherd WD and Kaplan S (1984) Localization of phospholipid biosynthetic enzyme activities in cell-free fractions derived from *Rhodopseudomonas sphaeroides*. J Biol Chem 259: 942–948

Campbell TB and Lueking DR (1983) Light-mediated regulation of phospholipid synthesis in *Rhodopseudomonas sphaeroides*. J Bacteriol 155: 806–816

Cashel M and Rudd KE (1987) The stringent response. In: Neidhardt FC, Ingraham JL, Low KB, Magasanik B, Schaechter M and Umbarger HE (eds) *Escherichia coli* and *Salmonella typhimurium*: Cellular and Molecular Biology, Vol 2, pp 1410–1438. American Society for Microbiology, Washington, DC

Cedergreen RA and Hollingsworth RI (1994) Occurrence of sulfoquinovosyl diacylglycerol in some members of the family *Rhizobiaceae*. J Lipid Res 35: 1452–1461

Cooper CL and Lueking DR (1984) Localization and characterization of the *sn*-glycerol-3-phosphate acyltransferase in *Rhodopseudomonas sphaeroides*. J Lipid Res 25: 1222–1232

Cooper CL, Boyce SG and Lueking DR (1987) Purification and characterization of *Rhodobacter sphaeroides* acyl carrier protein. Biochemistry 26: 2740–2746

Cronan JE and Rock CO (1987) Biosynthesis of membrane lipids. In: Neidhardt FC, Ingraham JL, Low KB, Magasanik B, Schaechter M and Umbarger HE (eds) *Escherichia coli* and *Salmonella typhimurium*: Cellular and Molecular Biology, Vol 1, pp 474–497. American Society For Microbiology, Washington, DC

Cruden DJ and Stanier RY (1970) The characterization of *Chlorobium* vesicles and membranes isolated from green bacteria. Arch Microbiol 72: 115–134

Deisenhofer J and Michel H (1989) The photosynthetic reaction center from the purple bacterium *Rhodopseudomonas viridis*. EMBO J 8: 2149–2170

Deisenhofer J and Norris JR (eds.) (1993) The Photosynthetic Reaction Center. Academic Press, San Diego

Deisenhofer J, Epp O, Miki K, Huber R and Michel H (1985) Structure of the protein subunits in the photosynthetic reaction centers from *Rhodopseudomonas viridis* at 3Å resolution. Nature 318: 618–624

Ditandy T and Imhoff JF (1993) Preparation and characterization of highly pure fractions of outer membrane, cytoplasmic and intracytoplasmic membranes from *Ectothiorhodospira mobilis*. J Gen Microbiol 139: 111–117

Donohue TJ, Cain BD and Kaplan S (1982a) Purification and characterization of an *N*-acylphosphatidylserine from *Rhodopseudomonas sphaeroides*. Biochemistry 21: 2765–2773

Donohue TJ, Cain BD and Kaplan S (1982b) Alterations in the phospholipid composition of *Rhodopseudomonas sphaeroides* and other bacteria induced by Tris. J Bacteriol 152: 595–606

Dryden SC and Dowhan W (1996) Isolation and expression of the *Rhodobacter sphaeroides* gene (*pgsA*) encoding phosphatidylglycerophosphate synthase. J Bacteriol 178: 1030–1038

Eaglesham ARJ, Ellis JM, Evans WR, Fleishman DE, Hungria M and Hardy RWF (1990) The first photosynthetic N$_2$-fixing *Rhizobium*: Characteristics. In: Gresshoff PM, Roth LE, Stacey G and Newton WL (eds) Nitrogen Fixation: Achievements and Objectives, pp 805–811. Chapman and Hall Limited, London

Fleishman DE, Evans WR and Miller IM (1995) Bacteriochlorophyll-containing *Rhizobium* species. In: Blankenship RE, Madigan MT and Bauer CE (eds) Anoxygenic Photosynthetic Bacteria, pp 123–136. Kluwer Academic Publishers, Dordrecht

Fraley RT, Lueking DR and Kaplan S (1978) Intracytoplasmic membrane synthesis in synchronous cell populations of *Rhodopseudomonas sphaeroides*: Polypeptide insertion into growing membrane. J Biol Chem 253: 458–464

Fraley RT, Lueking DR and Kaplan S (1979a) The relationship of intracytoplasmic membrane assembly to the cell division cycle in *Rhodopseudomonas sphaeroides*. J Biol Chem 254: 1980–1986

Fraley RT, Yen GSL, Lueking DR and Kaplan S (1979b) The physical state of the intracytoplasmic membrane of *Rhodopseudomonas sphaeroides* and its relationship to the cell division cycle. J Biol Chem 254: 1987–1991

Gabriel O (1987) Biosynthesis of sugar residues for glycogen, peptidoglycan, lipopolysaccharide, and related systems. In: Neidhardt FC, Ingraham JL, Low KB, Magasanik B, Schaechter M and Umbarger HE (eds) *Escherichia coli* and *Salmonella typhimurium*: Cellular and Molecular Biology, Vol 1, pp 504–511. American Society for Microbiology, Washington, DC

Gage DA, Huang ZH and Benning C (1992) Comparison of sulfoquinovosyl diacylglycerol from spinach and the purple bacterium *Rhodobacter sphaeroides* by fast atom bombardment tandem mass spectrometry. Lipids 27: 632–636

Gest H and Favinger JL (1983) *Heliobacterium chlorum*, an anoxygenic brownish green photosynthetic bacterium containing a 'new' form of bacteriochlorophyll. Arch Microbiol 136: 11–16

Golbeck JH (1993) Shared thematic elements in photochemical reaction centers. Proc Natl Acad Sci USA 90: 1642–1646

Gorchein A (1968) Studies on the structure of the ornithine-containing lipid from non-sulphur bacteria. Biochim Biophys Acta 152: 358–367

Gorchein A (1973) Structure of the ornithine-containing lipid from *Rhodopseudomonas sphaeroides*. Biochim Biophys Acta 306: 137–141

Güler S, Seeliger S, Härtel H, Renger G and Benning C (1996) A null mutant of *Synechococcus* sp. PCC7942 deficient in the sulfolipid sulfoquinovosyl diacylglycerol. J Biol Chem 271: 7501–7507

Hofmann M and Eichenberger W (1995) Biosynthesis of homoserine lipid (DGTS) in *Rb. sphaeroides*. Biol Chem Hoppe Seyler, 376, S52

Hoger JH, Tai SP and Kaplan S (1987) Membrane adenosine

triphosphatase in synchronous cultures of *Rhodobacter sphaeroides*. Biochim Biophys Acta 898: 70–80

Holo H, Broch-Due M and Ormerod JG (1985) Glycolipids and the structure of chlorosomes in green bacteria. Arch Microbiol 143: 94–99

Imhoff JF (1984) Sulfolipids in phototrophic purple nonsulfur bacteria. In: Siegenthaler PA and Eichenberger W (eds) Structure, Function and Metabolism of Plant Lipids, pp 175–178. Elsevier Science Publishers, Amsterdam

Imhoff JF (1988) Lipids, fatty acids and quinones in taxonomy and phylogeny of anoxygenic phototrophic bacteria. In: Olson JM, Ormerod JG, Amesz J, Stackebrandt E and Trüper HG (eds) Green Photosynthetic Bacteria, pp 223–232. Plenum Press, New York

Imhoff JF (1991) Polar lipids and fatty acids in the genus *Rhodobacter*. System Appl Microbiol 14: 228–234

Imhoff JF and Bias-Imhoff U (1995) Lipids, quinones and fatty acids of anoxygenic phototrophic bacteria. In: Blankenship RE, Madigan MT and Bauer CE (eds) Anoxygenic Photosynthetic Bacteria, pp 179–205. Kluwer Academic Publishers, Dordrecht

Imhoff JF, Kushner DJ, Kushwaha SC and Kates M (1982) Polar lipids in phototrophic bacteria of the *Rhodospirillaceae* and *Chromatiaceae* families. J Bacteriol 150: 1192–1201

Jensen MT, Knudsen J and Olson JM (1991) A novel aminoglycosphingolipid found in *Chlorobium limicola* f. *thiosulfatophilum* 6230. Arch Microbiol 156: 248–254

Joyard J, Block MA, Malherbe A, Maréchal E and Douce R (1993) Origin and synthesis of galactolipid and sulfolipid head groups. In: Moore TS Jr (ed) Lipid Metabolism in Plants, pp 231–258. CRC Press, Boca Raton

Kates M (1986) Influence of salt concentration on membrane lipids of halophilic bacteria. FEMS Microbiol Rev 39: 95–101

Kenyon CN (1978) Complex lipids and fatty acids of photosynthetic bacteria. In: Clayton RK and Sistrom WR (eds) The Photosynthetic Bacteria, pp 281–313. Plenum Publishing Corp, New York

Kenyon CN and Gray AM (1974) Preliminary analysis of lipids and fatty acids of green bacteria and *Chloroflexus aurantiacus*. J Bacteriol 120: 131–138

Klemme B, Klemme JH and San Pietro A (1971) PPase, ATPase and photophosphorylation in chromatophores of *Rhodospirillum rubrum*: Inactivation by phospholipase A: Reconstitution by phospholipids. Arch Biochem Biophys 144: 339–342

Knacker T, Harwood JL, Hunter CN and Russell NJ (1985) Lipid biosynthesis in synchronized cultures of the photosynthetic bacterium *Rhodopseudomonas sphaeroides*. Biochem J 229: 701–710

Knudsen E, Jantzen E, Bryn K, Ormerod JG and Sirevåg R (1982) Quantitative and structural characteristics of lipids in *Chlorobium* and *Chloroflexus*. Arch Microbiol 132: 149–154

Ladha JK, Pareek RP, So RB and Becker M (1990) Stem nodule symbiosis and its unusual properties. In: Gresshoff PM, Roth LE, Stacey G and Newton WL (eds) Nitrogen Fixation: Achievements and Objectives, pp 633–640. Chapman and Hall, New York

Lueking DR and Goldfine H (1975) *Sn*-glycerol-3-phosphate acyltransferase activity in particulate preparations from anaerobic, light-grown cells of *Rhodopseudomonas sphaer-*

oides: Involvement of acyl thiolester derivatives of acyl carrier protein in the synthesis of complex lipids. J Biol Chem 250: 8530–8535

Lueking DR, Fraley RT and Kaplan S (1978) Intracytoplasmic membrane synthesis in synchronous cell populations of *Rhodopseudomonas sphaeroides*: Fate of 'old' and 'new' membrane. J Biol Chem 253: 451–457

Minnikin DE and Abdolrahimzadeh H (1974) The replacement of phosphatidylethanolamine and acidic phospholipids by an ornithine-amide lipid and a minor phosphorus-free lipid in *Pseudomonas fluorescens* NCMB129. FEBS Lett 43: 257–260

Minnikin DE, Abdolrahimzadeh H and Baddiley J (1974) Replacement of acidic phospholipids by acidic glycolipids in *Pseudomonas diminuta*. Nature 249: 268–269

Olson JM, Shaw EK, Gaffeney JS and Scandella CJ (1983) A fluorescent aminolipid from a green photosynthetic bacterium. Biochemistry 22: 1819–1827

Olson JM, Shaw EK, Gaffeney JS and Scandella CJ (1984) *Chlorobium* aminolipid. A new membrane lipid from green sulfur bacteria. In: Sybesma C (ed) Advances in Photosynthesis Research, Vol 3, pp 139–142. Nijhoff/Junk, The Hague

Onishi JC and Niederman RA (1982) *Rhodopseudomonas sphaeroides* membranes: Alterations in phospholipid composition in aerobically and phototrophically grown cells. J Bacteriol 149: 831–839

Övermöhle M, Riedl I, Linscheid M and Heinz E (1995) Lipids of *Rhodopseudomonas viridis*: Experiments on structure and biosynthesis. Biol Chem Hoppe Seyler, 376, S55

Prince RC (1990) Bacterial photosynthesis: From photons to Δ*p*. In: Krulwich TA (ed) The Bacteria, Vol 12, pp 111–149. Academic Press, New York

Radcliffe CW, Broglie RM and Niederman RA (1985) Sites of phospholipid biosynthesis during induction of intracytoplasmic membrane formation in *Rhodopseudomonas sphaeroides*. Arch Microbiol 142: 136–140

Radcliffe CW, Steiner FX, Carman GM and Niederman RA (1989) Characterization and localization of phosphatidylglycerolphosphate and phosphatidylserine synthases in *Rhodobacter sphaeroides*. Arch Microbiol 152: 132–137

Radunz A (1969) Über das Sulfoquinovosyl-Diacylglycerin aus höheren Pflanzen, Algen und Purpurbakterien. Hoppe-Seyler's Z Physiol Chem 350: 411–417

Ratledge C and Wilkinson SG (1988) Microbial Lipids, Vol 1, Academic Press, San Diego, CA

Raymond JC and Sistrom WR (1969) *Ectothiorhodospira halophila*: A new species of the genus *Ectothiorhodospira*. Arch Microbiol 69: 121–126

Rossak M, Tietje C, Heinz E and Benning C (1995) Accumulation of UDP-sulfoquinovose in a sulfolipid-deficient mutant of *Rhodobacter sphaeroides*. J Biol Chem 270: 25792–25797

Russell NJ and Harwood JL (1979) Changes in the acyl lipid composition of photosynthetic bacteria grown under photosynthetic and non-photosynthetic conditions. Biochem J 181: 339–345

Sato N (1992) Betaine lipids. Bot Mag Tokyo 105: 185–197

Sato N and Murata N (1982) Lipid biosynthesis in blue-green algae, *Anabena variabilis*. I Lipid classes. Biochim Biophys Acta 710: 271–278

Schmid PC, Kumar VV, Weis BK and Schmid HHO (1991)

Phosphatidyl-Tris rather than *N*-acylphosphatidylserine is synthesized by *Rhodopseudomonas sphaeroides* grown in Tris-containing media. Biochemistry 30: 1746–1751

Shiba T and Harashima K (1986) Aerobic photosynthetic bacteria. Microbiol Sci 3: 376–378

Shimada K (1995) Aerobic anoxygenic phototrophs. In: Blankenship RE, Madigan MT and Bauer CE (eds) Anoxygenic Photosynthetic Bacteria, pp 105–122. Kluwer Academic Publishers, Dordrecht

Stackebrandt E, Murray RGE and Trüper HG (1988) *Proteobacteria* classis nov., a name for the phylogenetic taxon that includes the 'purple bacteria and their relatives.' Int J Syst Bacteriol 38: 321–325

Staehelin LA, Golecki JR and Drews G (1980) Supramolecular organization of chlorosomes (chlorobium vesicles) and of their membrane attachment sites in *Chlorobium limicola*. Biochim Biophys Acta 589: 30–45

Steiner S, Conti SF and Lester RL (1969) Separation and identification of the polar lipids of *Chromatium* strain D. J Bacteriol 98: 10–15

Steiner S, Sojka GA, Conti SF, Gest H and Lester RL (1970) Modification of membrane composition in growing photosynthetic bacteria. Biochim Biophys Acta 203: 571–574

Tai SP and Kaplan S (1984) Purification and properties of a phospholipid transfer protein from *Rhodopseudomonas sphaeroides*. J Biol Chem 259: 12178–12183

Tai SP and Kaplan S (1985) Intracellular localization of phospholipid transfer activity in *Rhodopseudomonas sphaeroides* and a possible role in membrane biogenesis. J Bacteriol 164: 181–186

Takacs BJ and Holt SC (1971) *Thiocapsa floridana*: A cytological, physical and chemical characterization. II. Physical and chemical characteristics of isolated and reconstituted chromatophores. Biochim Biophys Acta 233: 278–295

Thiemann B and Imhoff JF (1991) The effect of salt on the lipid composition of *Ectothiorhodospira*. Arch Microbiol 156: 376–384

Tindall BJ and Grant WD (1986) The anoxygenic phototrophic bacteria. Soc Appl Bacteriol Symp Ser 13: 115–155

Webb MS and Green BR (1991) Biochemical and biophysical properties of thylakoid acyl lipids. Biochim Biophys Acta 1060: 133–158

Woese CR (1987) Bacterial evolution. Microbiol Rev 51:221–271

Woese CR, Kandler O and Wheelis ML (1990) Towards a natural system of organisms: Proposal for the domains Archaea, Bacteria, and Eucarya. Proc Natl Acad Sci USA 87: 4576–4579

Yen GS, Cain BD and Kaplan S (1984) Cell-cycle-specific biosynthesis of the photosynthetic membrane of *Rhodopseudomonas sphaeroides*: Structural implications. Biochim Biophys Acta 777: 41–55

Chapter 6

The Physical Properties of Thylakoid Membrane Lipids and Their Relation to Photosynthesis

W. Patrick Williams
Division of Life Sciences, King's College London, Campden Hill, London W8 7AH, U.K.

Summary

The three main areas in which research into the physical properties of thylakoid membrane lipids has impacted on photosynthetic research are: lipid phase behavior and its relevance to membrane phase separations; lipid-protein interactions in the context of the structure and function of the main photosynthetic membrane complexes; and the fluidity properties of membranes in terms of thermal adaptation and the factors limiting rates of electron transport.

Many of our current ideas in these areas are based on a simple extrapolation of results obtained in studies carried out on purified thylakoid membrane lipids, or mixtures of such lipids. It has, however, become increasingly clear in recent years that such extrapolations are far from straightforward. This chapter is devoted to providing a review of the literature underlying these ideas in the context of what is currently known about the physical properties of thylakoid membrane lipids and their interaction with other membrane components.

P.-A. Siegenthaler and N. Murata (eds): Lipids in Photosynthesis: Structure, Function and Genetics, pp. 103–118.
© 1998 Kluwer Academic Publishers. Printed in The Netherlands.

I. Phase Behavior of Thylakoid Membrane Lipids

The main polar lipids found in thylakoid membranes are MGDG, DGDG, SQDG and PG. It is now thought that PC, which is often reported to be a thylakoid membrane component, is largely, if not entirely, restricted to the chloroplast envelope and that its presence in thylakoid membrane preparations reflects their contamination with chloroplast envelope membranes (Dorne et al., 1990). For convenience, discussion of lipid phase properties is divided into those of the purified lipids, binary lipid mixtures, total polar lipid extracts and native membranes.

A. Purified Lipids

MGDG is a non-bilayer lipid forming an Hex_{II} phase when dispersed in water at temperatures above its gel to liquid crystal phase transition temperature. The Hex_{II} phase, as illustrated in Fig. 1b, consists of cylindrical inverted micelles packed on an hexagonal lattice. Freeze-fracture and negative stain electron-micrographs showing these micelles are presented in Figs. 2a and 2b. In the gel state, MGDG, in common with all membrane lipids, forms a lamellar (L_β) phase. Most molecular species of MGDG transform directly from the L_β to the Hex_{II} phase on heating without going through a stable liquid crystalline lamellar (L_α) phase. A freeze-fracture electron-micrograph showing a sample of MGDG undergoing a transition between the bilayer and non-bilayer states is presented in Fig. 2c.

MGDG isolated from higher plant chloroplasts tends to be highly unsaturated. As such, it is in a liquid-crystal Hex_{II} phase at room temperature and converts to a gel phase only at temperatures well below 0 °C (Shipley et al., 1973). In order to avoid problems associated with low-temperature measurements, most of the studies on the phase behavior of MGDG have been carried out on more-saturated

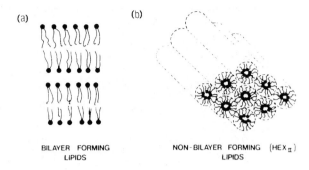

Fig. 1. Diagrammatic representation of the molecular organization of (a) the liquid-crystal lamellar phase of bilayer-forming lipids (b) the Hex_{II} phase formed by non-bilayer lipids.

molecular species that have higher transition temperatures.

Extensive measurements have been made on synthetic and semi-synthetic di-18:0 and di-16:0 MGDG (Sen et al., 1981a, 1983; Mannock et al., 1985a; Lis and Quinn, 1986; Mansourian and Quinn, 1987; Mannock and McElhaney, 1991). These studies reveal a complex pattern of transitions involving the formation on storage at low temperatures of one, or more, highly condensed lamellar crystal (L_c) phases. Their formation reflects the ability of MGDG to form regular hydrogen bond networks between the sugar headgroups of the lipid. Crystal phases of this type, it must be emphasized, are only found in the pure lipid and there is no evidence to suggest that such forms can exist in native membrane systems.

It has recently become clear, that the difficulties associated with the study of the phase behavior of highly unsaturated MGDG species, arise mainly from their high degree of unsaturation rather than the inherently low temperatures of their transitions. Calorimetric measurements indicate that molecular species of membrane lipids containing two polyenoic acyl chains, commonly found in chloroplast membranes, are characterized by extremely broad, low entropy gel to liquid-crystal transitions. In the case of di-18:3 MGDG, the molar entropy change is about 1.0 kcal/mol and the transition, which is centered at about –35 °C, spans nearly forty degrees (Sanderson and Williams, 1992). This contrasts with an entropy of about 8.0 kcal/mol and a transition at about 70 °C spanning about four degrees for di-18:0 MGDG. Apart from these differences, the pattern of phase changes occurring in di-18:3 MGDG, illustrated in Fig. 3, is very similar to that previously observed for saturated molecular species.

Abbreviations: DGDG – digalactosyldiacylglycerol; DMSO – dimethylsulfoxide; DPH – 1,6-diphenyl-2,3,5-hexatriene; DOPE – dioleoyl phosphatidylethanolamine; DSC – differential scanning calorimetry; MGDG – monogalactosyldiacylglycerol; Hex_{II} – inverted hexagonal phase; L_α – lamellar liquid crystal phase; L_β – lamellar gel phase; L_c – lamellar crystal phase; LHC-II – chlorophyll *a/b* light-harvesting protein; PC – phosphatidylcholine; PE – phosphatidylethanolamine; PG – phosphatidylglycerol; POPE – palmitoyloleoyl phosphatidylethanolamine; SQDG – sulfoquinovosyldiacylglycerol

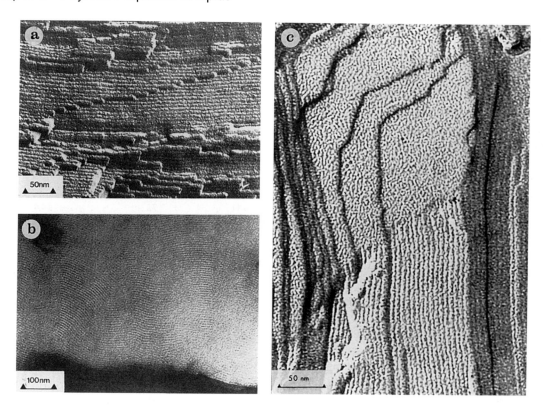

Fig. 2. Freeze-fracture (a) and negative stain (b) electronmicrographs of the Hex_{II} phase of MGDG showing the cylindrical inverted micelles and their central aqueous channel, (c) freeze-fracture electronmicrograph of an MGDG sample undergoing a transition between the gel and Hex_{II} phases.

Broad phase transitions of the type seen for di-18:3 MGDG, are a common feature of highly unsaturated lipids. Similar transitions are seen for dipolyenoic derivatives of PC (Keough and Kariel, 1987; Williams et al., 1993), PE (Williams et al., 1993) and DGDG (W. P. Williams, unpublished). Wide-angle X-ray diffraction measurements (Sanderson and Williams, 1993; Williams et al., 1993) indicate that the gel states of such lipids are much more disordered than those of more-saturated molecular species. This lack of order appears to be associated with difficulties in packing two highly unsaturated acyl chains into the regular hexagonal lattice typical of conventional L_β states. The packing constraints in molecular species containing a single polyenoic chain, while still significant, are much smaller. This is reflected in the intermediate breadth of the $L_\beta \rightarrow H_{II}$ transition phase of 1–16:0, 2–18:3 MGDG which is centered at about $-10\ °C$ and spans fifteen to twenty degrees (N. Tsvetkova, unpublished).

The phase behaviors of DGDG, PG and SQDG, have received much less attention than that of MGDG. Measurements performed on fully-saturated DGDG

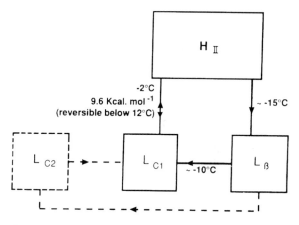

Fig. 3. Diagram illustrating the relationship between the different phases seen in di-18:3 MGDG. Solid lines indicate transitions observed in DSC measurements. Dashed lines indicate a possible second crystalline seen in fully saturated molecular species. Based on the work of Sanderson and Williams (1993).

(Sen et al., 1981a) indicate that it is a conventional bilayer-forming lipid. Apart from the occurrence of very broad, low enthalpy phase transitions of the type described above for MGDG, measurements performed on polyenoic molecular species show no unexpected features.

The phase properties of saturated and monoenoic fatty acid derivatives of PG have been extensively studied over the years (Tilcock, 1986). Most of this work, however, has been carried out on lipids prepared synthetically or isolated from animal tissues. Little work has been carried out on PG isolated from plant sources. In particular, no detailed studies of the phase behavior of the molecular species containing *trans*-Δ^3-hexadecenoic acid (16:1*tr*) residues, which is unique to photosynthetic membranes, have appeared. Very little information is available regarding the properties of SQDG. Shipley et al. (1973) and Mannock et al. (1985b), using X-ray diffraction, have shown that SQDG is again a bilayer-forming lipid.

A summary of the currently available data relating to the transition temperatures (T_c) and transition enthalpies (ΔH) of purified thylakoid membrane lipids of different degrees of saturation are presented in Tables 1 and 2. It is important, however, to remember that being negatively charged at neutral pH, the phase behavior of SQDG and PG are strongly influenced by the pH and the ionic composition of the suspending medium.

B. Binary Mixtures

The non-bilayer forming lipid MGDG and the bilayer-forming lipid DGDG typically account for about 70% of the polar membrane lipids of photosynthetic membranes. A number of studies of the properties of mixtures of these two lipids have been carried out. A detailed NMR study of the phase properties of MGDG/DGDG mixtures at low levels of hydration has been published by Brentel et al. (1985). The structural organization of fully-hydrated mixtures has been extensively studied by freeze-fracture electron microscopy (Sen et al., 1981b, 1982a,b, Sprague and Stahelin, 1984). These studies indicate that at low MGDG:DGDG ratios (below about 20 mol% MGDG) the non-bilayer lipid remains constrained within a bilayer conformation. As the proportion of MGDG in the mixture increases, spherical inverted micelles are first formed. At higher MGDG concentrations, these tend to fuse into cylindrical micelles occasionally isolated within a bilayer but more commonly in the form of separate aggregates. Further increases in the proportion of MGDG lead to the formation of cubic phases (see Fig. 4).

Table 1. Thermal data for lipid phase transitions for individual molecular species of the lipids found in photosynthetic membranes

Transition	Transition Temperature (°C)	Molar Enthalpy (kcal/mol)
MGDG		
di-18:0[a]		
$L_c \rightarrow H_{II}$	85	25.7
$L_\beta \rightarrow L_\alpha$	73.5	11.1
$L_\alpha \rightarrow H_{II}$	76	1.0
di-16:0[b]		
$L_c \rightarrow H_{II}$	72	14.4
$L_\beta \rightarrow H_{II}$	52	5.9
di-18:3[c]		
$H_{II} \rightarrow L_\beta$	$-23 - -50$	-1.0
$L_c \rightarrow H_{II}$	-2	9.6
DGDG		
di-18:0[d]		
$L_\beta \rightarrow L_\alpha$	51	–
PG		
di-16:0[e]		
$L_\beta \rightarrow L_\alpha$	40	–

Data from [a] Mannock and McElhaney (1991) using synthetic lipid [similar values based on semi-synthetic preparations have been reported by Sen et al. (1983), Mansourian and Quinn (1986); Lis and Quinn (1986)]; [b] Mannock et al.(1985a); [c] Sanderson and Williams (1992); [d] Sen et al. (1981); [e] Murata and Yamaya (1984).

Table 2. Thermal data for lipid phase transitions for lipid classes (mixed molecular species) isolated from photosynthetic membranes

Lipid Source/Class	Transition Temperature Range (°C)	Molar Enthalpy (kcal/mol)
Anacystis nidulans[a]		
MGDG		
$L_c \rightarrow H_{II}$	$38 -$	12.5
$L_\beta \rightarrow H_{II}$	$6 -$	2.8
DGDG		
$L_\beta \rightarrow L_\alpha$	$-16 - 11$	5.6
SQDG		
$L_\beta \rightarrow L_\alpha$	$-24 - 15$	4.4
PG		
$L_\beta \rightarrow L_\alpha$	$-3 - 13$	5.4
Vicia Faba[b]		
SQDG	$-32 - -19$	1.0

Data from [a] Mannock et al. (1985b) and [b] Sanderson (1992).

In the presence of excess water, the formation of non-bilayer structures is favored by increased temperature and the presence of soluble co-solutes such as glycerol, DMSO and ethylene glycol (Sen et al., 1982b). The driving force for non-bilayer structure formation in the presence of such co-solutes is an entropic effect associated with a tendency of the lipid to minimize its surface hydration area. The non-bilayer structures which are characterized by smaller surface areas per molecule than the corresponding lamellar phase are thus preferentially stabilized. A detailed combined X-ray diffraction/DSC study of these effects has been carried out for the non-bilayer forming lipid POPE (Sanderson et al., 1991). Similar effects are observed with isolated chloroplast membranes (Williams et al., 1992) and chloroplast lipid extracts (YashRoy, 1994) but no detailed study of the effects of these co-solutes on pure MGDG has yet appeared.

C. Total Polar Lipid Extracts

A large number of studies have been reported, particularly in the early literature, on the phase properties of total polar lipid extracts. These were mainly aimed at determining the temperature of any gel to liquid-crystal phase transition temperatures that might occur in such mixtures. The presence on the one hand of high concentrations of the non-bilayer lipid MGDG and on the other hand of the negatively charged lipids PG and SQDG, means that the structural organization and phase properties of

such extracts are highly dependent on the ionic strength and pH of the suspending medium. The importance of these factors has been ignored in most studies performed on total membrane lipid extracts and this must be taken into account when attempting to interpret the significance of the results reported in such studies.

Total polar extracts of thylakoid membranes disperse to form small unilamellar liposomes in distilled water but the presence of relatively low concentrations of ions (which shield the charges on the negatively charged lipids) or added protons (which alter their degree of ionization) tend to lead to extensive non-bilayer lipid separation and/or aggregation. In the presence of concentrations of ions typical of those found in vivo and/or standard chloroplast isolation media, the structural organization of total polar lipid extracts of thylakoid membranes is, predominantly non-bilayer (Gounaris et al., 1983a). Electronmicrographs illustrating typical structures found in such dispersions are presented in Figs. 4b and 4c. Comparison of these with the bilayer organization of native chloroplast membranes provides eloquent testimony to the dangers of extrapolating results based on studies of phase behavior of lipid extracts to the native membrane.

The other major problem with many of the phase-transition studies performed on total polar lipid extracts, is the choice of methods used to study the properties of the dispersions. The majority of such studies are based on the use of spin and/or fluorescence probes. While such measurements are

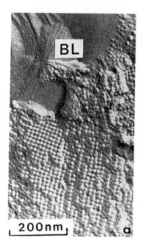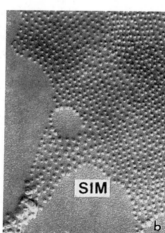

Fig. 4. Freeze-fracture electronmicrographs of aqueous dispersions of (a) a binary mixture of MGDG/DGDG (2:1 wt/wt) showing a mixture of bilayer (BL) and cubic phase lipid; (b and c) total polar lipid extracts of broad bean chloroplasts showing spherical inverted micelles (SIM) and cylindrical inverted micelles (CIM).

relatively easy to interpret in model systems containing a single, or at most two, well-defined lipid molecular species, the potential for misinterpretation in systems total polar lipid extracts of chloroplast membranes that may contain twenty to thirty different molecular species is very high (see reviews of Quinn and Williams 1985, 1990, and Williams and Quinn, 1987 for detailed discussions).

D. Native Membrane Systems

The scope for measurements of lipid phase properties in native membrane systems is very limited. Freeze-fracture electron microscopy has proved useful for picking up the formation of gel-phase lipid in the plasma and photosynthetic membranes of thermophilic cyanobacteria (Armond and Staehelin, 1979; Brand et al., 1979; Furtado et al., 1979; Verwer et al., 1979, Ono and Murata, 1982). The presence, at low temperatures, of smooth particle-free areas in such membranes clearly signals the presence of patches of phase-separated gel-phase lipid from which the intrinsic membrane proteins have been excluded (see Fig. 5). DSC (Furtado et al., 1979; Ono et al., 1983; Mannock et al., 1985b) and X-ray diffraction measurements (Tsukomoto et al., 1980) confirm the

formation of gel-phase lipid in such membranes.

Phase separations of this type are also seen in the membranes of non-photosynthetic thermophilic organisms and are reflections of cold-shock phenomena and are not, as is often suggested, directly related to chilling-sensitivity in higher plants. The early suggestion by Lyons and Raison that similar phase transitions occur in the bulk lipids of the mitochondria (Lyons, 1973) and chloroplasts (Raison, 1973) of chilling-sensitive plants has now been abandoned (see Quinn and Williams, 1978, 1983 and 1985 for discussions). It also appears that the related suggestion by Murata and Yamaya (1984) that chilling sensitivity in higher plants is associated with the phase-separation of highly saturated molecular species of PG is likely to be incorrect. The involvement of PG is not in dispute but recent work of Murata and his group (Gombos et al., 1994; refer also to Chapter 13, Gombos and Murata) now suggests that chilling-sensitivity is closely linked to the ability of plants to repair low-temperature photo-inhibitory damage rather than to membrane damage due to exposure to low temperature per se.

While low-temperature gel phase-separations in higher plant chloroplasts have never been convincingly demonstrated, freeze-fracture electron micros-

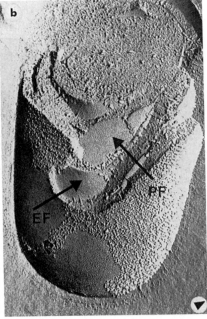

Fig. 5. Freeze-fracture electronmicrographs of *Anacystis nidulans* grown at 38 °C and thermally quenched from (a) 35 °C and (b) 15 °C. Formation of gel phase lipid in the sample quenched from 15 °C is indicated by the presence of particle-free patches in the exoplasmic (EF) and protoplasmic (PF) faces of the thylakoid membranes and the cytoplasmic membrane.

copy studies clearly indicate the occurrence of non-bilayer lipid phase-separations in such membranes in response to exposure to higher temperatures (Gounaris et al., 1983b, 1984) or the addition of high concentrations of soluble co-solutes (Williams et al., 1992). Typical examples of such phase-separations are shown in Fig. 6. The driving force for the phase-separations seen in the presence of co-solutes, as discussed in Section I.B, appears to be largely entropic.

The fact that the threshold temperature for non-bilayer phase separations in heat-treated chloroplasts appeared to coincide with that for the loss of PS II electron transport activity led to suggestions that non-bilayer phase-separations might be linked to the destabilization of the PS II light-harvesting apparatus and hence to the thermal stability of the photosynthetic apparatus (Gounaris et al., 1984a,b). Williams and Gounaris (1992), however, have since demonstrated that the presence of high concentrations of soluble co-solutes of the type that favor non-bilayer lipid phase separation, stabilize PS II oxygen-evolving cores against heat damage suggesting that the two phenomena are not connected.

II. Lipid-Protein Interactions

One of the most striking feature of the above studies is the large difference that exists between the phase-behavior of mixtures of chloroplast lipids and native photosynthetic membranes. The presence of proteins clearly plays a major part in imposing a bilayer configuration in the native membranes and demonstrates the importance of lipid-protein interactions in such membranes. The detailed nature of these interactions is, however, far from clear. For discussion purposes they are conveniently divided into two main categories; non-specific and specific interactions.

A. Non-Specific Interactions

In general, membranes with high protein contents such as chloroplast, mitochondrial and retinal rod membranes tend to have higher proportions of non-bilayer forming lipids. The ability of membrane proteins to suppress Hex_{II} formation has been demonstrated for a number of membrane proteins including the red blood cell protein glycophorin reconstituted in DOPE (Taraschi et al., 1982) and cytochrome c oxidase reconstituted in cardiolipin

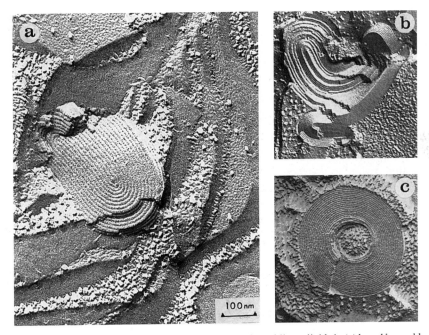

Fig. 6. Freeze-fracture electronmicrographs showing the phase separation of non-bilayer lipids in (a) broad bean chloroplasts incubated for 5 min at 40 °C (b) and (c) pea chloroplast membranes resuspended in 2.4 M sucrose. Samples were thermally quenched from room temperature. Data from Gounaris et al. (1983) and Williams et al. (1992).

(Rietveld et al., 1987). It is generally assumed, although it has never been directly demonstrated, that the photosynthetic light-harvesting proteins have a similar effect on MGDG.

Interactions of this type are envisaged as being non-specific in the sense that they reflect the existence of a preference for the lipids to accumulate on a time-average basis at the lipid-protein interface as a result of the hydrophillic/hydrophobic requirements of the interface rather than any specific interaction between the protein and the lipid. The high concentration of MGDG present in thylakoid membranes has often been suggested to be associated with the efficient sealing of the large protein complexes of the photosynthetic light-harvesting apparatus into the thylakoid membrane (Quinn and Williams, 1985; Murphy, 1986). Again, while an attractive idea, it still lacks direct experimental support.

B. Specific Interactions

The photosynthetic literature contains many articles advocating the idea of interactions between specific lipid classes and different membrane proteins or protein assemblies that are important to their proper function. Much of this work has been reviewed elsewhere (Webb and Green, 1991; Williams, 1994) and only the more recent studies will be referred to here. Broadly speaking, these studies fall into five categories; those based on detergent extraction experiments, reconstitution experiments and antibody studies, those involving direct physical measurements of lipid-protein interactions using spin-probe and saturation ESR techniques, and those using mutants.

1. Detergent Extraction Studies

The general approach adopted in studies of this type has been to use detergents to extract different membrane protein complexes from photosynthetic membranes, to analyze these extracts for residual lipid and to attribute any enrichment in a specific class of lipid to specific interactions that prevent its removal. The reader is referred to the detailed analysis of the early literature in this field presented by Webb and Green (1991). These authors point to the inherent inconsistency of many of the results obtained in such studies and conclude that much of this can be directly attributed to methodological factors: a view this reviewer strongly endorses.

Of the more recent studies, those carried out by Murata et al. (1990), Huner et al. (1992) and Trémolières et al. (1994) are probably the most interesting. These involve careful analyses of the lipid content of different types of PS II preparation. In all cases significant differences in lipid class composition and fatty acid saturation were observed with different fractions. However, as these authors acknowledge, it is difficult to know whether studies of this type reflect genuine lipid requirements or simply differences in the relative extractability of lipids.

2. Lipid Reconstitution Studies

A great deal of effort has been expended in attempts to determine specific lipid requirements for membrane proteins by means of lipid reconstitution studies. Again the reader is referred to the review of Webb and Green (1991) for a detailed analysis of the early literature. As Webb and Green correctly point out, little care was taken in many of these studies to distinguish between preparations in which the proteins are integrated into lamellar, Hex$_{II}$ phases or mixed detergent/lipid micelles. An even more basic criticism of such studies, is that even if due care is taken to avoid such problems, it is still impossible to judge whether a given lipid requirement reflects a specific in vivo lipid requirement or non-specific effects characteristic of the conditions existing in the isolated complex. One important piece of information that does seem to emerge from reconstitution studies is a requirement for PG for the trimerization of LHC-II. There is a long history of reports of the association of PG with oligomeric forms of LHC-II (Trémolières and Siegenthaler, 1997). This view is strengthened by the report of Nussberger et al. (1993) that release of PG from LHC-II by phospholipase action or the proteolytic cleavage of 49 amino acid residues from the N terminus prevents LHC-II trimerization. The same authors have reported a specific requirement for DGDG for the formation of two- and three-dimensional crystals of the trimeric complex.

3. Antibody Studies

A rather different approach to searching for specific lipid-protein interactions in photosynthetic membranes has been adopted by Radunz and his co-workers. They have isolated different PS II polypeptides using SDS-polyacrylamide gel electrophoresis and then made Western blots using antisera against the main photosynthetic membrane lipids to test for lipid binding.

Working with tobacco, they report the existence of MGDG interactions with most of the polypeptides of the PS II light-harvesting apparatus, including D_1 and D_2 polypeptides of the reaction center and the three polypeptides corresponding to the extrinsic proteins comprising the oxygen evolution center (Voss et al., 1992). They found no interaction, however, with a 66 kDa polypeptide which they identify as the heterodimer of the D_1 and D_2 polypeptides, which interacted instead with their DGDG and SQDG antisera. They have also reported interactions between antisera to all the main thylakoid lipids and the α- and β-subunits of CF_1 of spinach (Haase et al., 1993) and of PG with the D_1 polypeptide of the filamentous cyanobacteria *Oscillatoria chalybea* (Kruse et al., 1994).

In the case of the experiments on tobacco, it is difficult to imagine that MGDG can be specifically associated with so many PS II polypeptides, especially those of extrinsic membrane proteins. This result is more consistent with the view that the lipids were carried over as part of the detergent shell of the polypeptides. The apparent absence of MGDG, and the presence of other lipids, in the 66 kDa is very puzzling but, as mentioned briefly above, there are many reports of the association of specific lipid classes with photosynthetic proteins based on detergent studies.

The key question, as Radunz and his collaborators appreciate, is to distinguish between interactions of this type, which probably have little or no relevance to the native membrane, and specific interactions likely to be of structural significance. In an attempt to resolve this problem, they subjected their sedimented preparations to organic solvent extraction and periodate oxidation to try to remove, or degrade, weakly bound lipids. They also tried treating their cellulose nitrate blotting membranes with lipase. Only the lipase treatment was effective.

It is very difficult when using extremely sensitive non-quantitative techniques to be sure what such results really mean. The residual lipids are clearly strongly associated with the SDS-solubilized polypeptides. However, as with the detergent extraction studies referred to above, this does not necessarily mean that such associations exist in the native medium. The suggestion by Voss et al. (1992) that such lipids should be considered to be prosthetic groups of the polypeptides is hard to sustain.

4. Spin-Probe and Saturation Transfer ESR Studies

All intrinsic membrane proteins interact with the lipids in their immediate vicinity that constitute their boundary layer. In the case of photosynthetic membranes, which are particularly rich in protein, a major part of the membrane lipid fraction is located in such layers (Murphy, 1986; Joliot et al., 1992). As the environment of a lipid molecule in a boundary layer is necessarily different from that of a similar molecule in the bulk lipid phase, not surprisingly such lipids have different physical properties.

The existence of identifiable boundary layer lipids was first demonstrated in model systems by Jost and her co-workers (1973). Using ESR techniques, they showed, that beef-heart mitochondrial cytochrome oxidase is surrounded by an annulus of lipid molecules that are motionally constrained with respect to the bulk lipid when viewed on an ESR timescale ($\approx 10^{-7}$–10^{-9}s). These findings give rise to three key questions: (a) to what extent is exchange between the boundary layer and the bulk lipid phase constrained, (b) is there any element of selectivity in the composition of the bilayer, (c) to what extent, if any, does the enrichment of the boundary layer by a particular lipid species influence the structural organization and/or function of the protein in question.

In the case of thylakoid membranes, Li et al. (1989a,b, 1990) have demonstrated that nitroxide-labeled MGDG, PC and PG, all show appreciable motional restriction on the ESR time-scale (10^{-7}s) in thylakoid membranes. As illustrated in Fig. 7, they identified two components in their spectra; one associated with the motionally restricted lipid and the other with the more-fluid bulk lipid. Measurements of this type suggest that PG tends to concentrate in the boundary layer of one or more of the major intrinsic membrane complexes of photosynthetic membranes (Li et al., 1989b, 1990; Ivancich et al., 1994). The question of whether such an accumulation plays any part in determining the overall structure or function of the membrane system, however, remains unanswered.

A great deal of work has been carried out on lipid-protein interactions in biological membranes and it is now generally recognized that the constraints on lipid motion seen on the ESR timescale usually disappear when viewed on an NMR timescale (10^{-3} – 10^{-4} s) indicating that a relatively rapid exchange occurs between the boundary layer and the bulk

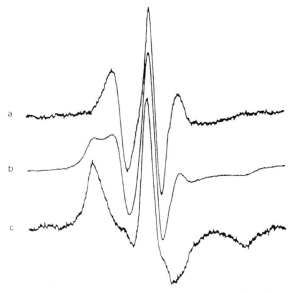

Fig. 7. ESR spectra of 14-PGSL spin-label intercalated into cucumber thylakoid membranes. The measured spectrum is shown in (b), the spectra corresponding to the fluid component (33% of total) and motionally restricted component (67% of total) are shown in (a) and (c) respectively. Data from Li et al. (1990).

phase (see Gennis, 1989 and references therein). In line with expectation, Li et al. (1989a) using saturation transfer ESR measurements have shown that the lipids in PS II-enriched membranes containing very low amounts of lipid (lipid/chlorophyll/protein, 16:17:67 w/w) have effective rotational correlation times of about 10^{-5} s.

Ivancich et al. (1994), however, have reported major spectral differences in experiments in which they change the order of addition of spin-labeled and excess unlabelled PC to PS II-enriched membrane fragments. They interpret this as indicating the existence of two sets of binding sites. One set (consisting of about 100 mol/820 kDa protein) appears to allow rapid exchange with bulk lipid and the other (consisting of about 50 mol/820 kDa) in which exchange is much slower or non-existent. The existence of the rapidly exchanging sites is easily understood but it is difficult to understand why there should be such a large number of what would appear to be effectively non-exchangeable lipid binding sites for PC, a non-thylakoid lipid, associated with PS II.

5. Studies Involving Mutants

The major development in recent years regarding the

determination of the importance of lipid-protein interactions in photosynthetic membranes, has undoubtedly been the increased availability of mutants lacking specific lipid molecular species. The importance of these mutants lies not so much in any information that they give about particular lipid-protein interactions but in the fact that they allow these interactions to be put into a more realistic perspective.

Particularly important in this context are the isolation of a fatty acid desaturase (*fad*) mutant of *Arabidopsis* that lacks 16:1*tr* PG (Browse and Sommerville, 1985; McCourt et al., 1985; refer also to Chapter 14, Vijayan et al.) and a *Rhodobacter sphaeroides* mutant totally lacking in SQDG (Benning et al., 1993; refer to Chapter 5, Benning). These mutants appear to show little structural or functional change from the wild-type organisms. This clearly indicates that whatever the role played by 16:1*tr* PG in the organization of chloroplast thylakoid membranes, and SQDG in the membranes of photosynthetic bacteria, these roles can be fulfilled by other lipids. This suggests that while general non-specific interactions of the type associated with lipid boundary layers may well be of importance, more specific interactions involving particular lipid molecular species are likely to be of less importance than is often argued.

Structural studies have revealed interesting differences in the detailed morphology of several *fad*-mutants of *Arabidopsis* (see Tsvetkova et al., 1994, 1995 and references therein). The relationship of these differences, if any, to lipid-protein interactions is, however, still a matter of debate and is clearly an area worthy of further study.

Grenier et al. (1990a) and Trémolières et al. (1991) have shown that exogenous 16:1*tr* PG, in the form liposomes added to the growth medium, is taken up by mutants of *Chlamydomonas reinhardtii* lacking this particular lipid. The lipid, or its fatty acids, is readily incorporated into the thylakoid membranes with a resulting increase in the proportion of appressed membranes. Great care has to be taken in interpreting such measurements as similar increases are seen in wild-type cells supplemented with unsaturated PC (Grenier et al., 1991). Nevertheless, the concept of the use of lipid supplementation experiments to investigate the role of lipids in determining structural changes in mutants is an intriguing possibility.

III. Membrane Fluidity Measurements

While the general concept of membrane fluidity is simple to understand, a number of difficulties arise when attempts are made to quantify the extent of fluidity. The lipid bilayer is an inherently anisotropic system. As such the value of the physical parameters that describe its properties vary with position. The motional fluidity of the lipid chains close to their headgroups, for example is clearly very different from that at the center of the bilayer.

A. Measurement of Membrane Fluidity

One approach to this problem is to measure membrane fluidity in terms of the lateral diffusion constants of given molecular species. Such measurements are, however, extremely difficult in practice. Methods are available for the measurement of rotational and linear diffusion rates of proteins using triplet probes and such techniques as fast recovery after photobleaching (FRAP). Lateral diffusion rates of lipids can be estimated from saturation transfer ESR and NMR measurements in model systems but these methods are difficult to apply to chloroplast membranes.

The approach normally adopted in photosynthesis studies, is to define fluidity in terms of physical parameters that describe the motion of spin or fluorescence probes either attached to the fatty acyl chains of membrane lipids, or free to move in the membrane phase. Two distinct sets of problems are associated with such measurements. The first lies with the relation between the measured parameter and the type of motion undergone by the probe and the second with the interpretation of probe measurements performed in systems as complex as the thylakoid membrane.

In general, the parameters measured in fluorescence and spin probe studies are influenced by two factors the range of the motion and the rate of the motion (see Stubbs, 1983 for a detailed discussion). The implications of this are conveniently illustrated in terms of the use of the fluorescence probe 1,6-diphenyl-2,3,5-hexatriene (DPH). DPH is a rod-shaped molecule with a high affinity for the hydrophobic regions of lipid bilayers. If it is excited using a beam of polarized light, any motion of the probe occurring in the time interval between the excitation and de-excitation will result in the depolarization of the fluorescence emission. This is reflected in a decrease in fluorescence anisotropy (r)

given by:

$$r = \frac{I_{/\!/} - I_{\perp}}{I_{/\!/} + 2I_{\perp}} \tag{1}$$

where $I_{/\!/}$ and $I_{\perp}$ are the fluorescence intensities measured parallel to, and perpendicular to, the plane of polarization of the exciting light.

The motion of DPH in the membrane is usually represented by a "wobbling in a cone" model. In terms of this model, the probe precesses within a cone of half-angle Θ at an average rotational velocity v. The problem is that steady state measurements do not distinguish between constraints imposed on the *range* of the probe, reflected in Θ, and the *rate* of the motion, reflected in v. Similar decreases in the value of r are seen if the value of Θ is decreased due to the presence of high concentrations of membrane proteins, as illustrated in Fig. 8, or the value of v decreases, as a consequence of viscosity changes due to changes in lipid saturation. This problem can be resolved by measuring the time-dependence of fluorescence depolarization in flash experiments (Kinosita and Ikegami, 1984; Gennis, 1989) but the equipment required is much more complex than that needed for simple steady-state measurements.

The second set of problems referred to above, lies in the extrapolation of the results obtained from probe studies carried out on simple model systems such as a liposomal dispersion of a single molecular species of lipid (in practice usually dipalmitoyl

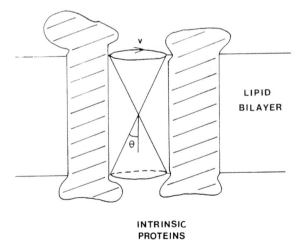

Fig. 8. Diagrammatic representation of the restrictions imposed on the motion of DPH in membranes containing high concentrations of intrinsic membrane proteins. See text for explanation.

phosphatidylcholine) to the much more hetero-geneous system making up a natural membrane. In the model systems, the probe exists in a uniform environment. In natural membranes, the probe can distribute itself between a variety of possible environments and it is almost impossible to be sure whether the probe is reporting on the properties of the membrane as a whole or a particular environment into which it has preferentially partitioned.

B. Relevance to Photosynthesis

There are three areas where fluidity and lateral diffusion measurements are of special interest in photosynthesis. These are the effects of growth temperature on membrane fluidity; the relative fluidity of the granal and stromal membranes of the thylakoid membrane; and the closely related question of the mobility of electron carriers in the plane of the photosynthetic membrane. It must be emphasized that when discussing these problems most authors have ignored the many theoretical and practical problems associated with measurements using spin and fluorescence probes. Conclusions based mainly, or entirely, on such evidence should, therefore, be viewed with extreme caution.

1. Temperature Adaptation

In common with many other biological organisms, plants appear to obey the homeoviscosity principle in that they tend to adjust the composition of their membranes so that their fluidity remains approxi-mately constant at their growth temperature (Raison et al., 1982; Mannock et al., 1985b). The way that this change is achieved, however, varies greatly. In some cases, the main factor is changes in lipid saturation either by de novo lipid synthesis or transacylation, in others the change is achieved by alterations in lipid/protein ratios.

2. Fluidity of Granal and Stromal Membranes

Measurements of the steady state value of the fluorescence anisotropy of DPH intercalated into granal and stromal membrane fractions suggest that the stromal membranes are appreciably more fluid than granal membranes (Ford et al., 1982; Barber et al., 1984). However, when a fuller analysis is carried out using time-resolved fluorescence anisotropy, in which the time dependence of fluorescence r(t) is

measured, it becomes clear that this apparent difference in fluidity is a reflection of a reduction in the half-angle of the rotation cone as a result of the higher concentration of protein in grana as opposed to stromal membranes (Ford and Barber, 1983).

3. Mobility of Electron Carriers

The question of the relative fluidities of the granal and stromal membranes is closely related to that of the mobility of the electron transport carrier plastoquinone. The presence of high concentrations of relatively immobile light-harvesting complexes in the granal membrane might be expected to restrict the mobility of electron carriers such as plastoquinone and as such have implications with respect to the relative locations of PS II units and the cytochrome $b_6 f$ complex.

Blackwell and her collaborators have used measurements of the quenching of pyrene fluores-cence by plastoquinone and its derivatives to estimate plastoquinone mobility in liposomes (Blackwell et al., 1987), proteoliposomes (Blackwell and Whit-marsh, 1990), thylakoid membranes and thylakoid membrane sub-fractions (Blackwell et al., 1994). Their main finding is that the mobility of plasto-quinone in protein-free liposomes appears to be about $1-3 \times 10^{-7}$ cm^2.s^{-1} as compared to $0.1-3 \times 10^{-9}$ cm^2.s^{-1} in the thylakoid fractions.

There are, as the authors take pains to point out, a number of theoretical uncertainties about the precise interpretation of their results. Their main conclusion is that the large reduction in quinone mobility in the native membrane as compared to the liposomal system is due to a hindering of free diffusion by the presence of relatively immobile intrinsic membrane proteins. This view is almost certainly sound and is strongly supported by their measurements on proteoliposome systems where the presence of intrinsic membranes occupying a calculated interfacial surface area of 16–26% resulted in a 10-fold reduction in quinone mobility (Blackwell and Whitmarsh, 1990).

These findings are in line with a range of practical and theoretical studies that suggest the conventional idea that small molecule diffusion in biological membranes can be adequately described as a random walk process on a two-dimensional lattice made up of identical points (the fluid sheet model) needs to be replaced by models in which due account is taken of the presence of membrane components that can block free access to such points (percolation models). Joliot and Lavergne and their co-workers, have estimated that the area fraction occupied in stacked membrane regions is 0.41–0.53, and have developed an interesting model for the thylakoid membrane in which long-range PQ diffusion is restricted by the presence of a fluctuating network of intrinsic membrane proteins (Joliot et al., 1992; Lavergne et al., 1992).

IV. Concluding Remarks

Physical studies performed on lipids have given rise to many important ideas in photosynthesis. The occurrence of phase transitions, the role of non-bilayer lipids, the possible importance of lipid-protein interactions and membrane fluidity are all important topics under active discussion in the photosynthetic literature. The aim of this chapter has been to provide a review of the literature underlying these topics while at the same time pointing out the many difficulties associated with extrapolating these ideas to the thylakoid membrane system.

It now seems that the importance of lipid phase transitions is much less than had originally been proposed. They are clearly of importance in 'cold-shock' phenomena of the type seen in the thermophilic cyanobacteria but their role in such phenomena as chilling-sensitivity is debatable. The precise role of non-bilayer lipids in membranes is still unclear. Early suggestions that they help seal proteins into membranes remain extremely plausible but detailed evidence is still missing. The idea that they play a major role in stabilizing the photosynthetic light-harvesting apparatus now seems highly questionable.

The large difference that exists between the phase behavior of mixtures of chloroplast membrane lipids and native photosynthetic membranes, discussed in Section I, clearly indicates that lipid-protein interactions play a major part in imposing a bilayer configuration in the native membranes. While there is little doubt regarding the significance of non-specific lipid-protein interactions, there is very little information available as to their precise nature. In strong contrast, at first sight, there appears to be an abundance of information relating to specific lipid-protein interactions. Unfortunately much of this work, as argued in Section II, provides information which relates to detergent solubilized complexes and cannot be reliably related back to the situation in the native

thylakoid membrane. The existence of a specific binding site for PG involved in the trimerization of LHC-II is, however, still an intriguing possibility.

Finally, a good deal of progress has been made over the years regarding gaining an understanding of the dynamic properties of photosynthetic membrane lipids. The emergence of a new generation of more useful models of membrane dynamics, discussed in Section III, is particularly welcome and is likely to have important implications both in terms of our understanding of temperature adaptation phenomena and the limiting factors in electron transport rates.

References

Armond PA and Staehelin LA (1979) Lateral and vertical displacement of integral membrane proteins during lipid phase transition in *Anacystis nidulans*. Proc Natl Acad Sci USA 76: 1901–1905

Barber J, Ford RC, Mitchell RAC and Millner P (1984) Chloroplast thylakoid membrane fluidity and its sensitivity to temperature. Planta 161: 375–380

Benning C (1998) Membrane lipids in anoxygenic photosynthetic bacteria. In: Siegenthaler PA and Murata N (eds) Lipids in Photosynthesis: Structure, Function and Genetics, pp 83–101. Kluwer, Dordrecht

Benning C, Beatty JT, Prince RC and Somerville CR (1993) The sulfolipid sulfoquinovosyldiacylglycerol is not required for photosynthetic electron transport in *Rhodobacter sphaeroides* but enhances growth under phosphate limitation. Proc Natl Acad Sci USA 90: 1561–1565

Blackwell MF and Whitmarsh J (1990) Effect of integral membrane proteins on the lateral mobility of plastoquinone in phosphatidylcholine proteoliposomes. Biophys J 58: 1259–1271

Blackwell MF, Gounaris, K, Zara SJ and Barber J (1987) A method for estimating lateral diffusion coefficients in membranes from steady-state fluorescence quenching studies. Biophys J 51: 735–744

Blackwell M, Gibas C, Gyax S, Roman D and Wagner B (1994) The plastoquinone diffusion coefficient in chloroplasts and its mechanistic implications. Biochim Biophys Acta 1183: 533–543

Brand J, Kirchanski SJ and Ramirez-Mitchell R (1979) Chill-induced morphological alterations in *Anacystis nidulans* as a function of growth temperature. Planta 145: 63–68

Brentel I, Selstam E and Lindblom G. (1985) Phase equilibria of plant galactolipids. The formation of a bicontinuous cubic phase. Biochim Biophys Acta 812: 816–826

Browse J, McCourt P and Somerville CR (1985) A mutant of *Arabidopsis* lacking a chloroplast-specific lipid. Science 227: 763–765

Dorne A-J, Block MA, Joyard J and Douce R (1990) Do thylakoids really contain phosphatidylcholine? Proc Natl Acad Sci USA 87: 71–74

Ford, RC and Barber J (1983) Time dependent decay and anisotropy of fluorescence from diphenyl hexatriene embedded in the chloroplast thylakoid membrane. Biochim Biophys Acta 722: 341–348

Ford RC, Chapman DJ, Barber J, Pedersen, JZ and Cox RP (1982) Fluorescence polarization and spin label studies of the fluidity of stromal and granal chloroplast membranes. Biochim Biophys Acta 681: 145–151

Furtado D, Williams WP, Brain APR and Quinn PJ (1979) Phase separations in membranes of *Anacystis nidulans* grown at different temperatures. Biochim Biophys Acta 555: 352–357

Gennis, RB (1989) Biomembranes, Molecular Structure and Function. Springer-Verlag, New York

Gombos Z and Murata N (1998) Genetically engineered modulation of the unsaturation of glycerolipids and its consequences in tolerance of photosynthesis to temperature stresses. In: Siegenthaler PA and Murata N (eds) Lipids in Photosynthesis: Structure, Function and Genetics, pp 249–262. Kluwer, Dordrecht

Gombos Z, Wada H and Murata N (1994) The recovery of photosynthesis from low-temperature photoinhibition is accelerated by the unsaturation of membrane lipids: A mechanism of chilling tolerance. Proc Natl Acad Sci USA 91: 8787–8791

Gounaris K, Sen A, Brain APR, Quinn PJ and Williams WP (1983a) The formation of non-bilayer structures in total polar lipid extracts of chloroplast membranes. Biochim Biophys Acta 728: 129–139

Gounaris K, Brain APR, Quinn PJ and Williams WP (1983b) Structural and functional changes associated with heat-induced phase-separation of non-bilayer lipids in chloroplast thylakoid membranes. FEBS Lett 153: 47–52

Gounaris K, Brain APR, Quinn PJ and Williams WP (1984) Structural reorganisation of chloroplast thylakoid membranes in response to heat-stress. Biochim Biophys Acta 766: 98–208

Grenier G, Guyon D, Roche O, Dubertret G and Trémolières A (1990) Modification of the membrane fatty acid composition of *Chlamydomonas reinhardtii* cultured in the presence of liposomes. In: Quinn PJ and Harwood JL (eds) Plant Lipid Biochemistry, Structure and Utilization, pp 367–369. Portland Press, London

Grenier G, Guyon D, Roche O, Dubertret G and Trémolières A (1991) Modification of the membrane fatty acid composition of *Chlamydomonas reinhardtii* cultured in the presence of liposomes.Plant Physiol Biochem 29: 429–440

Haase R, Unthan M, Couturier P, Radunz A and Schmid GH (1993) Determination of glycolipid, sulpholipid and phospholipids in the thylakoid membrane. Z Naturforsch 48C: 621-631

Huner NPA, Campbell D, Kroll M, Hayden DB, Myscich EM, Basalyga S and Williams JP (1992) Differential detergent stability of the major light-harvesting complex II in thylakoids isolated from monocotyledonous and dicotyledonous plants. Plant Physiol 99: 830–836

Ivancich A, Horvath LI, Droppa H, Horvath O and Farkas T (1994) Spin label study of lipid solvation of supramolecular photosynthetic complexes in thylakoids. Biochim Biophys Acta 1196: 51–56

Joliot P, Lavergne J. and Béal D (1992) Plastoquinone compartmentation in chloroplasts. 1. Evidence for domains with different rates of photo-reduction. Biochim Biophys Acta 1101: 1–12

Jost PC, Griffith OH, Capaldi RA and Vanderkooi G (1973) Evidence for boundary layer lipid in membranes. Proc Natl Acad Sci USA 70: 480–484

Keough KMW and Kariel N (1987) Differential scanning calorimetric studies of aqueous dispersions of phosphatidyl-cholines containing two polyenoic chains. Biochim Biophys Acta 902: 11–18

Kinosita K and Ikegami A (1984) Reevaluation of the wobbling dynamics of diphenylhexatriene in phosphatidylcholine and cholesterol/phosphatidylcholine membranes. Biochim Biophys Acta 769: 523–527

Kruse O, Radunz A and Schmid GF (1994) Phosphatidylglycerol and β-carotene bind onto the D_1 protein of the filamentous cyanobacterium *Oscillatoria chalybea*. Z Naturforsch 49C: 115–124

Lavergne J, Bouchaud J-P and Joliot P (1992) Plastoquinone compartmentation in chloroplasts. 1. Theoretical aspects. Biochim Biophys Acta 1101: 13–22

Li G, Horvath LI, Knowles PF, Murphy DJ and Marsh D (1989a) Spin-label saturation transfer ESR studies of protein-lipid interactions in Photosystem II-enriched membranes. Biochim Biophys Acta 987: 187–192

Li G, Knowles PF, Murphy DJ, Nishida I and Marsh D (1989b) Spin-label ESR studies of lipid-protein interactions in thylakoid membranes. Biochemistry 28: 7446–7452

Li G, Knowles PF, Murphy DJ and Marsh D (1990) Lipid-protein interactions in thylakoid membranes of chilling-resistant and -sensitive plants studied by spin label electron spin resonance spectroscopy. J Biol Chem 265: 16867–16872

Lis L and Quinn PJ (1986) A time-resolved X-ray study of a crystalline phase bilayer transition in a saturated mono-galactosyldiacylglycerol system. Biochim Biophys Acta 862: 81–86

Lyons JM (1973) Chilling injury in plants. Annu Rev Plant Physiol 24: 445–466

McCourt P, Browse J, Watson J and Arntzen CJ (1985) Analysis of photosynthetic antenna function in a mutant of *Arabidopsis thaliana* (L.) lacking trans-hexadecanoic acid. Plant Physiol 78: 853–858

Mannock D and McElhaney RN (1991) Differential calorimetry and X-ray diffraction studies of a series of synthetic β-D-galactosyl diacylglycerols. Biochem Cell Biol 69: 863–867

Mannock D, Brain APR and Williams WP (1985a) The phase behaviour of 1,2-diacyl-3-monogalactosyl-*sn*-glycerol derivatives. Biochim Biophys Acta 817: 289–298

Mannock DA, Brain APR and Williams WP (1985b) Phase behaviour of the membrane lipids of the thermophilic blue-green alga *Anacystis nidulans*. Biochim Biophys Acta 821: 153–164

Mansourian AR and Quinn PJ (1986) Phase properties of binary mixtures of monogalactosyldiacylglycerols differing in hydrocarbon chain substituents dispersed in aqueous systems. Biochim Biophys Acta 855: 169–178

Murata N and Yamaya J (1984) Temperature dependent phase behaviour of phosphatidylglycerols from chilling sensitive and chilling resistant plants. Plant Physiol 74: 1016–1024

Murata N, Higashi S-I and Fujimura Y (1990) Glycerolipids in various preparations of Photosystem II from spinach chloroplasts. Biochim Biophys Acta 1019: 261–268

Murphy D (1986) Structural properties and molecular organization of the acyl lipids of photosynthetic membranes. In: Staehelin LA and Arntzen CJ (eds) Encyclopedia of Plant Physiology,

Vol 19, Photosynthesis III: Photosynthetic Membranes and Light Harvesting Systems, pp 713–725. Springer-Verlag, Berlin

Nussberger S, Dörr K, Wang DN and Kühlbrandt W (1993) Lipid-protein interactions in crystals of plant light-harvesting complex. J Mol Biol 234: 347–356

Ono T-A and Murata N (1982) Chilling susceptibility of the blue-green alga *Anacystis nidulans*. Plant Physiol 69: 125–129

Ono T-A, Murata N and Fujita T (1983) Thermal analysis of membrane lipids from the blue-green algae *Anacystis nidulans* and *Anabaena variabilis*. Plant Cell Physiol 24: 635-639

Quinn PJ and Williams WP (1978) Plant lipids and their role in membrane function. Prog Biophys Molec Biol 34: 109–173

Quinn PJ and Williams WP (1983) The structural role of lipids in photosynthetic membranes. Biochim Biophys Acta 737: 223–266

Quinn PJ and Williams WP (1985) Environmentally induced changes in chloroplast membranes and their effects on photosynthetic function. In: Barber J (ed) Topics in Photosynthesis. Vol 6, pp 1–47. Elsevier, Amsterdam

Quinn PJ and Williams WP (1990) Structure and dynamics of plant membranes. In: Harwood J and Walton J (eds) Methods in Plant Biochemistry. Vol 4, pp 297–340. Academic Press, New York

Raison JK (1973) The influence of temperature-induced phase changes in the kinetics of respiratory and other membrane-associated enzyme systems. J Bioenerg 4: 285–309

Raison JK, Roberts JKM and Berry JA (1982) Correlations between the thermal stability of chloroplast (thylakoid) membranes and the composition and fluidity of their polar lipids upon acclimation of the higher plant *Nerium oleander* to growth temperature. Biochim Biophys Acta 688: 218–228

Rietveld A, Kenemade TJJM, Hak T, Verkleij AJ and de Kruijff B (1987) The effect of cytochrome *c* oxidase on lipid polymorphism of model membranes containing cardiolipin. Eur J Biochem 164: 137–148

Sanderson PW (1992) Membrane lipid phase behavior and its relationship to the stability of biological membranes. PhD thesis, Kings College, University of London, UK

Sanderson PW and Williams WP (1992) Low-temperature phase behaviour of the major plant leaf lipid monogalactosyldiacylglycerol. Biochim Biophys Acta 1107: 77–85

Sanderson PW, Lis LJ, Quinn PJ and Williams WP (1991) The Hofmeister effect in relation to membrane lipid phase stability. Biochim Biophys Acta 1067: 43–50

Sen A, Williams WP and Quinn PJ (1981a) The structure and thermotropic properties of pure 1,2-diacylgalactosylglycerols in aqueous systems. Biochim Biophys Acta 663: 380–389

Sen A, Williams WP, Brain APR, Dickens MJ and Quinn PJ (1981b) Formation of inverted micelles in dispersions of mixed galactolipids. Nature 293: 488–490

Sen A, Williams WP, Brain APR and Quinn PJ (1982a) Bilayer and non-bilayer transformations in aqueous dispersions of mixed sn-3-galactosyldiaclylglycerols isolated from chloroplasts. Biochim Biophys Acta 685: 297–306

Sen A, Brain APR, Quinn PJ and Williams WP (1982b) Formation of inverted lipid micelles in aqueous dispersions of mixed sn-3-galactosyldiacylglycerols induced by heat and ethylene glycol. Biochim Biophys Acta 686: 215–224

Sen A, Mannock DA, Collins DJ, Quinn PJ and Williams WP (1983) Thermotropic phase properties and structure of 1,2-disteraoylgalactosylglycerols in aqueous solutions. Proc R

Soc Lond B218: 349–364

Shipley GG, Green JP and Nichols BW (1973) The phase behaviour of monogalactosyl, digalactosyl and sulpho-quinovosyl diglycerides. Biochim Biophys Acta 311: 531–544

Sprague SG and Staehelin LA (1984) Effects of reconstitution method on the structural organisation of isolated chloroplast membranes. Biochim Biophys Acta 777: 306–322

Stubbs CD (1983) Membrane fluidity: structure and dynamics of membrane lipids. In: Campbell PN and Marshall RD (eds) Essays in Biochemistry. Vol 19, pp 1–39. Biochemical Society, London

Taraschi TF, de Kruijff B, Verkleij A and Van Echteld CJA (1982) Effect of glycophorin on lipid polymorphism a ^{31}P-NMR study. Biochim Biophys Acta 685: 153–161

Tilcock CPS (1986) Lipid polymorphism. Chem Phys Lipids 40: 109–125

Trémolières A and Siegenthaler PA (1998) Reconstitution of photosynthetic structures and activities with lipids. In: Siegenthaler PA and Murata N (eds) Lipids in Photosynthesis: Structure, Function and Genetics, pp 175–189. Kluwer, Dordrecht

Trémolières A, Roche O, Dubertret F, Guyon D and Garnier J (1991) Restoration of thylakoid appression by Δ^3-trans-hexadecanoic acid containing phosphatidylglycerol in a mutant of *Chlamydomonas reinhardtii*. Relationship with the regulation of excitation energy distribution. Biochim Biophys Acta 1059: 286–292

Trémolières A, Dainese P and Bassi R (1994) Heterogeneous lipid distribution among chlorophyll-binding proteins of Photosystem II in maize mesophyll chloroplasts. Eur J Biochem 221: 721–730

Tsvetkova NM, Brain APR and Quinn PJ (1994) Structural characteristics of thylakoid membranes of *Arabidopsis* mutants deficient in lipid fatty acid desaturation. Biochim Biophys Acta 1192: 286–292

Tsvetkova NM, Apostolova EL, Brain APR, Williams WP and Quinn PJ (1995) Factors influencing PS II particle array formation in *Arabidopsis thaliana* chloroplasts and the relationship of such arrays to the thermostability of PS II. Biochim Biophys Acta 1228: 201–210

Tsukomoto Y, Yeki T, Mitsui T, Ono T-A and Murata N (1980) Relationship between growth temperature of *Anacystis nidulans* and phase transition temperatures of its thylakoid membranes. Biochim Biophys Acta 602: 673-675

Vijayan P, Routabout JM and Browse J (1998) A genetic approach to investigating membrane lipid structure and photosynthetic function. In: Siegenthaler PA and Murata N (eds) Lipids in Photosynthesis: Structure, Function and Genetics, pp 263–285. Kluwer, Dordrecht

Verwer W, Ververgaert PJJT, Leunissen-Bijvelt J and Verkleij AJ (1978) Particle aggregation in the photosynthetic membranes of the blue-green alga *Anacystis nidulans*. Biochim Biophys Acta 504: 231–234

Voss R, Radunz A and Schmid GH (1992) Binding of lipids onto polypeptides of the thylakoid membrane. 1. Galactolipids and sulpholipids as prosthetic groups of core polypeptides of the Photosystem II complex. Z Naturforsch 47C: 406–415

Webb MS and Green BR (1991) Biochemical and biophysical properties of thylakoid acyl lipids. Biochim Biophys Acta 1060: 133–158

Williams WP (1994) The role of lipids in the structure and function of photosynthetic membranes. Prog Lipid Res 33: 119–127

Williams WP and Gounaris K (1992) Stabilisation of PS II-mediated electron transport in oxygen-evolving PS II core preparations by the addition of compatible co-solutes. Biochim Biophys Acta 1100: 92–97

Williams WP and Quinn PJ (1987) The phase behaviour of lipids in photosynthetic membranes. J Bioenerg Biomembr 19: 605-624

Williams WP, Brain APR and Dominy PJ (1992) Induction of non-bilayer lipid phase separations in chloroplast thylakoid membranes by compatible co-solutes and its relation to the thermal stability of Photosystem II. Biochim Biophys Acta 1099: 137–144

Williams WP, Sanderson PW, Cunningham BA, Wolfe DH and Lis LJ (1993) Phase behaviour of membrane lipids containing polyenoic acyl chains. Biochim Biophys Acta 1148: 285–290

YashRoy RC (1994) Destabilisation of lamellar dispersion of thylakoid membrane lipids by sucrose. Biochim Biophys Acta 1212: 129–133

Chapter 7

Molecular Organization of Acyl Lipids in Photosynthetic Membranes of Higher Plants

Paul-André Siegenthaler
Laboratoire de Physiologie végétale, Université de Neuchâtel,
Rue Emile Argand 13, CH-2007 Neuchâtel, Switzerland

P.-A. Siegenthaler and N. Murata (eds): Lipids in Photosynthesis: Structure, Function and Genetics, pp. 119–144.
© *1998 Kluwer Academic Publishers. Printed in The Netherlands.*

Summary

The extreme diversity of thylakoid acyl lipids and their unique (physico)chemical characteristics suggest that they are arranged, at the molecular level, as distinct membrane domains. Several experimental approaches, including fractionation of subchloroplast particles, separation of appressed and non-appressed regions of the thylakoids and purification of (chlorophyll-) protein complexes, show that acyl lipids are asymmetrically distributed within the plane of the membrane and that some of them (generally those more saturated than the bulk lipids) are tightly bound to proteins. However, conclusions based only on such evidence should be viewed with caution because the use of detergents to obtain subparticles or lipoprotein complexes may result in differential displacements of acyl lipids and pigments non-covalently bound to proteins. In this respect, the use of cyclodextrins as a new tool for the controlled lipid depletion of thylakoid membranes is discussed in some detail. These molecules are cyclic oligosaccharides consisting of six to eight glucopyranose units linked by α (1–4) bonds which adopt a torus shape and are able to bind a range of small guest molecules of poor water solubility (e.g., lipids) within their hydrophobic cavity to form a water soluble guest-cyclodextrin inclusion complex. The advantage of such an approach is to avoid the use of detergents. The detection of lipids by antibodies directed to individual lipids bound to the surface of the thylakoid membrane, to subchloroplast particles or to individual proteins is also presented.

The transmembrane distribution of acyl lipids in the thylakoid membrane has been extensively studied. We report here on the different experimental approaches, with special emphasis on the enzymatic one which consists of digesting lipids, stepwise and selectively, in the two membrane monolayers. The results indicate that the outer monolayer is highly enriched in MGDG and PG while the inner one contains high levels of DGDG, thus confirming the general sidedness of thylakoid membrane components. The modulation of the unsaturation/saturation ratio of acyl lipids by catalytic hydrogenation is also described in order to assess the effect of the unsaturation degree on thylakoid structures.

The problem of transfer of (galacto)lipids from their sites of synthesis (the envelope membranes) to the thylakoid network is an intriguing one. Recent data favor the hypothesis that the intrachloroplastic lipid export is primarily achieved by a mechanism involving transient fusions between inner envelope and thylakoid membranes. This new proposal points to the key role of chloroplast envelope membranes in the establishment of lipid asymmetry in thylakoid membranes. Finally, several paradigms of the thylakoid membrane molecular organization based on the known lipid topology and properties are discussed.

I. Introduction

Chloroplasts play a crucial role in sustaining life on earth by their dual property in performing the primary fixation of carbon and also in releasing oxygen for use in respiration. A key factor in these processes is the trapping of light energy by chlorophylls. Collection of light and its transformation into chemical energy occur in the thylakoid membrane

---

Abbreviations: DCPIP – dichlorophenolindophenol; DGDG – digalactosyldiacylglycerol; GC – gas chromatography; HPLC – high-performance liquid chromatography; LAH – lipolytic acyl hydrolase; LHC (LHCP) – light-harvesting chlorophyll *a/b* protein; MGDG – monogalactosyldiacylglycerol; PC – phosphatidylcholine; PG – phosphatidylglycerol; PS I – Photosystem I; PS II – Photosystem II; R – resistant; S – susceptible; SQDG – sulfoquinovosyldiacylglycerol; 16:1(3t) – *trans*-Δ^3-hexadecenoic acid; 18:3 – $\Delta^{9,12,15}$ *cis, cis, cis* octadecatrienoic acid (α-linolenic acid)

which is one of the most remarkable transducing systems in the biological world. In order for the light-dependent reactions to take place, a high degree of molecular organization of its constituents is needed. Several (chlorophyll-) protein complexes, involved in the capture of light, in photosynthetic electron flow and phosphorylation, are separated both in the plane of and across the membrane. Their asymmetric distribution confers vectorial properties, necessary for energy conservation, to the thylakoid membrane. Acyl lipids are the second major components of the thylakoid membrane. They consist of four classes (MGDG, DGDG, SQDG and PG) which are characterized by the occurrence of a great number of molecular species (Nishihara et al., 1980; Xu and Siegenthaler, 1996a). These lipid classes have been found also to be asymmetrically distributed, both laterally and transversally, and in this respect may well play specific and dynamic roles in photosynthetic functions (refer to Chapter 8 by Siegenthaler and Trémolières).

These four glycerolipid classes are found in all oxygen evolving photosynthetic organisms from the cyanobacteria to higher plants. This unique lipid composition might result from a common ancestor for all these organisms. Nevertheless, it is quite surprising that no other lipid has ever been introduced during the course of evolution. This suggests that a strong selection pressure exists which maintains such a uniform lipid composition.

Thus, two questions can be addressed: Does this special composition confer specific properties to the photosynthetic membrane? Do some of these lipids interact specifically with certain proteins of this membrane to insure an optimal photosynthetic efficiency? At present, there is no clear answer to these questions. However, the concept of a photosynthetic membrane in which only proteins and pigments play important functions, the lipidic matrix (i.e. the bulk lipids) being limited only to its passive support, is far from reality. Indeed, some specific lipid-protein interactions are now well documented. For instance, Mattoo et al. (1987) have demonstrated in *Spirodela* plants that one of the first acceptors of the PS II, the D1 protein (i.e. the target of urea/triazine-type herbicides) which has one of the highest turnover rates recorded so far for a membrane protein, is acylated (covalently) by palmitate to reach its correct place in the PS II reaction center. Furthermore, weak interactions may also display a great specificity. For example, PG containing the unusual fatty acid *trans*-Δ^3-hexadecenoic acid [16:1(3t)] seems to play a subtle but crucial role not only in the function but also in the biogenesis of the photosynthetic membrane. It tunes the correct organization of the light harvesting antennae (LHC) which, in situ, present an extremely complex structure (Dubacq and Trémolières, 1983; Nussberger et al., 1993; Hobe et al., 1994, 1995; Chapter 9, Trémolières and Siegenthaler).

Compared to all other biological membrane molecular architecture, the photosynthetic membrane is one of those showing the highest heterogeneity (Murphy, 1986a). It is characterized by a marked lateral heterogeneity of its protein components. Analyses of purified photosynthetic complexes show that lipids are not uniformly distributed and that certain lipids are tightly bound to particular protein complexes. Moreover, the photosynthetic membrane displays a marked transversal heterogeneity with a very different set of proteins on the intrathylakoidal (or lumenal) and stromal sides (Murphy, 1986a). It is now firmly established that lipids also display a marked transversal heterogeneity (Siegenthaler and Rawyler, 1986; Siegenthaler et al., 1988 and references therein).

On the other hand, one has to consider that the biogenesis of photosynthetic membranes is a highly complex process which is under the control of both chloroplastic and nuclear genomes. The chloroplastic genome codes for many plastidial proteins, e.g., the large subunit of the ribulose 1,5-bisphosphate carboxylase, and also several protein components involved in the light and electron transport reactions as well as in the CF_o-CF_1 ATP synthase complex. However, the contribution of the nuclear genome remains essential for the synthesis of certain thylakoid proteins such as the LHC which are synthesized in the cytoplasm, imported into the chloroplast, then assembled in the thylakoid membrane (Archer and Keegstra, 1990; Soll and Alefsen, 1993 and references therein). Recent reports show that lipid-protein interactions are involved in chloroplast protein import (Chapter 10, de Kruijff et al.).

It also appears that the rates of synthesis of the three main components of the photosynthetic membrane, i.e., the proteins, the pigments and the lipids, have to be strictly regulated to achieve the biogenesis of a functional thylakoid. For instance, if chlorophylls were not correctly associated with proteins and lipids, they would be photodestroyed very rapidly. In *Pharbitis nil* cotyledons, the synthesis of linolenic acid is induced only at high temperature; if the plant is grown at 12 °C, it cannot synthesize linolenic acid and the cotyledons remain yellow because the chlorophylls are immediately photo-degraded in the absence of this fatty acid (Tchang et al., 1985).

The purpose of this Chapter is to review comprehensively some of the recent advances made in elucidating the structure and the molecular arrangement of acyl lipids in the photosynthetic membrane of higher plants. Over the past fifteen years, several reviews have been published on these topics (Dubacq and Trémolières, 1983; Gounaris and Barber, 1983; Quinn and Williams, 1983; Barber and Gounaris, 1986; Gounaris et al., 1986; Murphy, 1986a; Siegenthaler and Rawyler, 1986; Trémolières,

1991; Webb and Green, 1991; Selstam and Widell-Wigge, 1993; Williams, 1994; Chapter 6, Williams). The functional aspects of lipids in photosynthetic membranes as well as the reconstitution of photosynthetic structures and activities with lipids will be dealt with in Chapters 8 (Siegenthaler and Trémolières) and 9 (Trémolières and Siegenthaler).

II. Unique Features of Acyl Lipids in Photosynthetic Membranes

It is generally accepted that the major role of acyl lipids in membranes is to form a bilayer which will act as a structural matrix that will limit the transmembrane movement of most hydrophilic solutes. The bilayer provides a so-called 'bulk lipid phase' which prevents non-specific protein-protein aggregation and a fluid state allowing protein diffusion and conformational change to occur. The above properties, however, could be easily fulfilled by any lipids. Thus, the extreme diversity of lipid molecular species and membrane-specific lipid composition cannot be only explained in these terms. Compared to other biomembrane lipids, thylakoid lipids display several interesting and unique features which justify special attention and may provide outstanding information on the functions of this particular type of membrane (Gounaris et al., 1986).

Firstly, the amount of thylakoid lipids is particularly high in the biosphere. Because of the huge amounts of photosynthetic membranes in algae and plants, the glycosylglycerides (MGDG and DGDG) are, in fact, the most abundant membrane lipids in the world.

Secondly, thylakoid lipid composition is unique in that, instead of the commonly encountered phospho-glycerides, the major components are glyco-sylglycerides (Table 1). Thylakoid membranes contain about 50 mol% of their acyl lipids as MGDG, 30% as DGDG and about 5% as SQDG, a sulfolipid which seems to be present mainly in photosynthetic membranes. The only significant phospholipid is PG (about 10% of total acyl lipids).

Thirdly, the two galactolipids are characterized by an exceptionally high content of α-linolenic acid (up to 95% for MGDG). In the so-called '16:3 plants', hexadecatrienoic acid is also found in MGDG. The acyl composition of the two acidic lipids is also rich in linolenate but in addition SQDG contains about 35% palmitate and PG contains palmitate and a unique acid, trans-Δ^3-hexadecenoic acid (Dubacq and Trémolières, 1983; Xu and Siegenthaler, 1996b). Thus, the thylakoid fatty acid composition is unique and gives rise to a great number of molecular species (Nishihara et al., 1980; Xu and Siegenthaler, 1996a). A comparison of the different molecular species of PG in spinach and squash thylakoid membranes is given in Table 2.

Fourthly, it is known that isolated diacyl lipids extracted from biological membranes can adopt on hydration either a lamellar or a hexagonal type II configuration. In this respect, thylakoid membranes from higher plants are notable in that the major glycolipid (MGDG) comprising half of the total lipid content may adopt, under defined conditions, such a non-bilayer configuration (Quinn and Williams, 1983; Chapter 6, Williams). This is in contrast to all other lipid classes present in thylakoids which will form typical bilayer structures. The potential to form non-bilayer structures is quite different in the two thylakoid monolayers. For instance, when taking into account

Table 1. Content of glycerolipid classes in thylakoid membranes of spinach (*Spinacia oleracea* L. cv. Nobel) leaf and of squash (*Cucurbita moschata* Durch. cv. Shirakikuza) cotyledons. Data are expressed as mean values ± SD (n = 3)

Lipid class	Spinach		Squash	
	nmol/mg Chl	mol%	nmol/mg Chl	mol%
MGDG	1262 ± 35	52.0 ± 1.3	1318 ± 83	53.2 ± 1.3
DGDG	758 ± 28	31.2 ± 1.0	761 ± 30	30.7 ± 0.8
SQDG	127 ± 10	5.2 ± 0.4	136 ± 1	5.5 ± 0.2
PG	281 ± 20	11.6 ± 0.8	261 ± 7	10.5 ± 0.6
PC	Tr		Tr	

Tr, trace amounts (< 0.5 mol %) of PC are due to contaminations of thylakoid preparations by chloroplast envelope membrane vesicles. Contaminants are generally higher in spinach than in squash thylakoids (Y. N. Xu and P. A. Siegenthaler, unpublished; Xu and Siegenthaler, 1996b)

Table 2. Comparison of the content of phosphatidylglycerol (PG) molecular species in thylakoid membranes from spinach (a chilling resistant plant) and squash (a chilling sensitive plant) (Xu and Siegenthaler, 1996a)

Molecular species of PG	Spinach nmol/mg Chl	mol %	Squash nmol/mg Chl	mol %
18:3/16:1(3t)	214.7	76.4	75.4	28.9
18:3/16:0	40.7	14.5	14.4	5.5
18:2/16:1(3t)	2.3	0.8	17.0	6.5
18:2/16:0	3.4	1.2	3.4	1.3
18:1/16:1(3t)	1.4	0.5	20.1	7.7
18:1/16:0	Tr	Tr	6.5	2.5
18:0/16:1(3t)	1.4	0.5	18.5	7.1
18:0/16:0	Tr	Tr	2.6	1.0
16:0/16:1(3t)	13.8	4.9	80.9	31.0
16:0/16:0	3.7	1.3	22.5	8.6

Tr amount < 0.5 mol %

only those acyl lipids which are present in one of the two thylakoid monolayers and calculating the non-bilayer/bilayer forming lipid ratios, one discovers that the outer monolayer of the prothylakoid membrane displays the ability to form lamellar structures while the outer monolayer of the thylakoid membrane should display nonlamellar configurations. The physiological significance of such a finding has been addressed (Rawyler et al., 1987) and discussed in terms of biogenesis and stability of the thylakoid membrane (Giroud and Siegenthaler, 1988).

Fifthly, the lipid composition of thylakoids is dominated, as mentioned above, by glycolipids rather than phospholipids. It has been reported that a mixture of both MGDG and DGDG form monolayers which are much more condensed than those formed by phospholipids with comparable fatty acyl chains (Bishop et al., 1980). As a consequence, the packing properties of total thylakoid lipids are determined essentially by the packing characteristics of the two galactolipids. Very strong head-group interactions between these molecules are likely to enhance the stability of the membrane and its impermeability. Possibly, in order to compensate for its high lipid stability, the thylakoid membrane contains an unusually high amount of α-linolenic acid in most of its lipids. An obvious explanation is that such fatty acids ensure that the membrane remains in a highly fluid state (Ford and Barber, 1983). This property may be required for optimizing the fluidity of the thylakoid membrane which is characterized by a

high protein-to-lipid ratio. Such a highly fluid system is required not just for the diffusion of lipophilic compounds (e.g., plastoquinones) but for the lateral diffusion of protein complexes (Allen, 1992).

Sixthly, the occurrence of *trans-*Δ^3-hexadecenoic fatty acid in PG molecules is doubtless a peculiarity of thylakoid membranes. The possible role of PG molecular species containing this unusual fatty acid is not fully understood but it has been proposed that they are implicated in the supramolecular organization and efficiency of the oligomeric form of LHCII (Dubacq and Trémolières, 1983; Nussberger et al., 1993), as well as in the phenomenon of chilling sensitivity in plants (Roughan, 1985; Murata and Nishida, 1990) and of the herbicide resistance of some plant species (Burke et al., 1982; Siegenthaler and Mayor, 1992).

Finally, as will be discussed in this chapter, the thylakoid membrane is characterized by a pronounced transmembrane asymmetry of all of its acyl lipids (Siegenthaler and Rawyler, 1986 and references therein; Rawyler et al., 1987; Siegenthaler and Giroud, 1986; Siegenthaler et al., 1988, 1989a) and, to a lesser degree, by a lateral heterogeneity, mainly of MGDG (Murphy, 1986a,b; Gounaris et al., 1986).

The extreme diversity of thylakoid acyl lipids and their unique (physico)chemical characteristics as well as their peculiar arrangement in the membrane strongly suggest that distinct domains and specific functions of acyl lipids exist in the thylakoid membrane.

III. Asymmetric Distribution and Molecular Organization of Acyl Lipids in Thylakoids

A. Lateral Heterogeneity

1. Methodological Strategies

Several attempts have been made to determine if, in a manner similar to proteins, acyl lipids are also asymmetrically distributed in the plane of the thylakoid membrane. The following approaches have been used: (a) Mild solvent extraction of freeze-dried membranes (Costes et al., 1972, 1978); (b) Fractionation of subchloroplast particles enriched in PS I or PS II activities (Eichenberger et al., 1977; Henry et al., 1983; Bednarz et al., 1988; Ouijja et al., 1988); (c) Separation of appressed and non-appressed regions of the thylakoids or granal and intergranal fractions (Tuquet et al., 1977; Gounaris et al., 1983; Henry et al., 1983; Murphy and Woodrow, 1983; Chapman et al., 1984; Bednarz et al., 1988); (d) Purification of one of the protein complexes found in the thylakoid membrane (Rawyler et al., 1980; Heinz and Siefermann-Harms, 1981; Trémolières et al., 1981; Rémy et al., 1982; Doyle and Yu, 1985; Gounaris and Barber, 1985; Pick et al., 1985; Sigrist et al., 1988; Murata et al., 1990; Trémolières et al., 1994); (e) Detection of lipids by using antibodies directed to individual lipids bound to the surface of the thylakoid membrane, to subchloroplast particles or to individual proteins (Radunz 1981; Voss et al., 1992b; Kruse and Schmid, 1995 and references therein; Makewicz et al., 1995). Only a few of these approaches will be discussed below.

2. Preferential Association of Lipids with Thylakoid Subfractions and Protein Complexes

The composition in acyl lipids and proteins of several thylakoid subfractions has been determined and tentatively related to particular function(s). The main conclusion of these studies is that acyl lipids are asymmetrically distributed within the plane of the thylakoid membrane, although Chapman et al. (1984) have questioned this distribution (apart from a very slight enrichment of MGDG in appressed membranes). It is worth emphasizing here, that the lateral heterogeneity of thylakoid proteins is much more pronounced than that of lipids. On the basis of such results it is hard to pin-point a specific functional role

of a particular lipid class. At best, one can establish a tenuous correlation between enrichment of a particular lipid and a special function, e.g., PG containing the *trans*-hexadecenoic acid, is preferentially associated with the PS II LHCP (Dubacq and Trémolières, 1983) which is mainly located in the appressed regions of the grana (Andersson and Anderson, 1980). Nevertheless, it is very likely that specific lipids are needed for the functional activities and/or the stabilization of certain thylakoid complexes, but such requirements should involve only a small amount of a particular lipid molecular species as suggested by Siegenthaler and Rawyler (1986) and recently substantiated by Nussberger et al. (1993) and Hobe et al. (1994). Already in 1985, Gounaris and Barber (1985) reported specific lipid-protein interactions in a purified photochemically active reaction center complex, isolated from PS II preparations of spinach chloroplasts by Triton X-100 solubilization and sucrose gradient. This complex, which is photochemically active, contains five polypeptides which are characteristic of the reaction center II and only two polar lipids associated with it, MGDG and SQDG, the former being the most abundant. These results point to several important features: (a) The reaction center is isolated as a lipoprotein-pigment complex; (b) The lipid composition of the complex is very uncharacteristic for the thylakoid bilayer since it contains only MGDG and SQDG with unusual low levels of unsaturation; (c) The PS II reaction center seems to be associated with particular lipid molecular species, e.g., palmitic, stearic and oleic fatty acids make up most of SQDG present; MGDG is also very different from that found in intact thylakoids with the total absence of hexadecatrienoic acid. In *Chlamydomonas reinhardii*, Sigrist et al. (1988) found that the apoproteins of LHCII are tightly bound to SQDG, 80% of its acyl chains being palmitic acid.

Also of interest is the finding by Pick et al. (1985) that SQDG is the only lipid to be associated with the coupling factor complex (CF_o-CF_1) of spinach and *Dunaliella salina*. This suggests that this acidic glycolipid is firmly bound to the ATP synthetase complex and may play a special role in the mechanism of energy coupling in photosynthetic systems. In contrast, the purified cytochrome b_6f complex contains only phospholipids i.e. no galactolipid and no SQDG (Doyle and Yu, 1985). The purified complex has little plastoquinol-cytochrome c reductase activity in the absence of added lipid.

Bednarz et al. (1988) compared the lipid composition of thylakoids and of photosynthetically active PS II particles from the tobacco mutant NC95. This mutant has the peculiarity of displaying green and yellow-green leaf areas. Green areas exhibit PS I and PS II reactions and have chloroplasts with normal morphological structure and granal/stromal thylakoid ratio. In contrast, chloroplasts from the yellow-green patches have only intergranal thylakoids and exhibit only PS I reactions. This feature offers a unique opportunity to compare directly, within one plant and without the use of detergents, the lipid composition of PS I with that of PS II. The lipids of PS II particles originating from chloroplasts with appressed thylakoids are composed of 37% glyco-lipids, 4% phospholipids, 5% carotenoids and 54% chlorophyll while the lipid composition of chloro-plasts with non-appressed thylakoids consist of 83% glycolipids, 14% phospholipids and only 0.5% carotenoids and 2% chlorophyll. Thus, thylakoids active only in Photosystem I contain more than twice the amount of glycolipids and four times more phospholipids than PS II active preparations. According to Bednarz et al. (1988), these results confirm the lateral heterogeneity of thylakoid lipids and suggest that the fluidity of the intergranal region is likely to be higher than that of the appressed lamellae. However, the fact that PS II particles are prepared with detergents limits the value of such a comparison.

More recently, Murata et al. (1990) have determined the fatty acid and glycerolipid composition of three types of preparation of PS II from spinach thylakoids, namely the PS II membrane, the PS II core complex and the PS II reaction center complex. The molecular ratios of lipid to photochemical reaction center II (P680) in these preparations were estimated to be about 150:1, 10:1, 1:1, respectively. It is noteworthy that only one molecule of MGDG, and possibly one molecule of PG, are necessary for the maintenance of the structural organization of the PS II reaction center complex and for the photochemical charge separation. Furthermore, the fatty acids of these two lipids are much more saturated (50% of the total fatty acids) than those in the bulk lipids of thylakoid and PS II membranes (10% of the total fatty acids). Since increased saturation of fatty acids results in a more rigid structure of lipid molecules, the occurrence of saturated fatty acids in the lipids suggests that they stabilize the conformation of the five polypeptides of this reaction center complex. On the other hand, the PS II core complex contains only three molecules each of MGDG, DGDG and PG per P680. Since this complex is highly active in photochemical charge separation and O_2 evolution, one can expect that either some or all of these lipid classes are necessary for the proper conformation of the PS II core complex, which is composed of about ten polypeptides. This result is in accordance with our previous finding that PG as well as MGDG are essential for sustaining PS II electron transport activity (Siegenthaler et al., 1989a; Rawyler and Siegenthaler, 1989).

These few examples show that the approach consisting of isolating subfractions or highly purified protein complexes from the thylakoid membrane may be useful in determining the lateral heterogeneity as well as the minimal specific associations between proteins and lipids necessary to maintain the structural integrity of a thylakoid particle or to sustain a particular function (for a review, see Williams, 1994; Chapter 8, Siegenthaler and Trémolières). However, this approach presents several drawbacks and limitations, and also contradicts some reconstitution experiments. Firstly, the tight association with or the enrichment of certain lipids in a thylakoid particle or a lipid-protein complex does not mean at all that these lipids are necessarily involved in the functional expression of these subchloroplast fractions or complexes. Secondly, the systematic use of detergents to obtain membrane subparticles or lipo-protein complexes may result in differential displacements of acyl lipids and pigments non-covalently bound to proteins (Helenius and Simons, 1975). Thirdly, Pick et al. (1985) reported that highly purified preparations of CF_o-CF_1 from thylakoids contain almost exclu-sively SQDG which is likely to be essential for the integrity of the enzyme. However, the same authors have demonstrated that reconstitution of purified CF_o-CF_1 with chloroplast glycolipids stimulates the ATPase activity of the complex and that MGDG is the only chloroplast lipid which by itself activates ATP hydrolysis (Pick et al., 1984; Pick et al., 1987), despite the fact that no evidence was found for tightly bound MGDG molecules in CF_o-CF_1 preparations. Conclusions based only on such evidence should, therefore, be viewed with caution.

B. Transmembrane Heterogeneity

1. Methodological Strategies

The transversal localization of lipids in biological

membranes, especially in erythrocyte membranes, has been carried out primarily by chemical or enzymatic modification, and by exchange or immunochemical procedures (Roelofsen and Zwaal, 1976; Etemadi, 1980). In thylakoid membranes from higher plants, the classical chemical modification of phospholipid aminogroups is unfortunately not possible because of the lack of such groups in the lipids present in the membrane. However, attempts have been made to allow PG to chemically react with the diazonium salt of sulphanilic acid, followed by reduction in the presence of NaB^3H_4 (Unitt and Harwood, 1985). Concerning the galactolipids, their distribution across the thylakoid membrane was investigated by tritium labeling of membrane vesicles of opposite sidedness (Sundby and Larsson, 1985). Tritium was introduced into the galactose headgroups of the lipids by oxidation with galactose oxidase or sodium periodate, and subsequent reduction with tritiated sodium borohydride (Sundby and Larsson, 1985; Rawyler et al., 1987). In these experiments, however, the results can be biased by non-specific labeling. Other approaches applied to thylakoids, such as immunological studies (Radunz, 1976, 1977, 1981; Radunz et al., 1984; Radunz and Schmid, 1989; Voss et al., 1992a,b; Haase et al., 1993; Makewicz et al., 1995), have also confirmed the heterogeneity of the thylakoid membrane lipids (see Section III.C). The enzymatic approach has been particularly rewarding in assessing the precise acyl lipid topography in the thylakoid membrane of higher plants.

This approach consists of treating thylakoid membranes under various conditions with lipolytic enzymes. The rationale of this approach has been discussed in detail (Siegenthaler, 1982; Rawyler and Siegenthaler, 1985; Siegenthaler and Giroud, 1986; Siegenthaler et al., 1987a, 1989a,b). This technique requires the accurate control of quite a number of factors including the structural integrity of thylakoids, many (physico)chemical parameters of the incubation medium, the amount and specificity of the lipolytic enzyme, the constant testing of the lipid content during hydrolysis, the influence of lipid products on lipid asymmetry, the sidedness of the membrane, etc. The advantage of this technique is not only to assess the transmembrane distribution of acyl lipids but, simultaneously, to draw relationships between the hydrolysis of certain lipids and the impairment of particular photochemical functions in the thylakoid membrane.

2. Transmembrane Distribution of Acyl Lipid Classes

Earlier studies have shown that a treatment by small amounts of the lipolytic acyl hydrolase (LAH) from potato results in a stepwise hydrolysis of all membrane acyl lipids. Galactolipids are the first lipids to be attacked by LAH, then the enzyme has access to ionic lipids (Rawyler and Siegenthaler, 1980). This has been suggested to be due to a microheterogeneity of lipids at the membrane surface and/or to surface pressure requirements differing for each substrate of the enzyme (Bishop, 1983).

In contrast to LAH, which is a non-specific enzyme, phospholipases (A_2, C or D) hydrolyze only phospholipid but not glycolipid molecules. From detailed kinetics of phospholipid hydrolysis at 0 °C in the presence of the above enzymes, we have found that PG is asymmetrically arranged across the thylakoid membrane. The molar outside/inside distribution of PG is $70 \pm 5/30 \pm 5$ in oat (Siegenthaler and Giroud, 1986), spinach (Siegenthaler et al., 1989a) and squash (Xu et al., 1994) thylakoid membranes (see Fig. 1). These results give a quantitative support to the qualitative estimation by immunological techniques of the distribution of PG in *Antirrhinum* thylakoid membranes (Radunz, 1981). The enrichment of PG in the outer leaflet was also found in thylakoids from barley, lettuce and pea chloroplasts (Unitt and Harwood, 1985). The transmembrane distribution of phospholipids is similar in oat prothylakoids (Siegenthaler and Giroud, 1986). These results indicate that the incorporation of newly synthesized chlorophyll-protein complexes does not modify the transmembrane arrangement of phospholipids and that phospholipid asymmetry is generated at an earlier stage of membrane biogenesis (Siegenthaler and Giroud, 1986; Section IV).

In addition to phospholipid distribution, the localization of galactolipids in the thylakoid membrane from spinach was also investigated (Rawyler and Siegenthaler, 1985; Rawyler et al., 1987). These studies have been carried out in resting membranes (in the dark) by measuring the kinetics of galactolipid hydrolysis catalyzed by the lipase from *Rhizopus arrhizus*. The molar outside/inside distribution is $65 \pm 3/35 \pm 3$ for MGDG and $15 \pm 2/85 \pm 2$ for DGDG (see Fig. 2B). The transversal distribution of MGDG in thylakoid membranes working under conditions of coupled or uncoupled electron transport is the same as in resting membranes

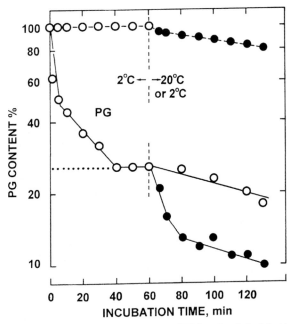

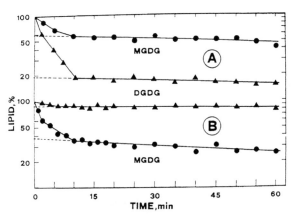

Fig. 1. Transmembrane distribution of PG in spinach thylakoid membrane as determined by the enzymatic approach. Time course of PG hydrolysis at 2 °C (○) and at 20 °C (●) of spinach thylakoids in the presence (—) or absence (- - -) of pancreatic phospholipase A₂. Results are expressed as semilog plots. The 100% value corresponds to 251 nmol/mg chlorophyll. Extrapolation (.....) to zero time of the plateau level reached at 2 °C after 40 to 60 min indicates the outer/inner molar ratio of PG. For more details, refer to Siegenthaler et al. (1989a).

Fig. 2. Transmembrane distribution of MGDG and DGDG in spinach thylakoid membrane from intact thylakoids and thylakoid inside-out vesicles. Time course of galactolipid hydrolysis at 10 °C for thylakoid inside-out vesicles (A) and at 2 °C for intact thylakoids (B) treated with the lipase from *Rhizopus arrhizus*. (●), MGDG; (▲), DGDG. The 100% values correspond to 286 ± 31 (n = 9) and to 155 ± 18 (n = 9) nmol/mg chlorophyll for MGDG and DGDG, respectively, in inside-out vesicles (A) and to 1375 and 675 nmol/mg chlorophyll for MGDG and DGDG, respectively, in intact thylakoids (B). For more details refer to Siegenthaler et al. (1988).

(Rawyler and Siegenthaler, 1989). A similar transmembrane distribution has been found in thylakoids from a great variety of temperate climate higher plants (Rawyler et al., 1987) as well as in oat prothylakoids (Giroud and Siegenthaler, 1988). From these results we have calculated the non-bilayer forming/bilayer forming lipid ratios. These ratios suggest that both monolayers of the oat prothylakoid and the inner monolayer of oat thylakoid membranes should display lamellar structures (e.g., ratio < 2.5). In contrast, the outer monolayer of the thylakoid membrane should form non-lamellar configurations (e.g., ratio > 2.5). Thus, it is concluded that the incorporation of chlorophyll-complexes into the nascent thylakoid membrane modifies neither the galactolipid nor the phospholipid transmembrane distribution. However, these complexes appear to be crucial in preserving a bilayer configuration of the greening membrane which would otherwise adopt non-lamellar structures (Giroud and Siegenthaler, 1988).

The galactolipid and phospholipid transmembrane distribution has been verified unambiguously in spinach by using thylakoid inside-out vesicles. Their galactolipid (Fig. 2) and phospholipid asymmetry is 'just' the opposite of that found in intact thylakoids (Siegenthaler et al., 1988; Duchêne et al., 1996). Although these results were expected, they are of importance in view of the earlier controversy concerning the relevance of using lipolytic enzymes to determine lipid asymmetry (Webb and Green, 1991). The distribution and main characteristics of acyl lipids in the outer and inner monolayer of the thylakoid membrane in spinach chloroplasts are illustrated in Fig. 3.

3. Transmembrane Distribution of Phosphatidylglycerol Molecular Species

As reported above, the outer monolayer of the thylakoid membrane is highly enriched in PG in spinach (Siegenthaler et al., 1989a), oat (Siegenthaler and Giroud, 1986), barley, lettuce and pea (Unitt and Harwood, 1985). These latter authors have found that in all three plant species tested, 16:1(3t), one of the main fatty acid of PG, is exclusively located in the

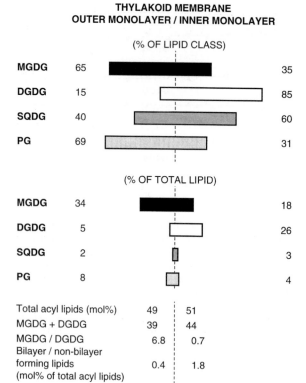

THYLAKOID MEMBRANE
OUTER MONOLAYER / INNER MONOLAYER

(% OF LIPID CLASS)

MGDG	65	35
DGDG	15	85
SQDG	40	60
PG	69	31

(% OF TOTAL LIPID)

MGDG	34	18
DGDG	5	26
SQDG	2	3
PG	8	4

Total acyl lipids (mol%)	49	51
MGDG + DGDG	39	44
MGDG / DGDG	6.8	0.7
Bilayer / non-bilayer forming lipids (mol% of total acyl lipids)	0.4	1.8

Fig. 3. Distribution and characteristics of acyl lipids in the outer and inner monolayer of the thylakoid membrane from spinach chloroplasts. Lipids are expressed as mol% of each lipid class (upper part) and mol% of total lipid (middle part). The composition in lipid classes expressed in nmol/mg Chl is given in Table 1. For more details refer to Siegenthaler et al. (1987b) and Giroud and Siegenthaler (1988).

outer monolayer. One consequence of this result is that no PG molecular species containing 16:1(3t) should be located in the inner monolayer. Spinach, lettuce, potato and squash thylakoid membranes contain ten different PG molecular species (Xu and Siegenthaler, 1996). In spinach (a chilling resistant plant) only three of them display values >1.2 mol %, i.e., 18:3/16:1(3t), 18:3/16:0 and 16:0/16:1(3t), whereas in squash (a chilling sensitive plant) nine of them display values >1.2 mol % (see Table 2). In spinach, we found that the relative amount of three main species, especially the 18:3/16:1(3t) one, which accounts for 76 mol %, is almost equally distributed within the two monolayers (Siegenthaler et al., 1992). These results indicate that the dependence of electron flow activity on inner PG (Siegenthaler et al., 1989a) does not lie on the identity of PG molecular species but rather in their interactions with specific inner proteins or protein domains.

C. Binding of Anti-Lipid Antibodies to the Thylakoid Membrane

1. Binding of Anti-Lipid Antibodies to the Surface of the Membrane

Radunz was among the first to use antibodies directed to native lipids to explore the molecular organization of lipids at the surface of the thylakoid membrane. To this end, specific antibodies to SQDG (Radunz and Berzborn, 1970), PG (Radunz, 1971), MGDG (Radunz, 1972) and DGDG (Radunz, 1976) were obtained by immunization of rabbits by lipids previously adsorbed on methylated bovine serum albumin. Control tests revealed that immunoglobulins are specifically directed to the polar head group moiety of the lipid but not to the fatty acid one. This type of interaction has to be kept in mind when interpreting serological results. Earlier work dealing with various subchloroplast fractions showed that the antigenic determinants of all thylakoid lipid classes could be localized, to various extents, at the outer and/or inner surfaces of the membrane. However, a mixture of anti-lipid antibodies covers only one fourth of the surface, indicating that lipids are likely to fill up the gaps between the protein molecules, thus facilitating, by means of their fluidity properties, conformational changes of proteins (Radunz, 1977, 1981).

2. Binding of Anti-Lipid Antibodies to the PS II Complex

Voss et al. (1992b) prepared PS II complexes from chloroplasts of wild type tobacco *Nicotiana tabacum* and from two chlorophyll mutants. The hydrophobic peptides of these complexes were analyzed for bound lipid molecules by western blot with monospecific anti-lipid antibodies. The antibodies directed to DGDG and SQDG react exclusively with the 66 kDa peptide which is the heterodimer D1/D2. In contrast, the antibody directed to MGDG does not react with the heterodimer but with all the other peptides of the PS II complex including the isolated D1 (32 kDa) and D2 (34 kDa) peptides, the two chlorophyll-binding peptides (43 and 48 kDa), the apoproteins of cytochrome b_{559} (10 and 4 kDa) and the peptides of the oxygen-evolving complex (33, 23 and 16 kDa). Various treatments of the PS II preparations (washings with unpolar and polar organic solvents, treatments with sodium periodate and the lipase from *Rhizopus arrhizus*) indicate that lipids are tightly bound and not just unspecifically adsorbed onto the peptides. Altogether, Voss et al. (1992b) concluded that: (1) Bound sulfolipids and digalactolipids are localized as prosthetic groups at the surface of the native D1/D2 heterodimer (66 kDa) of the PS II reaction center in such a way as to be accessible to their antibodies; (2) These bound lipids might, therefore, hold the dimer together; (3) MGDG binds to all the PS II peptides in decreasing order: LHCII > D1 and D2 > the chlorophyll binding peptides > the three extrinsic peptides of the oxygen evolving complex; (4) The tight binding of certain lipids to defined peptides of the PS II core has a functional significance, since treatments of thylakoid membranes or PS II particles with certain anti-lipid antibodies inhibit the electron transport reaction on the donor side of PS II (Chapter 8, Siegenthaler and Trémolières).

Furthermore, the intrinsic polypeptide D1, isolated from the PS II-preparation of the cyanobacterium *Oscillatoria chalybea* by SDS-gel electrophoresis, reacts, in the western blotting, only with the antibodies directed to PG and β-carotene, but not to MGDG and SQDG. The binding of PG to the D1-core polypeptide was confirmed by HPLC lipid and GC fatty acid analyses (Kruse et al., 1994). In this organism, the PG/D1-polypeptide molar ratio is about 20:1.

3. Binding of Anti-Lipid Antibodies to the PS I Complex

PS I preparations from the wild type tobacco *Nicotiana tabacum* var. JWB consist of the core complex I (CCI) and the light-harvesting protein complex I (LHCI). The core complex contains the two core polypeptides having the same apparent Mr of 66 kDa and several polypeptides of lower Mr (22, 20, 19, 17, 16 and 9 kDa). LHCI consists of 4 subunits (28, 26, 25 and 24 kDa). Using specific anti-lipid antibodies, Makewicz et al. (1995) were able to show that PS I preparations contain only MGDG and PG in the ratio 2:1. The two lipids differ with respect to binding strength to the PS I polypeptides. Organic solvent extraction and boiling experiments showed that PG is much more tightly bound to CCI than MGDG. In addition, these authors calculated that CCI contains a proportion of 27 molecules of MGDG and 42 of PG, while LHCI contains only MGDG. These results highlight the great hetero-

geneity of lipids, not only between PS II and PS I complexes but also within one complex [CCI(II) versus LHCI(II)].

4. Binding of Anti-Lipid Antibodies to the Coupling Factor

The analysis of the binding of antibodies to lipids onto the surface of the thylakoid membrane, before and after the removal of the CF$_1$-complex with sodium bromide, suggests that in the immediate vicinity of CF$_1$, SQDG and MGDG occur in higher concentrations and are likely to be arranged in domains (Haase et al., 1993). This means that the hydrophobic portion of the coupling factor spanning the thylakoid membrane (i.e., CF$_0$) is surrounded by MGDG, in particular in the outer monolayer of the membrane. The reason for this accumulation of MGDG molecules presumably lies in the highly unsaturated character of the fatty acids of this lipid, which might facilitate, due to higher fluidity of the surrounding membrane regions, conformational changes of the complex. In agreement with this concept is the recent finding that a depletion of MGDG in the outer monolayer of the thylakoid membrane is much more effective in decreasing photo- and dark phosphorylation than a depletion of PG which contains fewer unsaturated molecular species than MGDG (Siegenthaler and Vallino, 1995).

The hydrophilic CF$_1$-portion of the ATPase is composed of five subunits with the stoichiometry α_3, β_3, γ, δ and ε. The binding of glycolipids and phospholipids onto the subunits of CF$_1$ from spinach was determined by using monospecific antibodies and western blotting. The electroneutral galactolipids (MGDG and DGDG) as well as the two electronegative (anionic) lipids (SQDG and PG) are bound to the two large subunits α and β of the CF$_1$-complex. The small subunits δ and ε react with none of the antibodies tested. Occasionally, the γ-subunit appears to be marked by the MGDG antibody. These results show that the lipids forming the bilayer of the thylakoid membrane are also present and bound to the large subunits of the CF$_1$-complex which protrudes out of the membrane, into the stroma (Haase et al., 1993). Based on lipolytic treatments and methanol extraction experiments as well as the use of different antibodies against MGDG and SQDG, these authors concluded that: (1) Lipids are differently bound to the α- and β-subunits indicating that the accessibility of the lipid antigenic determinants on both subunits is different or that the distribution of the lipids on the two-subunits is different; (2) Since the lipid antibodies are not directed to the fatty acid region but to the polar head group region, one can assume that the lipids are linked via the fatty acid residues to the α- and β-subunits. Only after the decomposition of lipids by lipases (but not after methanol extraction) is the antigen-antibody interaction abolished.

D. Modulation of the Acyl Lipid Saturation by Catalytic Hydrogenation of Thylakoids

The concept of lateral separation of PS II and PS I raises questions concerning the electron transfer and energy distribution between the two photosystems. As a mobile redox carrier, plastoquinone links Photosystem II and the cytochrome $b_6 f$ complex. It has been proposed that this compound is located primarily within the fluid bilayer-midplane region of the thylakoid membrane and thus can move laterally at very high rates (Millner and Barber, 1984). In addition, the regulation of quantal distribution between PS I and PS II requires the physical movement of the LHC between appressed and non-appressed membranes (Barber, 1982; Allen, 1992). These concepts emphasize the importance of the thylakoid membrane lipid moiety as the matrix in which diffusion of certain membrane components has to occur. In this respect, the fluidity and packing properties of this membrane in response to various conditions (e.g., environmental stimuli such as temperature, light variations, etc.) are important factors to be considered. Furthermore, it has been proposed that thylakoid fluidity can be estimated by the lipid: protein ratio (Ford and Barber, 1983). This ratio can be manipulated artificially by incorporating either lipids or proteins into the thylakoid membrane (for a review see Siegenthaler and Rawyler, 1986).

Chemical modification of acyl lipids provides another powerful tool for studying their role in the structure and functions of thylakoid membranes. At present, the only manipulation which directly and exclusively changes the fatty acid core of membrane lipids is the catalytic hydrogenation of their double bonds (for an extensive review, see Quinn et al., 1989). This method was originally developed by Chapman and Quinn (1976) and further investigated in several laboratories to obtain finally a water-soluble catalyst Pd $(QS)_2$, a sulphonated alizarine complex of Pd(II). This catalyst has been used successfully to modulate in situ chloroplast membrane lipids by homogenous catalytic hydrogenation (Vigh et al., 1985). The catalyst itself does not affect the photosynthetic activity but causes an extensive loss of unsaturated fatty acids in the presence of hydrogen gas without affecting the double bonds of neutral lipids such as chlorophylls, carotenoids and plastoquinone (Szalontai et al., 1986). Poly-unsaturated fatty acids, namely linolenic acid, are hydrogenated at a higher rate than the monoenoic acids and converted into more saturated C_{18} acids (Vigh et al., 1985). Concerning the different lipid classes of thylakoids, their susceptibility to hydrogenation decreases in the order MGDG > DGDG > SQDG > PG (Horváth et al., 1986b). This effect could be due to a charge repulsion between the negatively-charged functional groups of SQDG and PG and the sulphonated alizarine groups on the palladium catalyst complex. Moreover, it has been found that progressive hydrogenation of these lipids in situ produces a change in their organization, e.g., single bilayer vesicles are converted to large aggregates of planar bilayer stacks in which the hydrocarbon chains are predominantly in the gel phase configuration (Horváth et al., 1986b). These authors proposed a mechanism for structural changes occurring in chloroplast membranes submitted to catalytic hydrogenation (Horváth et al., 1986a). Although during hydrogenation the basic thylakoid structure remains unchanged, the saturation of only 10% of fatty acyl double bonds already induces a definite decrease in the dimension of both thylakoids and loculi which arises from a thickening of the single membranes with a simultaneous decrease in the spacing between membranes. These changes might be accounted for by the alignment of the hydrocarbon chains of saturated lipids and the increased hydrophobicity of the membranes.

More recently, the water-soluble palladium complex was used as a probe to test the acyl lipid molecular organization of liposomes and thylakoids originating from triazine-resistant (R) and triazine-susceptible (S) lines of *Solanum nigrum* (Mayor et al., 1992). When thylakoids isolated from R- and S-biotype are submitted to catalytic hydrogenation, the double bond content of total lipids as well as that of individual lipids (MGDG, DGDG and SQDG) decreases appreciably, reaching its minimum after 5 min. However, great differences are observed between the saturation curves of the two biotypes. Although both plots are biphasic, i.e., a lag phase is followed by the main reaction, hydrogenation begins 50 s later in S- than in R-thylakoids resulting in a 30 to 40% difference in the extent of saturation between the two samples. Surprisingly, the rates of catalytic hydrogenation of PG are similar in both types of membranes. However, when liposomes, formed from thylakoid polar lipids originating from R- or S-biotype, are submitted to a similar hydrogenation procedure, R- and S-samples display identical saturation curves (Mayor et al., 1992). These findings can be interpreted as follows: (1) The abolition of the original membrane assembly (liposomes) results in equal accessibility of all lipids to the catalyst; (2) The factors rendering some glycolipid pools relatively less accessible to the catalyst in S-type membranes are inoperative in the case of PG. The differences in the apparent accessibility of the catalyst to membrane lipid substrates could be due to: (1) A differential effect of the catalyst itself on the microviscosity of the different bulk lipid populations; (2) The existence of lipid microdomains due to lateral heterogeneity in membranes; (3) A phase separation of non-bilayer lipids forming H_{II} phase (Mayor et al., 1992 and references therein). Further evidence to support differences in membrane architecture of S- and R-type thylakoids is provided by the lifetime distribution analysis of diphenylhexatriene (Mayor et al., 1992). For an apparently identical lipid class and fatty acid composition and fluidity of thylakoids, the greater environmental heterogeneity of the fluorophore in S-membranes may be attributed to stronger protein-lipid interactions. Intramembraneous domains of acyl lipids are likely to be present only in the atrazine-susceptible biotype, despite the fact that some of them are relatively well protected against the catalyst.

As mentioned above, the technique of catalytic hydrogenation provides a valuable tool in assessing the role of acyl lipids in the photosynthetic membrane (Quinn et al., 1989; Chapter 8, Siegenthaler and Trémolières). However, a major drawback of this method is the homogeneity property of the hydrogenation catalysis. Although the hydrogenation extent is different for each of the thylakoid acyl lipids, nevertheless they are apparently all randomly hydrogenated. The stepwise hydrogenation may reflect differences in the accessibility of the catalyst and/or selective interaction between the catalyst and lipid populations. Thus, any conclusions about the extent of hydrogenation and activity inhibition can be drawn regardless of the lipid population which has been hydrogenated.

E. Cyclodextrins: A New Tool for the Controlled Lipid Depletion of Thylakoid Membranes

Cyclodextrin (CD) molecules are cyclic, non-reducing oligosaccharides consisting of 6, 7 or 8 (n) glucopyranose units linked by α (1–4) bonds (Fig. 4A). They adopt a torus shape (Fig. 4B) and are able to bind a range of small guest molecules of poor water solubility within their hydrophobic cavity to form a water-soluble guest-CD inclusion complex (Fig. 4C) (Szejtli, 1988, 1990; Bekers et al., 1991). As shown in Fig. 4B, all secondary OH groups are located on the edge of the torus-like CD molecule, while all primary OH groups are on the other side. The lining of the cavity is formed by H atoms and glucosidic oxygen-bridge atoms. Therefore, this surface is slightly apolar which makes the CD molecule an empty capsule of molecular size. When it is filled with the molecule of another substance (e.g., a lipid) it forms an inclusion complex which is an entity comprising a host (CDs) and a guest (lipid) molecule, held only by physical forces, i.e., without covalent bonding. In aqueous solution the slightly apolar CD cavity is occupied by water molecules which are energetically unstable (polar-apolar interactions) and can therefore be readily substituted by lipid guest molecules which are less polar than water molecules. The essence of the molecular encapsulation is that one, two or three CD molecules contain one or more entrapped guest molecules (adapted from Szejtli, 1990). A variety of lipid molecules have been shown to form such complexes with cyclodextrins (Ohtani et al., 1989; Okada et al., 1989). So far, cyclodextrins have been used mainly in the biomedical and galenic fields (Szejtli, 1988, 1990; Irie et al., 1992; Nakanishi et al., 1992 and references therein).

Recently, we have explored the possibility of using cyclodextrins as lipid complexing agents for lipid depletion purposes in structural studies of thylakoid membranes as illustrated in Fig. 4C. According to Szejtli (1990), the inclusion complex is a highly dynamic depot maintaining a low and constant concentration of the slowly dissolved free guest. In aqueous CD solutions, the solubility of lipophilic substances increases. The extent of this solubility enhancement depends on the type and concentration of CD, temperature, etc., but mainly on the chemical structure of the substrate. In addition to this two-step process, lipid molecules may be directly removed from the membrane in a one- and fast-step, as

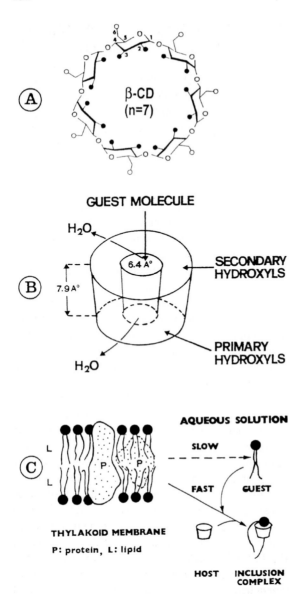

Fig. 4. Structure and relevant properties of cyclodextrins. Cyclodextrins (CDs) are cyclic, non-reducing oligosaccharides, where n corresponds to the number of glycopyranosidic units. α-CD [n = 6; diameter of the cavity: 5.2 Å; diameter of periphery: 14.6 Å; height of torus: 7.9 Å]; β-CD [n = 7; 6.4 Å; 15.4; 7.9 Å, respectively]; γ-CD [n = 8; 8.3 Å; 17.5 Å; 7.9 Å, respectively] (from Bekers et al., 1991). A: The structure and the numbering of the atoms of β-CD. B: Schematic representation of β-CD (adapted from Szejtli, 1990). C: Schematic illustration of the host-guest interaction (adapted from Szejtli, 1990). In the case of thylakoid membrane galactolipids, model system shows that only one acyl chain of the MGDG molecule can be introduced in the cavity of β-CD. This does not exclude, however, that more than one β-CD can be partners in the inclusion complex (see, for instance, Stoddart, 1992).

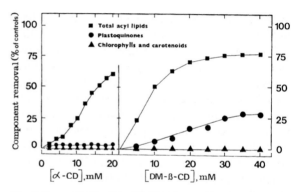

Fig. 5. Removal of total acyl lipids, plastoquinone, chlorophylls and carotenoids by increasing concentrations of α-cyclodextrin (α-CD) and DM-β-cyclodextrin (DM-β-CD) in spinach thylakoid membranes. After an incubation at 0 °C for 5 min, cyclodextrins were removed by washing and thylakoids analyzed. Total lipid and plastoquinone amounts in control thylakoid membranes were 2.5 and 0.1 μmol/mg chlorophyll, respectively. For further details, refer to Rawyler and Siegenthaler (1996).

illustrated in the scheme (Fig. 4C). To this end, α- and β-cyclodextrins as well as DM-β-cyclodextrin [heptakis-(2,6-di-O-methyl)-β-cyclodextrin] and HP-β-cyclodextrin [hydroxypropyl-β-cyclodextrin] have been used alone or in combination (Rawyler and Siegenthaler, 1996). The removal of lipids is characterized by the following features: (a) As shown in Fig. 5, both CDs removed acyl lipids, but the efficiency of the process, measured by the initial slope of the curve, is approximately three times higher with DM-β-CD than with α-CD. The extent of removal depends on both CD and thylakoid concentrations; (b) Chlorophylls and carotenoids are not extracted from thylakoids by both CDs whereas plastoquinones are removable, though only in small amounts, with a low efficiency (DM-β-CD > α-CD) (Fig. 5); (c) A treatment up to 20 mM α- or DM-β-CD, does not affect the protein content of thylakoid membranes as judged by the total amount of protein and by the content of two individual intrinsic polypeptides belonging to the light-harvesting (LHCIIα and LHCIIβ) and of two extrinsic polypeptides (α- and β-subunits of the coupling factor CF_1 complex) measured in control and CD-treated membranes; (d) For all lipid classes, the extent of depletion is higher at 0 °C than at 20 °C; (e) The presence of $MgCl_2$ reduces the removal of PG and SQDG but does not affect galactolipid depletion levels; (f) Stable lipid depletion levels in thylakoid membranes are reached after 5 to 10 min of cyclodextrin treatment at 0 °C; (g) Of the four CDs tested, only three (α-CD, β-CD, and DM-β-CD) promote lipid depletion whereas one (hydroxypropyl-β-CD) fails completely to do so; (h) As illustrated in Fig. 6, α-CD displays a higher selectivity towards lipid classes than does DM-β-CD, but in both cases the removal order was SQDG $\geq$ PG $\geq$ MGDG $\geq$ DGDG; (i) Interestingly, when α- and β-CDs are used together, their effect is additive for PG and SQDG, but synergistic for galactolipids; (j) Finally, α-CD shows a preference for those lipids containing 16-carbon acyl chains whereas DM-β-CD is essentially insensitive to the fatty acid composition of the lipids.

In conclusion, CD-mediated lipid removal provides a valuable and versatile tool to achieve controlled and specific lipid depletions in biological membranes—thus avoiding the use of either detergents or lipolytic enzymes.

IV. Origin of Lipid Heterogeneity in Thylakoids

A. Role of Chloroplast Envelope Membranes

In higher plant chloroplasts, the thylakoid membrane network develops an extended area in order to maximize its absorbance capacity and, consequently, its overall efficiency for photon capture, energy transduction and energy conservation. This is reflected, at the molecular level, by the fact that at least 90% of the chloroplast acyl lipids belong to the thylakoids whereas the remaining 10% are shared between the two envelope membranes (Joyard and Douce, 1976). However, thylakoids are essentially deprived of the ability to synthesize lipids (Dorne et al., 1990) and must therefore rely on the chloroplast envelope for supply of membrane lipids.

The two envelope membranes greatly contribute to the synthesis and renewal of the main acyl lipids of the chloroplast, namely of galactolipids. Whatever the chloroplastic or cytoplasmic origin of its diacylglycerol backbone, MGDG is formed by a galactosyl transfer between UDP-galactose and diacylglycerol, catalyzed by the UDP-galactose: diacylglycerol galactosyltransferase or MGDG synthase (Joyard et al., 1991). The final step of DGDG synthesis has been assigned to the galactolipid: galactolipid galactosyltransferase (Heemskerk et al., 1990). Whereas the pathways leading to the synthesis of galactolipids and to their subsequent desaturation have been extensively studied (Heemskerk and Wintermans, 1987; Joyard and Douce, 1987;

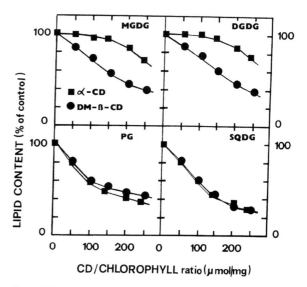

Fig. 6. Changes in the relative amounts of individual diacyl lipid classes of CD-treated thylakoid membranes. Incubation was at 0 °C for 30 min. The chlorophyll concentration was 50 µg/ml with α-CD and 75 µg/ml with DM-β-CD and the 100% values of the controls were 1300 (MGDG), 685 (DGDG), 260 (PG) and 206 (SQDG) nmol lipid/mg chlorophyll, respectively. For further details, refer to Rawyler and Siegenthaler (1996).

Schmidt and Heinz, 1990; Joyard et al., 1991), the problem of the transfer of (galacto)lipids from their sites of synthesis (the envelope membranes) to the thylakoid network has received much less attention.

Experimental evidence suggesting that (galacto) lipid molecules are exported from the envelope to the thylakoid membranes was first provided by Williams et al. (1979) and later confirmed (Joyard et al., 1980; Bertrams et al., 1981). However, apart from its occurrence and its rapidity, no information is available on the detailed aspects of this export process. More recently, this problem has been tackled by two groups using different approaches. Morré et al. (1991) have used a cell-free system to show that MGDG can be transferred from isolated envelope vesicles to immobilized thylakoid membranes in an ATP- and temperature-dependent fashion. In our laboratory, we have studied lipid export using intact chloroplasts supplied with UDP-[^{14}C]galactose, [1-^{14}C]acetate and [^{14}C]glycerol 3-phosphate. The transfer in organello of newly synthesized lipid molecules from inner envelope to thylakoid membranes, as well as their subsequent transbilayer distribution in these membranes, have been studied in intact chloroplasts (Rawyler et al., 1992, 1995).

At this point, it is worth mentioning that there are

lines of evidence suggesting that thylakoids might also be the site of glycerolipid synthesis (Sandelius and Selstam, 1984; Omata and Murata, 1986; Slabas and Smith, 1988) and desaturation of fatty acids (Ohnishi and Thompson Jr., 1991; Mustardy et al., 1996). At present, however, it is not known whether both mechanisms (i.e. the transfer of lipids from chloroplast envelope to thylakoids and the in situ lipid synthesis and desaturation in thylakoids) coexist and to what extent they do.

B. Lipid Export from Chloroplast Envelope to Thylakoid Membranes

1. Possible Mechanisms

If one assumes that thylakoids have to import their lipids (or part of them) from the chloroplast envelope, one can envisage the lipid transfer as the result of at least four possible basic mechanisms (Rawyler et al., 1992): (a) The inner envelope membrane buds off and releases vesicles which move through the stroma and fuse with the thylakoid membranes; (b) A lipid transfer protein extracts a lipid molecule from the stroma-facing monolayer of the inner envelope membrane, carries it through the stroma and inserts it into the stroma-facing monolayer of the thylakoid membrane; (c) A partial fusion is triggered between the stroma-facing monolayers of both inner envelope membrane and thylakoid membrane allowing lipid molecules to diffuse laterally within these monolayers. In the case of MGDG, the inverted micelles at the fusion locus (Siegel, 1986a,b) should prevent the contents of the two other monolayers from being mixed; (d) A complete fusion occurs between inner envelope and thylakoid membranes, connecting both their stroma-facing monolayers and their lumen-facing monolayers together, thus allowing an unimpeded lateral diffusion of lipid molecules (or of other components) within the whole bilayers.

2. Strategy for Studying Lipid Export in Intact Chloroplasts

For the sake of clarity, the methodological aspects for studying the lipid export process in organello are outlined in Fig. 7 (Rawyler et al., 1992, 1995). In the first four steps (1–4, Fig. 7), intact chloroplasts are provided with a suitable radioactive precursor (acetate or glycerol-3-phosphate, UDP-galactose) together

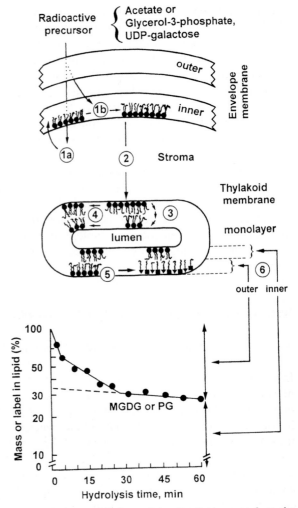

Fig. 7. Strategy used for studying the lipid export from the chloroplast envelope to the thylakoid membrane (see explanations in the text). For details, refer to the text and to Rawyler et al. (1992, 1996).

with all the necessary cofactors and are allowed to synthesize membrane lipids (Heemskerk and Wintermans, 1987) either from the de novo pathway (step 1a: labeled acetate or glycerol-3-phosphate) or by glycosylation of preexisting lipids (step 1b: labeled UDP-galactose). Some of the newly made lipids are exported (step 2) from their site of synthesis (envelope membrane) to their final destination (the thylakoid). When present in the thylakoid membrane, these lipids may undergo transbilayer movement (step 3) and may be redistributed between the outer (stroma-facing) and the inner (lumen-facing) monolayers. Lipids may also diffuse laterally (step 4). Since a method is available to prepare pure, envelope-free thylakoids from intact chloroplasts (Rawyler et al., 1992), a comparative determination of the amount of lipid radioactivity in the intact plastids and in the corresponding thylakoid membranes will give direct information on the extent, rate and selectivity of the lipid export process. Furthermore, purified thylakoids can be treated by lipolytic enzymes (Rawyler and Siegenthaler, 1985; Siegenthaler et al., 1989a). For instance, both the lipase from *R. arrhizus* and the phospholipase A_2 will generate a free fatty acid (single-tailed arrow) and the corresponding lyso-lipid (single tailed square) from galactolipids and phosphatidylglycerol (double-tailed circles), respectively (step 5). Then a comparative sidedness analysis of the lipid mass and label distribution between both monolayers of the thylakoid membranes can be performed (step 6). The presence of radioactive lipids in either monolayer of the thylakoid membrane will depend not only on the transfer mechanism involved, but also on the topology of lipid synthesis in envelope membranes and on the possible occurrence of lipid translocation activities across the thylakoid bilayer. Keeping the temperature around 0 °C can essentially prevent transmembrane movement of lipids between the two monolayers (Rawyler and Siegenthaler, 1985; Siegenthaler et al., 1989a). At this stage, the amount of radioactivity associated with lyso-MGDG, lyso-DGDG or lyso-PG will therefore represent the amount of exported MGDG, DGDG or PG that is located in the outer monolayer of thylakoid membranes at the end of the incubation period of intact chloroplasts with the radioactive precursor, while the residual parent lipids (unlabeled) correspond to the inner pools. This strategy helps in identifying which export mechanisms, involving stromal intermediates (vesicles, lipid transfer protein) or not (semi- or complete fusion), are operative in plastids.

3. Characterization and Involvement of Lipid Export from Chloroplast Envelopes in Thylakoid Lipid Asymmetry

Intact chloroplasts, isolated from mature and young spinach, young pea and mature lettuce leaves have been incubated either in the dark with UDP-[^{14}C]galactose or in the light with [^{14}C]acetate or [^{14}C]glycerol 3-phosphate. The former precursor is used to follow the fate of MGDG and DGDG made from polyunsaturated preexisting diacylglycerol, whereas the latter ones are used to follow the fate of

MGDG and PG, respectively, after the de novo synthesis. The essential data are summarized below (Rawyler et al., 1992, 1995): (a) Purified (envelope-free) thylakoids contain radiolabelled MGDG, DGDG and PG, indicating that these molecules are exported from the inner envelope; (b) The amounts exported are strictly proportional to the amounts synthesized, provided that the necessary substrates are not limiting; (c) Lipid export is class selective; under the conditions used, as much as 50–80% of MGDG, 87% of PG and 20–30% of DGDG synthesized are exported to thylakoids; (d) The MGDG/DGDG ratio is 7 in intact chloroplasts and 18 in thylakoids, suggesting a preferential export of MGDG; (e) However, within the MGDG class labeled from [^{14}C]acetate, there is hardly any selectivity in the export of its various molecular species; (f) For MGDG, the proportionality coefficient, which represents the proportion of chloroplast lipid synthesized used for export towards thylakoids, is higher in chloroplasts from young leaves than from mature leaves, and higher in spinach chloroplasts than in pea and lettuce chloroplasts; (g) In most cases, the transmembrane distribution of labeled lipids in thylakoids matches closely the corresponding distribution of mass, regardlesss of plant age and species and of incubation time and temperature. In some cases, however, small but significant differences occur between the label and mass transbilayer distributions of MGDG (labeled molecules more inwardly oriented), DGDG and PG (more outwardly oriented).

4. A Key Role for Chloroplast Envelopes in the Establishment of Lipid Asymmetry in Thylakoid Membranes

The above data (Rawyler et al., 1992, 1995) show that the intrachloroplastic export of acyl lipids from their initial biosynthesis site (the inner envelope membrane) to their final destination site (the thylakoid membrane) is a general process which occurs in various higher plants and involves most of the lipid classes of the inner chloroplast envelope membrane. In the last step of the export process, those lipid molecules now located in thylakoid membranes adopt a transmembrane distribution that is either definitive or (particularly when desaturation is still in progress) liable to further adjustments.

A peculiarity of the lipid transfer process is its class selectivity which is expressed by the fact that certain lipids (e.g., MGDG and PG) are exported towards thylakoids in clear preference to others (e.g., DGDG and probably SQDG). This suggests that the topography of lipid-synthesizing enzymes at the envelope level is involved in this selectivity.

The data which are available as yet and which have been discussed thoroughly (Rawyler et al., 1995) favor the hypothesis that the intrachloroplastic lipid export is primarily achieved by a mechanism involving transient complete fusions between inner envelope and thylakoid membranes, although a vesicular transfer cannot be excluded. This new proposal points to the key role of chloroplast envelope membranes in the establishment of MGDG, DGDG and PG (and likely SQDG) asymmetry in thylakoid membranes. Thylakoid lipid asymmetry is primarily preestablished in the chloroplast envelope by the topography of its lipid-synthesizing enzymes, together with the occurrence of relatively fast lateral diffusion and translocation rate of the newly synthesized lipids. Transient fusion between inner envelope and thylakoid membranes would allow lipid export by lateral diffusion and it would build the observed lipid asymmetry in thylakoids.

C. A Model for the Origin of Lipid Asymmetry in Thylakoid Membranes

A model showing the fundamental role played by chloroplast envelope membranes in the establishment of acyl lipid asymmetry across the thylakoid membrane is illustrated in Fig. 8.

The outer envelope membrane with its two monolayers (a,b) is separated from the inner envelope membrane with its two monolayers (d,e) by the intermembrane space (c). Because of the cytoplasmic origin of UDP-galactose (Murphy, 1982a) and also the impermeability of the inner envelope membrane to this compound, we have proposed (Rawyler et al., 1995) that galactosylation of the diacylglycerol DAG_1 (synthesized de novo through the Kornberg-Pricer pathway) by the UDP-galactose: diacylglycerol galactosyl transferase (Joyard and Douce, 1976) takes place at the cytoplasmic side of the inner envelope

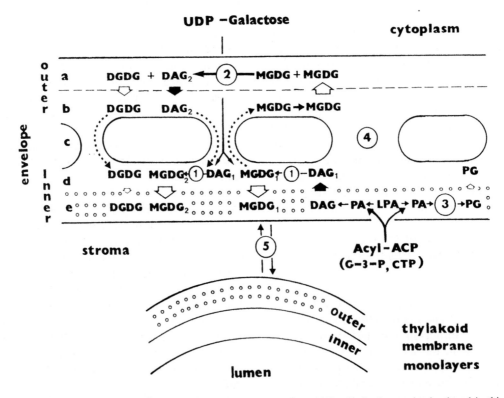

Fig. 8. A model for the origin of lipid asymmetry in thylakoid membranes and for the fundamental role played in this process by chloroplast envelope membranes (see explanations in the text). Adapted from Rawyler et al. (1996).

membrane to yield $MGDG_1$. This molecule may then undergo an inwardly oriented transbilayer movement at a relatively high, non-limiting rate (large open arrow), possibly in a quasi-one step transfer (Simbeni et al., 1990). $MGDG_1$ may also diffuse laterally until it reaches a contact site (4) which enables it to enter (van Venetië and Verkleij, 1982) and equilibrate (Simbeni et al., 1990) in the outer envelope membrane. From the known localization (Dorne et al., 1982) of the galactolipid: galactolipid galactosyltransferase (2), a dismutation of MGDG molecules is then assumed to occur at the cytoplasmic side of the outer envelope membrane to yield DGDG (Heemskerk et al., 1990) and DAG_2. After its synthesis, DGDG must first be translocated inwardly, before diffusing across contact sites (4) to reach the cytoplasmic side of the inner membrane, where it accumulates because of a slow rate (small open arrow) of transbilayer movement to the stroma side of this membrane. DAG_2 is assumed to be translocated very rapidly (large black arrows) across the outer envelope bilayer and to diffuse via contact sites (4) towards the cytoplasmic side of the inner envelope membrane where it is eventually galactosylated (1) to $MGDG_2$ and inwardly translocated as described for MGDG. Desaturases, not represented in the model, are expected to function in the inner envelope membrane (Joyard et al., 1991). Lastly, the assembly of phosphatidic acid (PA) from lysophosphatidic acid (LPA) requires acyl residues of stromal origin and, accordingly, takes place at the stroma side of the inner membrane (Joyard and Douce, 1977; Block et al., 1983; Andrews et al., 1985). Since the other cofactors required for PG synthesis (glycerol 3-phosphate and CTP) are available in the stroma (Dyer, 1984), the subsequent steps (3) leading to the formation of PG (Mudd and de Zacks, 1981; Andrews and Mudd, 1985) are assumed to be also localized at the stroma side of the inner envelope membrane, where PG accumulates preferentially because of a slow rate (small open arrow) of outward transbilayer movement.

The overall result is the synthesis-dependent establishment of a lipid asymmetry in the inner envelope membrane, with the stroma side being primarily enriched in PG and (to an extent which depends on its desaturation level) in MGDG and with the cytoplasmic side enriched in DGDG. By analogy with MGDG synthesis, one may even suggest that SQDG synthesis (Seifert and Heinz, 1992) occurs at the cytoplasmic side of the inner envelope membrane, thus giving rise to a marked enrichment of this monolayer in SQDG, provided that its translocation rate is slow. Lipid export from the inner envelope membrane to thylakoids (5) would then be achieved by transient fusion events (Heber and Heldt, 1981; Carde et al., 1982; Rawyler et al., 1992) connecting both stroma-facing monolayers (dotted zones) together and both complementary monolayers (lumenal for thylakoids and cytoplasmic for inner envelope membrane) together. Possible mechanisms for such fusion events have been described (Siegel, 1986a,b).

V. Search for a Paradigm of the Thylakoid Membrane Molecular Organization Based on Lipid Topology and Properties

Each type of membrane is characterized by the amount and kind of its proteins and lipids. The membrane components are integrated into a functional whole that transcends the sum properties of its parts. The structural arrangement of the membrane components is therefore of primary concern to an understanding of membrane function; conversely, knowledge of the functional activities may help to conceive the membrane structure.

When considering the numerous models which have been published in the literature on the molecular organization of the thylakoid membrane, it is surprising that only a few researchers have attempted to represent the acyl lipids according to their topological distribution and associated functions. In the now classical fluid mosaic model of the membrane all lipids are alike, as represented by a polar head group and two apolar tails. However, it is known that the thylakoid membrane contains four lipid classes including about 11 main lipid molecular species and 27 minor ones (Nishihara et al., 1980; Xu and Siegenthaler, 1996) which display lateral and transversal heterogeneities, specific interactions with intrinsic membrane complexes and/or proteins (Section III) as well as appreciable motional restriction on the ESR time-scale (Li et al., 1989; Ivancich et al., 1994). Consequently, any model of the thylakoid membrane should take into consideration the diversity and the known properties of the lipids.

One of the first models of the thylakoid membrane involving lipids at the molecular level has been proposed by Weier and Benson (1967). According to this model, the bilayer structure is formed by a juxtaposition of lipoprotein subunits. The components facing the aqueous spaces (the stroma, the loculi and the fret channels) consist of relatively hydrophilic groups of proteins and associated surfactant lipids, while their hydrophobic groups are buried within the partitions between the two rows of subunits. These authors envisage a two-dimensional array of hydrophobic protein subunits associated with lipids. The interior of the lipoprotein sheets of the partition is hydrophobic, while the exterior presents hydrophilic regions of the protein chains and the polar head groups of the bound lipids. According to this model the phytyl chains of the chlorophyll molecules are intercalated within the internal hydrophobic region of intrinsic proteins. In contrast, Anderson (1975) proposed that the phytyl chains are placed adjacent to the outside perimeter of the hydrophobic portions of the two major intrinsic protein complexes, in contact with proteins on one side and lipids on the other, and thus are part of the boundary lipids of the two chlorophyll-protein complexes.

However, as noted by Murphy (1982b), these first models fail to represent the distribution and nature of the different lipid classes and do not take into account the deformation of the lipid bilayer nor the nature of the tightly curved margins of the thylakoid sacs. Moreover, the relatively huge amounts of non-bilayer forming MGDG associated with the thylakoid membrane indicate that non-planar and non-bilayer regions should play an important role in both the structure and function of such a membrane. This author has calculated that the proportion of lipid-occupied membrane is 21.3% and 12.5% in the outer and inner margins, respectively, and 66.2% in the flat regions. Around the thylakoid margin a high degree of curvature is imposed upon the bilayer, which then becomes differentiated into an inner-facing concave surface and an outer-facing convex surface. Packing constraints favor the presence of cone-shaped lipids, such as MGDG (i.e., characterized by a small polar head group and a relatively bulky hydrophobic region), on the concave face and wedge-shaped lipids such as DGDG on the convex face. Murphy (1982b, 1986b) has also proposed a schematic representation of the deformations in the bilayer structure of thylakoids induced by inverted micelles. Large hydrophobic integral membrane proteins may be stabilized within a completely hydrophobic structure formed from inverted micelles sandwiched between the two halves of the lipid bilayer. The inverted micelles consist of cone-shaped lipids, such as

MGDG. On the other hand, the lipid bilayer, which is extensively deformed into regions of convex curvature, should be stabilized by wedge-shaped lipids such as DGDG, PG or SQDG. The known distribution of acyl lipids in the outer monolayer of the thylakoid membrane, expressed as percent of total lipids (Fig. 3), suggests that these three wedge-shaped lipids may play an equal role in this respect (Siegenthaler et al., 1987). A scheme showing a tentative distribution of phospholipids in the thylakoid membrane as revealed by phospholipase treatments has also been reported (Siegenthaler, 1982).

Based on the observation that the addition of SQDG to mixtures of MGDG and DGDG induces the appearance of paracrystalline arrays of 80–100 Å lipidic particles, Sakai et al. (1983) proposed that SQDG may occupy the region in the leaflet opposite the MGDG inverted micelles or the region on the convex side of the bulge or cusp in the corresponding models of lipidic particles. According to these authors, the orderly arranged lipidic particles containing negatively charged SQDG may be the basis for the orderly arrangement of other molecules functioning in photosynthesis in the thylakoid membrane.

More recently, we have proposed a model of the PS II region of the thylakoid membrane with special emphasis on the lipid distribution (Siegenthaler et al., 1995). This model, inspired from that of Jansson (1994) for proteins, is based essentially on the lipid composition of various PS II (Murata et al., 1990) and purified LHCII (Trémolières et al., 1994) preparations. The model also takes into consideration the transmembrane distribution of MGDG, DGDG and SQDG (Rawyler et al., 1987) and of PG and its role in the PS II electron flow activity (Siegenthaler et al., 1989a). In brief, the basic concepts leading to the topology of acyl lipids in the PS II region of the thylakoid membrane are as follows (Siegenthaler et al., 1995): (a) The bulk lipid phase of the membrane is essentially constituted of highly unsaturated galactolipids which represent about 40 mol % of the total lipids in each monolayer; (b) The other minor lipids (PG and SQDG) and a few galactolipid molecules are associated with proteins. These lipids are much more saturated than the bulk lipids; (c) The lipids which are tightly bound to proteins might be located, for instance, between the transmembrane spanning domains (α-helices) of the protein, near a chlorophyll molecule. This is the case for the lipids of the reaction center (D1 and D2) and core (CP43 and 47) proteins of Photosystem II as well as those encountered in the major LHCII antenna proteins (Lhcb1-3). These lipids may have a precise functional role; (d) Other lipid populations which are also closely associated with proteins are involved in the structural organization of the PS II complex. For instance, galactolipids having an intermediate saturation seem to play an important role in the attachment of the minor LHCII antenna proteins (Lhcb4-6) to the core complex (Trémolières et al., 1994). On the other hand, PG appears to be involved in the assembly of monomers into LHCII trimers (Hobe et al., 1995); (e) In addition, several results on the function of PS II in relation to lipids (refer to Chapter 8, Siegenthaler and Trémolières), reconstitution experiments (refer to Chapter 9, Trémolières and Siegenthaler), can bring further information on the topology of lipids. The best example concerns the PG molecules. Only two small subpopulations (3 mol % each) of PG which are located in the inner monolayer have a recognized functional role in the PS II electron transport. All the other inner and outer (sub)populations of this lipid (94 mol %) presumably play a structural role in the PS II region (Siegenthaler et al., 1989a).

When building a model of the thylakoid membrane or of certain of its domains, one has to remember that the membrane is not a static structure but a highly dynamic one which responds to environmental and physiological stimuli. The cohesion of its components is insured by specific protein-protein and protein-lipid interactions as well as non-specific van der Waal's attractions between the molecules constitutive of the membrane. In addition to these long recognized types of interactions, one has to take into consideration recent observations showing ion-induced adhesion between DGDG vesicles (Webb et al., 1988) which is probably a consequence of the formation of specific complexes between the lipid head groups and cations (Menikh and Fragata, 1993). These new findings have been discussed by Stys (1995) in terms of the current theories (i.e., the surface-charge and the molecular recognition between proteins) on the segregation of Photosystems I and II and the interlamellar interactions (stacking).

Acknowledgments

I thank all my collaborators who made essential contributions to the work described in this chapter: Drs. C. Giroud, J.-P. Mayor, A. Rawyler and Y.N. Xu, Mrs S. Duchêne, M. Meylan Bettex, J. Smutny and J. Vallino, as well as Mrs. C. Bettinelli for expert assistance in the preparation of the manuscript. I would also like to acknowledge Prof. W. Eichenberger, Prof. E. Stutz, Drs. A. Trémolières and L. Bovet for critically reading the manuscript, Prof. E. Heinz for his valuable advice and Prof. R. Deschenaux for helping me to visualize the lipid-cyclodextrin inclusion complex with model systems. Thanks are also due to Mrs. D. Jones Siegenthaler for improvement of the English in this chapter and to my wife, Madeleine, for her help and patience. The work from my laboratory was supported by the Swiss National Science Foundation and the University of Neuchâtel (Switzerland).

References

Allen JF (1992) Protein phosphorylation in regulation of photosynthesis. Biochim Biophys Acta 1098: 275–335

Anderson JM (1975) The molecular organization of chloroplast thylakoids. Biochim Biophys Acta 416: 191–235

Andersson B and Anderson JM (1980) Lateral heterogeneity in the distribution of chlorophyll-protein complexes of the thylakoid membranes of spinach chloroplasts. Biochim Biophys Acta 593: 427–440

Andrews J and Mudd JB (1985) Phosphatidylglycerol synthesis in pea chloroplasts. Pathway and localization. Plant Physiol 79: 259–265

Andrews J, Ohlrogge JB and Keegstra K (1985) Final step of phosphatidic acid synthesis in pea chloroplasts occurs in the inner envelope membrane. Plant Physiol 78: 459–465

Archer EK and Keegstra K (1990) Current views on chloroplast protein import and hypotheses on the origin of the transport mechanism. J Bioenerg Biomem 22: 789–810

Barber J (1982) Influence of surface charges on thylakoid structure and function. Annu Rev Plant Physiol 33: 261–295

Barber J and Gounaris K (1986) What role does sulpholipid play within the thylakoid membrane? Photosynth Res 9: 239–249

Bednarz J, Radunz A and Schmid GH (1988) Lipid composition of Photosystem I and II in the tobacco mutant *Nicotiana tabacum* NC 95. Z Naturforsch 43c: 423–430

Bekers O, Uijendaal EV, Beijnen JH, Bult A and Underberg WJM (1991) Cyclodextrins in the pharmaceutical field. Drug Development and Industrial Pharm 17: 1503–1549

Bertrams M, Wrage K and Heinz E (1981) Lipid labelling in intact chloroplasts from exogenous nucleotide precursors. Z Naturforsch 36c: 62–70

Bishop DG (1983) Functional role of plant membrane lipids. In: Thompson WW, Mudd JB and Gibbs M (eds) Biosynthesis and Function of Plant Lipids, pp. 81–103. Am Soc Plant Physiol, Rockville, USA

Bishop DG, Kenrick JR, Bayston JH, MacPherson AS and Johns SR (1980) Monolayer properties of chloroplast lipids. Biochim Biophys Acta 602: 248–259

Block MA, Dorne AJ, Joyard J and Douce R (1983) Preparation and characterization of membrane fractions enriched in outer and inner envelope membranes from spinach chloroplasts. I. Electrophoretic and immunochemical analyses. J Biol Chem

258: 13273–13280

Burke JJ, Wilson RF and Swafford R (1982) Characterization of chloroplasts isolated from triazine-susceptible and triazine-resistant biotypes of *Brassica campestris* L. Plant Physiol 70: 24–29

Carde JP, Joyard J and Douce R (1982) Electron microscopic studies of envelope membranes from spinach plastids. Biol Cell 44: 315–324

Chapman D and Quinn PJ (1976) A method for the modulation of membrane fluidity: Homogeneous catalytic hydrogenation of phospholipids and phospholipid-water model biomembranes. Proc Natl Acad Sci US 73: 3971–3975

Chapman DJ, De Felice J and Barber J (1984) Lipids at sites of quinone and herbicide interaction with the photosystem two pigment-protein complex of chloroplast thylakoids. In: Siegenthaler PA and Eichenberger W (eds) Structure, Function and Metabolism of Plant Lipids, pp 457–464. Elsevier/North-Holland Biomedical Press, Amsterdam, New York

Costes C, Bazier R and Lechevallier D (1972) Rôle structural des lipides dans les membranes des chloroplastes de Blé. Physiol Vég 10: 291–317

Costes C, Bazier R, Baltscheffsky H and Hallberg C (1978) Mild extraction of lipids and pigments from *Rhodospirillum rubrum* chromatophores. Plant Sci Lett 12: 241–249

de Kruijff B, Pilon R, van't Hof R and Demel R (1998) Lipid-protein interactions in chloroplast protein import. In: Siegenthaler PA and Murata N (eds) Lipids in Photosynthesis: Structure, Function and Genetics, pp 191–208. Kluwer Academic Publishers, Dordrecht

Dorne AJ, Block MA, Joyard J. and Douce R (1982) The galactolipid:galactolipid galactosyltransferase is located on the outer surface of the outer membrane of the chloroplast envelope. FEBS Lett 145: 30–34

Dorne AJ, Joyard J and Douce R (1990) Do thylakoids really contain phosphatidylcholine? Proc Natl Acad Sci USA 87: 71–74

Doyle MF and Yu CA (1985) Preparation and reconstitution of a phospholipid deficient cytochrome b_6-*f* complex from spinach chloroplasts. Biochem Biophys Res Commun 131: 700–706

Dubacq JP and Trémolières A (1983) Occurrence and function of phosphatidylglycerol containing Δ^3-*trans*-hexadecenoic acid in photosynthetic lamellae. Physiol Vég 21: 293–312

Duchêne S, Smutny J and Siegenthaler PA (1996) Topological and photosynthetic functional sidedness of phospholipids in spinach thylakoid membrane, inside-out and right-side-out vesicles. Experientia 52: S08–12

Dyer TA (1984) The chloroplast genome: Its nature and role in development. In: Baker NR and Barber J (eds) Chloroplast Biogenesis, pp 23–69. Elsevier, Amsterdam

Eichenberger W, Schaffner JC and Boschetti A (1977) Characterization of proteins and lipids of Photosystem I and II particles from *Chlamydomonas reinhardi*. FEBS Lett 84: 144–148

Etemadi A-H (1980) Membrane asymmetry. A survey and critical appraisal of the methodology. II. Methods for assessing the unequal distribution of lipids. Biochim Biophys Acta 604: 423–475

Ford RC and Barber J (1983) Time-dependent decay and anisotropy of fluorescence from diphenylhexatriene embedded in the chloroplast thylakoid membrane. Biochim Biophys Acta 722: 341–348

Giroud C and Siegenthaler PA (1988) Development of oat prothylakoids into thylakoids during greening does not change transmembrane galactolipid asymmetry but preserves the thylakoid bilayer. Plant Physiol 88: 412–417

Gounaris K and Barber J (1983) Monogalactosyldiacylglycerol: the most abundant polar lipid in nature. Trends Biochem Sci 8: 378–381

Gounaris K and Barber J (1985) Isolation and characterisation of a Photosystem II reaction centre lipoprotein complex. FEBS Lett 188: 68–72

Gounaris K, Sundby C, Andersson B and Barber J (1983) Lateral heterogeneity of polar lipids in the thylakoid membranes of spinach chloroplasts. FEBS Lett 156: 170–174

Gounaris K, Barber J and Harwood JL (1986) The thylakoid membranes of higher plant chloroplasts. Biochem J 237: 313–326

Haase R, Unthan M, Couturier P, Radunz A and Schmid GH (1993) Determination of glycolipids, sulfolipid and phospho-lipids in the thylakoid membrane. Z Naturforsch 48c: 623–631

Heber U and Heldt HW (1981) The chloroplast envelope: Structure, function, and role in leaf metabolism. Annu Rev Plant Physiol 32: 139–168

Heemskerk JWM and Wintermans JFGM (1987) Role of the chloroplast in the leaf acyl-lipid synthesis. Physiol Plant 70: 558–568

Heemskerk JWM, Storz T, Schmidt RR and Heinz E (1990) Biosynthesis of digalactosyldiacylglycerol in plastids from 16:3 and 18:3 plants. Plant Physiol 93: 1286–1294

Heinz E and Siefermann-Harms D (1981) Are galactolipids integral components of the chlorophyll-protein complexes in spinach thylakoids? FEBS Lett 124: 105–111

Helenius A and Simons K (1975) Solubilization of membranes by detergents. Biochim Biophys Acta 415: 29–79

Henry LEA, Mikkelsen JD and Møller BL (1983) Pigment and acyl lipid composition of Photosystem I and II vesicles and of photosynthetic mutants in barley. Carlsberg Res Commun 48: 131–148

Hobe S, Prytulla S, Kühlbrandt W and Paulsen H (1994) Trimerization and crystallization of reconstituted light-harvesting chlorophyll *a/b* complex. EMBO J 13: 3423–3429

Hobe S, Kuttkat A, Förster R and Paulsen H (1995) Assembly of trimeric light-harvesting chlorophyll *a/b* complex in vitro. In: Mathis P (ed) Photosynthesis: From Light to Biosphere, Vol 1, pp 47–52. Kluwer Academic Publishers, Dordrecht

Horváth G, Droppa M, Szito T, Mustardy LA, Horváth LI and Vigh L (1986a) Homogeneous catalytic hydrogenation of lipids in the photosynthetic membrane: Effects on membrane structure and photosynthetic activity. Biochim Biophys Acta 849: 325–336

Horváth G, Mansourian AR, Vigh L, Thomas PG, Joo F and Quinn PJ (1986b) Homogeneous catalytic hydrogenation of the polar lipids of pea chloroplasts *in situ* and the effects on lipid polymorphism. Chem Phys Lipids 39: 251–264

Irie T, Fukunaga K and Pitha J (1992) Hydroxypropyl-cyclodextrins in parenteral use. I: Lipid dissolution and effects on lipid transfers in vitro. J Pharm Sci 81: 521–523

Ivancich A, Horváth LI, Droppa M, Horváth G and Farkas T (1994) Spin label EPR study of lipid solvation of supramolecular photosynthetic protein complexes in thylakoids. Biochim Biophys Acta 1196: 51–56

Jansson S (1994) The light-harvesting chlorophyll *a/b*-binding proteins. Biochim Biophys Acta 1184: 1–19

Joyard J and Douce R (1976) Préparation et activités enzymatiques de l'envelope des chloroplastes d'Epinard. Physiol Vég 14: 31–48

Joyard J and Douce R (1977) Site of synthesis of phosphatidic acid and diacylglycerol in spinach chloroplasts. Biochim Biophys Acta 486: 273–285

Joyard J and Douce R (1987) Galactolipid synthesis. In: Stumpf PK (ed) The Biochemistry of Plants. A Comprehensive Treatise. Vol 9: Lipids: Structure and Function, pp 215–274. Academic Press Inc, New York

Joyard J, Douce R, Siebertz HP and Heinz E (1980) Distribution of radioactive lipids between envelopes and thylakoids from chloroplasts labelled in vivo. Eur J Biochem 108: 171–176

Joyard J, Block MA and Douce R (1991) Molecular aspects of plastid envelope biochemistry. Eur J Biochem 199: 489–509

Kruse O and Schmid GH (1995) The role of phosphatidylglycerol as a functional effector and membrane anchor of the D1-core peptide from Photosystem II-particles of the cyanobacterium *Oscillatoria chalybea*. Z Naturforsch 50c: 380–390

Kruse O, Radunz A and Schmid GH (1994) Phosphatidylglycerol and β-carotene bound onto the D1-core peptide of Photosystem II in the filamentous cyanobacterium *Oscillatoria chalybea*. Z Naturforsch 49c: 115–124

Li G, Knowles PF, Murphy DJ, Nishida I and Marsh D (1989) Spin label saturation transfer ESR studies of lipid-protein interactions in thylakoid membranes. Biochemistry 28: 7446–7452

Makewicz A, Radunz A and Schmid GH (1995) Structural modifications of the photosynthetic apparatus in the region of Photosystem I in *Nicotiana tabacum* as a consequence of an increased CO_2-content of the atmosphere. Z Naturforsch 50c: 511–520

Mattoo AK and Edelman M (1987) Intramembrane translocation and posttranslational palmitoylation of the chloroplast 32-Kd herbicide-binding protein. Proc Natl Acad Sci USA 84: 1497–1501

Mayor JP, Török Z, Siegenthaler PA and Vigh L (1992) Differential accessibility of a hydrogenation catalyst to acyl lipids in thylakoid membranes from atrazine-resistant and -susceptible *Solanum nigrum* biotypes. Plant Cell Physiol 33: 209–215

Menikh A and Fragata M (1993) Fourier-transform infrared spectroscopic study of ion-binding and intramolecular interactions in the polar head of digalactosyldiacylglycerol. Eur Biophys J 22: 249–258

Millner PA and Barber J (1984) Plastoquinone as a mobile redox carrier in the photosynthetic membrane. FEBS Lett 169: 1–6

Morré DJ, Morré JT, Morré SR, Sundqvist C and Sandelius AS (1991) Chloroplast biogenesis. Cell-free transfer of envelope monogalactosylglycerides to thylakoids. Biochim Biophys Acta 1070: 437–445

Mudd JB and DeZacks R (1981) Synthesis of phosphatidylglycerol by chloroplasts from leaves of *Spinacia oleracea* L. (spinach). Arch Biochem Biophys 209: 584–591

Murata N and Nishida I (1990) Lipids in relation to chilling sensitivity of plants. In: Wang CY (ed) Chilling Injury of Horticultural Crops, pp 181–199. CRC Press, Boca Raton, FL

Murata N, Higashi S-I and Fujimura Y (1990) Glycerolipids in various preparations of Photosystem II from spinach chloroplasts. Biochim Biophys Acta 1019: 261–268

Murphy DJ (1982a) The biosynthesis of photosynthetic membrane lipid precursors in higher plants and its integration with photosynthetic carbon assimilation. In: Wintermans JFGM and Kuiper PJC (eds) Biochemistry and Metabolism of Plant Lipids, pp 51–56. Elsevier Biomedical Press, Amsterdam

Murphy DJ (1982b) The importance of non-planar bilayer regions in photosynthetic membranes and their stabilisation by galactolipids. FEBS Lett 150: 19–26

Murphy D (1986a) The molecular organisation of the photosynthetic membranes of higher plants. Biochim Biophys Acta 864: 33–94

Murphy D (1986b) Structural properties and molecular organization of the acyl lipids of photosynthetic membranes. In: Staehelin LA and Arntzen CJ (eds) Encyclopedia of Plant Physiology, Vol 19, Photosynthesis III: Photosynthetic Membranes and Light Harvesting Systems, pp 713–725. Springer-Verlag, Berlin

Murphy D and Woodrow IE (1983) Lateral heterogeneity in the distribution of thylakoid membrane lipid and protein components and its implications for the molecular organisation of photosynthetic membranes. Biochim Biophys Acta 725: 104–112

Mustardy L, Los DA, Gombos Z, Tsvetkova N, Nishida I and Murata N (1996) Intracellular distribution of fatty acid desaturases in cyanobacterial cells and higher-plant chloroplasts. In: Williams JP, Khan MU and Lem NW (eds) Physiology, Biochemistry and Molecular Biology of Plant Lipids, pp 87–89. Kluwer Academic Publishers, Dordrecht

Nakanishi K, Nadai T, Masada M and Miyajima K (1992) Effect of cyclodextrins on biological membrane. II. Mechanism of enhancement on the intestinal absorption of non-absorbable drug by cyclodextrins. Chem Pharm Bull 40: 1252–1256

Nishihara M, Yokota K and Kito M (1980) Lipid molecular species composition of thylakoid membranes. Biochim Biophys Acta 617: 12–19

Nussberger S, Dörr K, Wang DN and Kühlbrandt W (1993) Lipid-protein interactions in crystals of plant light-harvesting complex. J Mol Biol 234: 347–356

Ohnishi M and Thompson Jr GA (1991) Biosynthesis of the unique *trans*-Δ^3-hexadecenoic acid component of chloroplast phosphatidylglycerol: evidence concerning its site and mechanism of formation. Arch Biochem Biophys 288: 591–599

Ohtani Y, Irie T, Uekama K, Fukunaga K and Pitha J (1989) Differential effects of α-, β- and γ-cyclodextrins on human erythrocytes. Eur J Biochem 186: 17–22

Okada Y, Koizumi K, Ogata K and Ohfuji T (1989) Inclusion complexes of lipids with branched cyclodextrins. Chem Pharm Bull 37: 3096–3099

Omata T and Murata N (1986) Glucolipid synthesis activities in cytoplasmic and thylakoid membranes from the cyanobacterium *Anacystis nidulans*. Plant Cell Physiol 27: 485–490

Ouijja A, Farineau N, Cautrel C and Guillot-Salomon T (1988) Biochemical analysis and photosynthetic activity of chloroplasts and Photosystem II particles from a barley mutant lacking chlorophyll *b*. Biochim Biophys Acta 932: 97–106

Pick U, Gounaris K, Admon A and Barber J (1984) Activation of the CF_0-CF_1, ATP synthase from spinach chloroplasts by chloroplast lipids. Biochim Biophys Acta 765: 12–20

Pick U, Gounaris K, Weiss M and Barber J (1985) Tightly bound sulpholipids in chloroplast CF_0-CF_1. Biochim Biophys Acta

808: 415–420

Pick U, Weiss M, Gounaris K and Barber J (1987) The role of different thylakoid glycolipids in the function of reconstituted chloroplast ATP synthase. Biochim Biophys Acta 891: 28–39

Quinn PJ and Williams WP (1983) The structural role of lipids in photosynthetic membranes. Biochim Biophys Acta 737: 223–266

Quinn PJ, Joo F and Vigh L (1989) The role of unsaturated lipids in membrane structure and stability. Prog Biophys Mol Biol 53: 71–103

Radunz A (1971) Phosphatidylglycerin-antiserum und seine Reaktionen mit Chloroplasten. Z Naturforsch 26b: 916–919

Radunz A (1972) Lokalisierung des Monogalaktosyldiglycerids in Thylakoidmembranen mit serologischen Methoden. Z Naturforsch 27b: 822–826

Radunz A (1976) Localization of the tri- and digalactosyl diglyceride in the thylakoid membrane with serological methods. Z Naturforsch 31c: 589–593

Radunz A (1977) Binding of antibodies onto the thylakoid membrane. II. Distribution of lipids and proteins at the outer surface of the thylakoid membrane. Z Naturforsch 32c: 597–599

Radunz A (1981) Application of antibodies in the analysis of structural configuration of thylakoid membranes. Ber Deutsch Bot Ges 94: 477–489

Radunz A and Berzborn R (1970) Antibodies against sulphoquinovosyl-diacyl glycerol and their reactions with chloroplasts. Z Naturforsch 25b: 412–419

Radunz A and Schmid GH (1989) Comparative immunological studies on the CF_1-complex in mutants of N. tabacum, exhibiting different capacities for photosynthesis and photorespiration and different chloroplast structures. Z Naturforsch 44c: 689–697

Radunz A, Bader KP and Schmid GH (1984) Serological investigations of the function of galactolipids in the thylakoid membrane. Z Pflanzenphysiol 114: 227–231

Rawyler A and Siegenthaler PA (1980) Role of lipids in functions of photosynthetic membranes revealed by treatment with lipolytic acyl hydrolase. Eur J Biochem 110: 179–187

Rawyler A and Siegenthaler PA (1985) Transversal localization of monogalactosyldiacylglycerol and digalactosyldiacylglycerol in spinach thylakoid membranes. Biochim Biophys Acta 815: 287–298

Rawyler A and Siegenthaler PA (1989) Change in the molecular organization of monogalactosyldiacylglycerol between resting and functioning thylakoid membranes. Involvement of the CF_o-CF_1-ATP synthetase. Biochim Biophys Acta 975: 283–292

Rawyler A and Siegenthaler PA (1996) Cyclodextrins: A new tool for the controlled lipid depletion of thylakoid membranes. Biochim Biophys Acta 1278: 89–97

Rawyler A, Henry LEA and Siegenthaler PA (1980) Acyl and pigment lipid composition of two chlorophyll-proteins. Carlsberg Res Commun 45: 443–451

Rawyler A, Unitt MD, Giroud C, Davies H, Mayor JP, Harwood JL and Siegenthaler PA (1987) The transmembrane distribution of galactolipids in chloroplast thylakoids is universal in a wide variety of temperate climate plants. Photosynth Res 11: 3–13

Rawyler A, Meylan M and Siegenthaler PA (1992) Galactolipid export from envelope to thylakoid membranes in intact chloroplasts. I. Characterization and involvement in thylakoid

lipid asymmetry. Biochim Biophys Acta 1104: 331–341

Rawyler A, Meylan-Bettex M and Siegenthaler PA (1995) (Galacto)lipid export from envelope to thylakoid membranes in intact chloroplasts. II. A general process with a key role for the envelope in the establishment of lipid asymmetry in thylakoid membranes. Biochim Biophys Acta 1233:123–133

Rémy R, Trémolières A, Duval JC, Ambard-Bretteville F and Dubacq JP (1982) Study of the supramolecular organization of light-harvesting chlorophyll protein (LHCP). FEBS Lett 137: 271–275

Roelofsen B and Zwaal RFA (1976) The use of phospholipases in the determination of asymmetric phospholipid distribution in membranes. In: Korn ED (ed) Methods in Membrane Biology, Vol 7, pp 147–177. Plenum Press, New York

Roughan PG (1985) Phosphatidylglycerol and chilling sensitivity in plants. Plant Physiol 77: 740–746

Sakai WS, Yamamoto HY, Miyazaki T and Ross JW (1983) A model for chloroplast thylakoid membranes involving orderly arrangements of negatively charged lipidic particles containing sulphoquinovosyldiacylglycerol. FEBS Lett 158: 203–207

Sandelius AS and Selstam E (1984) Localization of galactolipid biosynthesis in etioplasts isolated from dark-grown wheat (Triticum aestivum L.). Plant Physiol 76: 1041–1046

Schmidt H and Heinz E (1990) Desaturation of oleoyl groups in envelope membranes from spinach chloroplasts. Proc Nat Acad Sci USA 87: 9477–9480

Seifert U and Heinz E (1992) Enzymatic characteristics of UDP-sulfoquinovose: diacylglycerol sulfoquinovosyl transferase from chloroplast envelopes. Bot Acta 205: 197–205

Selstam E and Widell-Wigge A (1993) Chloroplast lipids and the assembly of membranes. In: Sundqvist C and Ryberg M (eds) Pigment-Protein Complexes in Plastids: Synthesis and Assembly, pp 241–277. Academic Press, Inc., San Diego

Siegel DP (1986a) Inverted micellar intermediates and the transitions between lamellar, cubic, and inverted hexagonal lipid phases. I. Mechanism of the $L_\alpha \leftrightarrow H_{II}$ phase transitions. Biophys J 49: 1155–1170

Siegel DP (1986b) Inverted micellar intermediates and the transitions between lamellar, cubic, and inverted hexagonal lipid phases. II. Implications for membrane-membrane interactions and membrane fusion. Biophys J 49: 1171–1183

Siegenthaler PA (1982) Transmembrane distribution and function of lipids in spinach thylakoid membranes: rationale of the enzymatic modification method. In: Wintermans JFGM and Kuiper PJC (eds) Biochemistry and Metabolism of Plant Lipids, pp 351–358. Elsevier Biomedical Press, Amsterdam

Siegenthaler PA and Giroud C (1986) Transversal distribution of phospholipids in prothylakoid and thylakoid membranes from oat. FEBS Lett 201: 215–220

Siegenthaler PA and Mayor JP (1992) Changes in the binding and inhibitory properties of urea/triazine-type herbicides upon phospholipid and galactolipid depletion in the outer monolayer of thylakoid membranes. Photosynth Res 31: 57–68

Siegenthaler PA and Rawyler A (1986) Acyl lipids in thylakoid membranes: distribution and involvement in photosynthetic functions. In Staehelin LA and Arntzen CJ (eds) Encyclopedia of Plant Physiology, Vol 19, Photosynthesis III, pp 693–705. Springer Verlag, Berlin

Siegenthaler PA and Trémolières A (1998) Role of acyl lipids in the function of photosynthetic membranes in higher plants. In: Siegenthaler PA and Murata N (eds) Lipids in Photosynthesis:

Structure, Function and Genetics, pp 145–173. Kluwer Academic Publishers, Dordrecht

Siegenthaler PA and Vallino J (1995) Effect of a selective depletion of acyl lipids on the light and dark phosphorylation in spinach thylakoid membranes. In: Mathis P (ed) Photosynthesis: From Light to Biosphere, Vol III, pp 225–228. Kluwer Academic Publishers, Dordrecht

Siegenthaler PA, Rawyler A and Giroud C (1987) Spatial organization and functional roles of acyl lipids in thylakoid membranes. In: Stumpf PK, Mudd JB and Nes WD (eds) The Metabolism, Structure and Function of Plant Lipids, pp 161–168. Plenum Press, New York and London

Siegenthaler PA, Sutter J and Rawyler A (1988) The transmembrane distribution of galactolipids in spinach thylakoid inside-out vesicles is opposite to that found in intact thylakoids. FEBS Lett 228: 94–98

Siegenthaler PA, Rawyler A and Smutny J (1989a) The phospholipid population which sustains the uncoupled non-cyclic electron flow activity is localized in the inner monolayer of the thylakoid membrane. Biochim Biophys Acta 975: 104–111

Siegenthaler PA, Rawyler A and Mayor JP (1989b) Structural and functional aspects of acyl lipids in thylakoid membranes from higher plants. In: Biacs PA, Gruiz K, and Kremer T (eds) Biological Role of Plant Lipids, pp. 171–180. Akadémiai Kiado, Budapest and Plenum Publishing Corporation, New York and London

Siegenthaler PA, Xu Y, Bovet L, Smutny J and Rawyler A (1992) Molecular organization and functions of acyl lipids in spinach chloroplast membranes. In: Cherif A, Ben Miled-Daoud D, Marzouk B, Smaoui A and Zarrouk M (eds) Metabolism, Structure and Utilization of Plant Lipids, pp 285–288. Centre National Pédagogique, Tunisie

Siegenthaler PA, Xu Y, Smutny J, Meylan-Bettex M, Vallino J and Rawyler A (1995) Thoughts concerning a new paradigm of the Photosystem II region of the thylakoid membrane based on lipid structure and function. In: Kader JC and Mazliak P (eds) Plant Lipid Metabolism, pp 170–172. Kluwer Academic Publishers, Dordrecht

Sigrist M, Zwillenberg C, Giroud C, Eichenberger W and Boschetti A (1988) Sulfolipid associated with the light-harvesting complex associated with Photosystem II apoproteins of *Chlamydomonas reinhardii*. Plant Sci 58: 15–23

Simbeni R, Paltauf F and Daum G (1990) Intramitochondrial transfer of phospholipids in the yeast, *Saccharomyces cerevisiae*. J Biol Chem 265: 281–285

Slabas AR and Smith CG (1988) Immunological localization of acyl carrier protein in plants and *Escherichia coli*: Evidence for membrane association in plants. Planta 175: 145–152

Soll J and Alefsen H (1993) The protein import apparatus of chloroplasts. Physiol Plant 87: 433–440

Stoddart F (1992) Cyclodextrins, off-the-shelf components for the construction of mechanically interlocked molecular systems. Angew Int Ed Engl 31: 846–848

Stys D (1995) Stacking and separation of Photosystem I and Photosystem II in plant thylakoid membranes: A physico-chemical view. Physiol Plant 95: 651–657

Sundby C and Larsson C (1985) Transbilayer organization of the thylakoid galactolipids. Biochim Biophys Acta 813: 61–67

Szalontai B, Droppa M, Vigh L, Joo F and Horváth G (1986) Selectivity of homogeneous catalytic hydrogenation in saturation of double bonds of lipids in chloroplast lamellae. Photobiochem Photobiophys 10: 233–240

Szejtli J (1988) Cyclodextrin Technology. In: Davies JED (ed) Topics in Inclusion Science, pp 1–306. Kluwer Academic Publishers, Dordrecht

Szejtli J (1990) The cyclodextrins and their applications in biotechnology. Carbohydrate Polymers 12: 375–392

Tchang F, Connan A, Robert D and Trémolières A (1985) Effect of light and temperature on α-linolenic acid biosynthesis in developing cotyledons of *Pharbitis nil*. Physiol Vég 24: 361–371

Trémolières A (1991) Lipid-protein interactions in relation to light energy distribution in photosynthetic membrane of eukaryotic organisms. Role of trans- Δ^3 - hexadecenoic acid-containing phosphatidylglycerol. Trends Photochem Photobiol 2: 13–32

Trémolières A and Siegenthaler PA (1998) Reconstitution of photosynthetic structures and activities with lipids. In: Siegenthaler PA and Murata N (eds) Lipids in Photosynthesis: Structure, Function and Genetics, pp 175–189. Kluwer Academic Publishers, Dordrecht

Trémolières A, Dubacq JP, Ambard-Bretteville F and Rémy R (1981) Lipid composition of chlorophyll-protein complexes. FEBS Lett 130: 27–31

Trémolières A, Dainese P and Bassi R (1994) Heterogenous lipid distribution among chlorophyll-binding proteins of Photosystem II in maize mesophyll chloroplasts. Eur J Biochem 221: 721–73

Tuquet C, Guillot-Salomon T, de Lubac M and Signol M (1977) Granum formation and the presence of phosphatidylglycerol containing *trans*-Δ^3-hexadecenoic acid. Plant Sci Lett 8: 59–64

Unitt MD and Harwood JL (1985) Sidedness studies of thylakoid phosphatidylglycerol in higher plants. Biochem J 228: 707–711

Van Venetië R and Verkleij AJ (1982) Possible role of non-bilayer lipids in the structure of mitochondria. A freeze-fracture electron microscopy study. Biochim Biophys Acta 692: 397–405

Vigh L, Joo F, Droppa M, Horváth LI and Horváth G (1985) Modulation of chloroplast membrane lipids by homogeneous catalytic hydrogenation. Eur J Biochem 147: 477–481

Voss R, Radunz A and Schmid GH (1992a) Glycolipids are prosthetic groups of polypeptides of the reaction center complex of Photosystem II. In: Argyroudi-Akoyunoglou JH (ed) Regulation of Chloroplast Biogenesis, pp 417–422. Plenum Press, New York

Voss R, Radunz A and Schmid GH (1992b) Binding of lipids onto polypeptides of the thylakoid membrane. I. Galactolipids and sulpholipid as prosthetic groups of core peptides of the Photosystem II complex. Z Naturforsch 47c: 406–415

Webb MS and Green BR (1991) Biochemical and biophysical properties of thylakoid acyl lipids. Biochim Biophys Acta 1060: 133–158

Webb MS, Tilcock CPS and Green BR (1988) Salt-mediated interactions between vesicles of the thylakoid lipid digalactosyldiacylglycerol. Biochim Biophys Acta 938: 323–333

Weier TE and Benson AA (1967) The molecular organization of chloroplast membranes. Amer J Bot 54: 389–402

Williams JP, Simpson EE and Chapman DJ (1979) Galactolipid

synthesis in *Vicia faba* leaves. Plant Physiol 63: 669–673

Williams WP (1994) The role of lipids in the structure and function of photosynthetic membranes. Prog Lipid Res 33: 119–127

Williams WP (1998) The physical properties of thylakoid membrane lipids and their relation to photosynthesis. In: Siegenthaler PA and Murata N (eds) Lipids in Photosynthesis: Structure, Function and Genetics, pp 103–118 Kluwer Academic Publishers, Dordrecht

Xu YN and Siegenthaler PA (1996a) Phosphatidylglycerol molecular species of photosynthetic membranes analyzed by HPLC: theoretical considerations. Lipids 31: 223–229

Xu YN and Siegenthaler PA (1996b) Effect of non-chilling temperature and light intensity during growth of squash cotyledons on the composition of thylakoid membrane lipids and fatty acids. Plant Cell Physiol 37: 471–479

Xu YN, Meylan M and Siegenthaler PA (1994) Effect of phospholipase A_2 on the rate and extent of phosphatidylglycerol molecular species hydrolysis in spinach and squash thylakoids. Experientia 50: S23–15

Chapter 8

Role of Acyl Lipids in the Function of Photosynthetic Membranes in Higher Plants

Paul-André Siegenthaler
Laboratoire de Physiologie végétale, Université de Neuchâtel,
Rue Emile Argand 13, CH-2007 Neuchâtel, Switzerland

Antoine Trémolières
Université de Paris sud, Centre d'Orsay, Institut de Biotechnologie des Plantes,
Bâtiment 630, F-91405 Orsay, France

Summary

The great diversity of thylakoid acyl lipids and their unique (physico)chemical characteristics as well as their peculiar topology in the membrane strongly suggest that specific functions and distinct domains of acyl lipids exist in the thylakoid membrane.

P.-A. Siegenthaler and N. Murata (eds): Lipids in Photosynthesis: Structure, Function and Genetics, pp. 145–173.
© *1998 Kluwer Academic Publishers. Printed in The Netherlands.*

To investigate the functional significance of thylakoid lipid composition, several approaches have been adopted which are presented in this chapter. The functional alterations occurring in aged thylakoids and during leaf senescence can be attributed, at least in part, to a release of free fatty acids and/or to the loss or modification of the parent lipids which are essential for the functioning of the thylakoid membrane. The effects of aging are strikingly similar to those of exogenous free unsaturated fatty acids. Controlled lipolytic treatment of thylakoid membranes is another approach allowing the stepwise digestion of specific acyl lipids, first in the thylakoid outer monolayer, then in the inner one. These topological lipid depletions show that only certain lipid populations (e.g. located in one of the two membrane monolayers) are able to sustain photochemical reactions. The removal of thylakoid acyl lipids by cyclodextrins (cyclic oligosaccharides consisting of six to eight glucopyranose units linked by $\alpha(1-4)$bonds) represents a potentially interesting depletion technique, since no detergents are used as in the preparation of subchloroplast particles and no free fatty acids and lysolipids are produced as in the enzymatic approach. The modulation of thylakoid membrane fluidity by homogenous catalytic hydrogenation in situ shows quite clearly the importance of the lipid unsaturation for optimal photochemical reactions. In addition, alteration in membrane fluidity may be the first signal in the perception of temperature changes in the plant environment. Another interesting approach is to use antibodies directed to specific membrane lipids and measure the photochemical reactions which are impaired in thylakoids. Lipids are also involved in the mode of action of herbicides by mediating the accessibility of these compounds to their binding site, for instance to the Q_B protein level. Finally, a number of *Arabidopsis* and *Chlamydomonas* mutants have been characterized as being deficient in desaturases or glycerol-3-phosphate acyltransferases. The availability of these mutants provides another tool in determining the physiological consequences of variations in lipid unsaturation, though information concerning photosynthetic activity is still rather scarce. All these approaches present advantages and drawbacks.

In spite of considerable effort, the structure/function relationship of acyl lipids in the thylakoid membrane remains ambiguous and elusive. This is due probably to the fact that, in contrast to proteins, lipids have by themselves no recognized catalytic properties. They allow the maintenance of an appropriate conformation and orientation of proteins which may express their function only in the presence of specific lipids. This function is expected to require only a few specific lipid molecules as it appears to be the case for phosphatidylglycerol containing *trans*-Δ^3-hexadecenoic acid in the formation of the trimeric chlorophyll *a/b* light-harvesting protein of Photosystem II and the development of grana stacks. In the literature, there is no clear distinction between the role of these specific lipids and the general physicochemical properties of the membrane bulk lipids which are likely to be involved in structural aspects of the membrane such as the mode of herbicide action, chilling injury, photoinhibition and, in general, responses to environmental changes.

I. Introduction

The photosynthetic membranes of higher plant chloroplasts consist of extensive planar lamellar regions bound by tightly curved margins and enclosing an inner aqueous compartment, the lumen. This network of membranes can be divided into appressed, non-appressed and marginal regions (Murphy, 1986a, b). Appressed regions contain most of the thylakoid PS II complex, including its LHCII antennae, while non-appressed regions are enriched in PS I and the CF_o-CF_1 complex responsible for ATP formation. The cytochrome b_6f complex appears to be equally distributed between both appressed and non-appressed membranes. In addition, there is a so-called mobile subpopulation of LHCII which, in its phosphorylated form, is dissociated from PS II and diffuses into the non-appressed membranes in order to facilitate energy transfer to PS I. The marginal regions ought to be free of protein owing to steric

Abbreviations: BSA – bovine serum albumin; CD – cyclodextrin; DCCD – N, N'-dicyclohexylcarbodiimide; DCMU – 3-(3′, 4′-dichlorophenyl)-1, 1-dimethylurea; DCPIP – dichlorophenolindophenol; DGDG – Digalactosyldiacylglycerol; FCCP – carbonylcyanide 4-trifluoromethoxyphenylhydrazone; LAH – lipolytic acyl hydrolase; LHC (LHCP) – chlorophyll *a/b* light-harvesting protein; LRa – lipase from *Rhizopus arrhizus*; MGDG – monogalactosyldiacylglycerol; PC – phosphatidylcholine; PG – phosphatidylglycerol; PLA$_2$, PLC or PLD – phospholipase A$_2$, C or D; PS I – Photosystem I; PS II – Photosystem II; $Q_A(Q_B)$ – primary (secondary) quinone acceptor of PS II; S_o, S_1, S_2, S_3, S_4 – five states of the oxygen-evolving system, according to the Kok S state Model; SQDG – sulfoquinovosyldiacylglycerol; 16:1(3t) – *trans*-Δ^3-Hexadecenoic acid; 18:3 – α-linolenic acid constraints imposed upon the bulky protein complexes by the tight curvature of the membrane (Murphy, 1982). Coupled non-cyclic electron transport in thylakoid membranes is achieved through a spatial organization of chlorophyll-protein complexes linked together by protein and non-protein redox components. These constituents are characterized by an asymmetric distribution in both the lateral and transversal planes of the thylakoid membrane (Andersson et al., 1985; Murphy, 1986a; Wollenberger et al., 1994), thus conferring vectorial properties to the membrane.

All these membrane components are embedded in a lipid matrix consisting of several lipid classes (MGDG, DGDG, SQDG and PG) and molecular species (Nishihara et al., 1980; Xu and Siegenthaler, 1996). The molecular organization of these lipids in the thylakoid membrane as well as the origin of their asymmetric transmembrane distribution have been dealt with in Chapter 7 (Siegenthaler).

Several recent studies strongly suggest that thylakoid acyl lipids are not merely an inert matrix into which the various membrane proteins are inserted. Indeed, their strong interactions with many membrane proteins as well as their pronounced transverse heterogeneity suggest that these lipids play a dynamic and essential role not only in the structure but also in the function of the photosynthetic apparatus. However, one has to point out that the results related to the functional aspects of thylakoid lipids are often ambiguous because these molecules, in contrast to proteins, have by themselves no recognized catalytic properties. Rather, they allow the maintenance of an appropriate conformation and orientation of proteins which may express their function (only) in the presence of specific lipids. It is therefore often difficult to dissociate the functional from the structural role of lipids (Siegenthaler and Rawyler, 1986).

The purpose of this chapter is to review comprehensively some of the historical and more recent advances made in elucidating the functional role of acyl lipids in the photosynthetic membrane of higher plants. Over the past fifteen years, several reviews have been published on this topic (Bishop, 1983; Dubacq and Trémolières, 1983; Quinn and Williams, 1983; Barber and Gounaris, 1986; Gounaris et al., 1986; Siegenthaler and Rawyler, 1986; Siegenthaler et al., 1987b, 1989b; Webb and Green, 1991; Williams, 1994, 1998; Somerville, 1995; Gombos and Murata, 1998; Trémolières and

Siegenthaler, 1998; Vijayan et al., 1998). Here, we will present evidence showing that several photosynthetic functions depend, to varying extents, on the presence, the nature and the topology of certain acyl lipid classes or molecular species in the thylakoid membrane.

II. Methodological Approaches

The study of the molecular organization of acyl lipids and of their involvement in thylakoid membrane function requires suitable (bio)chemical tools aimed at altering the lipid content and the lipid class and fatty acid composition of these membranes. Several approaches are designed for lipid depletion and/or modification. They include: (1) Thylakoid aging in vitro (Siegenthaler and Rawyler, 1977; Dupont and Siegenthaler, 1985); (2) Solvent extraction of freeze-dried membranes (Costes et al., 1972, 1978; Krupa and Baszynski, 1975); (3) Mild solubilization of membrane components by detergents (Murphy, 1986a,b; Eckert et al., 1987; Siefermann-Harms et al., 1987; Murata et al., 1990; Trémolières et al., 1994). The last two approaches are often followed by reconstitution with synthetic or native lipids; (4) Controlled lipolytic treatment of thylakoids by specific enzymes (phospho- and galactolipases, acyl hydrolases), often followed by the removal of hydrolysis products by defatted bovine serum albumin (Rawyler and Siegenthaler, 1981a,b, 1985, 1989; Siegenthaler and Rawyler, 1986 and references therein; Rawyler et al., 1987; Giroud and Siegenthaler, 1988; Horváth et al., 1989; Droppa et al., 1995; Siegenthaler and Vallino, 1995); (5) Removal of membrane acyl lipids by cyclodextrins (Rawyler and Siegenthaler, 1996). Other approaches involve: (6) Lipid enrichment (i.e. modification of lipid/protein ratio) of thylakoids by fusion with liposomes of known composition (Hoshina, 1979; Siegel et al., 1981; Millner et al., 1983; see also Chapter 9, Trémolières and Siegenthaler); (7) Lowering the unsaturation of acyl chains by in situ catalytic hydrogenation of membrane lipids (Vigh et al., 1985; Horváth et al., 1986a,b, 1987; Gombos et al., 1988); (8) The use of anti-lipid antibodies as reagents for the localization and functional characterization of lipids (Radunz et al., 1984a,b; Voss et al., 1992a,b); (9) Comparison of chilling-sensitive and -resistant plants as well as wild and mutant plants that are deficient in certain lipid metabolic steps and therefore in certain lipid molecular species (Murata and Nishida, 1990; Somerville and Browse, 1991; Somerville, 1995); (10) Biogenesis and physiological approaches (see Section VII); (11) Comparison of herbicide treatment in control and lipid-depleted membranes (Section VIII).

The above-mentioned lipid depletion techniques (2–5) have advantages but also various drawbacks. For instance, solvent extraction is restricted to relatively non-polar lipids because the solvents usually employed are themselves non-polar. Detergent treatment suffers from a lack of specificity since it invariably removes proteins and pigments in addition to membrane lipids; moreover, undesirable detergent molecules can displace and/or substitute for lipids within the membrane. Lipolytic enzymes may alter the membrane structure but also change the qualitative membrane lipid composition by generating new products (e.g., free fatty acids, lysolipids and diacylglycerols). These products, if not removed (Siegenthaler et al., 1987a), can lead to changes in the remaining lipid-protein interactions. Though several cyclodextrins are available that display various cavity diameters and properties, none of them are specific in removing only one lipid class (Rawyler and Siegenthaler, 1996). Furthermore, liposome-mediated lipid enrichment of thylakoids is not easily controllable and prone to back-extraction of thylakoid lipids into liposomes. Catalytic hydrogenation in situ always leaves some catalyst molecules within the membrane. Also, the decrease in the double bond index is often accompanied by the formation of artifactual *trans*-unsaturated isomers. In practice, there is no technique currently available which modifies or removes lipid molecules from thylakoid membranes without introducing foreign compounds and/or altering the pigment and protein content of these membranes. This may explain why completely different approaches have been proposed recently, e.g. the use of chilling-resistant and -sensitive plant species or plant mutants affected in their lipid metabolism.

III. Aging of Thylakoids In Vitro and Effects of Free Fatty Acids on Photochemical Reactions

An important feature of the synthesis of chloroplast membrane lipids is the strict exclusion of free fatty acids from any of the participating cell compartments.

It follows that any free fatty acids, mostly linolenic acid, found in the chloroplast in vivo or in vitro is almost certainly the product of lipolysis (Galliard, 1980; Thomas, 1986). Sastry and Kates (1964) described an enzyme from *Phaseolus* leaves which they called galactolipase (EC 3.1.1.26) and which causes rapid damage to membrane lipids. The enzyme can hydrolyze a range of acyl lipids in addition to galactolipids and is therefore more appropriately referred to as a non-specific lipid acyl hydrolase (LAH). The enzyme from spinach (Helmsing, 1967) and runner bean (Helmsing, 1969; Burns et al., 1977) leaves has been purified and its properties established. An interesting feature of the enzyme reaction is its autocatalytic nature, since fatty acids, mostly linolenic acid, released by hydrolysis enhance further activity by acting as a physiological detergent (Galliard, 1971). Thus, lipolytic enzymes may cause rapid damage to membrane acyl lipids and may be of physiological significance in response to senescence (Gepstein, 1988) as well as to pathological infection (Slusarenko et al., 1991) and physical damage (Galliard, 1980).

A. Effect of Leaf Senescence and Thylakoid Aging In Vitro

During senescence, chloroplasts and thylakoids from green leaves of higher plants gradually change their (ultra)structure and lose their photosynthetic abilities (Gepstein, 1988). Disappearance of chlorophyll is one of the most prominent phenomena of senescence, and eventually the rate of chlorophyll degradation is usually considered to be a reliable criterion of leaf senescence and a measure of the age-related deterioration of the photosynthetic capacity (Matile, 1992 and references therein). Loss of photosynthetic competence during senescence is attributable, at least in part, to impairment of photosynthetic electron transport in the thylakoid membrane. Jenkins and Woolhouse (1981) observed that in senescing bean leaves electron flow through PS II and PS I decreases, but that the rate of non-cyclic electron transport declines to a greater degree than the activities of either of the photosystems. This has prompted the proposal that it is largely an impairment of electron flow between the two photosystems that limits the availability of photosynthetic reducing power with advancing senescence. According to Holloway et al. (1983), the rate limiting step is the transfer of electron from plastohydroquinone to the cytochrome b_6f

complex. An attractive hypothesis is that during senescence there is a decrease in thylakoid membrane fluidity that limits the ability of electron carriers to interact properly (Yamamoto et al., 1981). However, unlike plasma and microsomal membranes, thylakoid membranes do not sustain a decrease in bulk lipid fluidity with advancing age (Thompson, 1988 and references therein).

The involvement of acyl lipids in chloroplast senescence is far from being elucidated and often controversial. However, several changes or phenomena that might alter photochemical reactions have been called upon (Thomas, 1986; Gepstein, 1988; Thompson, 1988; Matile, 1992): (1) Quantitative and qualitative changes in lipid composition; (2) Membrane lipid phase transition from a liquid-crystalline to a gel phase resulting in the formation of gel phase lipid domains in the membranes which are responsible for decreased fluidity and ultimately membrane leakiness and loss of permeability. Moreover, the formation of gel phase lipid may alter the whole pattern of protein organization: (3) Free radicals that accumulate in aging tissues and are known to have deleterious effects on membranes, may initiate lipid peroxidation and in turn induce membrane rigidification; (4) Senescence-associated breakdown of acyl lipids by lipolytic enzymes has also been invoked to explain impairment of photochemical reactions (Gepstein, 1988; Matile, 1992 and references therein). However, as pointed out by Matile (1992), senescence-induced lipid acyl hydrolase which could explain the rapid loss of bulk acyl lipids and hence the decline of photochemical activities has not been established so far. In our opinion, such a straightforward correlation is highly improbable since it has been demonstrated that certain photochemical activities (e.g. non-cyclic electron flow, photo- and dark phosphorylation) are sustained by very small lipid population pools which are arranged in domains either in the inner or in the outer monolayer of the thylakoid membrane (Siegenthaler et al., 1989a,b; Siegenthaler and Vallino, 1995; see also Section IV.C). Moreover, the role of a few PG molecules in the trimerization process of LHCII is well documented (see also Chapter 9, Trémolières and Siegenthaler).

As senescent chloroplasts, isolated thylakoids are unstable and alter their structure and photochemical reactions, even when stored at 0 °C. During aging in vitro, thylakoids undergo irreversible swelling (Siegenthaler, 1972) accompanied by a lack of light-induced shrinkage and a decrease in the chlorophyll protein complexes concomitant with changes in the acyl lipid composition (Henry et al., 1982). Interestingly, aging of isolated spinach thylakoids results also in an acid shift of the pH optimum for electron flow through both photosystems or through either photosystem alone (Siegenthaler and Rawyler, 1977 and references therein). This phenomenon was shown to be related to an alteration of the thylakoid membrane integrity, as attested by the dissipation of proton gradient, i.e. a decrease in the proton pump activity and ΔpH (see Fig. 1 for further explanations, refer to Section III.B) and by a gradual shift of the phase transition to lower temperatures in aged thylakoid membranes (van Ginkel and Fork, 1881). Upon aging, thylakoid membranes display a dramatic decrease in the content of cytochrome b_{559}HP (high potential) which is not concomitant with a corresponding increase in one of the lower potential forms of the cytochrome. In contrast, cytochrome f appears to be rather resistant towards the aging process whilst the extent of reducibility of cytochrome b_{563} (i.e. b_6) is higher in aged than in freshly isolated thylakoids (Dupont and Siegenthaler, 1985). Furthermore, cytochrome breakdown and bleaching of pigments (chlorophylls a and b, and β-carotene) are obviously two related phenomena, probably sharing some common step(s), because scavengers of free radicals greatly slow down the decrease in both cytochrome and pigment contents (Dupont and Siegenthaler, 1986).

The functional alterations occurring in aged thylakoids can be attributed to a release of endogenous free fatty acids (mainly linolenic acid) following hydrolysis of acyl lipids, mainly galactolipids (Siegenthaler and Rawyler, 1977; Siegenthaler et al., 1981). Indeed, the effects of aging in vitro on photochemical reactions are strikingly similar to those of exogenous free unsaturated fatty acids. The inhibition of photochemical reactions during aging can also be the result of the loss of the parent lipids themselves which are essential for the functioning of the thylakoid membrane (for further details, see Section IV.C).

B. Effect of Free Fatty Acids

The release of free fatty acids and lysolipids during the enzymatic reaction catalyzed by LAH may contribute to structural and functional alterations of cells and organelles. For instance, it was experi-

Intact thylakoid

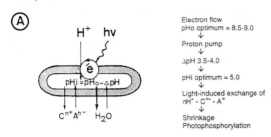

Thylakoid deteriorated by fatty acids or aged in vitro

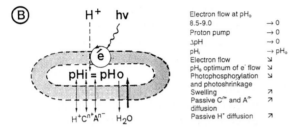

Fig. 1. Scheme illustrating the relationships between the shift in pH optimum of electron flow toward acidity, ΔpH and pHi of thylakoids in intact thylakoids (A) and in fatty acid-treated or aged thylakoids (B). For more details, refer to the text (Section III.B) and to Siegenthaler and Depéry (1976).

mentally verified that exogenous polyunsaturated fatty acids (e.g., linolenic acid) enhance thylakoid swelling in vitro and inhibit light-induced shrinkage (Siegenthaler, 1972). Furthermore, they cause a sequential inhibition of PS II and PS I electron transport and of the associated photophosphorylation. The site of inhibition by unsaturated fatty acids lies on the oxidizing side of PS II, probably between the hypothetical carriers Y2 and Y3, as illustrated in Fig. 2. Interestingly, Mn^{2+} ions are able to restore the electron flow activity, most likely by creating a shunt bypassing the fatty acid inhibition site (Siegenthaler, 1974).

Golbeck et al. (1980) extended these results by showing that the effects of linolenic acid on photosynthetic electron transport reactions are localized at a site on the donor side of PS I and at two functionally distinct sites in PS II. At the first site of PS II, there is a time-dependent loss of the loosely bound pool of Mn implicated in the water-splitting mechanism. Mn^{2+} was found to be most effective in preventing linolenic acid effects. At the second PS II site, the photochemical charge separation is rapidly inhibited by linolenic acid as evidenced by the high initial fluorescence yield. When thylakoids are washed

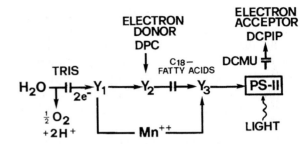

Fig. 2. Schematic representation of the action sites of C_{18}-unsaturated fatty acids, Tris washing, 1, 5-diphenylcarbazide (DPC), $MnCl_2$ and DCMU in the Photosystem II electron transport (H_2O or DPC $\rightarrow$ 2, 6-dichlorophenol indophenol:DCPIP) in thylakoids. Y_1, Y_2 and Y_3 are unknown electron carriers (adapted from Siegenthaler, 1974).

after treatment with linolenic acid, the fluorescence returns to its original low value and there is a resumption of artificial donor activity from diphenylcarbazide to 2, 6-dichlorophenolindophenol (DCPIP). In PS I, an increase in the electron transport rate occurs at low 18:3/chlorophyll molar ratios, followed by a decline in rate at high ratios. The increase may result from an alteration of membrane structure that allows greater reactivity of the artificial donors (e.g. DCPIP) with plastocyanin. The decrease is due to loss of plastocyanin from the membrane (Golbeck et al., 1980).

Unsaturated fatty acids also cause a shift in the pH optimum for electron flow activity toward acidity (Fig. 1). Low concentrations of linolenic acid which display an uncoupling effect do not shift the pH optimum for both PS II and PS I electron flow systems whilst higher concentrations having no uncoupling effect do shift the pH optimum. It was postulated that the acid shift of the pH optimum and the inhibition of electron flow activity by fatty acids are due to deterioration of the thylakoid membrane causing an increase in the permeability of the membrane to water and protons combined with an inhibition of the light-induced proton uptake mechanism (Siegenthaler, 1973). This hypothesis is based on the fact that the rate of electron flow is controlled not only by the degree of energy coupling but also by the internal pH of the thylakoid compartment, as reported by Avron's group (Rottenberg et al., 1972). A thorough study of the relations between the shift in pH optimum of several types of electron flow activity, ΔpH [i.e., the difference between the external (pHo) and internal (pHi) pH] across the thylakoid membrane and pHi is in full accordance with this hypothesis (Siegenthaler and

Depéry, 1976). The data and the proposed interpretation are summarized in Fig. 1. In intact thylakoids, light initiates the proton pump mechanism which creates a ΔpH between the outer and inner space of the membrane (Fig. 1A). This results in a lowering of the pHi. In the light, pHo 8.5–9.0 corresponds to pHi close to 5, both pH values being optimal for electron transport (Rottenberg et al., 1972). Under these normal conditions, the uptake of protons is accompanied by an exchange of cations and anions as well as by a shrinkage of thylakoids and maximum photophosphorylation. Thus, the electron flow activity seems to be delicately controlled by the proton pump and ΔpH. In thylakoids treated by fatty acids or aged in vitro (Fig. 1B), the membrane becomes deteriorated in such a way that the proton pump and the resulting ΔpH diminish, the extent of the decreases depending upon the 18:3/chlorophyll molar ratio or aging time of isolated thylakoids. The pHi values increase and eventually reach pHo. These values are no longer optimal for electron flow activity. Since the size of ΔpH is lowered in treated thylakoids, the pHo optimum for electron flow is shifted toward the acidic side and both photophosphorylation and light-induced shrinkage diminish. Under these conditions, the fluxes of cations and anions are altered in such a way that swelling occurs and light-induced shrinkage decreases (Siegenthaler and Depéry, 1976).

IV. Depletion of Thylakoid Acyl Lipids

A. Solvent Extraction of Freeze-Dried Membranes

Mild extraction of lipids from freeze-dried membranes has been used by several authors for determining the molecular associations between the photosynthetic membrane components. Costes et al. (1972) made a careful study of the extractibility of the different lipids in wheat chloroplast membranes by various organic solvents. They showed that each lipid displays a double heterogeneity in the distribution between the organic fractions and the acyl chain composition, and explained this compartmentation on the basis of interactions (hydrophobic forces, ionic and hydrophobic bonds) between lipids themselves and lipids and proteins. Interestingly, Costes et al. (1978) found that in *Rhodospirillum rubrum* chromatophores, PG is the only lipid to be isosoluble with bacteriochlorophyll suggesting that

this lipid might stabilize the pigments associated with the photosystem of chromatophores. These early findings have been confirmed recently by reconstitution experiments (see Chapter 9, Trémolières and Siegenthaler and references therein). The extraction approach by organic solvents may provide useful information on the structural role and organization of lipids in the thylakoid membrane but is not particularly adequate for determining the specific lipid requirement involved in photosynthetic functions (Krupa and Baszinski, 1975).

B. Mild Solubilization of Membrane Components by Detergents

Thylakoid subfractions and (chlorophyll) protein complexes prepared in the presence of detergents have been extensively employed over these past 30 years (Berthold et al., 1981; Heinz and Siefermann-Harms, 1981; Dunahay et al., 1984; Ikeuchi et al., 1985; Murata et al., 1990; Trémolières et al., 1994) for determining the minimal specific associations between proteins and lipids necessary to ensure the structural integrity and/or to sustain a particular function of thylakoid particles. The reader is referred to reviews for discussion of the literature concerning the structural and functional aspects of these particles (Dunahay et al., 1984; Webb and Green, 1991; Williams, 1994). This approach presents several drawbacks and limitations which are discussed in Section III.A.2 of Chapter 7 (Siegenthaler).

C. Controlled Lipolytic Treatment of Thylakoid Membranes

1. Early Contributions

The enzymatic approach consists of treating thylakoid membranes under various conditions with lipolytic enzymes. This allows one not only to assess the transmembrane distribution of acyl lipids (see Chapter 7, Siegenthaler), but simultaneously, to draw relationships between the modification or depletion of certain lipids which are topologically defined and the alteration of particular photochemical functions in the thylakoid membrane. In early works, the conclusions are often ambiguous (for a review, see Siegenthaler and Rawyler, 1986 and references therein). At least three reasons might explain the lack of clear-cut conclusions. Firstly, most authors did not recognize clearly that lipolytic enzymes first degrade bulk lipids which do not support specific function. Indeed, part of the bulk lipids can be hydrolyzed without affecting directly and significantly a function. Secondly, the fact that the thylakoid membrane consists of a lipid bilayer in which the outer and inner monolayers are mutually independent up to a certain degree, was not considered. Thirdly, in most studies, detailed kinetics of lipid hydrolysis and concomitant alterations of functions are lacking. Finally, the hydrolysis products such as lysoderivatives and free fatty acids were often not removed from the membrane, thus preventing a clear distinction between the effect of acyl lipid hydrolysis itself and the breakdown products on the measured function. These products, namely free fatty acids, are known to interact with some photosynthetic components of the membrane and impair photochemical functions (Siegenthaler, 1972, 1973, 1974; Hoshina and Nishida, 1975; Siegenthaler and Depéry, 1977; Siegenthaler and Rawyler, 1977; Golbeck et al., 1980). Thus, the removal of fatty acids and lyso-lipids by bovine serum albumin (BSA) is a pre-requisite requirement for studying the role of lipids in thylakoid membrane function (Siegenthaler et al., 1987a,b).

In early works, lipolytic acyl hydrolases (LAH) partially purified from either bean leaves (Anderson et al., 1974; Shaw et al., 1976; Michalski and Kaniuga, 1980; Krupa, 1982, 1983), spinach leaves (Krupa and Baszynski, 1975) or potato tubers (Hirayama and Matsui, 1976; Rawyler and Siegenthaler, 1980) were used. Although lipolytic treatments have marked effects on photosynthetic electron transport, simple correlations, either temporal or quantitative between lipid hydrolysis and loss of activity, could not be strictly established. In addition, results did not allow a functional role to be attributed to a specific lipid because of the non-specificity of LAH.

In contrast to LAHs, the use of phospholipases offers the advantage of selectively degrading the phospholipid class of thylakoids (PG) and of lowering the amount of membrane lipid hydrolyzed (i.e. phospholipids account for no more than 15% of the total membrane acyl lipids), thereby limiting damages to the bilayer structure. Several studies show that a PLA_2 treatment which alters 60–80% of PG inhibits non-cyclic and PS II electron flows (Hirayama and Matsui, 1976; Rawyler and Siegenthaler, 1981a; Jordan et al., 1983). In the presence of BSA (which binds lipid breakdown products), the hydrolysis of

lipids by PLA_2 is similar, but the activities are not significantly altered (Hirayama and Nomotobori, 1978; Duval et al., 1979). This indicates that the hydrolysis products exert an inhibitory effect which can be relieved through BSA binding. Moreover, the extent of lipid degradation roughly accounts for the PG population which is located in the outer leaflet of the thylakoid membrane, thus suggesting that the presence of residual PG in the inner monolayer may be the explanation for the persistence of PS II activity.

Interestingly, Duval et al. (1979) showed that PLA_2-treatment of thylakoids dissociates the oligomeric into the monomeric form of LHCP and that the reconstitution kinetics are more rapid with PG containing 16:1(3t) than with any other type of lipids (Rémy et al., 1982; Dubacq and Trémolières, 1983). This is in agreement with the conclusion of Krupa (1984) and Horváth et al. (1989) that PG is essential for the stabilization of oligomeric structures of LHC complex. Moreover, several authors proposed that PLA_2 treatment blocks the electron flow somewhere between the entry point of 1,5-diphenylcarbazide and plastoquinones (Rawyler and Siegenthaler, 1981a; Thomas et al., 1985) or, more precisely, at a site located in the electron transfer step from pheophytin to Q, possibly by altering the relative distance and/or orientation between these two components (Jordan et al., 1983). This interpretation is reminiscent of that proposed by Golbeck et al. (1980) concerning the action site of added linolenic acid. Another feature which has also to be taken into consideration is that exposure of thylakoids to PLA_2 leads to membrane lipid phase separation and irreversible formation of non-bilayer lipid structures (Thomas et al., 1985). These structural changes are accompanied by an inhibition of PS II-mediated electron transport and a stimulation of PS I-mediated transport (Rawyler and Siegenthaler, 1981b; Thomas et al., 1985). According to Jordan et al. (1983) the stimulation of PS I electron transport activity is not due to increased spillover but rather to an increased rate of diffusion or accessibility of the donors which feed electrons into PS I.

This literature survey shows that definite conclusions cannot be drawn from early results. In our opinion, the limitations of the enzymatic approach at that time was mainly due to a failure of applying the above-mentioned basic concepts which are *sine qua non* conditions for obtaining valuable results on the functionality of topologically distinct lipids.

2. Role of Phospholipids

Conditions can be chosen under which phospholipid depletion occurs in a stepwise fashion, first in the outer, then in the inner monolayer of the thylakoid membrane (Siegenthaler et al., 1987a; 1989a; refer also to Chapter 7, Siegenthaler). Our results show the occurrence of at least three phospholipid populations with respect to uncoupled non-cyclic electron flow activity (Fig. 3): (a) A first one, corresponding to a complete phospholipid depletion (phase A) in the outer monolayer (i.e. about 69% of the total PG) leads to less than 20% inhibition of the electron flow activity; (b) A second one, corresponding to a partial phospholipid depletion (phase B, C and D) in the inner monolayer (i.e. about 19% of PG) causes more than 80% inhibition of the activity; (c) A third PG population (12%), which is inaccessible to PLA$_2$ and presumably localized in the inner monolayer, is not involved in the electron flow activity. As illustrated in Fig. 3 (right panel), the inner PG population consists of distinct subpopulation pools (B to D), some of which are very efficient in sustaining the electron flow activity e.g., 3% of PG in pool C supports 40% of the activity. These inner subpopulation pools have been evidenced by a stepwise delocalization of inner PG molecules through outward

transbilayer movement. These pools inhibit the activity in a non-linear fashion, suggesting the occurrence of interactions of different types and strengths between certain PG molecules and the other interacting membrane components, e.g. proteins (Siegenthaler et al., 1989b). During phospholipid hydrolysis, the osmotic responsiveness of thylakoids towards sorbitol remains unaltered, indicating that the structure of thylakoids is not impaired by the enzymatic treatment. Altogether, these results point to the non-equivalent role of PG in the inner and outer monolayers of the thylakoid membrane in supporting the electron flow activity (Siegenthaler et al., 1987b, 1989b).

The importance of inner PG population in sustaining the uncoupled non-cyclic electron flow activity has been substantiated and confirmed by using grana membranes (i.e. thylakoid inside-out vesicles) obtained by the batch procedure of Andreasson et al. (1988). Indeed, a depletion of PG in the outer leaflet of such vesicles inhibits most of the uncoupled non-cyclic electron flow activity (Duchêne et al., 1995). Moreover, it was shown recently that a depletion of PG in the outer thylakoid monolayer (about 62 mol %) does not alter the dark-phosphorylation activity, as defined and measured by Jagendorf and Uribe (1966). A further depletion

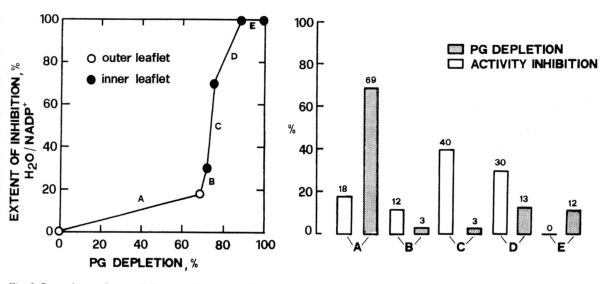

Fig. 3. Dependence of uncoupled non-cyclic electron flow activity on various phosphatidylglycerol (sub)population molecules in the outer and inner thylakoid membrane monolayers. PG depletion refers to either depletion per se (steps A and D) or to delocalization of inner PG to the outer monolayer (steps B and C). Left side graph: Inhibition extent of the electron flow activity (percentage of control) as a function of PG depletion; Right side graph: Comparative visualization of these two parameters. See explanations in the text and in Siegenthaler et al. (1989a,b).

of PG in the inner monolayer (from 62 to about 83 mol%) inhibits only 25% of the activity (Siegenthaler and Vallino, 1995).

In order to gain further insight into the role of phospholipids in the structural-functional relationship of thylakoid membranes, a few authors studied the effect of digestion of phospholipid molecules in the presence of PLC and D (Siegenthaler et al., 1984; Droppa et al., 1995). Such treatments remove the soluble polar head moiety of the phospholipid molecules, keeping intact the sn-1 and sn-2-positioned fatty acyl chains in the outer monolayer of the membrane. The results reported by Droppa et al. (1995) show that the progressive removal of phosphoglycerol from PG located at the outer surface of the membrane affects differentially the electron transport activity as well as the fluorescence induction and thermoluminescence characteristics of PLC-treated thylakoids. A treatment by low concentrations of PLC (corresponding to about 10% of phospho-glycerol removal) results in no alteration of electron transport but in a fast reduction of Fv which can be restored by DCMU. These results indicate that the electron flow from Q_A^- is intensified by the polar head group removal. Higher concentrations of the enzyme (corresponding to about 40% of polar head group removal in PG) induce strong reduction in both electron flow and Fv while only part of Fv can be restored by DCMU. This indicates that the polar head group of PG interacts not only with the acceptor side of PS II but also affects electron transfer before the inhibiting site of DCMU. The location of this second inhibitory site was investigated by measuring the thermoluminescence characteristics of PLC-treated thylakoids (Droppa et al., 1995). The fact that the thermoluminescence B band (originating from the $S_2 S_3 Q_B^-$ charge recombination) is abolished faster than the Q band (originating from the $S_2 S_3 Q_A^-$ charge recombination) confirms the conclusion that the removal of the polar head group of PG causes a rapid oxidation of Q_B^-. Furthermore, the decrease of Q band and the concomitant increase of C band (originating from tyrosine $D^+ Q_A^-$ recombination) upon PLC treatment may indicate that Q_A^- is capable for charge recombination but of the positive charges of S_2 and S_3 states are no longer available for this process. This makes possible that the positive charge stored on tyrosine D participates in the recombination reaction resulting in the emission of band C (Demeter et al., 1993).

Altogether, the above results show that PG molecules, as attested by depletion experiments and amputation of certain portions of the molecule, display specific roles in controlling the required thylakoid membrane organization that maintains the appropriate electron transport and associated reactions. However, no data are yet available as to the molecular species of PG which are involved in these processes.

3. Role of Galactolipids

The changes occurring in the molecular organization of MGDG when a resting thylakoid membrane is energized upon illumination were studied by Rawyler and Siegenthaler (1989). The transversal distribution of MGDG in thylakoid membranes working under coupled (addition of ADP and inorganic phosphate) or uncoupled (addition of an uncoupler such as FCCP) electron transport conditions is the same as in resting (darkness) membranes, e.g., the outer/inner molar ratio is about $62 \pm 3/38 \pm 3$ (see Chapter 7, Siegenthaler). These results indicate that, if changes in the MGDG organization occur during the dark/light transition, they cannot be detected at the level of the transmembrane distribution of this lipid and that they are probably more subtle and, consequently, more difficult to estimate. Taking advantage of the fact that many lipolytic enzymes are sensitive to lipid packing, the lipase from *Rhizopus arrhizus* (LRa) was used to probe relative changes in the packing pressure of its main substrate, MGDG, in thylakoid membranes. First, it was checked that when thylakoid membranes are catalytically hydrogenated according to Vigh et al. (1985) or enriched in cholesterol (two conditions under which the packing pressure of the membrane is increased), the rate of MGDG hydrolysis by the lipase is markedly diminished. Furthermore, compared to the dark control, the initial rate of MGDG hydrolysis does not change when thylakoid membranes are incubated in the light under various conditions such as basal, uncoupled (+FCCP) or DCMU-inhibited electron flow from H_2O to methyl-viologen. In contrast, the rate of MGDG hydrolysis is markedly accelerated when the membranes are allowed to work under phosphorylating (+ADP + Pi) conditions. A pretreatment of thylakoids with NaBr (CF_1 removal), DCCD (CF_0 block) or phlorizin (CF_1 block) strongly decreases the rate of MGDG hydrolysis compared to that of the dark control. Under coupled conditions of electron flow, the fast hydrolysis rate of MGDG can be reduced within 5 s by switching off the light or by adding DCMU or FCCP. The above results are interpreted in terms of reversible changes in MGDG packing in the outer monolayer of the thylakoid membrane that depends on the functional status of the coupling factor CF_oCF_1. The decreased MGDG packing generated in phosphorylating thylakoids may be associated with an increase in membrane fluidity, thereby facilitating the long-range diffusional processes required for optimal photosynthetic electron transport (Rawyler and Siegenthaler, 1989).

The above results suggest that the packing of MGDG may be intimately related to the rate of photophosphorylation. It was therefore of interest to study the effect of a selective depletion of this thylakoid major acyl lipid on the photo- and dark-phosphorylation (Siegenthaler and Vallino, 1995). Provided that free fatty acids and lyso-MGDG resulting from the enzymatic hydrolysis by the LRa-treatment are removed, a complete depletion of MGDG in the outer monolayer (65 mol % of the total MGDG) impairs completely both activities (Fig. 4). Under these conditions, MGDG is still present in the inner monolayer. Furthermore, the inhibition of the activities as a function of MGDG depletion varies stepwise, indicating that topologically distinct MGDG pools have different accessibility to the enzyme in the thylakoid membrane. Furthermore, the fact that both photo- and dark-phosphorylation activities are impaired in a similar fashion indicates that MGDG depletion in the outer monolayer does not alter the basal electron flow, which is indeed the case and is therefore not responsible for the observed inhibition. Thus, in contrast to PG, outer MGDG is absolutely required for the functioning of the ATP synthase machinery in the thylakoid membrane (Siegenthaler and Vallino, 1995), as already suggested by reconstitution experiments (Pick et al., 1984, 1987).

D. Removal of Thylakoid Acyl Lipids by Cyclodextrins

Cyclodextrin (CD) molecules are cyclic oligo-saccharides consisting of 6, 7 or 8 glucopyranose units linked by α (1–4) bonds (Szejtli, 1990; Bekers et al., 1991). They adopt a torus shape and are able to bind a range of small guest molecules (e.g., lipids) of poor water solubility within their hydrophobic cavity to form water soluble guest-CD inclusion complexes. The structure and the main properties of cyclodextrins as well as the host (CD)-guest (lipids of the thylakoid membrane) possible interactions are described in Chapter 7 (Siegenthaler and references therein).

The CD-induced depletion of lipids from thylakoid membranes has important implications for their functions (Rawyler and Siegenthaler, 1996). Spinach thylakoids, treated with a mixture of α- and β-CD (5 mM each at 0 °C for 10 min), showed a 35%

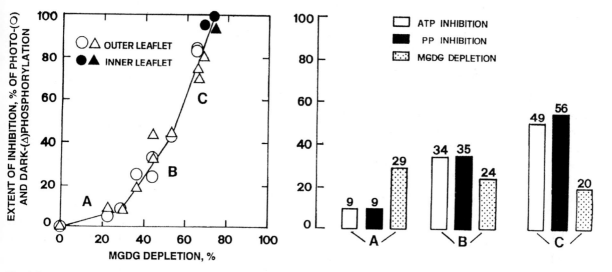

Fig. 4. Dependence of photo- (PP) and dark-phosphorylation (ATP) on various MGDG (sub)population molecules in the outer and inner thylakoid membrane monolayers. Left side graph: Inhibition extent of photo- (○, ●) and dark (△, ▲) phosphorylation (percentage of control) as a function of MGDG depletion. Right side graph: Comparative visualization of these three parameters. For further details, refer to Siegenthaler and Vallino (1995).

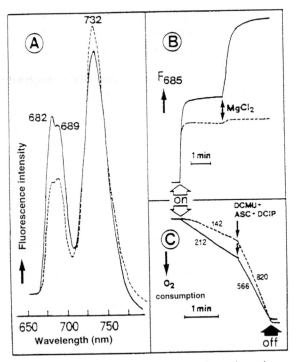

Fig. 5. Effect of a treatment of thylakoid membranes by a mixture of α- and β-cyclodextrin on (A) the 77 K fluorescence emission spectrum, (B) the kinetics of the Mg^{2+}-induced increase in chlorophyll fluorescence (F_{685}) at 20 °C, and (C) the electron transport activities through both PS II + I or through PS I alone. Under the conditions used, the lipid removal (expressed as mol % of each lipid class) is for MGDG (30), DGDG (32), PG (46) and SQDG (58); this corresponds to a depletion of 35% of the total lipids. In (A) and (B), fluorescence was excited at 480 ± 10 nm. In (C) numbers on the curves express the oxygen consumption rates in μmol (mg chlorophyll)$^{-1}$h^{-1}. Thylakoids were suspended in a medium (50 μg chlorophyll/ml) and incubated at 0 °C for 10 min in the absence (——) or in the presence (-----) of both α-CD and β-CD together (5 mM each). Reprinted from Rawyler and Siegenthaler (1996) with kind permission of Elsevier Science-NL, 1055 KV Amsterdam, The Netherlands.

subsequent failure of Mg^{2+} to increase F_{685} value in the delipidated membranes (Fig. 5B). Finally, the behavior of the electron flow activity is also affected by the lipid depletion (Fig. 5C). The first phase of the trace, reflecting NH_4Cl-uncoupled electron flow activity from H_2O to methylviologen, i.e., through both PS II + I, shows an inhibition (−33%) in lipid-depleted thylakoids, whereas the second phase (DCMU + ascorbate + DCPIP), reflecting Photosystem I-specific electron transport, shows a stimulation (+45%). Though further experiments need to be carried out to interpret these results, the following lines of thoughts can be considered: (1) Though CD-induced lipid depletion does not greatly alter the general balance between bilayer-forming (DGDG, PG, SQDG) and non-bilayer-forming (MGDG) lipids in thylakoid membranes, it may however change the barrier properties of membranes, as it removes part of their sealing material; (2) In addition, the fluidity of the thylakoid membrane may be altered by CD treatment, as the lipid/protein ratio, believed to control the fluidity of thylakoid membranes (Ford et al., 1982; see also Webb and Green, 1991 and references therein), is decreased; (3) Also, lipid depletion is expected to decrease the relative distances between the supramolecular complexes of thylakoid membranes by bringing them into closer proximity; (4) All these CD-induced structural modifications have consequences on the thylakoid membrane functions, the most obvious being the alteration of the distribution of excitation energy between the two photosystems (Rawyler and Siegenthaler, 1996).

V. Modulation of Thylakoid Membrane Fluidity

It is well established that both the composition and the nature of membrane lipids are of crucial importance in the function of membrane proteins (Quinn and Williams, 1983). Chloroplast membranes are characterized by the presence of large proportions of polyunsaturated lipids. The major acyl residue is linolenic acid which accounts for at least 70% of the total fatty acid content of higher plant chloroplast lamellae. However, the functional significance of the highly unsaturated nature of chloroplast lipids has not been fully elucidated. It has been suggested that fatty acid unsaturation is required for maintaining the appropriate membrane fluidity over a wide range

depletion of their total lipids but an unaltered polypeptide profile with respect to control membranes. Under these conditions, the lipid removal (expressed as mol % of each lipid class) is for MGDG (30%), for DGDG (32%), for PG (46%) and for SQDG (58%). The 77 K fluorescence spectrum of the treated sample shows that the $F_{732}/F_{682-689}$ emission ratio is significantly increased from 1.4 in the control to 2.4 in treated membranes and that the relative height of the F_{682} and F_{689} peaks is reversed (Fig. 5A). Moreover, the intensity of fluorescence emission at 20 °C measured in the presence of DCMU as a function of time is characterized by a lower level of the initial fluorescence signal (F_{685}) and by a of environmental temperatures and to induce chilling resistance and cold hardiness (Quinn and Williams, 1983; Webb and Green, 1991). Moreover, the lipid composition is believed to play an important role in the molecular organization of the thylakoid membrane by determining the cooperative interactions between its functional components and complexes (see Chapter 7, Siegenthaler). These components undergo rotational and lateral movements within the membrane and are consequently dependent on the fluidity of the lipid matrix. Membrane fluidity is primarily determined by the length and the unsaturation degree of the alkyl chains of the constituent lipids, as well as by the relative composition of its lipids and proteins, and by the temperature. Fluidity which can be defined as the degree of unhindered mobility experienced by the acyl chains of lipid molecules, is commonly believed to be associated with the degree of thylakoid lipid unsaturation. However, as pointed out and discussed by Webb and Green (1991) there is no consistent correlation between unsaturation and either order parameter or diffusion rate.

The fluidity of the thylakoid membrane can be modulated by varying the lipid/protein ratio and by homogeneous catalytic hydrogenation in situ.

A. Modification of the Lipid/Protein Ratio

The reader is referred to the reviews of Siegenthaler and Rawyler (1986) and Webb and Green (1991) for a discussion of the early literature. The experimental approaches consist of artificially manipulating the lipid/protein ratio by incorporating lipids (liposomes) or proteins (such as melittin) into thylakoid membranes and measuring photochemical functions. Though the results are often contradictory, they can be interpreted in terms of alterations of the spatial relationship between membrane components upon lipid enrichment, and restrictions in lipid molecule motion upon addition of protein, i.e., a decrease in the overall membrane fluidity. However, one has to remember that the relationship between the lipid/protein ratio and membrane fluidity is still a subject of controversial discussion.

Plants seem to be able to adjust the lipid/protein ratio of their membrane according to the function they have to carry out and probably also in response to variations in temperature. For instance, in photosynthetic thylakoid membranes, non-appressed (stromal) lamellae display a higher lipid/protein ratio and higher fluidity than appressed (granal) lamellae though both types of thylakoids have nearly the same degree of unsaturation (Murphy and Woodrow, 1983). This finding is in accordance with the role of stromal thylakoids in allowing the diffusion of mobile phosphorylated LHCII towards Photosystem I (Allen, 1992). As another example, the fadC mutant of *Arabidopsis*, deficient in activity of the chloroplast n-6 desaturase, accumulates high levels of 16:1 and 18:1 lipids and has a correspondingly reduced level of polyunsaturated lipids (Hugly et al., 1989). The altered lipid composition of the mutant has pronounced effects on the chloroplast ultrastructure, e.g., a 48% decrease in the amount of appressed thylakoid membranes. Furthermore, the mutant is characterized by a net loss of 36% of the thylakoid membranes per chloroplast and a corresponding reduction in chlorophyll. Interestingly, the reduction of the unsaturation degree is partly compensated by an increase in lipid/protein ratio with the result that non-cyclic electron flow and PS I rates are unimpaired, although there is a 20% decrease in PS II activity. Though there are very few reports comparing lipid unsaturation and lipid/protein ratios, it nevertheless appears that the plant can modulate both parameters in thylakoid membranes in such a way as to ensure optimal functions (refer also to the review by Webb and Green, 1991).

B. Lowering the Unsaturation of Thylakoid Lipids by Catalytic Hydrogenation

The $Pd(QS)_2$ catalyst used for thylakoid hydrogenation offers a suitable tool for studying the role of unsaturation/saturation levels of lipids in the regulation of membrane fluidity and photosynthetic activities (Quinn et al., 1989 and see Chapter 7, Siegenthaler). This technique makes it possible to decrease the number of double bonds of fatty acids while retaining the preexisting architecture of the membrane (Restall et al., 1979). Homogeneous catalytic hydrogenation in chloroplast membrane lipids was pioneered by the Szeged's group (Vigh et al., 1985). Basic findings showed that: (1) The catalyst causes an extensive loss of unsaturated fatty acids in the presence of H_2 gas without affecting the double bonds of neutral lipids such as chlorophylls, carotenoids and plastoquinone (Szalontai et al., 1986); (2) The polyunsaturated fatty acids are hydrogenated at a faster rate than the monoenoic acids; (3) During hydrogenation the orientational ordering of membrane lipids, as measured with the C-12 positional isomer of spin-labeled stearic acid, displays a slight increase; (4) Most important, the catalyst can be removed easily after washing of thylakoids and in itself does not alter photosynthetic activities; (5) Progressive saturation of double bonds of lipids (e.g., decrease in fluidity) primarily inhibits the electron transport between the two photosystems followed by the inhibition of PS II electron flow, while PS I electron transport activity is not inhibited even by 50% fatty acid hydrogenation (Vigh et al., 1985). This has led to the suggestion that the unsaturation level of acyl chains lipids plays a crucial role by ensuring the lateral mobility of plastoquinone between PS II and PS I (Horváth et al., 1986a), in contrast to the fact that even in the case of 30% hydrogenation, no change in the reduction rate of flash-oxidized cytochrome f is observed (Gombos et al., 1988).

The effect of lipid hydrogenation was further investigated on the PS II reactions by studying and comparing the characteristics of fluorescence induction and thermoluminescence (Hideg et al., 1986). Hydrogenation increases the level of intermediate fluorescence (Fi) without affecting significantly Fo and the maximal level of fluorescence (Fm). This might indicate an enhancement of Q_A^- in the hydrogenated samples similar to that found in the presence of DCMU and, moreover, a limited transfer of electrons from Q_A toward the plastoquinone pool. At both low and high light excitations, the progressive hydrogenation concomitantly increases the value $\Delta Fi / \Delta F_{max}$ (Fi–Fo/F$_{max}$–Fo) from 0.2 up to 0.5 confirming a decreased electron transport between Q_A and Q_B. However, thermoluminescence (TL) experiments reveal that the B band of TL which is generated by charge recombination of $S_2Q_B^-$ and $S_3Q_B^-$ redox couples, is present and even increased by saturation of double bonds of lipids. It is concluded that the Q_B component can be charged partly in hydrogenated thylakoids but cannot be reoxidized since the electron transfer between the two photosystems is blocked (Vigh et al., 1985).

Based on fluorescence induction kinetics measurements of thylakoids, Horváth et al. (1987) proposed a model which might help to visualize the effect of hydrogenation with respect to PS II organization and function (Fig. 6). Hydrogenation decreases the rate of electron transfer between Q_A and Q_B, disconnects the chlorophyll a/b LHCII-peripheral, effectively converting a PS IIα unit into a PS IIβ unit and free

LHCII in hydrogenated thylakoids. Thus, the saturating level of fatty acids may play an important role in regulating the association of the peripheral chlorophyll a/b light harvesting complex II with PS IIβ, and determining the PS II heterogeneity within PS IIα and PS IIβ units (Horváth et al., 1987).

As expected, both cyclic and non-cyclic phosphorylation show significant changes upon saturation of the double bonds of fatty acyl residues of membrane lipids (Horváth et al., 1986a). The capacity of non-cyclic phosphorylation as well as the uncoupling extent (± methylamine) are decreased gradually by progressive hydrogenation, and no ATP is formed when about 35% of double bonds of lipids are saturated. Horváth et al. (1986a) found that the cyclic phosphorylation appears to be more resistant to lipid hydrogenation suggesting that either hydrogenation affects cyclic electron flow less than non-cyclic electron flow or that the sites of proton translocation, driving the two types of phosphorylation, are located in different regions of the photosynthetic membranes.

In all the above-cited experiments, one has to remember that a progressive hydrogenation of the lipids in situ produces changes in the organization of the lipids. Single bilayer vesicles are converted to large aggregates of planar bilayer stacks in which the hydrocarbon chains are predominantly in the gel phase configuration (Horváth et al., 1986b). Furthermore, thickening of single membranes with a simultaneous decrease in the spacing between membranes is observed (Horváth et al., 1986a). These changes might be accounted for by the alignment of the hydrocarbon chains of saturated lipids and the increased hydrophobicity of the membranes. However, the orientational pattern of chlorophyll a molecules is not altered by saturating up to 50% fatty acyl double bonds, indicating that the energy-transfer processes amongst the chlorophyll molecules remains functional after hydrogenation (Thomas et al., 1986).

Under physiological conditions, the extent of hydrogenation of membrane lipids is, however, much smaller than that obtained in early studies by hydrogenation in situ. Under these latter conditions, a large proportion of fatty acids are converted to saturated fatty acids which might produce a solid (gel) phase in the thylakoid membrane and, consequently, alter its function. Thus, only small extent of hydrogenation in situ should by considered for the understanding of the effect of lipid (un)saturation on in vivo photosynthetic activities.

Exposure of higher plant cells to heat stress results

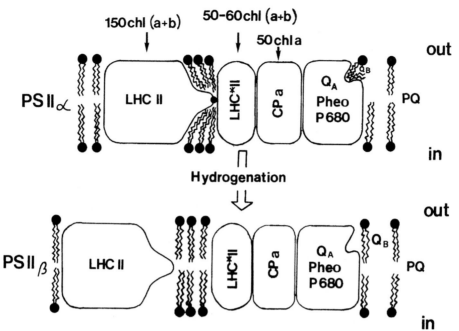

Fig. 6. Tentative model showing the effect of lipid hydrogenation on the PS II structure and function. PS II consists of the reaction center (P_{680}, Pheophytin, Q_A and Q_B) and, attached to it, the core chlorophyll a (CPa) and the LHCII-intrinsic (LHC*II) antenna. The association of PS IIβ with the LHCII peripheral antennae (LHCII) results in the formation of PS IIα. Note that the hydrogenation treatment dissociates the LHCII peripheral antennae (leading to the formation of LHCIIβ) and inhibits the $Q_A \rightarrow Q_B$ interaction (for details, see text). Redrawn from Horvath et al. (1987) with kind permission of the authors.

in irreversible damage to photosynthesis prior to impairment of other cell functions. PS II-mediated electron transport and photophosphorylation rates are particularly susceptible to high temperatures. However, PS I-supported electron transport is stimulated rather than inhibited at the threshold temperature at which the damage to PS II reactions occurs (Thomas et al., 1986a). The loss of PS II activity is accompanied by a sharp increase in chlorophyll a fluorescence from PS I (Vigh et al., 1989 and references therein). These data have been interpreted in terms of the blockage of PS II reaction centers and a dissociation of the peripheral LHC of PS II. The observed increase in DCPIP$_2$-supported PS I activity has been attributed to the thermal stress-induced formation of new acceptor sites for DCPIP$_2$ within the cytochrome b_6f complex, more particularly at the reducing site of the Rieske FeS center. It was also suggested that the stimulation of PS I activity is related to the phase separation of non-bilayer forming lipids in heat-stressed membranes (Vigh et al., 1989 and ref. therein). Employing the technique of homogenous catalytic hydrogenation, Thomas et al. (1986) observed that hydrogenation of lipids reduces the tendency of heated membranes to destack and

vesiculate at high temperatures. Measurements of chlorophyll *a* fluorescence emission and the thermal properties of hydrogenated thylakoids suggest that the hydrogenation process also leads to an increase in the thermal stability of pigment-protein complexes of the PS II light harvesting apparatus. Furthermore, a selective saturation of *cis* double bonds of lipid alkyl chains of thylakoids results in a marked increase in the threshold temperatures at which both the thermal damage of PS II-mediated electron transport and formation of non-bilayer lipid phase occur (Thomas et al., 1986). Heat-induced stimulation of DCPIPH$_2$-supported PS I electron transport can also be controlled by the level of unsaturated lipids in the thylakoid membrane, e.g., lipid saturation above 35% totally prevents the appearance of any increase in electron flow to PS I upon heat treatment (Vigh et al., 1989).

It has already been pointed out that fatty acid unsaturation is required to maintain the appropriate fluidity of the thylakoid membrane in order to optimize the cooperative interactions of its functional components. However, this very complex molecular organization which has to be, within certain limits, highly dynamic, appears to be very fragile when the environmental factors (temperature, light, etc.) are beyond the normal physiological conditions. This raises the question of how eukaryotic cells maintain the physical structure of their membrane lipid bilayers within tolerable limits. Yet little is known about how this is achieved and what the cellular sensors concerned are (Maresca and Cossins, 1993). However, a clue to what is involved comes from recent data obtained by Vigh et al. (1993). In the blue-green alga *Synechocystis* (PCC 6803), these authors showed that the catalytic hydrogenation of a small pool of plasma membrane fatty acids activates the transcription of the *desA* gene which encodes an acyl-lipid desaturase. This enzyme acts at the $\Delta 12$ position of fatty acids of membrane lipids (Wada et al., 1990). Interestingly, the effects of hydrogenation and a decrease in temperature both contribute to increasing the level of *desA* transcript. Thus, it appears that the first signal in the perception of a temperature change is an alteration in the fluidity of the plasma membrane in *Synechocystis* (for more details see Chapter 13, Gombos and Murata and references therein).

VI. Immunological Approach

A. Involvement of Glycolipids in the Electron Transport Activity

Antibodies directed to glyco- and phospholipids have been shown to be useful tools in understanding not only the topology of these lipids in the thylakoid membrane (for a review, see Chapter 7, Siegenthaler) but also the role lipids play in photosynthetic reactions (Radunz et al., 1984a). In order to establish relationships between treatment with antibodies against lipids of the thylakoid membrane and the corresponding effects on photosynthetic functions, several requirements have to be fulfilled among which: (a) Thylakoid membranes have to be placed under osmotically active conditions and (b) Physiological reaction conditions have to be such as to offer a high binding affinity to the antibodies (Voss et al. 1992a). The inhibition of any activity, e.g., of the electron transport, can be interpreted by assuming that the binding of antibodies induces a conformational change of certain electron transport components, thus leading to the observed inhibition. This requires, of course, that lipids antigenic determinants are bound to these components. By means of the western blot procedure, Voss et al.

(1992b) were able to show that glycolipids (MGDG, DGDG and SQDG) are bound onto various polypeptides of the PS II complex.

The effect of monospecific antibodies directed to galactolipids (Radunz et al., 1984a), phospholipids (Radunz, 1984) and SQDG (Radunz et al., 1984b) has been investigated in the photosynthetic electron transport of higher plant thylakoids. All the above antibodies inhibit, though to varying extents, the photoreduction of DCPIP with water as the native electron donor as well as the reduction of anthraquinone-2-sulfonate with the electron donor couple DCPIP/ascorbate (see Fig. 7). The degree of inhibition of the galactolipid and SQDG antibodies in the region of Photosystem I depends on the pH (maximum inhibition in the pH range 6.8 to 7.5) and temperature (maximum inhibition between 20 and 30 °C) of the assay medium. Around 10 °C the antibodies cause no or only minimal inhibition. From earlier circular dichroism experiments it was concluded that hydrophobic interactions occur in vitro between SQDG and a 25 kDa polypeptide. The binding of antibodies against SQDG (or other lipids) may lead to conformational changes of the proteins involved in the electron flow or light collection, thus causing photosynthetic inhibition (Radunz et al., 1984a and references therein).

The sites of inhibition of antibodies against lipids are localized on the donor side of PS II as well as of PS I, as illustrated in Fig. 7. The binding of these antibodies results only in a partial blocking of the electron transport reactions. This observation is likely to be attributed to the granal structure of the chloroplasts, in which large regions of the partitions are not accessible to antibodies. Another explanation would be that only part of the lipids are accessible to antibodies on the stroma side of the thylakoid membrane. In addition, considerable size differences exist between antigen- and antibody molecules. These differences might lead to the consequence that one antibody, once bound to the rather small antigen molecule, protects other lipid molecules in its immediate vicinity.

B. Involvement of Phosphatidylglycerol in Photosystem II Structure and Function

Recently, Kruse et al. (1994) isolated by SDS-gel electrophoresis the D1-peptide from PS II-preparation of the filamentous cyanobacterium *Oscillatoria chalybea*. On a western blot, the intrinsic peptide

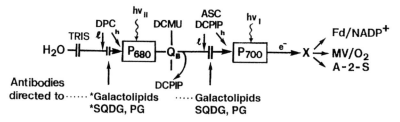

Fig. 7. Photosynthetic electron transport scheme showing the approximate sites of inhibition of the antibodies directed to galactolipids (MGDG and DGDG), SQDG and PG. *, the inhibition of the electron flow activity is observed only after Tris treatment; A-2-S: anthraquinone-2-sulfonate; l (low) and h (high) concentrations of l, 5-diphenylcarbazide (DPC) and ascorbate (ASC) + DCPIP. Scheme redrawn from Radunz et al. (1984a,b) and Radunz (1984).

reacts with the antibodies against PG but not with antibodies against galactolipids. The binding of PG onto the D1-peptide (in the ratio 20:1) has been confirmed by lipid analysis (Kruse et al., 1994). As antibodies are directed towards the glycerophosphate residue but not towards the fatty acids, the above-mentioned authors concluded that PG is bound to the peptide via its fatty acid tail and that the presence of a negative charge in the molecule favors the binding of cations like Mg^{2+} and Ca^{2+} which may also contribute to maintaining the structural and functional integrity of the lipoprotein complex. This hypothesis may explain the stimulatory effects on photosynthetic reactions (Fragata et al., 1991). The fact that incubation of PS II-preparation with phospholipase A_2 inhibits 25% of the activity that can be fully restored by adding back PG, confirms that bound PG may have a functional role (Kruse et al., 1994).

Kruse and Schmid (1995) reported on the role of PG as a functional effector for charge separation at the D1/D2-heterodimer and as a membrane anchor for the D1-core peptide. Purified PS II particle preparations from *O. chalybea* were treated with phospholipase A_2, then supplemented with lipid emulsions. The resulting inhibition of the electron transport activity is relieved only by PG molecules which are added back to the preparation. These authors proposed a model in which PG is a functional effector for the optimal conformation of the D1 protein in the PS II core complex (Fig. 8). PG molecules are unusually tightly bound to the D1 protein by hydrophobic interactions, though a covalent binding seems improbable. The localization of PG binding sites is substantiated by trypsin treatment of the D1 protein and analysis of the obtained oligopeptides with HPLC and immunoblotting. The binding sites are likely to be confined to the hydrophobic amino acid section between arginine 27 and 225, which is known to be the membrane anchor

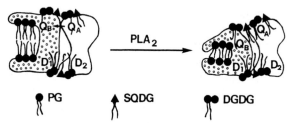

Fig. 8. Tentative model showing the effect of a phospholipase A_2 (PLA$_2$) treatment on the structure and function (electron transport activity) at the D1/D2 heterodimer in the reaction center of PS II. Bound sulfolipid (SQDG) and digalactolipid (DGDG) molecules are localized as prosthetic groups at the surface of the native D1/D2 heterodimer, holding the dimer together. Phosphatidylglycerol (PG) has to be considered as essential for the orientation and stabilization of the D1-protein and cannot be replaced by other lipids because of its high specificity; its role can be compared with that of a functional effector. Though the type of binding between PG and D1 remains unknown, one assumes that ionic interactions and van der Waal's forces in hydrophobic pockets or clefts might be the principle binding systems between D1 and PG. After PLA$_2$ treatment, the cleft is closed and the function of the lipo-protein complex obliterated. Adapted from Kruse and Schmid (1995) and from the data reported by Voss et al. (1992b).

of D1. This has led to the conclusion that the anionic PG plays an important role in the PS II complex, by binding onto the intrinsic D1 protein, as an effector for charge separation, as well as a stabilizer in the thylakoid membrane structure (Kruse and Schmid, 1995).

VII. Physiological Approach

Environmental and physiological changes leading to adaptive and regulatory processes in the photosynthetic apparatus and simultaneously in its lipid composition were reported to occur in at least two types of events.

A. Effect of Phosphatidylglycerol Unsaturation on Chilling Sensitivity

The first feature concerns the acclimation of plants to low, but not freezing temperatures. The mechanism by which such an adaptation takes place, has been carefully studied by the group of Murata (see Chapter 13, Gombos and Murata). In brief, when plants are submitted to low temperatures some species are still able to perform photosynthesis and preserve the integrity of their photosynthetic apparatus. These plants were therefore called chilling-resistant plants. The other species lose their capacity to perform photosynthesis and their structures are irreversibly damaged at low temperatures. These plants are called chilling-sensitive plants. The fatty acid unsaturation of one chloroplast lipid appears to be directly related to this adaptation, i.e., phosphatidylglycerol. Plants able to desaturate this lipid at the sn-1 position are chilling-resistant while those unable to do it are chilling-sensitive. The Japanese group has most effectively demonstrated the involvement of this lipid by genetically manipulating the unsaturation of PG in both higher plants and cyanobacteria (Wada et al., 1990; Murata et al., 1992). It was also shown that in *Synecchocystis* PCC 6803, a decrease in the temperature induces the expression of desaturase genes leading to an increase in the unsaturation level (Los et al., 1993). This process allows the cyanobacterium to perform photosynthesis under low temperature conditions. The mechanism by which such a regulation occurs is related to the fluidity of the membrane, in particular to the degree of unsaturation of PG (Moon et al., 1995). In addition, it was found to be linked to the photoinhibition process by controlling the renewal of the D1 protein (Gombos et al., 1994; Moon et al., 1995). For instance, when the saturation degree of lipids is too high, degraded forms of D1 accumulate in the membrane under strong light conditions and newly synthesized D1 proteins are unable to be rapidly and correctly inserted in the PS II complex. The PS II is thus severely and irreversibly photodamaged because in such a rigid membrane, the efficiency of the turnover of PS II reaction center D1 protein is impaired, most probably at the posttranslational level (Kanervo et al., 1995).

B. Role of Phosphatidylglycerol in Functional Organization of Thylakoid Membranes

The second feature also concerns phosphatidyl-glycerol which is involved in the biogenesis of the main light-harvesting antenna complex (LHCII) and the development of appressed membranes in the chloroplast. In photosynthetic eukaryotic organisms, the thylakoid membrane contains PG which is characterized by high levels of 16:1(3t) fatty acid esterified at the *sn*-2 position. It was demonstrated that this lipid plays a crucial role in the supra-molecular organization of the LHCII by inducing the trimerization of the antenna complex. This trimer-ization seems to be in turn a prerequisite for the formation of the grana stacks. Correlations between the presence of this lipid [the PG 16:1(3t)], the formation of the trimeric LHCII and the development of the grana stacks have been observed in several physiological and genetical situations (for more details refer to Chapter 9, Trémolières and Siegenthaler).

Huner et al. (1987, 1989) studied the relations between the lipid composition, the supramolecular organization of LHCII and the acclimation to low temperature in monocotyledon plants. A number of rye varieties are able to acclimate to low temperatures while preserving their photosynthetic activity, even at temperatures just below 0 °C. These authors observed that the percentage of 16:1(3t) in PG decreased markedly during the acclimation process in parallel with a progressive decrease in the temperature and shortening of the photoperiod. Such a correlation has to be considered with caution since it was shown that changes in the level of 16:1(3t)-PG following cold treatment greatly depend on the stage of cotyledon or leaf maturity (Xu and Siegenthaler, 1996, 1997). This change in lipid composition was correlated with changes in the ratio between trimeric and monomeric LHCII, as evidenced by electro-phoretic separation of photosynthetic complexes under mild denaturing conditions (Huner et al., 1987, 1989). In fact, the level of trimeric LHCII decreases with a concomitant increase in the level of monomeric LHCII in low temperature-acclimated plants which display a reduced percentage of 16:1(3t) in PG. On the other hand, the acquired resistance to low temperature does not affect the ability of thylakoids to regulate light energy distribution between PS II and PS I. Indeed, the kinase, involved in the phosphorylation of the mobile LHCII allowing this antenna to move from the PS II to the PS I region, was active in these resistant plants but not in plants unable to acclimate to low temperature. It was suggested that, in these latter plants, either the kinase cannot reach the substrate or the phosphorylated molecules cannot reach the PS I because of increasing viscosity of the membrane (Krupa et al., 1987; Moll et al., 1987; Coughlan and Hind, 1987). Assuming that PG 16:1(3t) is attached to the PS II complex and plays a key role in the stability of the trimeric forms of LHCII (see Chapter 9, Trémolières and Siegenthaler), one can envisage that, in monocotyledon plants acclimated to low temperatures, the decrease in the content of PG-16:1(3t) also diminishes the affinity between the PS II region and the mobile antennae thus favoring their movement toward the PS I region.

Another correlation between the content of 16:1(3t) in PG and the level of trimerization, as well as the extent of grana stacking was depicted in mutants of *Chenopodium* resistant to triazine (Gasquez et al., 1985). These mutants appear spontaneously in natural populations of *Chenopodium* and display a mutation at the D1 protein in such a way that this protein loses its affinity to triazine (Gressel, 1982). But this mutation also decreases the efficiency of the electron transfer between the two photosystems. As a secondary effect, however, the plants improve the light energy collection by increasing the oligomerization of the LHCII with a concomitant increase in the percentage of the PG-16:1(3t) (Pillai and St. John, 1981; Burke et al., 1982; Vaughn and Duke, 1984; Lemoine et al., 1986; Trémolières et al., 1988). The same kind of adaptation occurs in *Lemna minor* plants treated with only sublethal doses of atrazine where an increase in both the level of PG-16:1(3t) and grana stacks were observed (Laroche et al., 1989). It was also reported that in the chloroplasts of bundle sheath cells of C4 plants, both PG-16:1(3t) (Guillot-Salomon et al., 1973) and trimeric LHCII are absent (Bassi, 1985). Similar results were found by Lemoine et al. (1982). In chloroplasts of dark-grown pine cotyledons which still contain chlorophyll, both the PG-16:1(3t) and trimeric LHCII are absent, though polypeptides of LHCII accumulate in the monomeric form. When such dark-grown seedlings are placed under continuous light, trimeric LHCII and 16:1(3t) in PG accumulate. On the other hand, the light-harvesting efficiency increases in parallel with the degree of trimerization.

VIII. Role of Lipids in the Mode of Action of Herbicides

Well documented literature is available on the different modes of action of herbicides (Ashton and Crafts,

1981; Fedtke, 1982; Baker and Percival, 1991). Herbicides have been classified according to their known mode of action (Pfister and Urbach, 1983). While several herbicides act primarily on photosynthetic electron transport at the level of PS II, others can affect cell division and elongation, growth hormone regulation, pigment and lipid metabolism (Harwood, 1991) or membrane structure (e.g., via lipid peroxidation). However, in addition to their main effect on a primary target, herbicides often exert secondary effects which also contribute to the death of weeds.

In the thylakoid membrane, most studies are concerned with the interaction of herbicides with membrane proteins (Pfister and Urbach, 1983; Bowyer et al., 1991 and references therein; Percival and Baker, 1991). Since two routes are available for herbicides to reach their target(s), an aqueous one and a lipoidal one, and since most herbicides are lipophilic, it is surprising that little attention has been paid to the possible involvement of membrane lipids in the expression of the toxic character of these compounds. Studies with dichlorophenoxy acetic acid (2,4-D) have indicated that the transfer of 2,4-D from a hydrophilic phase to a hydrophobic phase was increased to a much greater extent by polar (PC, MGDG) than by apolar lipids (triacylglycerols); moreover, the rate of 2,4-D exchange decreased at low pH suggesting that lipids favor the transfer of the anionic form of 2,4-D (Smith, 1972). Several studies (Talbert and Camper, 1983; Stidham et al., 1985) suggest that herbicide, partitioning within the hydrophobic core of the bilayer, disturbs the packing of acyl chains to various extents according to their chemical nature.

A potentially useful avenue has been to explore the properties of the lipid matrix of thylakoid membranes isolated from weed biotypes which are susceptible (S) or resistant (R) to herbicides such as atrazine. In several species (*Brassica*, *Amaranthus*, *Poa*, *Chenopodium*), the amount of polar lipids (expressed on a chlorophyll basis) is greater in resistant(R)- than in susceptible(S)-biotypes (Burke et al., 1982; Lemoine et al., 1986; Chapman et al., 1985). This may be due to the fact that, in R-biotypes, a greater percentage of grana contains a larger number of thylakoids per granum (Vaughn and Duke, 1984; Mattoo et al., 1984). However, chloroplasts isolated from atrazine-R horseweed (*Conyza canadensis*) show a lower amount of polar lipids (relative to chlorophyll) than that of S-biotypes (Lehoczki et al., 1985). In addition, R-biotypes contain a higher proportion of MGDG and a lower proportion of DGDG as compared to S-biotypes (Pillai and St-John, 1981; Burke et al., 1982; Lehoczki et al., 1985; Chapman et al., 1985; Mayor, 1991). Other polar lipids do not show appreciable differences between the two biotypes. Due to an enrichment in linoleic acid and to a lower level of palmitic acid in their glycolipids, chloroplast total lipids exhibit a higher degree of unsaturation in R-biotypes (Lehoczki et al., 1985; Pillai and St-John, 1981; Mayor, 1991). However, PG is enriched in 16:1(3t) acid in thylakoids from atrazine R-biotypes (Burke et al., 1982) as well as in appressed thylakoid membranes (enriched in PS II) from R-biotypes (Chapman et al., 1985). Although minor differences in lipid composition between R- and S-biotypes may contribute to triazine resistance, it is more likely that they reflect secondary alterations in membrane organization associated with changes in the relative levels of pigment-protein complexes, as suggested by Chapman et al. (1985). Therefore, it is not only in terms of differential lipid composition but probably more in terms of lipid molecular organization that an involvement of membrane lipids in the herbicide resistance should be envisaged. The physical state of liposomes prepared with chloroplast lipids extracted from R- and S-biotypes was investigated using fluorescence polarization techniques (Lehoczki et al., 1985). These authors reported that the lipids from R-biotypes yielded liposomes which are more fluid than those prepared with lipids of S-biotypes. Another interesting observation is that following lipid enrichment of pea thylakoid membranes, the H_2O to methylviologen electron transport shows a decreased sensitivity to DCMU (Millner et al., 1983).

Taken together, these data indicate that thylakoid membrane lipids are likely to play a role in the mode of action of herbicides. An unanswered question is whether lipids are also involved in the phenomenon of herbicide resistance. Some of the above results suggest, however, that lipids may mediate the accessibility of herbicides to their binding site(s) at the Q_B protein level (Lehoczki et al., 1985).

A. Model Systems

Unilamellar vesicles were prepared from various synthetic or natural lipids by the technique of Batzri and Korn (1973), and their thermotropic behavior was studied by turbidimetry or by fluorescence techniques (Lee, 1975; Mayor, 1991). In the latter case, chlorophyll was included in the lipids at a lipid/probe ratio of 50:1. Lipids chosen were dimyristoyl PC (DMPC), dipalmytoyl PC (DPPC), distearoyl PC (DSPC), as well as hydrogenated MGDG, DGDG, and total (depigmented) thylakoid lipids. The influence of DCMU and atrazine on the thermotropic behavior of these vesicles was studied, together with that of compounds known to act as fluidity-altering agents (cholesterol, benzyl alcohol). The partition coefficients of DCMU and atrazine between vesicles made of different lipid classes and the surrounding water phase was also determined (Mayor, 1991).

The effects of DCMU were generally more important than those of atrazine, whatever the lipid class or the acyl chain length in a given lipid class. The most salient feature of the herbicide treatments was a decrease in the main transition temperature of the different lipids together with a decrease in the enthalpy of the transition. On the other hand, when the gel to liquid-crystalline transition was essentially flattened by inclusion of 20 mol % cholesterol in the lipid vesicles, both herbicides restored partly the transition, atrazine being more efficient than DCMU in this case. The effect of a given concentration of DCMU on the thermotropic behavior of DPPC vesicles was essentially similar to that of a 50 times higher concentration of benzyl alcohol. These results suggest that the fluidity of vesicles made of saturated lipids is increased by herbicides. It appears that these two herbicides penetrate the bilayer to different depths and to different extents, as also suggested by the difference between their partition coefficient (200 for DCMU and 5 for atrazine, in DPPC vesicles; 290 for DCMU and 5 for atrazine in saturated thylakoid lipid vesicles) (Mayor, 1991).

B. Thylakoid Membranes

Thylakoid membranes were isolated from spinach and from atrazine-resistant (R) or susceptible (S) leaves of *Solanum nigrum*. The effects of DCMU and atrazine were investigated on several parameters (Mayor, 1991): (a) the I_{50} dependence on the osmolarity and temperature; (b) the thylakoid packed volume dependence on the herbicide concentration; (c) the electron flow activity dependence on the temperature; (d) the effect of cholesterol enrichment of thylakoid membrane on the sensitivity of electron flow to herbicides. Results led to the following conclusions (Mayor, 1991): DCMU and atrazine interact quite differently with the thylakoid membrane. Their partition coefficients between thylakoid membrane and the water phase seem to reflect those measured in model systems (see above). The degree of lateral compression (and particularly the degree of lipid packing) of the thylakoid membrane seems to gear the accessibility of herbicides to their binding site. This is especially the case for atrazine. It also appears that at least in *S. nigrum*, the resistance trait is linked to the chilling sensitivity and reciprocally. Since chilling resistance or sensitivity has been correlated with different properties of the lipid phase of thylakoid membrane (Murata and Nishida, 1990), it may well be that herbicide resistance show a similar correlation (Mayor, 1991).

Another approach was to remove lipids from the outer monolayer of thylakoid membrane by a controlled hydrolysis of phospholipids by PLA_2 or galactolipids by LRa in the presence of BSA, and then to determine both I_{50} and the binding pattern (scatchard plots) of herbicides to these lipid-depleted thylakoid membranes using ^{14}C-labeled DCMU or atrazine (Siegenthaler and Mayor, 1992). As already mentioned, these lipid depletions have only moderate inhibitory effects on electron flow provided that the inner monolayer is still intact. The main conclusion of these experiments was that at least in the S-biotype, some acyl lipids (not yet fully identified) of the outer monolayer interact with the D1 protein (herbicide-binding protein) so as to maintain it in a conformation that allows an optimal compromise between binding affinity and inhibitory power of herbicides, thereby supporting our contention that membrane lipids play a role not only in the mode of action of herbicides, but probably also in the phenomenon of herbicide resistance (Siegenthaler and Mayor, 1992).

IX. Functional Studies of Mutants Affected in Lipid Composition

A high degree of unsaturation of membrane lipids is crucial for survival of plants at low temperature. In this respect, thylakoid membrane lipids are highly unsaturated in order to form a liquid crystalline phase and, therefore, sustain maximal photosynthetic function within the broadest possible range of temperatures. A number of mutants of *Arabidopsis thaliana* and *Chlamydomonas* have been characterized as being deficient in desaturases or glycerol-

3-phosphate acyltransferases (Somerville and Browse, 1991; Sato et al., 1996; see also Chapter 4, Wada and Murata). The availability of these mutants provides another tool in determining the physiological consequences of variations in lipid unsaturation (see also Section V.B). However, information concerning photosynthetic activity in these mutants is rather scarce.

Browse et al. (1986) and McCourt et al. (1987) described a mutant of *A. thaliana* (*fad* D renamed *fad* 7 according to Ohlrogge and Browse, 1995) which exhibits a large reduction of both 18:3 and 16:3 acyl groups in all membrane lipids, including chloroplast membrane lipids. The fluidity of the thylakoid membranes are not significantly altered by the mutation. Compared to the wild type, the *fad* 7 mutant displays unaltered photosynthetic capacity, including CO_2 fixation, whole chain electron transport, PS II and PS I activities, PS II/PS I ratio as revealed by the 77 K F_{685}/F_{734} ratio and the mechanism leading to redistribution of absorbed energy between the two Photosystems. Thus, a partial reduction in the level of trienoic acyl groups in chloroplast membranes results in no significant changes in the structure of the chlorophyll-protein complex nor in the efficiency of energy capture and transfer (Browse et al., 1986). Similar observations were made with the *A. thaliana fad* B (renamed *fad* 5) mutant which is deficient in the activity of the chloroplast $\omega 9$ desaturase converting 16:0 (at the sn-2 position) of MGDG to 16:1 (Kunst et al., 1989b). The chloroplast membranes of the *fad* 5 mutants are more resistant than those of the wild type to thermal inactivation of the photosynthetic electron transport.

The *fad* C (renamed *fad* 6) mutant from *A. thaliana* is deficient in the chloroplast $\omega 6$ desaturase, which normally desaturates 16:1 and 18:1. As a consequence, the proportion of polyunsaturated fatty acids in thylakoids is reduced to about 50% of that of wild type levels with a concomitant accumulation of 16:1 and 18:1 lipids (Browse et al., 1989). The altered lipid composition of the mutant has pronounced effects on chloroplast ultrastructure (e.g., 48% decrease in the amount of appressed membranes) and a corresponding reduction in chlorophyll and thylakoid protein content, electron transport rates and thermal stability of thylakoids (Hugly et al., 1989). These results corroborate the conclusions reached in catalytic hydrogenation experiments, that lipid unsaturation may function in maintaining membrane properties required for normal electron transport rates (Vigh et al., 1985; see also Section V.B). In addition, the increase in 77 K fluorescence at 685 nm (corresponding to LHCII) in the *fad* 6 mutant thylakoids is consistent with the concept that a decrease in lipid unsaturation may cause partial dissociation of peripheral-LHCII from the PS II complex, as suggested by Horváth et al. (1987; see also Fig. 6) and a lower efficiency of excitation energy transfer from LHCII to PS I and/or of some structural impairment of the peripheral antenna of PS I in the mutant (Hugly et al., 1989).

Hugly and Somerville (1992) compared the growth and development of two lipid mutants (*fad* B and *fad* C) and the wild type of *Arabidopsis* at low temperature. Results provide evidence that the high level of chloroplast lipid polyunsaturation is required for chloroplast biogenesis and contributes to the low temperature fitness of the organism. Based on freeze-fracture electron microscopy, Tsvetkova et al. (1994) reached the same conclusion that the decreased level of lipid fatty acid unsaturation affects the ability of the lipid matrix to mediate the assembly of components in the chloroplast membrane, in particular the ability to promote functional oligomeric assemblies of components of the photosynthetic apparatus. Rabatoul and Browse (1995) further investigated the role of lipid polyunsaturation in a mutant of *A. thaliana* lacking both α-linolenic (18:3) and hexadecatrienoic (16:3) acids, but containing high level of linoleic acid (65 mol %) in its leaf lipids. This mutant carries mutation in the three genes *fad* 3, 7 and 8 that mediate the synthesis of trienoic fatty acids. These authors demonstrated that trienoic fatty acids are not critical for photosynthesis at normal temperature but enhance the tolerance of photosynthesis to low temperatures (refer also to Chapter 14, Vijayan et al.).

A quite interesting mutant, designated *fad* A (renamed *fad* 4), which is supposed to lack the specific desaturase which converts 16:0 to 16:1(3t) at the *sn*-2 position of the glycerol, was used to test the role of 16:1(3t) in the formation of LHCII oligomer as well as in the functional association of LHCII and the photochemical reaction center (McCourt et al., 1985). The mutant lacks 16:1(3t) and has a corresponding increase in 16:0 but is otherwise indistinguishable from the wild type in fatty acid composition or morphology (Browse et al., 1985). Comparison of the electrophoretic separation patterns of the chlorophyll-protein complexes reveals that the mutant has more labile oligomeric forms of

the P_{700}-chlorophyll a-protein complex (CP1) and LHCII (LHCP[1]). The absence of LHCP[1] in the mutant mimics the results obtained by phospholipase A_2 treatment of thylakoids (Rémy et al., 1982). Obviously, PG containing 16:1(3t) stabilizes LHCP[1] against SDS-mediated dissociation. However, the *fad* 4 mutant and the wild type have indistinguishable rates of electron transport (Browse et al., 1985) and PS II photochemical efficiency i.e., the energy transfer between LHCII and the PS II reaction center, and between the photosystems (McCourt et al., 1985).

Another very interesting *Arabidopsis* mutant, *fab* 1 (for fatty acid biosynthesis), was recently described (Wu et al., 1994; Wu and Browse, 1995; Wu et al., 1997). In this mutant, the biochemical defect appears to be a reduction in the activity of the condensing enzyme (KAS II: 3-ketoacyl-acyl-carrier protein synthase II) responsible for the elongation of 16:0 to 18:0. As a consequence, this mutant is characterized by increased proportions of 16:0 in all of the major membrane lipids of the leaf tissue (Wu et al., 1994). The increased proportion of 16:0 (from 20 to 41 mol %) in thylakoid PG of *fab* 1 mutant results in elevated level of high-melting-point PG molecular species. More recently, Wu et al. (1997) extended this characterization by sampling leaf material from wild-type and *fab* 1 mutant *Arabidopsis* plants grown for 14 days at 22 °C before they were transferred to 2 °C (0 day). After transfer of the *fab* 1 plants to 2 °C for more than 7 days, the Fv/Fm ratio declines dramatically, in contrast to that of the wild-type plants which stabilizes at a value of about 0.7. Measurements of the quantum yield closely parallels the Fv/Fm values for both the mutant and wild-type plants. Wu et al. (1997) attributed the substantial collapse of photosynthesis that occurs in the *fab* 1 plants to a severe disruption of the photosystems (or at least of PS II). This decline in photosynthetic capacity is accompanied by reduction in chlorophyll content and the amount of chloroplast glycerolipids as well as by an extensive disruption of the thylakoid and chloroplast structure in the mutant. It is noteworthy that despite the almost complete loss of photosynthetic function and the destruction of photosynthetic machinery, *fab* 1 plants retained a substantial capacity for recovery following transfer to 22 °C. These results provide further demonstration of the importance of lipid unsaturation in chloroplast membrane for the proper growth and development of plants at low temperature (Wu et al., 1997).

Another class of *Arabidopsis* mutants that lack

activity for the first enzyme of the chloroplastic pathway, glycerol-3-phosphate acyltransferase, is due to a single nuclear mutation at the *act* 1 locus (Kunst et al., 1988, 1989a). This mutation effectively converts a 16:3 plant into a 18:3 plant. As a consequence, the *act* 1 mutants show a greatly reduced level of 16:3 acyl groups, characteristic of chloroplastic MGDG and a corresponding increase in C_{18}-fatty acids. The changes in leaf lipid composition did not significantly affect growth or development of the mutant under standard conditions. However, these changes were accompanied by a significant decrease in the ratio of appressed to nonappressed membranes, that was associated with an altered distribution of excitation energy transfer from antenna chlorophyll to PS II and PS I complexes. Interestingly, measurements of temperature-induced fluorescence yield enhancement suggest an increased thermal stability of the photosynthetic apparatus of the mutant (Kunst et al., 1989a).

Dörmann et al. (1995) also adopted a genetic approach to study the biosynthesis and function of galactolipids in higher plants. They isolated a mutant of *A. thaliana* that is deficient in DGDG. This mutant carries a recessive nuclear mutation at a single locus designated *dgd* 1. A strong reduction of the relative amount of DGDG (from 16.0 in the mutant to 1.2 mol% in the wild type) was observed as well as a reduction in the 16:3 fatty acid level and a concomitant increase in 18:3 in MGDG. No specific lipid compensated for the loss of DGDG. As a consequence, the ratio of the non-bilayer-forming (MGDG) to the bilayer-forming lipids (PG, SQDG, DGDG) is increased from 1.9 in the wild type to 4.5 in the *dgd* 1 mutant. The mutation resulted in stunted growth, pale leaf color, a huge increase in the total length of thylakoid membranes per plastid and a slight reduction in the ratio of appressed to nonappressed membranes. According to Dörmann et al. (1995) the increase in the curvature of thylakoid membranes in the mutant might be the direct consequence of the relative increase in the ratio of the non-bilayer-forming to bilayer-forming lipids. Compared to the wild type the *dgd* 1 mutant displays a reduction in the photosynthetic quantum yield and in total chlorophyll content as well as a decrease stability of LHCII during electrophoresis.

Similar experiments were carried out with *Chlamydomonas* mutants which were defective in the synthesis of chloroplast-specific lipids, PG (Dubertret et al., 1994) and SQDG (Sato et al., 1996)

or impaired in fatty acid desaturation (Sato et al., 1996). Dubertret et al., 1994 concluded that 16:1(3t)-PG is involved in the biogenesis of LHCII. However, chloroplasts of some orchids are devoid of 16:1(3t)-PG but are nevertheless able to trimerize LHCII and form grana stacks (Selstam and Krol, 1995; Chapter 11, Selstam). Sato et al. (1995) described a mutant of *C. reinhardtii* (hf-2 mt$^+$) defective in the synthesis of SQDG. The mutant shows reduced PS II activity with little effect on PS I activity, as compared with the parent (137c mt$^+$). According to these authors, SQDG is responsible for PS II activity by associating with the core and light-harvesting complex of PS II.

From all the mutant studies, it can be concluded that: (1) A high degree of lipid polyunsaturation in thylakoid membranes is required for chloroplast biogenesis, namely for promoting functional oligomeric assembly of components of the photosynthetic apparatus; (2) A high level of trienoic fatty acids in chloroplast lipids is not required, within certain limits, to support normal levels of the photosynthetic activities associated with thylakoid membranes; (3) In contrast, reduced levels of polyunsaturated lipids have generally pronounced effects on chloroplast ultrastructure, thylakoid membrane protein and chlorophyll content; (4) Lipid unsaturation directly affects the thermal stability of photosynthetic membranes; (5) Polyunsaturation of lipids contributes to the low-temperature fitness of the plant; (6) The involvement of PG in the formation of trimeric LHCII and then grana stacks is still a debatable question, as well as the exact role 16:1(3t) (refer also to Chapters 9, Trémolières and Siegenthaler; 14, Vijayan et al.).

Acknowledgments

P.A.S. thanks all his collaborators who made essential contributions to the work described in this Chapter: Drs C. Giroud, J.P. Mayor, A. Rawyler and Y. N. Xu, Mrs F. Depéry, S. Duchêne, M. Meylan Bettex, J. Smutny and J. Vallino, as well as Mrs C. Bettinelli for assistance in the preparation of the manuscript. Acknowledgments are due also to Drs. J. M. Ducruet, M. Fragata, G. Horváth, P. Matile, A. Radunz and L. Vigh who kindly provided reprints of their work, and to Mrs D. Jones Siegenthaler for improving the English manuscript. The work from the Laboratoire de Physiologie végétale was supported by the Swiss National Science Foundation and the University of Neuchâtel (Switzerland).

References

Allen JF (1992) Protein phosphorylation in regulation of photosynthesis. Biochim Biophys Acta 1098: 275–335

Anderson MM, McCarty RE and Zimmer EA (1974) The role of galactolipids in spinach chloroplast lamellar membranes. I. Partial purification of a bean leaf galactolipid lipase and its action on subchloroplast particles. Plant Physiol 53: 699–704

Andersson B, Sundby C, Åkerlund H-E and Albertsson P-A (1985) Inside-out thylakoid vesicles. An important tool for the characterization of the photosynthetic membrane. Physiol Plant 65: 322–330

Andreasson E, Svensson P, Weibull C and Albertsson P-A (1988) Separation and characterization of stroma and grana membranes—evidence for heterogeneity in antenna size of both Photosystem I and Photosystem II. Biochim Biophys Acta 936: 339–350

Ashton FM and Crafts AS (eds) (1981) Mode of Action of Herbicides, 2nd edition. Wiley-Interscience Publishers, New York

Baker NR and Percival MP (eds) (1991) Herbicides. Topics in Photosynthesis, Vol 10. Elsevier Science Publishers, Amsterdam

Barber J and Gounaris K (1986) What role does sulpholipid play within the thylakoid membrane? Photosynth Res 9: 239–249

Bassi R (1985) Spectral properties and polypeptide composition of the chlorophyll-protein from thylakoids of granal and agranal chloroplasts of maize (Zea mays L.). Carlsberg Res Commun 50: 127–143

Batzri S and Korn ED (1973) Single bilayer liposomes prepared without sonication. Biochim Biophys Acta 298: 1015–1019

Bekers O, Uijendaal EV, Beijnen JH, Bult A and Underberg WJM (1991) Cyclodextrins in the pharmaceutical field. Drug Development and Industrial Pharmacy 17: 1503–1549

Berthold DA, Babcock GT and Yocum CF (1981) A highly resolved, oxygen-evolving Photosystem II preparation from spinach thylakoid membranes. FEBS Lett 134: 231–234

Bishop DG (1983) Functional role of plant membrane lipids. In: Thompson WW, Mudd JB and Gibbs M (eds) Biosynthesis and Function of Plant Lipids, pp 81–103. Am Soc Plant Physiol, Rockville

Bowyer JR, Camilleri P and Vermaas WFJ (1991) Photosystem II and its interaction with herbicides. In: Baker NR and Percival MP (eds), Topics in Photosynthesis, Vol 10, Herbicides, pp 27–85. Elsevier Science Publishers, Amsterdam

Browse J, McCourt P and Somerville C (1985) A mutant of Arabidopsis lacking a chloroplast-specific lipid. Science 227: 763–765

Browse J, McCourt P and Somerville C (1986) A mutant of Arabidopsis deficient in $C_{18:3}$ and $C_{16:3}$ leaf lipids. Plant Physiol 81: 859–864

Browse J, Kunst L, Anderson S, Hugly S and Somerville C (1989) A mutant of Arabidopsis deficient in the chloroplast 16:1/18:1 desaturase. Plant Physiol 90: 522–529

Burke JJ, Wilson RF and Swafford R (1982) Characterization of chloroplasts isolated from triazine-susceptible and triazine-resistant biotypes of Brassica campestris L. Plant Physiol 70: 24–29

Burns DD, Galliard T and Harwood JL (1977) Catabolism of sulpholipid by an enzyme from the leaves of Phaseolus multiflorus. Biochem Soc Trans 5: 1302–1304

Chapman DJ, De Felice J and Barber J (1985) Characteristics of chloroplast thylakoid lipid composition associated with resistance to triazine herbicides. Planta 166: 280–285

Costes C, Bazier R and Lechevallier D (1972) Rôle structural des lipides dans les membranes des chloroplastes de Blé. Physiol Vég 10: 291–317

Costes C, Bazier R, Baltscheffsky H and Hallberg C (1978) Mild extraction of lipids and pigments from Rhodospirillum rubrum chromatophores. Plant Sci Lett 12: 241–249

Coughlan SJ and Hind G (1987) A protein kinase that phosphorylates light-harvesting complex is autophosphorylated and is associated with Photosystem II. Biochemistry 26: 6515–6521

Demeter S, Goussias C, Bernat G, Kovacs L and Petrouleas V (1993) Participation of the g=1.9 and g=1.82 EPR forms of the semiquinone-iron complex, Q_A^-.Fe of Photosystem II in the generation of the Q and C thermoluminescence bands, respectively. FEBS Lett 336: 352–356

Dörmann P, Hoffmann-Benning S, Balbo I and Benning C (1995) Isolation and characterization of an Arabidopsis mutant deficient in the thylakoid lipid digalactosyl diacylglycerol. Plant Cell 7: 1801–1818

Droppa M, Horváth G, Hideg E and Farkas T (1995) The role of phospholipids in regulating photosynthetic electron transport activities: treatment of thylakoids with phospholipase C. Photosynth Res 46: 287–293

Dubacq JP and Trémolières A (1983) Occurrence and function of phosphatidylglycerol containing Δ3-trans-hexadecenoic acid in photosynthetic lamellae. Physiol Vég 21: 293–312

Dubertret G, Mirshahi A, Mirshahi M, Gérard-Hirne C and Trémolières A (1994) Evidence from in vivo manipulations of lipid composition in mutants that the Δ³-trans-hexadecenoic acid-containing phosphatidylglycerol is involved in the biogenesis of the light-harvesting chlorophyll a/b-protein complex of Chlamydomonas reinhardtii. Eur J Biochem 226: 473–482

Duchêne S, Smutny J and Siegenthaler PA (1995) Transmembrane distribution of phospholipids in spinach thylakoid inside-out vesicles and involvement of outer and inner monolayer phospholipids in the photosynthetic electron flow activity. In: Lopez-Pérez MJ, Delgado C and Cebrian-Pérez JA (eds) 9th International Conference on Partitioning in Aqueous Two-Phase Systems. Advances in the Uses of Polymers in Cell Biology, Biotechnology and Environmental Sciences, Abstract Nr P23, University of Zaragoza, Spain

Dunahay TG, Staehelin LA, Seibert M, Ogilvie PD and Berg SP (1984) Structural, biochemical and biophysical characterization of four oxygen-evolving Photosystem II preparations from spinach. Biochim Biophys Acta 764: 179–193

Dupont J and Siegenthaler PA (1985) Alterations in the cytochrome composition of spinach thylakoid membranes induced by aging in vitro. J Plant Physiol 119: 347–357

Dupont J and Siegenthaler PA (1986) A parallel study of pigment bleaching and cytochrome breakdown during aging of thylakoid membranes. Plant Cell Physiol 27: 473–484

Duval JC, Trémolières A and Dubacq JP (1979) The possible role of transhexadecenoic acid and phosphatidylglycerol in light reactions of photosynthesis. FEBS Lett 106: 414–418

Eckert H-J, Toyoshima Y, Akabori K and Dismukes GC (1987) The involvement of lipids in light-induced charge separation in the reaction center of Photosystem II. Photosynth Res 14: 31–41

Fedke C (ed) (1982) Biochemistry and Physiology of Herbicide Action. Springer Verlag, Berlin

Ford RC, Chapman DJ, Barber J, Pedersen JZ and Cox RP (1982) Fluorescence polarization and spin-label studies of the fluidity of stromal and granal chloroplast membranes. Biochim Biophys Acta 681: 145–151

Fragata M, Strzalka K and Nénonéné EK (1991) MgCl$_2$-induced reversal of oxygen evolution decay in Photosystem II particles incubated with phosphatidylglycerol vesicles at high lipid/Photosystem II ratio. J Photochem Photobiol 11: 329–342

Fragata M, Menikh A and Nénonéné EK (1994) Functional and structural aspects of the thylakoid lipids in oxygen evolution in Photosystem II. Trends Photochem Photobiol 3: 201–210

Galliard T (1971) Enzymic deacylation of lipids in plants. The effects of free fatty acids on the hydrolysis of phospholipids by the lipolytic acyl hydrolase of potato tubers. Eur J Biochem 21: 90–98

Galliard T (1980) Degradation of acyl lipids: Hydrolytic and oxidative enzymes. In: Stumpf PK and Conn EE (eds) The Biochemistry of Plants. Vol 4, pp 85–116. Academic Press, New York

Gasquez J, Al Mourmar A and Darmency H (1985) Triazine herbicide resistance in Chenopodium album. Pestic Sci 16: 392–396

Gepstein S (1988) Photosynthesis. In: Noodén LD and Leopold AC (eds) Senescence and Aging in Plants, pp 85–109. Academic Press Inc, New York

Giroud C and Siegenthaler PA (1988) Development of oat prothylakoids into thylakoids during greening does not change transmembrane galactolipid asymmetry but preserves the thylakoid bilayer. Plant Physiol 88: 412–417

Golbeck JH, Martin IF and Fowler CF (1980) Mechanism of linolenic acid-induced inhibition of photosynthetic electron transport. Plant Physiol 65: 707–713

Gombos Z and Murata N (1998) Genetically engineered modulation of the unsaturation of glycerolipids and its consequences in tolerance of photosynthesis to temperature stresses. In: Siegenthaler PA and Murata N (eds) Lipids in Photosynthesis: Structure, Function and Genetics, pp 249–262. Kluwer Academic Publishers, Dordrecht

Gombos Z, Barabas K, Joo F and Vigh L (1988) Lipid saturation induced microviscosity increase has no effect on the reducibility of flash-oxidized cytochrome f in pea thylakoids. Plant Physiol 86: 335–337

Gombos Z, Wada H and Murata N (1994) The recovery of photosynthesis from low-temperature photoinhibition is accelerated by the unsaturation of membrane lipids: A mechanism of chilling tolerance. Proc Natl Acad Sci USA 91: 8787-8791

Gounaris K, Barber J and Harwood JL (1986) The thylakoid membranes of higher plant chloroplasts. Biochem J 237: 313–326

Gressel J (1982) Triazine herbicide interaction with a 32 000 Mr thylakoid protein—Alternative possibilities. Plant Sci Lett 25: 99–106

Guillot-Salomon T, Douce R and Signol M (1973) Rapport entre l'évolution ultrastructurale des feuilles de plantules étiolées de Maïs soumises à l'action de la lumière et la synthèse de nouvelles molécules de phosphatidylglycerol. Plant Sci Lett 1: 43–47

Harwood JL (1991) Herbicides affecting chloroplast lipid synthesis. In: Baker NR and Percival MP (eds) Herbicides, pp

209–246. Elsevier Science Publishers, Amsterdam

Heinz E and Siefermann-Harms D (1981) Are galactolipids integral components of the chlorophyll-protein complexes in spinach thylakoids? FEBS Lett 124: 105–111

Helmsing PJ (1967) Hydrolysis of galactolipids by enzymes in spinach leaves. Biochim Biophys Acta 144: 470–472

Helmsing PJ (1969) Purification and properties of galactolipase. Biochim Biophys Acta 178: 519–533

Henry LEA, Strasser RJ and Siegenthaler PA (1982) Alteration in the acyl lipid composition of thylakoids induced by aging and its effect on thylakoid structure. Plant Physiol 69: 531–536

Hideg E, Rozsa Z, Vaas I, Vigh L and Horvàth G (1986) Effect of homogeneous catalytic hydrogenation of membrane lipids on luminescence characteristics of the Photosystem II electron transport. Photobiochem Photobiophys 12: 221–230

Hirayama O and Matsui T (1976) Effects of lipolytic enzymes on the photochemical activities of spinach chloroplasts. Biochim Biophys Acta 423: 540–547

Hirayama O and Nomotobori JL (1978) Preparation and characterization of phospholipid-depleted chloroplasts. Biochim Biophys Acta 502: 11–16

Holloway PJ, Maclean DJ and Scott KJ (1983) Rate-limiting steps of electron transport in chloroplasts during ontogeny and senescence of barley. Plant Physiol 72: 795–801

Horváth G, Droppa M, Szító T, Mustárdy LA, Horváth LI and Vigh L (1986a) Homogeneous catalytic hydrogenation of lipids in the photosynthetic membrane: Effects on membrane structure and photosynthetic activity. Biochim Biophys Acta 849: 325–336

Horváth G, Mansourian AR, Vigh L, Thomas PG, Joó F and Quinn PJ (1986b) Homogeneous catalytic hydrogenation of the polar lipids of pea chloroplasts in situ and the effects on lipid polymorphism. Chem Physics of Lipids 39: 251–264

Horváth G, Melis A, Hideg E, Droppa M and Vigh L (1987) Role of lipids in the organization and function of Photosystem II studied by homogeneous catalytic hydrogenation of thylakoid membranes in situ. Biochim Biophys Acta 891: 68–74

Horváth G, Droppa M, Hideg E and Rozsa Z (1989) The role of phospholipids in regulating photosynthetic electron transport activities: treatment of chloroplasts with phospholipase A$_2$. J Photochem Photobiol, B: Biology 3: 515–527

Hoshina S (1979) Restoration of lysolecithin-induced inhibition of the Hill reaction in spinach chloroplasts by the addition of lecithin. Plant Cell Physiol 20: 1107–1116

Hoshina S and Nishida K (1975) Photoswelling and light-inactivation of isolated chloroplasts. II. Functional and structural changes in isolated chloroplasts under the influence of lysolecithin. Plant Cell Physiol 16: 475–484

Hugly S and Somerville C (1992) A role for membrane lipid polyunsaturation in chloroplast biogenesis at low temperature. Plant Physiol 99: 197–202

Hugly S, Kunst L, Browse J and Somerville C (1989) Enhanced thermal tolerance of photosynthesis and altered chloroplast ultrastructure in a mutant of Arabidopsis deficient in lipid desaturation. Plant Physiol 90: 1134–1142

Huner NPA, Krol M, Williams JP, Maissan E, Low PS, Roberts D and Thompson JE (1987) Low temperature development induces a specific decrease in trans-Δ3-hexadecenoic acid content which influences LHCII organization. Plant Physiol 84: 144–152

Huner NPA, Williams JP, Maissan E, Mysscich EG, Krol A, Laroche E and Singh J (1989) Low temperature-induced decrease in trans-Δ3-hexadecenoic acid content is correlated with freezing tolerance in cereals. Plant Physiol 89: 144–152

Ikeuchi M, Yuasa M and Inoue Y (1985) Simple and discrete isolation of an O_2-evolving PS II reaction center complex retaining Mn and the extrinsic 33 kDa protein. FEBS Lett 185: 316–322

Jagendorf AT and Uribe E (1966) ATP formation caused by acid-base transition of spinach chloroplasts. Proc Natl Acad Sci USA 55: 170–177

Jenkins GI and Woolhouse HW (1981) Photosynthetic electron transport during senescence of the primary leaves of Phaseolus vulgaris L. II. The reactivity of Photosystems one and two and a note on the site of reduction of ferricyanide. J Exp Bot 32: 989–997

Jordan BR, Chow WS and Baker AJ (1983) The role of phospholipids in the molecular organization of pea chloroplast membranes. Effect of phospholipid depletion on photosynthetic activities. Biochim Biophys Acta 725: 77–86

Kanervo E, Aro EM and Murata N (1995) Low unsaturation level of thylakoid membrane lipids limits turnover of the D1 protein of Photosystem II at high irradiance. FEBS Lett 364: 239–242

Krupa Z (1982) The action of lipases on chloroplast membranes. I. The release of plastocyanin from galactolipase-treated thylakoid membranes. Photosynth Res 3: 95–104

Krupa Z (1983) The action of lipase on chloroplast membranes. II. Polypeptide patterns of bean galactolipase- and phospholipase A_2-treated thylakoid membranes. Photosynth Res 4: 229–239

Krupa Z (1984) The action of lipases on chloroplast membranes. III. The effect of lipid hydrolysis on chlorophyll-protein complexes in thylakoid membranes. Photosynth Res 5: 177–184

Krupa Z and Baszynski T (1975) Requirement of galactolipids for Photosystem I activity in lyophilized spinach chloroplasts. Biochim Biophys Acta 408: 26–34

Krupa Z, Huner NPA, Williams JP and Maissan E (1987) Development at cold hardening temperatures. Plant Physiol 84: 19–24

Kruse O and Schmid GH (1995) The role of phosphatidylglycerol as a functional effector and membrane anchor of the D1-core peptide from Photosystem II-particles of the cyanobacterium Oscillatoria chalybea. Z Naturforsch 50c: 380–390

Kruse O, Radunz A and Schmid GH (1994) Phosphatidylglycerol and β-carotene bound onto the D1-core peptide of Photosystem II in the filamentous cyanobacterium Oscillatoria chalybea. Z Naturforsch 49c: 115–124

Ksenzenko VM, Zhukov EA, Romanenko VG and Molchanov MI (1994) Properties of Photosystem II pigment-lipid-protein complexes from bean chloroplast thylakoid membranes. Biochemistry 59: 271–276

Kunst L, Browse J and Somerville C (1988) Altered regulation of lipid biosynthesis in a mutant of Arabidopsis deficient in chloroplast glycerol-3-phosphate acyltransferase activity. Proc Natl Acad Sci USA 85: 4143–4147

Kunst L, Browse J and Somerville C (1989a) Altered chloroplast structure and function in a mutant of Arabidopsis deficient in plastid glycerol-3-phosphate acyltransferase activity. Plant Physiol 90: 846–853

Kunst L, Browse J and Somerville C (1989b) Enhanced thermal tolerance in a mutant of Arabidopsis deficient in palmitic acid unsaturation. Plant Physiol 91: 401–408

Laroche A, Beaumont G and Grenier G (1989) Analyse des lipides des chloroplastes isolés de Lemna minor en présence d'une concentration subléthale d'atrazine. Plant Physiol Biochem 27: 93–97

Lee AG (1975) Segregation of chlorophyll a incorporated into lipid bilayers. Biochemistry 14: 4397–4402

Lehoczki E, Pölös E, Laskay G and Farkas T (1985) Chemical compositions and physical states of chloroplast lipids related to atrazine resistance in Conyza canadensis L. Plant Sci 42: 19–24

Lemoine Y, Dubacq JP and Zabulon G (1982) Changes in light-harvesting capacities and Δ^3-trans-hexadecenoic acid in dark and light-grown Picea abies. Physiol Vég 20: 487–503

Lemoine Y, Dubacq JP, Zabulon G and Ducruet JM (1986) Organization of the photosynthetic apparatus from triazine resistant and susceptible biotypes of several plant species. Can J Bot 64: 2999–3007

Los D, Horváth I, Vigh L and Murata N (1993) The temperature-dependent expression of desaturase gene desA in Synecchocystis PCC 6803. FEBS Lett 318: 57–60

Maresca B and Cossins AR (1993) Fatty feedback and fluidity. Nature 365: 606–607

Matile P (1992) Chloroplast senescence. In: Barber N and Thomas H (eds) Crop Photosynthesis: Spatial and Temporal Determinants, pp 413–440. Elsevier, Amsterdam

Mattoo AK and Edelman M (1987) Intramembrane translocation and posttranslational palmitoylation of the chloroplast 32-Kd herbicide-binding protein. Proc Natl Acad Sci USA 84: 1497–1501

Mattoo AK, St John JB and Wergin WP (1984) Adaptive reorganization of protein and lipid components in chloroplast membranes associated with herbicide binding. J Cell Biochem 24: 163–175

Mayor JP (1991) Rôle des lipides diacylés dans l'interaction des herbicides photosynthétiques avec les membranes thylacoïdales de Solanum nigrum L. et Spinacia oleracea L. Thesis, University of Neuchâtel, Switzerland

McCourt P, Browse J, Watson J, Arntzen CJ and Somerville CR (1985) Analysis of photosynthetic antenna function in a mutant of Arabidopsis thaliana (L.) lacking trans-hexadecenoic acid. Plant Physiol 78: 853–858

McCourt P, Kunst L, Browse J and Somerville CR (1987) The effects of reduced amounts of lipid unsaturation on chloroplast ultrastructure and photosynthesis in a mutant of Arabidopsis. Plant Physiol 84: 353–360

Michalski WP and Kaniuga Z (1980) Photosynthetic apparatus in chilling-sensitive plants. VII. Comparison of the effect of galactolipase treatment of chloroplasts and cold-dark storage of leaves on photosynthetic electron flow. Biochim Biophys Acta 589: 84–99

Millner PA, Grouzis JP, Chapman DJ and Barber J (1983) Lipid enrichment of thylakoid membranes. I. Using soybean phospholipids. Biochim Biophys Acta 722: 331–340

Moll BA, Eilmann M and Steinback KE (1987) Phosphorylation of thylakoid proteins of Oryza sativa. Characterization and effects of chilling temperature. Plant Physiol 83: 428–433

Moon BY, Higashi SI, Gombos Z and Murata N (1995) Unsaturation of the membrane lipids of chloroplasts stabilizes the photosynthetic machinery against low-temperature photoinhibition in transgenic tobacco plants. Proc Natl Acad Sci USA 92: 6219–6223

Murata N and Nishida I (1990) Lipids in relation to chilling sensitivity of plants. In: Wang CY (ed) Chilling Injury of Horticultural Crops, pp 181–199. CRC Press, Boca Raton

Murata N, Higashi S-I and Fujimura Y (1990) Glycerolipids in various preparations of Photosystem II from spinach chloroplasts. Biochim Biophys Acta 1019: 261–268

Murata N, Ishizaki-Nishizaki O, Higashi H, Tasaka Y and Nishida I (1992) Genetically engineered alteration in the chilling sensitivity of plants. Nature 356: 710–713

Murphy DJ (1982) The importance of non-planar bilayer regions in photosynthetic membranes and their stabilisation by galactolipids. FEBS Lett 150: 19–26

Murphy D (1986a) The molecular organisation of the photosynthetic membranes of higher plants. Biochim Biophys Acta 864: 33–94

Murphy D (1986b) Structural properties and molecular organization of the acyl lipids of photosynthetic membranes. In: Staehelin LA and Arntzen CJ (eds) Encyclopedia of Plant Physiology, New Series, Vol 19, Photosynthesis III: Photosynthetic Membranes and Light Harvesting Systems, pp 713–725. Springer-Verlag, Berlin,

Murphy D and Woodrow IE (1983) Lateral heterogeneity in the distribution of thylakoid membrane lipid and protein components and its implications for the molecular organisation of photosynthetic membranes. Biochim Biophys Acta 725: 104–112

Nénonéné EK and Fragata M (1990) Effects of pH and freeze-thaw on photosynthetic oxygen evolution of Photosystem II particles incorporated into phosphatidylglycerol bilayers. J Plant Physiol 136: 615–620

Nishihara M, Yokota K and Kito M (1980) Lipid molecular species composition of thylakoid membranes. Biochim Biophys Acta 617: 12–19

Ohlrogge J and Browse J (1995) Lipid biosynthesis. Plant Cell 7: 957–970

Percival MP and Baker NR (1991) Herbicides and photosynthesis. In: Baker NR and Percival MP (eds) Herbicides, Topics in Photosynthesis, Vol 10, pp 1–26. Elsevier Science Publishers, Amsterdam

Pfister K and Urbach W (1983) Effects of biocides and growth regulators: Physiological basis. In: Lange OL, Nobel PS, Osmond CB and Ziegler H (eds) Encyclopedia of Plant Physiology, New Series, vol. 12D, pp 329–391. Springer Verlag, Berlin

Pick U, Gounaris K, Admon A and Barber J (1984) Activation of the CF_o-CF_1, ATP synthase from spinach chloroplasts by chloroplast lipids. Biochim Biophys Acta 765: 12–20

Pick U, Weiss M, Gounaris K and Barber J (1987) The role of different thylakoid glycolipids in the function of reconstituted chloroplast ATP synthase. Biochim Biophys Acta 891: 28–39

Pillai P and St John JB (1981) Lipid composition of chloroplast membranes from weed biotypes differentially sensitive to triazine herbicides. Plant Physiol 68: 585–587

Quinn PJ and Williams WP (1983) The structural role of lipids in photosynthetic membranes. Biochim Biophys Acta 737: 223–266

Quinn PJ, Joó F and Vigh L (1989) The role of unsaturated lipids in membrane structure and stability. Prog Biophys Molec Biol 53: 71–103

Radunz A (1984) Serological investigations on the function of phospholipids in the thylakoid membrane. In: Sybesma C (ed),

Advances in Photosynthesis Research, Vol III, pp 151–154. Martinus Nijhoff/Dr W Junk Publishers, The Hague

Radunz A, Bader KP and Schmid GH (1984a) Serological investigations of the function of galactolipids in the thylakoid membrane. Z Pflanzenphysiol 114: 227–231

Radunz A, Bader KP and Schmid GH (1984b) Influence of antisera to sulfoquinovosyl diglyceride and to β-sitosterol on the photosynthetic electron transport in chloroplasts from higher plants. In: Siegenthaler PA and Eichenberger W (eds) Structure, Function and Metabolism of Plant Lipids, pp 479–484. Elsevier Science Publishers, Amsterdam

Rawyler A and Siegenthaler PA (1981a) Transmembrane distribution of phospholipids and their involvement in electron transport, as revealed by phospholipase A_2 treatment of spinach thylakoids. Biochim Biophys Acta 635: 348- 358

Rawyler A and Siegenthaler PA (1981b) Regulation of Photosystem I electron flow activity by phosphatidylglycerol in thylakoid membranes as revealed by phospholipase treatment. Biochim Biophys Acta 638: 30–39

Rawyler A and Siegenthaler PA (1985) Transversal localization of monogalactosyldiacylglycerol and digalactosyldiacyl-glycerol in spinach thylakoid membranes. Biochim Biophys Acta 815: 287–298

Rawyler A and Siegenthaler PA (1989) Change in the molecular organization of monogalactosyldiacylglycerol between resting and functioning thylakoid membranes. Involvement of the CF_o-CF_1-ATP synthetase. Biochim Biophys Acta 975: 283–292

Rawyler A and Siegenthaler PA (1996) Cyclodextrins: a new tool for the controlled lipid depletion of thylakoid membranes. Biochim Biophys Acta 1287: 89–97

Rawyler A, Unitt MD, Giroud C, Davies H, Mayor JP, Harwood JL and Siegenthaler PA (1987) The transmembrane distribution of galactolipids in chloroplast thylakoids is universal in a wide variety of temperate climate plants. Photosynth Res 11: 3–13

Rémy R, Trémolières A, Duval JC, Ambard-Bretteville F and Dubacq JP (1982) Study of the supramolecular organization of light-harvesting chlorophyll protein (LHCP). FEBS Lett 137: 271–275

Restall CJ, Williams WP, Percival MP, Quinn PJ and Chapman D (1979) The modulation of membrane fluidity by hydro-genation processes. III. The hydrogenation of biomembranes of spinach chloroplasts and a study of the effect of this on photosynthetic electron transport. Biochim Biophys Acta 555: 119–130

Rottenberg H, Grunwald T and Avron M (1972) Determination of ΔpH in chloroplasts. 1. Distribution of [^{14}C]methylamine. Eur J Biochem 55: 54–63

Routaboul JM and Browse J (1995) A role for membrane lipid trienoic fatty acids in photosynthesis at low temperatures. In: Matthis P (ed) Photosynthesis: From Light to Biosphere, Vol IV, pp 861–864. Kluwer Academic Publishers, Dordrecht

Sastry and Kates (1964) Hydrolysis of monogalactosyl and digalactosyl diglyceride by specific enzymes in runner-bean leaves. Biochemistry 3: 1280–1287

Sato N, Sonoike K, Tsuzuki M and Kawaguchi A (1995) Impaired Photosystem II in a mutant of Chlamydomonas reinhardtii defective in sulfoquinovosyl diacylglycerol. Eur J Biochem 234: 16–23

Sato N, Sonoike K, Tsuzuki M and Kawaguchi A (1996) Photosynthetic characteristics of a mutant of Chlamydomonas *reinhardtii* impaired in fatty acid desaturation in chloroplasts. Biochim Biophys Acta 1274: 112–118

Selstam E (1998) Development of thylakoid membranes with respect to lipids. In: Siegenthaler PA and Murata N (eds) Lipids in Photosynthesis: Structure, Function and Genetics, pp 209–224. Kluwer Academic Publishers, Dordrecht

Selstam E and Krol M (1995) Trimeric LHCII plants lacking 16:1t in phosphatidylglycerol. In: Mathis P (ed.) Photosynthesis: From Light to Biosphere, Vol 1, pp 95–98. Kluwer Academic Publishers, Dordrecht

Shaw AB, Anderson MM and McCarty RE (1976) Role of galactolipids in spinach chloroplast lamellar membranes. Plant Physiol 57: 724–729

Siefermans-Harms D, Ninnemann H and Yamamoto HY (1987) Reassembly of solubilized chlorophyll-protein complexes in proteolipid particles—comparison of monogalactosyldiacylglycerol and two phospholipids. Biochim Biophys Acta 892: 303–313

Siegel CO, Jordan AE and Miller KR (1981) Addition of lipids to the photosynthetic membrane: Effects on membrane structure and energy transfer. J Cell Biol 91: 113–125

Siegenthaler PA (1972) Aging of the photosynthetic apparatus. IV. Similarity between the effects of aging and unsaturated fatty acids on isolated spinach chloroplasts as expressed by volume changes. Biochim Biophys Acta 275: 182–191

Siegenthaler PA (1973) Change in pH dependence and sequential inhibition of photosynthetic activity in chloroplasts by unsaturated fatty acids. Biochim Biophys Acta 305: 153–162

Siegenthaler PA (1974) Inhibition of Photosystem II electron transport in chloroplasts by fatty acids and restoration of its activity by Mn^{2+}. FEBS Lett 39: 337–340

Siegenthaler PA (1998) Molecular organization of acyl lipids in photosynthetic membranes of higher plants. In: Siegenthaler PA and Murata N (eds) Lipids in Photosynthesis: Structure, Function and Genetics, pp 119–144. Kluwer Academic Publishers, Dordrecht

Siegenthaler PA and Depéry F (1976) Influence of unsaturated fatty acids in chloroplasts. Shift of the pH optimum of electron flow and relations to ΔpH, thylakoid internal pH and proton uptake. Eur J Biochem 61: 573–580

Siegenthaler PA and Depéry F (1977) Aging of the photosynthetic apparatus. VI. Changes in pH dependence of ΔpH, thylakoid internal pH and proton uptake and relationships to electron transport. Plant Cell Physiol. 18: 1047–1055

Siegenthaler PA and Mayor JP (1992) Changes in the binding and inhibitory properties of urea/triazine-type herbicides upon phospholipid and galactolipid depletion in the outer monolayer of thylakoid membranes. Photosynth Res 31: 57–68

Siegenthaler PA and Rawyler A (1977) Aging of the photosynthetic apparatus. V. Change in pH dependence of electron transport and relationships to endogenous free fatty acids. Plant Sci Lett 9: 265–273

Siegenthaler PA and Rawyler A (1986) Acyl lipids in thylakoid membranes: distribution and involvement in photosynthetic functions. In: Staehelin LA and Arntzen CJ (eds) Encyclopedia of Plant Physiology, vol. 19, Photosynthesis III, pp 693–705. Springer Verlag, Berlin

Siegenthaler PA and Vallino J (1995) Effect of a selective depletion of acyl lipids on the light and dark phosphorylation in spinach thylakoid membranes. In: Mathis P (ed) Photosynthesis: From Light to Biosphere, Vol III, pp 225–228.

Kluwer Academic Publishers, Dordrecht

Siegenthaler PA, Rawyler A and Henry LEA (1981) A new type of correlation between changes in lipid composition and loss of electron transport activities during aging in vitro. In: Akoyunoglou G (ed) Photosynthesis II. Electron Transport and Photophosphorylation, pp 167–174. Balaban International Science Services, Philadelphia

Siegenthaler PA, Smutny J and Rawyler A (1984) Involvement of hydrophilic and hydrophobic portions of phospholipid molecules in photosynthetic electron flow activities. In: Siegenthaler PA and Eichenberger W (eds) Structure, Function and Metabolism of Plant Lipids, pp 475–478. Elsevier Science Publishers, Amsterdam

Siegenthaler PA, Smutny J and Rawyler A (1987a) Involvement of distinct populations of phosphatidylglycerol and phosphatidylcholine molecules in photosynthetic electron-flow activities. Biochim Biophys Acta 891: 85–93

Siegenthaler PA, Rawyler A and Giroud C (1987b) Spatial organization and functional roles of acyl lipids in thylakoid membranes. In: Stumpf PK, Mudd JB and Nes WD (eds) The Metabolism, Structure and Function of Plant Lipids, pp 161–168. Plenum Press, New York

Siegenthaler PA, Rawyler A and Ṣmutny J (1989a) The phospholipid population which sustains the uncoupled non-cyclic electron flow activity is localized in the inner monolayer of the thylakoid membrane. Biochim Biophys Acta 975: 104–111

Siegenthaler PA, Rawyler A and Mayor JP (1989b) Structural and functional aspects of acyl lipids in thylakoid membranes from higher plants. In: Biacs PA, Gruiz K, and Kremer T (eds) Biological Role of Plant Lipids, pp 171–180. Akadémiai Kiado, Budapest and Plenum Publishing Corporation, New York

Slusarenko AJ, Croft KP and Voisey CR (1991) Biochemical and molecular events in the hypersensitive response of bean to *Pseudomonas syringae* pv. *phaseolicola*. In: Smith CJ (ed) Proceedings of the Phytochemical Society of Europe. Biochemistry and Molecular Biology of Plant-Pathogen Interactions, pp 126–143. Clarendon Press, Oxford

Smith AE (1972) Lipid influence on 2,4-D transport and accumulation. Weed Sci 20: 45–48

Somerville C (1995) Direct tests of the role of membrane lipid composition in low-temperature-induced photoinhibition and chilling sensitivity in plants and cyanobacteria. Proc Natl Acad Sci USA 92: 6215–6218

Somerville C and Browse J (1991) Plant lipids: metabolism, mutants, and membranes. Science 252: 80–87

Stidham MA, Siedow JN, McIntosh TJ, Porter NA and Moreland DE (1985) Effects of phenylamide herbicides on the physical properties of phosphatidylcholine membranes. Biochim Biophys Acta 812: 721–730

Szalontai B, Droppa M, Vigh L, Joó F and Horváth G (1986) Selectivity of homogeneous catalytic hydrogenation in saturation of double bonds of lipids in chloroplast lamellae. Photobiochem Photobiophys 10: 233–240

Szejtli J (1990) The cyclodextrins and their applications in biotechnology. Carbohydrate Polymers 12: 375–392

Talbert DM and Camper ND (1983) Herbicide effects on liposome leakage. Weed Sci 31: 329–332

Thomas H (1986) The role of polyunsaturated fatty acids in senescence. J Plant Physiol 123: 97–105

Thomas PG, Brain APR, Quinn PJ and Williams WP (1985) Low

pH and phospholipase A$_2$ treatment induce the phase-separation of non-bilayer lipids within pea chloroplast membranes. FEBS Lett 183: 161–166

Thomas PG, Quinn PJ and Williams WP (1986a) The origin of photosystem-I-mediated electron transport stimulation in heat-stressed chloroplasts. Planta 167: 133–139

Thomas PG, Dominy PJ, Vigh L, Mansourian AR, Quinn PJ and Williams WP (1986b) Increased thermal stability of pigment-protein complexes of pea thylakoids following catalytic hydrogenation of membrane lipids. Biochim Biophys Acta 849: 131–140

Thompson JE (1988) The molecular basis for membrane deterioration during senescence. In: Noodén LD and Leopold AC (eds) Senescence and Aging in Plants, pp 51–83. Academic Press, New York

Trémolières A and Siegenthaler PA (1998) Reconstitution of photosynthetic structures and activities with lipids. In: Siegenthaler PA and Murata N (eds) Lipids in Photosynthesis: Structure, Function and Genetics, pp 175–189. Kluwer Academic Publishers, Dordrecht

Trémolières A, Darmecy H, Gasquez J, Dron M and Connan A (1988) Variation of *trans*-hexadecenoic content in two triazine resistant mutants of *Chenopodium album* and their susceptible progenitor. Plant Physiol 86: 967–970

Trémolières A, Dainese P and Bassi R (1994) Heterogenous lipid distribution among chlorophyll-binding proteins of Photosystem II in maize mesophyll chloroplasts. Eur J Biochem 221: 721–73

Tsvetkova NM, Brain APR and Quinn PJ (1994) Structural characteristics of thylakoid membranes of *Arabidopsis* mutants deficient in lipid fatty acid desaturation. Biochim Biophys Acta 1192: 263–271

Van Ginkel G and Fork DC (1981) The effect of ageing of spinach chloroplasts in vitro on the phase transition temperatures of thylakoid membranes. Photobiochem Photobiophys 2: 239–243

Vaughn KC and Duke SO (1984) Ultrastructural alterations to chloroplast in triazine-resistant weed biotypes. Physiol Plant 62: 510–520

Vigh L, Joo F, Droppa M, Horváth LI and Horváth G (1985) Modulation of chloroplast membrane lipids by homogeneous catalytic hydrogenation. Eur J Biochem 147: 477–481

Vigh L, Gombos Z, Horváth I and Joo F (1989) Saturation of membrane lipids by hydrogenation induces thermal stability in chloroplast inhibiting the heat-dependent stimulatin of Photosystem I-mediated electron transport. Biochim Biophys Acta 979: 361–364

Vigh L, Los DA, Horváth I and Murata N (1993) The primary signal in the biological perception of temperature: Pd-catalyzed hydrogenation of membrane lipids stimulated the expression of the *desA* gene in *Synechocystis* PCC6803. Proc Natl Acad Sci USA 90: 9090–9094

Vijayan P, Routaboul JM and Browse J (1998) A genetic approach to investigating membrane lipid structure and photosynthetic function. In: Siegenthaler PA and Murata N (eds) Lipids in Photosynthesis: Structure, Function and Genetics, pp 263–285. Kluwer Academic Publishers, Dordrecht

Voss R, Radunz A and Schmid GH (1992a) Glycolipids are prosthetic groups of polypeptides of the reaction center complex of Photosystem II. In: Argyroudi-Akoyunoglou JH (ed) Regulation of Chloroplast Biogenesis, pp 417–422. Plenum Press, New York

Voss R, Radunz A and Schmid GH (1992b) Binding of lipids onto polypeptides of the thylakoid membrane. I. Galactolipids and sulpholipid as prosthetic groups of core peptides of the Photosystem II complex. Z Naturforsch 47c: 406–415

Wada H and Murata N (1998) Membrane lipids in cyanobacteria. In: Siegenthaler PA and Murata N (eds) Lipids in Photosynthesis: Structure, Function and Genetics, pp 65–81. Kluwer Academic Publishers, Dordrecht

Wada H, Gombos Z and Murata N (1990) Enhancement of chilling tolerance of a cyanobacterium by genetic manipulation of fatty acid desaturation. Nature 347: 200–203

Webb MS and Green BR (1991) Biochemical and biophysical properties of thylakoid acyl lipids. Biochim Biophys Acta 1060: 138–158

Williams WP (1994) The role of lipids in the structure and function of photosynthetic membranes. Prog Lipid Res 33: 119–127

Williams WP (1998) The physical properties of thylakoid membrane lipids and their relation to photosynthesis. In: Siegenthaler PA and Murata N (eds) Lipids in Photosynthesis: Structure, Function and Genetics, pp 103–118. Kluwer Academic Publishers, Dordrecht

Wollenberger L, Stefansson H, Yu S-G and Albertsson P-A (1994) Isolation and characterization of vesicles originating from the chloroplast grana margins. Biochim Biophys Acta 1184: 93–102

Wu J and Browse J (1995) Elevated levels of high-melting-point phosphatidylglycerols do not induce chilling sensitivity in an *Arabidopsis* mutant. Plant Cell 7: 17–27

Wu J, James DW Jr, Dooner HK and Browse J (1994) A mutant of *Arabidopsis* deficient in the elongation of palmitic acid. Plant Physiol 106: 143–150

Wu J, Lightner J, Warwick N and Browse J (1997) Low-temperature damage and subsequent recovery of *fab*1 mutant *Arabidopsis* exposed to 2 °C. Plant Physiol 113: 347–356

Xu YN and Siegenthaler PA (1996a) Phosphatidylglycerol molecular species of photosynthetic membranes analyzed by HPLC: theoretical considerations. Lipids 31: 223–229

Xu YN and Siegenthaler PA (1996b) Effect of non-chilling temperature and light intensity during growth of squash cotyledons on the composition of thylakoid membrane lipids and fatty acids. Plant Cell Physiol 37: 471–479

Xu YN and Siegenthaler PA (1997) Low temperature treatments induce an increase in the relative content of both linolenic and Δ^3-*trans*-hexadecenoic acids in thylakoid membrane phosphatidylglycerol of squash cotyledons. Plant Cell Physiol 38: 611–618

Yamamoto Y, Ford RC and Barber J (1981) Relationship between thylakoid membrane fluidity and the functioning of pea chloroplasts. Effect of cholesterol hemisuccinate. Plant Physiol 67: 1069–1072

Chapter 9

Reconstitution of Photosynthetic Structures and Activities with Lipids

Antoine Trémolières
*Institut de Biotechnologie des Plantes, Université Paris-sud XI, Bâtiment 630,
F-91405 Orsay Cédex, France*

Paul-André Siegenthaler
*Laboratoire de Physiologie végétale, Université de Neuchâtel, Rue Emile Argand 13,
CH-2007 Neuchâtel, Switzerland*

Summary

This chapter presents a comprehensive overview of in vitro reconstitution experiments of photosynthetic activities and/or structures with lipids. Special attention has been focused on the biogenesis of the main light-harvesting chlorophyll-protein complex, the LHCII. One crucial step in this process consists of inducing specifically the trimerization of the monomeric protein-pigment complex by phosphatidylglycerol. It is an example of a highly specific interaction between one lipid and one protein family, since only one molecule of phosphatidylglycerol recognizing a short consensus sequence from the 16 to 21 amino acids of the N-terminal part of the apoprotein is able to induce a tridimensional configuration of the apoproteins allowing the trimerization of LHCII complex.

The second part of the chapter describes experiments in which trans-Δ3-hexadecenoic acid containing phosphatidylglycerol is targeted in vivo into the photosynthetic membrane of *Chlamydomonas reinhardtii* mutants lacking this lipid. This reconstitution induces not only the trimerization of LHCII but also the development of grana stacks and the restoration of light energy distribution. These in vivo experiments not only fully confirm the results obtained from in vitro experiments but also demonstrate that the formation of

P.-A. Siegenthaler and N. Murata (eds): Lipids in Photosynthesis: Structure, Function and Genetics, pp. 175–189.
© 1998 Kluwer Academic Publishers. Printed in The Netherlands.

appressed membranes strongly depends on the trimerization of LHCII. The in vivo approach also emphasizes the possible role played by the trans-Δ3-hexadecenoic acid, a fatty acid which is exclusively found in the phosphatidylglycerol of eukaryotic photosynthetic organisms.

Finally, the importance of the N-terminal part of the LHCII apoprotein in the differentiation of stacked and unstacked regions and in the lateral heterogenous organization of the photosynthetic membrane is discussed in relation to lipid-protein-pigment interactions.

I. Introduction

Among the great variety of techniques used to approach the functions of lipids in the photosynthetic membrane, one of the more conclusive is to demonstrate that the deficiency in one lipid class results in the impairment of some photosynthetic activities (Chapter 8, Siegenthaler and Trémolières) or structures (Chapter 7, Siegenthaler) and to restore a normal phenotype by reintroducing the lacking lipids. Transgenic approaches have been successfully used by Wada et al. (1990) who demonstrated that the introduction of desaturase genes in cyanobacteria restores photosynthetic activity at low temperature. Several groups (Murata et al., 1992; Wolter et al., 1992) have also manipulated the chilling tolerance of higher plants by changing the fatty acid composition of the phosphatidylglycerol in photosynthetic membranes by transferring genes coding for glycerol-3-phosphate acyltransferase (Chapter 13, Gombos and Murata). Although transformation experiments are a powerful tool for determining the family of genes involved in a given process, they do not give a precise understanding of the cascade of interactions occurring after the genetic expression. To overcome this limitation a dual approach has been developed. First, a genetic one, which consists of producing mutants (by classical random mutation or genetic engineering) lacking the ability to synthesize one of the numerous lipids involved in the lipid metabolism (Somerville and Browse, 1991; Chapter 14, Vijayan et al.). A second approach consists of developing a technique allowing the lacking lipid to be targeted into the correct membrane of the growing organism. This latter approach was developed recently

by using *Chlamydomonas reinhardtii* mutants (Garnier et al., 1990).

The comparison of these two approaches (i.e. reconstitution assays 'in vitro' and 'in vivo') has led to the conclusion that lipids may play a precise role in the biogenesis and functional organization of the thylakoid membrane.

II. In Vitro Approaches to Lipid-Protein Interactions in the Photosynthetic Membrane

A great number of strategies for the functional reassembly of membrane proteins into liposomes have been described in the literature (Rigaud et al., 1995). It is out of the scope of this review to describe all these approaches. However, in the experiments described below, the technique used for reconstitution will be mentioned in each case. Reconstitution of thylakoid components into artificial membranes has been reviewed by Ryrie (1986).

A. In Vitro Reconstitution of Photosynthetic Activities

Murata and Sato (1978) were among the first to study the absorption spectra of chlorophyll *a* in aqueous dispersions of lipids from photosynthetic membranes. They concluded that the presence of both galactolipids and charged lipids are necessary to reconstruct the state of chlorophyll *a* dissolved in the lipid phase in thylakoid membranes.

In 1979, van Ginkel observed an increase in O_2 and H^+ uptake with ascorbate/dichlorophenol indophenol as electron donor. This occurred when isolated PS I particles were mixed with lipidic vesicles prepared by sonication of soybean lecithin, although stronger effects were observed with Triton X-100 micelles. Larkum and Anderson (1982) used digitonin to extract PS II or PS I center complex from spinach chloroplasts and reconstituted them with LHCII incorporated in liposomes made of a total lipid extract from spinach chloroplasts. Light excitation of

Abbreviations: DGDG – digalactosyldiacylglycerol; LHCII – light-harvesting chlorophyll *a/b* protein complex; MGDG – monogalactosyldiacylglycerol; PAGE – polyacrylamide gel electrophoresis; PC – phosphatidylcholine; PG – phosphatidylglycerol; PG-16:1(3t) – phosphatidylglycerol containing trans-Δ3-hexadecenoic acid; PLA_2 – phospholipase A_2; PS I – Photosystem I; PS II – Photosystem II; SQDG – sulfoquinovosyldiacylglycerol chlorophyll *b* at 475 nm stimulates the fluorescence emission from both the PS II and the P700-chlorophyll *a* protein complex when LHCII is reconstituted with either of these complexes, thus demonstrating energy transfer between LHCII and PS I or PS II complexes in liposomes. No evidence was found of an energy transfer from the PS II to the PS I-complex in the same proteo-liposome preparation. Furthermore, Siefermann-Harms et al. (1982,1987) reported that only MGDG and no other thylakoid lipid can restore energy transfer from LHC to the photosystems in Triton X-100-solubilized membranes. This suggests that, in the presence of MGDG, chlorophyll-protein complexes reassemble to supramolecular structures similar to those in intact thylakoid membranes. In oxygen-evolving PS II particles which have lost preferentially DGDG during Triton X-100 solubilization, oxygen evolution is stimulated by the addition of a total polar lipid extract of thylakoid membrane. The same effect was achieved with isolated DGDG or PC (Gounaris et al., 1983). In spinach P700-enriched particles, energy transfer efficiency is highly increased only when lipids such as MGDG and PC are added together (Ikegami, 1983). The recovery of O_2 evolution of cholate-treated thylakoids induced by the simultaneous addition of the 17 and 23 kDa proteins is enhanced by a further addition of thylakoid total lipids up to about 75% of the non-depleted original broken thylakoids (Akabori et al., 1984). Murphy et al. (1984) reported that the reconstitution of LHCII with PS II complex in soybean phosphatidylcholine liposomes results in a large increase of PS II activity. Interestingly, thylakoid PS II particles display an increase in O_2 evolution when incorporated into PG vesicles (Fragata et al. 1991). Nénonéné and Fragata (1990) suggested that the PG-induced stimulation of oxygen evolution in the acidic side of pH results from the undissociated state of the hydroxyl group of the phosphoglyceryl moiety of PG, which is susceptible to hydrogen-bond formation with the carboxyl or amine groups in the protein backbone of the PS II polypeptides (see also Fragata et al., 1994). In similar experiments, Matsuda and Butler (1983) reported that a preparation of O_2-evolving PS II particles, which was selected on the basis of having a relatively low rate of O_2 evolution, contained very little high-potential cytochrome b_{559} and a less-than-normal amount of variable yield fluorescence. However, in a liposome preparation, consisting of DGDG and phosphatidylcholine, these particles showed considerably more high-potential cytochrome b_{559}, an almost normal amount of variable yield fluorescence and a substantially greater rate of O_2 evolution than in the original particles.

On the other hand, the cytochrome b_6f complex, which has been depleted of plastoquinone (PQ) and lipids, no longer functions as a plastoquinol-plastocyanin oxidoreductaz . It can, however, be reconstituted with PQ and exogenous lipids (DGDG, PG and PC) but not with MGDG and SQDG. Neither PQ nor lipid alone fully reconstitutes the electron transport in the depleted complex (Chain, 1985). In this kind of experiment the main problem is that some lipids are so firmly bound to photosynthetic complexes that it is difficult to deal with a fully delipidated preparation. Indeed, in an elegant work using antibodies against photosynthetic lipids, Vos et al. (1992) observed that MGDG, DGDG and SQDG are so tightly bound to the purified peptides of the PS II core that they are still present after a transfer of the peptides from polyacrylamide gel to nitrocellulose membranes. These authors concluded that galacto-lipids and the sulfolipid can be considered as prosthetic groups of the core polypeptide of the PS II.

B. Interaction Between Photosynthetic Membranes or Complexes and Liposomes

It has been shown that stacking of isolated thylakoids or aggregation of isolated LHCII particles can be reversibly induced in vitro by the addition of cations (e.g. by $MgCl_2$) and that the N-terminal parts of the polypeptides of the LHCII play a major role in this process (Izawa and Good, 1966; Staehelin, 1976). McDonnel and Staehelin (1980) reconstituted isolated LHCII in soybean lecithin liposomes and showed that cation-mediated aggregation of LHCII particles occurs in this reconstituted system. They concluded that the adhesion between LHCII particles is mediated by hydrophobic interactions and that cations are needed to neutralize surface charges on the particles. Ryrie and Fuad (1982) also reported that proteo-liposomes reconstituted with LHCP and the chloroplast diacyl lipids aggregate markedly in the presence of cations but vesicles containing LHCP prepared from trypsin-treated thylakoids do not. Siegel et al. (1981) showed that when isolated thylakoids are mixed with liposomes of phosphatidylcholine, these exogenous lipids can be integrated in the membrane after a freezing and thawing treatment. In 1985, Sprague et al. isolated LHCII and PS II from an octylglucoside-containing sucrose gradient after solubilization of the thylakoid membrane with Triton X-100 and octylglucoside, and reconstituted these complexes in the presence of DGDG or PC liposomes. They observed that in such proteoliposomes, membrane adhesion can be induced by Mg^{2+}. In the above experiments, however, no clear-cut specificity of lipids could be evidenced.

C. Insertion of Isolated LHCII in Developing Thylakoids

Day et al. (1984) studied thylakoids from intermittent-light-grown barley plants, which are characterized by a deficiency in LHCII complex. In these chloroplasts there are no true grana stacks but only limited appressed lamellae. Isolated thylakoids display limited but significant Mg^{2+}-induced membrane appression. When LHCII was re-incorporated by a sonication-freeze-thaw procedure, the reconstituted membrane, unlike the parent ones, exhibited extensive membrane appression. Darr and Arntzen (1986) developed a method for reinsertion of isolated LHCII into thylakoid membranes deficient in this complex. Thylakoids prepared from *Hordeum vulgare* leaves grown first in darkness then in intermittent light (a treatment giving partly etiolated leaves) are largely depleted in LHCII. Then the insertion of exogenous LHCII in the membrane during an in vitro assay results in changes of spectral and fluorescence properties of the membrane. The exogenous LHCII can be inserted by a freezing-thawing cyclic process or by dialysis with the detergent octylpolyoxyethylene. Both processes are equally efficient but, interestingly, the authors observed that the rate of in vitro insertion of LHCII was not very efficient when compared to the rate of in vivo insertion, i.e. in membranes of whole plants submitted to a continuous light treatment. They concluded that the membrane used as LHCII recipient lacks several important components which are only synthesized in continuous light, such as a lipid (or a special form of this lipid) which is exclusively found in the thylakoids of eukaryotic photosynthetic organisms after continuous light treatment (Dubacq and Trémolières, 1983). Recently, the group of Paulsen (Kuttkat et al., 1995) studied the insertion and binding with pigments of recombinant or mutagenized LHCII into isolated thylakoids. They showed that only the LHCII which is able to trimerize can be correctly stabilized in the membrane as

indicated by the sensitivity of the complex to protease (see Section II.F for further explanations). However, the functional role of lipids has not been considered in this work.

D. Reconstitution of Monomeric LHCII

An in vitro reconstitution system was developed by Plumey and Schmidt (1987). Thylakoid polypeptides (or thylakoids) from tobacco leaves were denatured by heating, then mixed with different proportions of chlorophylls and carotenoids and submitted to three cycles of freezing (6 to 12 h at −20 °C) and thawing (15 min at 20 °C). Then the mixture was submitted to lithiumdodecylsulfate/PAGE. Under these conditions, a green band corresponding to a monomeric pigment-protein complex appears which displays both spectral properties and a pigment-protein ratio which have the same characteristics as those of the native monomeric LHCII. In addition, these authors showed that chlorophylls *a* and *b* and xanthophylls are necessary for reconstitution, while β-carotene is not. This is in accordance with analyses of carotenoids in isolated photosynthetic complexes showing that β-carotene is essentially restricted to the PS II core (Bassi et al., 1993). However, since reconstitution with delipidated membrane and purified pigments reduces the reconstitution yield of LHCII, four lipids (MGDG, DGDG, PG and PC) were independently added to the reaction mixture but only an adverse effect was observed. Since lipids are not found in the reconstituted complex, the authors concluded that glycerolipids are not integral structural components of the monomeric LHCII. Nevertheless, as the amount of lipids remaining in the purified LHCII is very low, e.g. around two molecules of MGDG, two molecules of DGDG and one molecule of PG per polypeptide (Trémolières et al., 1994), this conclusion cannot be definitely accepted at this stage.

Three years later, Paulsen et al. (1990), using the same method as Plumey and Schmidt (1987), succeeded in reconstituting a monomeric LHCII by mixing chlorophylls and carotenoids with a polypeptide of LHCII which had been overproduced in *E. coli* transformed with a cab gene from pea without adding any lipids. Two groups (Paulsen and Hobe, 1992 and Cammarata and Schmidt, 1992) used the system of expression of cab gene in *E. coli* to produce mutagenized LHCII and thus determined what part of the polypeptide is necessary for reconstitution. The general conclusion is that the first

61 N-terminal and the last ten C-terminal amino acid residues do not play any role in pigment binding. However, all the three hydrophobic membrane spanning α-helix domains are necessary and only a small number of amino acid residues play an essential role in each border's regions between the N- and C-hydrophilic domains and the respective adjacent hydrophobic domains. Furthermore, it was observed by Paulsen and co-workers that the mutagenized LHCII either forms complexes with both chlorophylls and xanthophylls (e.g. their spectroscopic properties are very similar to those of the LHCII complex isolated from thylakoids) or does not form any stable complexes at all. This observation led to the conclusion that the stabilization of LHCII-pigment complexes is highly synergistic rather than based on individual pigment-binding sites provided by the protein.

Based on the above experiments, it was concluded that lipids do not play a role in the formation of the monomeric LHCII. Nevertheless, this conclusion was recently questioned. Indeed, when monomeric forms of LHCII are obtained by PLA_2 or chymotrypsin-trypsin treatments of native trimeric LHCII from pea, PG is degraded. Circular dichroism experiments show that, when monomers have lost their native PG, two bands at 640 and 661 nm are lacking. These are normally present in monomers obtained by treatments of trimeric LHCII by non-ionic detergent, i.e. under conditions where PG remains linked to the complex. These observations show that some chlorophyll a and b molecules are not correctly bound to the complex in the absence of native PG (Nussberger et al., 1994; Peterman et al., 1996) suggesting that this lipid could play a structural role in the organization of the monomer itself.

E. Lipid-Induced Oligomerization of LHCII

It has long been observed that, when solubilized photosynthetic membranes are analyzed by gel electrophoresis under mild detergent conditions, LHCII is resolved mainly into two green bands. In higher plants, the first one migrates at around 25 kDa, the second one at around 72 kDa. It is now admitted that the lower molecular weight band contains all the set of polypeptides related to the LHCII in a dissociated state and, consequently, represents monomeric LHCII. The band migrating at the higher apparent molecular weight is thought to represent an oligomeric state of the organization of the LHCII

polypeptides linked together by non-covalent interactions. For a long time the state of aggregation was not precisely determined. It is now well established that it is an oligomer consisting of three monomers (Kühlbrandt and Wang, 1991). The term 'oligomer' will be kept here to emphasize the fact that the exact structure of the LHCII complex in the membrane is not definitely elucidated and that we do not know whether all polypeptides of LHCII play the same role in the aggregation process. The idea that lipids and, in particular, PG could be involved in the oligomerization process first came from the observation that the LHCII oligomer was significantly enriched in PG as compared to the monomer (Trémolières et al., 1981). Rémy et al. (1982) and Krupa (1984) showed a marked and specific decrease in the oligomeric form of LHCII by digesting isolated thylakoids with PLA_2, a treatment which detaches about 70% of the 16:1(3t) from the PG molecule. In such PLA_2-treated thylakoid membranes and after washing to remove the excess PLA_2 as well as free fatty acids and lyso-PG, fluorescence properties are affected. The Fmax is decreased by about 15% and the efficiency of light collection at low light intensities is lowered (Rémy et al., 1982). Many in vitro reconstitution experiments were undertaken by Rémy et al. (1984) demonstrating that an oligomeric LHCII can be reconstituted by simply mixing and shaking at room temperature isolated monomeric LHCII with liposomes of PG-16:1(3t). The absence of cations in the medium is an absolute requirement for reconstitution. When the monomeric form of LHCII is first digested at the N-terminal side, no reconstitution occurs. All thylakoid lipids can induce the reconstitution in vitro except MGDG which is a non-bilayer forming lipid. PG-16:1(3t) is always the most efficient one. When 16:1(3t) is replaced by palmitic acid in the PG, the reconstitution occurs more slowly showing that 16:1(3t), which is found only in the PG of eukaryotic photosynthetic organisms (Dubacq and Trémolières, 1983), confers a higher affinity of PG for the LHCII. The importance of this special fatty acid was strengthened by reconstitution experiments in which the LHCII from rye chloroplasts, first isolated in its oligomeric state, then treated by PLA_2, leads to a large dissociation of the oligomer into its monomers (Krupa et al., 1992). The lipid-depleted LHCII was then mixed with lipids and sonicated under nitrogen for 2 h at 10 °C. Under these conditions, only PG containing 16:1(3t) but not PG containing palmitate induced reoligomerization.

F. Lipid-Induced Trimerization of Recombinant and Mutagenized LHCII

Hobe et al. (1994) succeeded in trimerizing LHCII monomers obtained from the purified apoprotein overexpressed in *E. coli*, then mixed with pigments and lipids. The mixture was sonicated for 1 min, then frozen at −196 °C and finally thawed in a water bath at 22 °C. Trimeric and monomeric LHCII were separated by centrifugation on a sucrose gradient. Both native and reconstituted LHCII trimers exhibit circular dichroism signals in the visible range, that are not seen in native or reconstituted monomeric LHCII. This indicates that some changes in the orientation of the pigments might occur during the trimerization process. Reconstituted trimers readily form two-dimensional crystals identical to crystals of the native trimeric complex. Hobe et al. (1995) explored the identity of the polypeptide region involved in the interactions with lipids. The first 15 amino acid residues on the N-terminal part of LHCII can be removed without loosing the lipid-induced trimerization ability. Inspection by punctual deletion or replacement of an amino acid residue using mutagenesis revealed that tryptophan in 16 and tyrosine in 17 (two aromatic amino acids) as well as arginine (a basic amino acid) in 21 positions are essential for the formation of LHCII trimers. These three amino acids cannot be removed or replaced by any other amino acids without the loss of the ability to trimerize with lipids. These amino acids are conserved in virtually all known sequences of LHCII apoproteins and the sequence WYXXXR from 16 to 21 was termed 'trimerization motif' and supposed to interact with PG (Hobe et al., 1995). Interestingly, it is a very small portion of the LHCII, located in the region where the polypeptide emerges outside the thylakoid toward the stroma, which seems to be involved (see Fig. 1). This site of interaction with the lipids is located very near the phosphorylation site. PG-16:1(3t) was also found to be the most efficient lipid for inducing trimerization. The molecular ratio of lipid to protein in the reconstituted trimers was determined (A. Trémolières, S. Hobe and H. Paulsen, unpublished) and was found to be near one, a value which is similar to that found in purified native LHCII (Trémolières et al., 1994). Another amino acid residue, at the carboxyl side of the polypeptide, tryptophan 222, was also shown to be involved in the trimerization, but no interaction with lipids has been

evidenced. A hypothetical scheme showing the interaction between PG and the trimerization site is presented in Fig. 1. It is proposed that this interaction results in changes in the three-dimensional configuration at the N-terminal end which allows the stabilization and the trimerization of the complex to occur.

G. Importance of Lipids in the Crystallization of LHCII

Kühlbrandt et al. (1994) succeeded in producing an atomic map at 3.4 Å resolution of LHCII from pea by electron crystallography. However, crystallization of the LHCII was only possible when starting from trimeric and never from monomeric LHCII. Furthermore, Nussberger et al. (1993) found that two different thylakoid lipids are specifically involved. DGDG binds to the isolated complex but can be removed by mild detergent treatment and anion-exchange chromatography. Removal of this lipid renders the complex unable to form two- or three-dimensional crystals. The ability to crystallize is completely restored by the addition of DGDG at a ratio of about four molecules of DGDG per polypeptide for three-dimensional crystallization. This suggests the existence of several binding sites at the periphery of the trimeric complexes. Two-dimensional crystals of purified protein grown in the presence of DGDG are more highly ordered than those obtained from the unfractionated complex. The second lipid involved is PG which binds more firmly than DGDG and cannot be removed with non-ionic detergent. Delipidation of LHCII can be achieved either by PLA_2 treatment or by proteolytic cleavage of the first 49 amino acid residues at the N-terminal side. Both treatments dissociate the native trimeric complex into monomers. This indicates that PG is directly involved in the formation of trimers and is indispensable for two- or three-dimensional crystallization. Both lipids are therefore present in two- and three-dimensional crystals and have distinct roles in the establishment of the structure of the complex. These lipids have not been localized on the atomic map which is presently badly resolved in the N-terminal region. The lipid composition of crystals from the native trimeric LHCII was determined and found to consist of two molecules of MGDG, two molecules of DGDG and one molecule of PG (A. Trémolières and W. Kühlbrandt, unpublished),

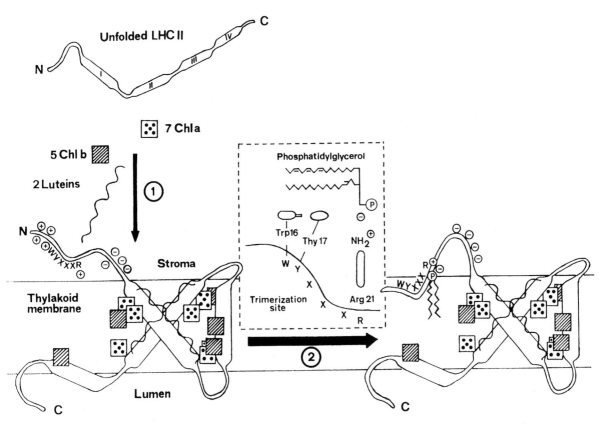

Fig. 1. Scheme illustrating the possible interaction between phosphatidylglycerol and the trimerization site of LHCII. (1) LHCII polypeptide(s) bind seven molecules of chlorophyll *a*, five molecules of chlorophyll *b* and two molecules of xanthophyll forming the monomeric chlorophyll-protein light-harvesting complex (Kühlbrandt et al., 1994). This reaction probably occurs in the membrane but monomeric LHCII remains largely accessible to trypsin or thermolysine degradation. (2) PG-16:1 (3t) binds the WYXXXR motif from the 16 to 21 amino acid residues of the N-terminal side bringing a negative charge in the 'tridimensional' configuration of the complex. As most of the PG-16:1 (3t) is localized in the stromatic lipid monolayer (Siegenthaler et al., 1989), it is likely that this interaction optimizes the orientation of the N-terminal side of the LHCII. It is suggested that the interaction of the negative charges of PG with the N-terminal sequence of LHCII, which is very rich in positive charges, will help the polypeptide sequence to be buried in the lipidic matrix by electrostatic interactions. In this configuration, only the negative charges brought by the extra thylakoidal segment between amino acid residues 22 and 53 will remain accessible in the stroma. Grana stacking will be induced when Mg^{2+} interacts with these charges (see Fig. 5). (3) Though the trimerization motif is involved in the formation of trimeric LHCII, another amino acid residue near the C-terminal side of LHCII, the tryptophan 222 (which can be replaced, for example in *Chlamydomonas* by a phenylalanine) is also likely to be involved in the trimerization process, but no interaction with lipids occurs at this place.

confirming previous analyses on LHCII, purified by different methods (Trémolières et al., 1994). A detailed comparison has been made by dichroic steady-state spectroscopic properties at 77K of several trimeric and monomeric forms of the LHCII (Nussberger et al., 1994; Peterman et al., 1996). The absorption and dichroism spectra indicate that in trimeric LHCII, the chlorophyll *b* absorption region is centered around 649 nm and is composed of at least five sub-bands near 640, 647, 649, 642 and 656 nm. The chlorophyll *a* absorption region is centered around 670 nm and is composed of at least five bands near 661, 668, 671, 673 and 676 nm. In monomeric LHCII the chlorophyll *b* bands near 647 and 652 and the chlorophyll *a* bands near 668 and 673 are absent in the circular dichroism spectrum. It is noteworthy that in the reconstitution assay starting from the recombinant LHCII (Hobe et al., 1994), the same differences in the circular dichroism spectra were found between trimeric and monomeric LHCII. A configuration in which pigments of the same nature are located in the monomers becomes excitonically coupled in the trimer and could explain this result (Nussberger et al., 1994).

III. In Vivo Approaches to Lipid-Protein Interactions by Targeting Lipids into Photosynthetic Membranes

One main limitation of apprehending lipid-protein interactions by in vitro techniques lies in the difficulty to ascertain that the observed interactions are really occurring in situ, i.e. in the membrane of the living organism. The possibility that proteins and lipids undergo artifactual relations can never be completely ruled out.

One way of overcoming this limitation is to use, as a model system, the unicellular green alga *Chlamydomonas reinhardtii* which presents many genetic and physiological advantages for the isolation of mutants. In its vegetative stage of growth, this alga remains in an haploid state and expresses all mutations. On the other hand, the rate of spontaneous mutations is exceptionally high. Furthermore, this alga presents the great advantage of growing under heterotrophic conditions allowing the study of non-photosynthetic mutants. As the first step, mutants affected in the synthesis of a given lipid were isolated (Maroc et al., 1987). This was achieved by random mutagenesis, but an approach by gene tagging can also be envisaged in the future. The second step consists of targeting the lacking lipids into the photosynthetic membrane by growing algae in the presence of liposomes made up with this lipid. It was observed that lipids from liposomes efficiently enter the cells by a mechanism which remains, however, to be elucidated. When a foreign glycerolipid enters the cell, fatty acids are very rapidly detached by endogenous lipases or acyltransferases and redistributed (or catabolyzed) within intracellular lipid species (Grenier et al., 1991). However, when the concentration of liposomes is sufficiently high, a part of the lipids escapes to this metabolic process and can be inserted and stabilized in a target membrane. Selected mutants of *Chlamydomonas* lacking PG-16:1(3t) were grown for 40 h in the presence of liposomes made up of this lipid at a concentration of 0.1 mg/mL, such a concentration having no effect on the growth rate. PG-16:1(3t) is very well incorporated in the photosynthetic membrane (Table 1) and only in this membrane (data not shown). These results which confirm those obtained by the in vitro approach strongly suggest that PG-16:1(3t) plays a key role in the biogenesis of LHCII. Furthermore, the in vivo approach should be

Table 1. Phosphatidylglycerol content of the wild type (WT), mf_1 and mf_2 mutants *Chlamydomonas reinhardtii* as well as the mf_2 mutant supplemented with phosphatidylglycerol containing 16:1(3t) during growth (mf_2 + PG) (from Dubertret et al., 1994)

	WT	mf_1	mf_2	mf_2 + PG
	% of total lipids			
Total PG	13.6	3.2	2.3	9.0
	% of PG fatty acids			
16:1(3t) in PG	29.3	0	0	20.0

useful in obtaining new insight into some aspects of the trimerization process and the thylakoid membrane biogenesis (Dubertret et al., 1994).

A. Role of PG-16:1(3t) in the Biogenesis and Trimerization of the LHCII

Among the set of mutants devoid of PS II reaction centers, two of them were found to show intriguing fluorescence properties (Maroc et al., 1987). Generally, a PS II-lacking mutant shows an abnormally high level of fluorescence because the main light-harvesting antennae cannot transmit energy to the missing PS II reaction center and reemit their energy as fluorescence (Wollman and Delepaire, 1984). On the contrary, these two special mutants show a lower fluorescence yield than the wild type during the induction phase and were therefore called *mf* mutants (*mf* for minimum fluorescence) (Garnier et al., 1990). Interestingly, they were found to be lacking in PG-16:1(3t) and to contain only PG molecular species devoid of this fatty acid, the total PG content being reduced to 1/3 to 1/4 of that found in the wild type (Table 1). The 16:1(3t) fatty acid which esterifies specifically the *sn*-2 position of the glycerol in PG was for a long time suspected to play an important function in relation to the antenna organization (Trémolières, 1991). Interestingly, these two mutants were found to be almost completely devoid of the trimeric LHCII, though the level of the monomeric LHCII was quite normal, e.g. only a very weak increase in the chlorophyll *a/b* ratio was recorded when the alga culture reached the plateau phase. A more detailed study showed that the polypeptide pattern of the LHCII is markedly disturbed in these two mutants. Among the four main polypeptides found in the LHCII (called in *Chlamydomonas* p11, p13, p16 and p17) only the p11 is relatively little affected while the content of the three others is significantly decreased. On the other hand, polypeptides immunologically related to the LHCII accumulate at low molecular weights. It was therefore suspected that, in mutants lacking PG-16:1(3t), LHCII polypeptides cannot be arranged and stabilized as a trimeric state and, therefore, are susceptible to be degraded by intracellular proteases. In order to prove that the absence of the lipid is responsible for this deficiency, PG-16:1(3t) was reincorporated into the photosynthetic membranes of the living mutants as previously described. Under these conditions, the trimeric LHCII largely reforms (Fig. 2) and a quite normal polypeptide pattern is restored in LHCII. This restoration only occurs with PG-16:1(3t) but not, for example, with PG-containing palmitic instead of 16:1(3t) acid at the *sn*-2 position of the glycerol. On the other hand, when the lipid was incorporated into the membrane under conditions where synthesis of LHCII polypeptides is inhibited by cycloheximide, the restoration of the ability to stabilize the trimeric LHCII does not occur (Garnier et al., 1990; Dubertret et al., 1994).

In conclusion, PG plays a crucial role in the LHCII trimerization process. Moreover, 16:1(3t) confers special properties to the PG molecule allowing high affinity interactions with some specific sites in the chlorophyll-protein complex. These in vivo results are in accordance with the reconstitution experiments reported above (see Section II).

B. Relations Between the Trimerization of LHCII, Grana Stacking and Light Energy Distribution in Photosynthetic Membranes

Correlations between the presence of PG-16:1(3t) and the trimeric form of LHCII, and the formation of appressed membranes in the chloroplast were reported under several physiological and genetic situations (see Trémolières, 1991). For example, it was found that the two mutants of *Chlamydomonas* devoid of PG-16:1(3t) and of the trimeric LHCII are almost completely unable to form grana stacks. When these mutants were supplemented in vivo with PG-16:1(3t), not only trimeric LHCII (Fig. 2) but also grana stacks reappear in the living cells showing clearly that these two levels of organization are linked together and depend on the presence of PG-16:1(3t) (Trémolières et al., 1991; see Fig. 3). A schematic diagram illustrates the sequential events leading to the formation of the monomeric form of LHCII and the

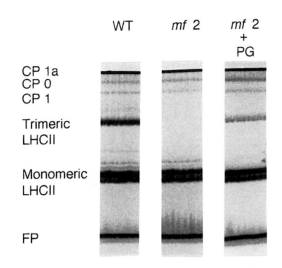

Fig. 2. Role of phosphatidylglycerol in the trimerization of LHCII monomers. Non-denaturing SDS/PAGE separation of octylglucoside-solubilized chlorophyll-protein complexes of the wild type of *C. reinhardtii* (WT), of the mutant lacking PG-16:1(3t) (mf_2) and of the same mutant grown for 40 h in the presence of liposomes containing PG-16:1(3t) (mf_2+PG). Cp1a, Cp0, Cp1: PS I complexes. FP: free pigments (from Dubertret et al., 1994).

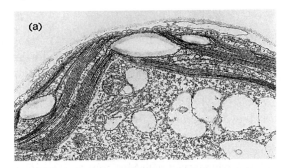

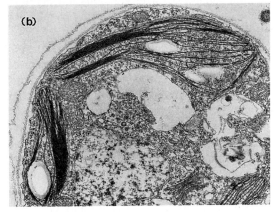

Fig. 3. Ultrastructure of *Chlamydomonas reinhardtii* cells of a mutant lacking PG-16:1(3t) grown for 40 h in the absence (a) or in the presence of liposomes made up with PG-16:1(3t) (b). A significant increase in the extent of grana stacks is observed after the incorporation of PG-16:1(3t) (from Dubertret et al., 1994).

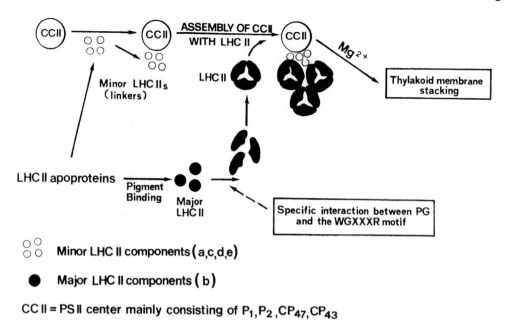

Minor LHC II components (a,c,d,e)

Major LHC II components (b)

CC II = PS II center mainly consisting of $P_1, P_2, CP_{47}, CP_{43}$

Fig. 4. Schematic diagram illustrating the sequential events leading to the formation of LHCII trimers. Note that PG is likely to be involved in a specific interaction with the WGXXXR motif which induces the LHCII trimerization. For the nomenclature of minor (a,c,d,e) and major (b) LHC II components refer to Dreyfuss and Thornber (1994a). The scheme is adapted from Dreyfuss and Thornber (1994a) and Trémolières et al. (1996).

trimerization of the LHCII complex (Fig. 4). PG is likely to be involved in a specific interaction with the WGXXXR motif of the N-terminal part of the major LHCII component which induces LHCII trimerization (see also Fig. 1). Another hypothetical scheme shows how the exposure of the negative charges of the apoprotein segment (between the amino acid residues 29 and 53) toward the stroma side of the membrane can interact with Mg^{2+} ions and therefore induce the appression of two thylakoid membranes (Fig. 5).

On the other hand, a detailed study of the regulation of light energy distribution in several mutants of *Chlamydomonas* has led to the following findings (Wollman and Delepaire, 1984; Delepaire and Wollman, 1985): (1) Mutants lacking only the PS II center (FI39) are able to perform the state II-state I transition in relation to the state of reduction of the plastoquinone pool; (2) The fluorescence emission spectra of these mutants at low temperature shows that when the pool of plastoquinone is chemically reduced (e.g. by sodium dithionite) only one main fluorescence peak (712 nm) appears in the PS I region, indicating that most of the light energy is transmitted to this photosystem. However, when the

pool of plastoquinone is oxidized, despite the lack of PS II, an important peak of fluorescence (682 nm) appears emitting exactly at the wavelength characteristic of LHCII. This shows that, in these mutants, LHCII can be reversibly attached to or detached from the PS II center, according to the state of reduction of the plastoquinone pool. This is likely to occur via a phosphorylation and dephosphorylation process of the LHCII, as previously described by Allen et al. (1981); (3) In the two mutants (mf_1 and mf_2) lacking PG-16:1(3t) this ability to perform the state II-state I transition in relation to the plastoquinone pool reduction state is almost completely abolished but can be significantly restored when PG-16:1(3t) is reintroduced in vivo into the photosynthetic membrane (Garnier et al., 1990; Trémolières et al., 1991).

It can be concluded that, in the absence of PG-16:1(3t), the polypeptides of the LHCII cannot be stabilized in the trimeric state and, consequently, grana stacks cannot be formed. Under these conditions, LHCII polypeptides are degraded and, consequently, loose their site of phosphorylation and the state II-state I transition ability (Dubertret et al., 1994).

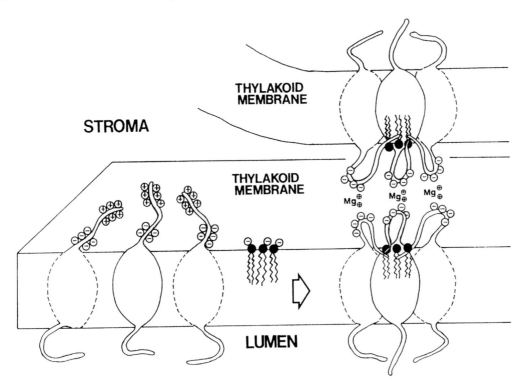

Fig. 5. Hypothetical scheme showing how Mg^{2+} ions induce the appression of two thylakoid membranes, i.e. thylakoid stacking.

IV. Conclusions and Perspectives

The dual approach for studying the lipid function based on reconstitution experiments in vitro and in vivo gives a new insight into the complexity of the organization of the light collecting apparatus around the PS II and the subtle relationship between lipids, proteins and pigments occurring during the biogenesis of the photosynthetic membrane. The very low lipid/protein ratio in the purified trimeric LHCII excludes the possibility that lipids act as a cement between polypeptides. It is more likely that they play a role in establishing the correct tertiary structure allowing the proper interaction between the LHCII components. We can envisage that even if monomeric LHCII can be formed without any lipids, it is nevertheless highly probable that lipids bind monomeric LHCII before trimerization in order to get an adequate tridimensional structure of the apoprotein. Another feature concerns the supramolecular organization of the LHCII. All experiments reported here prove that the trimeric organization of the LHCII polypeptides constitutes the basic ultrastructural and functional arrangement of the antennae around the PS II; however, the definitive

supramolecular organization could be even more complex, as proposed by Bassi and Dainese (1992), Kühlbrandt (1994), Dreyfuss and Thornber (1994a,b).

Another important question concerns the exact role played by the different polypeptides of the LHCII coded by the multigenic cab family. From the in vitro experiments using recombinant LHCII, it is clear that a homotrimer can be produced in vitro. Thus, the different polypeptides could play the same role in the trimerization process since all polypeptides encountered in the monomeric LHCII are also found in the trimer, in the same ratio. But the in vitro trimerization process could be quite different from the process occurring during the biogenesis of photosynthetic membranes in living organisms. For example, the high specificity as for PG-16:1(3t) observed in vivo was not as strict in vitro. Even if the crystallographic appearance found after the in vitro reconstitution is similar to that of the native LHCII (Kühlbrandt et al., 1994), one can envisage that, in vivo, the assembly of the different components might occur in a different physiological context. For example, it was found that when photosynthetic pigments, proteins and lipids are not synthesized in a correct dynamic equilibrium, the components which are not rapidly enough stabilized in their definitive structure, can be degraded by proteases (Bennett, 1981) or lipases. One might postulate that a more sophisticated mechanism occurs in vivo in which each polypeptide of the LHCII could play a specific function. The observation that in the *Chlamydomonas* mutants devoid of PG-16:1(3t) all polypeptides of the LHCII are not affected to the same extent by the lipid deficiency could be interpreted in this way.

Finally, it can be noted that the cab gene family is known to produce polypeptides with a very high level of homology, especially in the intrathylakoid region containing the three hydrophobic α-helices spanning the membrane, while some divergences are found in the N-terminal part of the LHCII (Green et al., 1991; Paulsen, 1994). However, it is this part of the polypeptide that is obviously involved in the interaction with the lipid and, consequently, in the trimerization and the formation of grana stacks. It can also be observed that the extent of appressed membranes is largely variable. They can be more or less important, in the number or in the size of stacked regions, not only when changing the physiological conditions (Boardman et al., 1975) or genotypes (Lemoine et al., 1986) but also in different eukaryotic photosynthetic species (Lemoine et al., 1982; Bassi, 1985). It seems quite clear that these distinct types of organization are related to the maximum efficiency of light harvesting in different ecophysiological situations, and also to the evolution of each species. We are probably far from a definitive understanding of the adaptation of the photosynthetic apparatus to the various physiological situations. The regulation process for light energy distribution between the two photosystems via the phosphorylation of the mobile antennae is likely to work in a highly different environment, as suggested by the elegant work of Huner et al. (1987; 1989) on the adaptation of rye to low temperatures. The genetic variability of the N-terminal part of the LHCII polypeptides could reflect these adaptive processes.

Acknowledgments

We would like to acknowledge Dr. Jacques Garnier, who died in 1992 and was one of the first to search for mutants affected in lipid metabolism in *Chlamydomonas*; Mr. Guy Dubertret for his constant and competent collaboration in the field of membrane protein biochemistry; Miss Annick Bennardo-Connan for excellent technical assistance in reconstitution experiments; Miss Odile Roche for her electron microscopic pictures; Mr. Roland Boyer for the photographs; Prof. Waldemar Eichenberger for critical reading of the manuscript; Mrs. Christiane Bettinelli for expert assistance in the preparation of the manuscript; Mrs. Delia Jones Siegenthaler for improvement of the English manuscript and Mrs. Jana Smutny for drawing Figs. 1, 4 and 5.

References

Akabori K, Imaoka A and Toyoshima Y (1984) The role of lipids and 17-kDa protein in enhancing the recovery of O_2 evolution in cholate-treated thylakoid membranes. FEBS Lett 173: 36–40

Allen JF, Bennet J, Steinback E and Arntzen CC (1981) Chloroplast protein phosphorylation couples plastoquinone redox state to distribution of excitation energy between photosystems. Nature 291: 25–29

Bassi R (1985) Spectral properties and polypeptide composition of the chlorophyll-protein complexes from thylakoids of granal and agranal chloroplast of maize (*Zea mays* L.). Carlsberg Res Commun 50: 127–143

Bassi R and Dainese P (1992) A supramolecular light-harvesting complex from chloroplast Photosystem II membranes. Eur J Biochem 204: 317–326

Bassi R, Pineau B, Dainese P and Marquardt J (1993) Carotenoid-binding proteins of Photosystem II. Eur J Biochem 212: 297–303

Bennett J (1981) Biosynthesis of the light-harvesting chlorophyll *a/b* protein. Polypeptide turnover in darkness. Eur J Biochem 118: 61–70

Boardman HK, Björkman O, Anderson JM, Goodchild DJ and Thorne SW (1975) Photosynthesis adaptation of higher plants to light intensity. Relationship between chloroplast structure, composition of the photosystems and photosynthetic rates. In: Avron M (ed) Proceedings of the 3rd International Congress on Photosynthesis, Vol II, pp 1809–1827. Elsevier, Amsterdam

Chain RK (1985) Involvement of plastoquinone and lipids in electron transport reactions mediated by the cytochrome b_6-f complex isolated from spinach. FEBS Lett 180: 321–325

Cammarata KV and Schmidt GW (1992) In vitro reconstitution of a light-harvesting gene product: deletion mutagenesis and analyses of pigment binding. Biochemistry 31: 2779–2789

Darr SC and Arntzen CJ (1986) Reconstitution of the light harvesting chlorophyll *a/b* pigment protein complex into developing chloroplast membranes using a dialyzable detergent. Plant Physiol 80: 931–937

Day DA, Ryrie IJ and Fuad N (1984) Investigations of the role of the main light-harvesting chlorophyll-protein complex in thylakoid membrane. Reconstitution of depleted membranes from intermittent-light-grown plants with the isolated complex. J Cell Biol 97: 163–172

Delepaire P and Wollman FA (1985) Correlation between fluorescence and phosphorylation changes in thylakoid membranes of *Chlamydomonas* in vivo. A kinetic analysis.

Biochim Biophys Acta 809: 277–286

Dreyfuss BW and Thornber JP (1994a) Assembly of the light-harvesting complexes (LHCs) of Photosystem II. Monomeric LHC IIb complexes are intermediates in the formation of oligomeric LHC IIb complexes. Plant Physiol 106: 829–839

Dreyfuss BW and Thornber JP (1994b) Organization of the light-harvesting complex of Photosystem I and its assembly during plastid development. Plant Physiol 106: 841–848

Dubacq JP and Trémolières A (1983) Occurrence and function of phosphatidylglycerol containing Δ3-trans-hexadecenoic acid in photosynthetic lamellae. Physiol Vég 21: 293–312

Dubertret G, Mirshahi A, Mirshahi M and Trémolières A (1994) Evidence from in vivo manipulation of lipid composition in mutants that the Δ3-trans-hexadecenoic acid-containing phosphatidylglycerol is involved in the biogenesis of the light-harvesting chlorophyll *a/b*-protein complex of *Chlamydomonas reinhardtii*. Eur J Biochem 226: 473–482

Fragata M, Strzalka K and Nénonéné EK (1991) MgCl$_2$-induced reversal of oxygen evolution decay in Photosystem II particles incubated with phosphatidylglycerol vesicles at high lipid/Photosystem II ratio. J Photochem Photobiol 11: 329–342

Fragata M, Menikh A and Nénonéné EK (1994) Functional and structural aspects of the thylakoid lipids in oxygen evolution in Photosystem II. Trends Photochem Photobiol 3: 201–210

Garnier J, Wu B, Maroc D, Guyon J and Trémolières A (1990) Restoration of both an oligomeric form of the light-harvesting antenna CPII and of a fluorescence state II-state I transition by Δ^3-trans-hexadecenoic acid-containing phosphatidylglycerol in cells of a mutant of *Chlamydomonas reinhardtii*. Biochim Biophys Acta 1020: 153–162

Gombos Z and Murata N (1998) Genetically engineered modulation of the unsaturation of glycerolipids and its consequences in tolerance of photosynthesis to temperature stresses. In: Siegenthaler PA and Murata N (eds) Lipids in Photosynthesis: Structure, Function and Genetics, pp 249–262. Kluwer Academic Publishers, Dordrecht

Gounaris K, Whitford D and Barber J (1983) The effect of thylakoid lipids on an oxygen-evolving Photosystem II preparation. FEBS Lett 163: 230–234

Green BR, Pichersky E and Kloppstech K (1991) Chlorophyll *a/b*-binding proteins: An extended family. Trends Biochem Sci 16: 181–186

Grenier G, Guyon D, Roche O, Dubertret G and Trémolières A (1991) Modification of the membrane fatty acid composition of *Chlamydomonas reinhardtii* cultured in the presence of liposomes. Plant Physiol Biochem 29: 429–440

Hobe S, Prytulla S, Kühlbrandt W and Paulsen H (1994) Trimerization and crystallization of reconstituted light-harvesting chlorophyll *a/b* complex. EMBO J 13: 3423–3429

Hobe S, Förster R, Kingler J and Paulsen H (1995) N-proximal sequence motif in light-harvesting chlorophyll *a/b* binding protein is essential for the trimerization of light-harvesting chlorophyll *a/b* complex. Biochemistry 43: 10224–10228

Huner NPA, Krol M, Williams JP, Maissan E, Low PS, Roberts D and Thompson JE (1987) Low temperature development induces a specific decrease in trans-Δ3-hexadecenoic acid content which influences LHCII organization. Plant Physiol 89: 12–18

Huner NPA, Williams JP, Maissan EE, Myscich EG, Krol M, Laroche A and Singh J (1989) Low temperature-induced decrease in *trans*-Δ^3-hexadecenoic acid content is correlated

with freezing tolerance in cereals. Plant Physiol 89: 144–150

Ikegami I (1983) Reconstitution of antenna in P-700-enriched particles from spinach chloroplasts. Biochim Biophys Acta 722: 492–497

Izawa S and Good NE (1966) Effects of salts and electron transport on the conformation on isolated chloroplasts. II: Electron microscopy. Plant Physiol 14: 544–552

Krupa Z (1984) The action of lipase on chloroplast membranes. III. The effect of lipid hydrolysis on chlorophyll-protein complexes in thylakoid membranes. Photosynth Res 5: 177–184

Krupa Z, Williams JP, Khan MU and Huner NPA (1992) The role of acyl lipids in reconstitution of lipid-depleted light-harvesting complex II from cold-hardened and nonhardened rye. Plant Physiol 100: 931–938

Kühlbrandt W (1994) Structure and function of the plant light-harvesting complex, LHC-II. Current Opinion in Structural Biology 4: 519–528.

Kühlbrandt W and Wang DN (1991) Three-dimensional structure of plant light-harvesting complex determined by electron crystallography. Nature 350: 130–134

Kühlbrandt W, Wang DN and Fujiyoshi Y (1994) Atomic model of plant light-harvesting complex by electron crystallography. Nature 367: 614–621

Kuttkat A, Grimm R and Paulsen H (1995) Light-harvesting chlorophyll *a/b* binding protein inserted into isolated thylakoids binds pigments and is assembled into trimeric light-harvesting complex. Plant Physiol 189: 1267–1300

Larkum AWD and Anderson JM (1982) The reconstitution of a Photosystem II protein complex, P-700-chlorophyll *a*-protein complex and light-harvesting chlorophyll *a/b*-protein. Biochim Biophys Acta 679: 410–421

Lemoine Y, Dubacq JP and Zabulon G (1982) Changes in light-harvesting capacities and Δ3-hexadecenoic acid content in dark and light-grown *Picea abies*. Physiol Vég 20: 487–503

Lemoine Y, Dubacq JP, Zabulon G and Ducruet JM (1986) Organization of the photosynthetic apparatus from triazine resistant and susceptible biotypes of several plant species. Can J Bot 64: 2999–3007

Maroc J, Trémolières A, Garnier J and Guyon D (1987) Oligomeric form of the light-harvesting chlorophyll *a/b* complex CP II, phosphatidylglycerol, Δ3-trans-hexadecenoic acid and energy transfer in *Chlamydomonas reinhardtii* wild type and mutants. Biochim Biophys Acta 893: 91–99

Matsuda H and Butler WL (1983) Restoration of high-potential cytochrome *b*-559 in Photosystem II particles in liposomes. Biochim Biophys Acta 725: 320–324

McDonnel A and Staehelin LLA (1980) Adhesion between liposomes mediated by the chlorophyll *a/b* light-harvesting complex isolated from chloroplast membranes. J Cell Biol 84: 40–56

Murata N and Sato N (1978) Studies on the absorption spectra of chlorophyll *a* in aqueous dispersions of lipids from the photosynthetic membranes. Plant Cell Physiol 19: 401–410

Murata N, Ishizaki-Nishizawa O, Higashi S, Hayashi H, Tasaka Y and Nishida I (1992) Genetically engineered alteration in the chilling sensitivity of plants. Nature 356: 710–713

Murphy DJ, Crowther D and Woodrow IE (1984) Reconstitution of light-harvesting chlorophyll-protein complexes with Photosystem II complexes in soybean phosphatidylcholine liposomes. FEBS Lett 165: 151–155

Nénonéné EK and Fragata M (1990) Effects of pH and freeze-thaw on photosynthetic oxygen evolution of Photosystem II particles incorporated into phosphatidylglycerol bilayers. J Plant Physiol 136: 615–620

Nussberger S, Dörr K, Wang DN and Kühlbrandt W (1993) Lipid-protein interaction in crystals of plant light-harvesting complex. J Mol Biol 234: 347–356

Nussberger S, Dekker JP, Kühlbrandt W, van Bolhuis BM, van Grondelle R and van Amerongen H (1994) Spectroscopic characterization of three different monomer forms of the main chlorophyll *a/b* binding protein from chloroplast membranes. Biochemistry 33: 14775–14783

Paulsen H (1994) Chlorophyll *a/b*-binding proteins. Photochem Photobiol 62: 367–382

Paulsen H and Hobe S (1992) Pigment-binding properties of mutant light-harvesting chlorophyll *a/b*-binding protein. Eur J Biochem 205: 71–76

Paulsen H, Rümler U and Rüdiger W (1990) Reconstitution of pigment-containing complexes from light-harvesting chlorophyll *a/b*-binding protein overexpressed in *Escherichia coli*. Planta 181: 204–211

Peterman EJG, Hobe S, Calkoen F, van Grondelle R, Paulsen H and van Amerongen H (1996) Low-temperature spectroscopy of monomeric and trimeric forms of reconstituted light-harvesting chlorophyll *a/b* complex. Biochim Biophys Acta 1273: 171–174

Plumey FG and Schmidt GW (1987) Reconstitution of chlorophyll *a/b* light-harvesting complexes: Xanthophyll-dependent assembly and energy transfer. Proc Nat Acad Sci USA 84: 146–150

Rémy R, Trémolières A, Duval JC, Ambard-Bretteville F and Dubacq JP (1982) Study of the supramolecular organization of the oligomeric light-harvesting chlorophyll-protein (LHCP): Conversion of the oligomeric form into monomeric one by phospholipase A_2 and reconstitution with liposomes. FEBS Lett 137: 271–275

Rémy R, Trémolières A and Ambard-Bretteville F (1984) Formation of oligomeric light-harvesting chlorophyll *a/b* protein by interaction between its monomeric form and liposomes. Photobiochem Photobiophys 7: 267–276

Rigaud JL, Pitard B and Levy D (1995) Reconstitution of membrane proteins into liposomes: application to energy-transducing membrane proteins. Biochim Biophys Acta 1231: 223–246

Ryrie IJ (1986) Reconstitution of thylakoid components into artificial membranes. In: Staehelin LA and Arntzen CJ (eds) Encyclopedia of Plant Physiology, New Series, Vol 19, Photosynthesis III. Photosynthetic Membranes and Light Harvesting Systems, pp 675–682, Springer-Verlag, Berlin

Ryrie IJ and Fuad N (1982) Membrane adhesion in reconstituted proteoliposomes containing the light-harvesting chlorophyll *a/b*-protein complex: the role of charged surface groups. Arch Biochem Biophys 214: 475–488

Siefermann-Harms D, Ross JW, Kaneshiro KH and Yamamoto HY (1982) Reconstitution by monogalactosyldiacylglycerol of energy transfer from light-harvesting chlorophyll *a/b*-protein complex to the photosystems in Triton X-100-solubilized thylakoids. FEBS Lett 149: 191–196

Sieferman-Harms D, Ninnemann H and Yamamoto HY (1987) Reassembly of solubilized chlorophyll-protein complexes in proteolipid particles: comparison of monogalactosyldiacyl-

glycerol and two phospholipids. Biochim Biophys Acta 892: 303–313

Siegel GO, Jordan AE and Miller KR (1981) Addition of lipid to the photosynthetic membrane: effects on membrane structure and energy transfer. J Cell Biol 91: 113–125

Siegenthaler PA (1998) Molecular organization of acyl lipid in higher plant photosynthetic membranes. In: Siegenthaler PA and Murata N (eds) Lipids in Photosynthesis: Structure, Function and Genetics, pp 119–144. Kluwer Academic Publishers, Dordrecht

Siegenthaler PA and Trémolières A (1998) Role of acyl lipids in the function of higher plant photosynthetic membranes. In: Siegenthaler PA and Murata N (eds) Lipids in Photosynthesis: Structure, Function and Genetics, pp 145–173. Kluwer Academic Publishers, Dordrecht

Siegenthaler PA, Rawyler A and Smutny J (1989) The phospholipid population which sustains the uncoupled non-cyclic electron flow activity is localized in the inner monolayer of the thylakoid membrane. Biochim Biophys Acta 975: 104–111

Somerville C and Browse J (1991) Plant lipids: Metabolism, mutants, and membranes. Science 252: 80–87

Sprague SG, Camm EL, Green BR and Staehelin LA (1985) Reconstitution of light-harvesting complexes and Photosystem II cores into galactolipids and phospholipids liposomes. J Cell Biol 100: 552–557

Staehelin LA (1976) Reversible particle movements associated with unstacking and restacking of chloroplasts membranes in vitro. J Cell Biol 28: 278–315

Trémolières A (1991) Lipid-protein interactions in relation to light energy distribution in photosynthetic membrane of eukaryotic organism. Role of trans-Δ3-hexadecenoic acid-containing phosphatidylglycerol. Trends Photochem Photobiol 2: 13–32

Trémolières A, Dubacq JP, Ambard-Bretteville F and Rémy R (1981) Lipid composition of chlorophyll-protein complexes. Specific enrichment in trans-hexadecenoic acid of an oligomeric form of light-harvesting chlorophyll *a/b* protein. FEBS Lett 130: 27–31

Trémolières A, Roche O, Dubertret G, Guyon D and Garnier J (1991) Restoration of thylakoid appression by Δ3-trans-hexadecenoic acid-containing phosphatidylglycerol in a mutant of *Chlamydomonas reinhardtii*. Relationships with the regulation of excitation energy distribution. Biochim Biophys Acta 1059: 286–292

Trémolières A, Dainese P and Bassi R (1994) Heterogenous lipid distribution among chlorophyll-binding proteins of Photosystem II in maize mesophyll chloroplasts. Eur J Biochem 221: 721–730

Trémolières A, Dubertret G and El Maani A (1996) In vivo manipulation of lipid composition in mutants of *Chlamydomonas* affected in phosphatidylglycerol metabolism: a tool to study lipid-protein interactions during trimerization of the LHCII. In: Schmitt R, Egemann P and Harris E (eds) Seventh International Conference on The Cell and Molecular Biology of *Chlamydomonas*, S–24. Regensburg University, Germany

Van Ginkel G (1979) Photoreactivity of isolated Photosystem I upon combination with artificial lipid membranes or Triton X-100 micelles. Photochem Photobiol 30: 397–404

Vijayan P, Routaboul JM and Browse J (1998) A genetic approach to investigating membrane lipid structure and photosynthetic function. In: Siegenthaler PA and Murata N (eds) Lipids in Photosynthesis: Structure, Function and Genetics, pp 263–285. Kluwer Academic Publishers, Dordrecht

Vos R, Radunz A and Schmidt CD (1992) Binding of lipids onto polypeptides of thylakoid membrane. I. Galactolipids and sulpholipid as prosthetic groups of core peptides of the Photosystem II complex. Z Naturforsch 47: 406–415

Wada H, Gombos Z and Murata N (1990) Enhancement of chilling tolerance of a cyanobacterium by genetic manipulation of fatty acid desaturation. Nature 347: 200–203

Wollman FA and Delepaire P (1984) Correlation between changes in light energy distribution and changes in thylakoid membrane polypeptide phosphorylation in *Chlamydomonas reinhardtii*. J Cell Biol 98: 1–7

Wolter FP, Schmidt R and Heinz E (1992) Chilling sensitivity of *Arabidopsis thaliana* with genetically engineered membrane lipids. EMBO J 11: 4685–4692

Chapter 10

Lipid-Protein Interactions in Chloroplast Protein Import

Ben de Kruijff[*], Rien Pilon[1], Ron van 't Hof[2] and Rudy Demel[*]

Department of Biochemistry of Membranes, Center for Biomembranes and Lipid Enzymology, Institute of Biomembranes, Utrecht University, Padualaan 8, 3584 CH Utrecht, The Netherlands

Summary

Chloroplasts rely for their biogenesis on the import of cytosolically synthesized proteins. These proteins carry N-terminal transit sequences which are responsible for the organelle specific import. The hypothesis is presented that specific interactions between the transit sequence and chloroplast specific lipids are involved in envelope passage of the precursor proteins. Experimental evidence obtained in model systems on the transit sequence-lipid interaction in support of this hypothesis is reviewed. The specificity of the transit sequence-lipid interaction and its consequences for the structure of the interacting components are translated into a model for the early steps in chloroplast protein import. The model predicts that the combination of specific lipid-protein and protein-protein interactions are responsible for an efficient import process.

---

[1] Present address: Department of Molecular and Cell Biology, University of California, Berkeley, USA
[2] Present address: Center for Protein Technology, TNO/WAU, The Netherlands

P.-A. Siegenthaler and N. Murata (eds): Lipids in Photosynthesis: Structure, Function and Genetics, pp. 191–208.

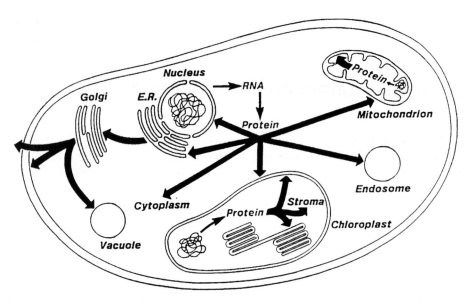

Fig. 1. A schematic representation of a plant cell in which the main protein transport routes are indicated.

I. Introduction

Proteins function at distinct sites in a cell, yet they are primarily synthesized in one compartment, the cytosol. For a concise collection of reviews, see Neupert and Lill (1992). This implies that newly synthesized proteins have to be transported to their final destination. Protein transport often involves membrane passage which is a fascinating event because large, often polar polypeptide chains of variable composition, have to move through an ultra-thin membrane without compromising the other essential barrier function of that membrane.

Within nucleated cells such as plant cells several main protein transport routes exist (Fig. 1). Cytosolically synthesized proteins directed into the endoplasmic reticulum enter the secretion pathway. Vesicular transport can carry these proteins further to compartments such as the Golgi system, the plasma

membrane, the vacuole and the outside world (Rapoport, 1992). Newly synthesized proteins can also be directed into the nucleus or stay in the cytosol. Of particular interest for this chapter is the transport route into plastids like chloroplasts, which are the most prominent organelles found in the plant cell. Chloroplasts are complex structures bounded by an envelope consisting of an outer and an inner membrane (Douce and Joyard, 1990). Within the stroma the thylakoid membrane system is present, which harbors the photosynthetic apparatus (Hall and Rao, 1987). Chloroplasts contain their own genome and have the capacity to synthesize proteins (Ellis, 1981; Sugiura, 1992). However, the coding capacity of the DNA in chloroplasts is much too limited to account for all chloroplast proteins (Umesono and Ozeki, 1987). Therefore, chloroplast biogenesis strictly depends on import of proteins synthesized in the cytosol. The same is true for mitochondria which also have a genome of insufficient coding capacity and therefore depend on protein import for growth (Douglas et al., 1986). Within the chloroplast extensive protein sorting processes have to take place because each intra-organellar compartment has its own specific set of proteins coming both from the cytosol as well as from synthesis within the stroma. It is the aim of this contribution to briefly describe the current insight into protein import into chloroplasts with special emphasis on protein passage through the envelope

Abbreviations: apoFd – apoferredoxin; C_{12}-maltose dodecyl maltose; C_{12}-PN – dodecyl phosphocholine; DGDG – digalactosyldiglyceride; DOPG – dioleoyl phosphatidylglycerol; $\Delta\Pi$ – surface pressure increase; Fd – Ferredoxin; H_{II} – inverted hexagonal phase; IM – inner membrane; MGDG – mono-galactosyldiglyceride; OM – outer membrane; Π – surface pressure; Π_i – initial surface pressure; PC – phosphatidylcholine; PG – phosphatidylglycerol; P-glycol dodecyl phosphoglycol; PI – phosphatidylinositol; preFd – precursor of ferredoxin; SQDG – sulfoquinovosyldiglyceride; trFd – transit peptide of ferredoxin; WT – wild type and to give an overview of our own research on the role specific chloroplast lipids might play in this process.

At some points a comparison will be made with mitochondrial protein import because these routes show similarities and only in plant cells operate simultaneously and yet with high fidelity. For recent comprehensive reviews on chloroplast protein import the reader is referred to De Boer and Weisbeek (1991) and Keegstra (1989).

A. Protein Import into Chloroplasts

The envelope forms the barrier for proteins imported into the chloroplast stroma. There appears to be one general import pathway for such proteins. The key features of this pathway are illustrated in Fig. 2 which pictures the import of ferredoxin (Fd) a protein which fulfills its function in electron transport during photosynthesis on the stromal side of the thylakoid membrane system (Hall and Rao, 1977). Fd follows the general import route and is synthesized in the cytosol as a precursor preferredoxin (preFd) carrying an N-terminal extension called the transit sequence (Smeekens et al., 1985). This transit sequence contains all information for chloroplast specific import. For instance gene fusion experiments revealed that it can direct foreign proteins into the chloroplast (Smeekens et al., 1987). Both chloroplast and mitochondrial precursor proteins are posttranslationally imported

into the organelle (Hay et al., 1984; Umesono and Ozeki, 1987). In vitro experiments using isolated chloroplasts revealed that preFd is efficiently imported without the aid of cytosolic factors (Pilon et al., 1992a). This might be a special property of this relatively small and polar precursor protein. More hydrophobic precursors seem to require cytosolic proteinaceous factors for import (Waegemann et al., 1990). Such helper proteins might have a chaperone type of function to maintain (or acquire) a translocation competent conformation; alternatively or in addition they could have a targeting function. PreFd is by itself an unstructured protein (Pilon et al., 1992a) and therefore might bypass involvement of cytosolic proteins. Envelope passage can be dissected into several steps. Binding of the precursor to the outer surface of the outer membrane (OM) is dependent on the transit sequence (Friedman and Keegstra, 1989) and requires low concentration of ATP (Olsen et al., 1989) and proteinaceous components on the envelope (Cornwell and Keegstra, 1987). Pretreatment of chloroplasts with proteases strongly reduces precursor binding and import (Friedman and Keegstra, 1989). What happens to the precursor between the outer surface of the outer membrane and the inner surface of the inner membrane (IM) is largely a mystery. Envelope passage of the precursor requires ATP which appears to be consumed in the stroma (Theg et al., 1989). Most likely several proteinaceous components are

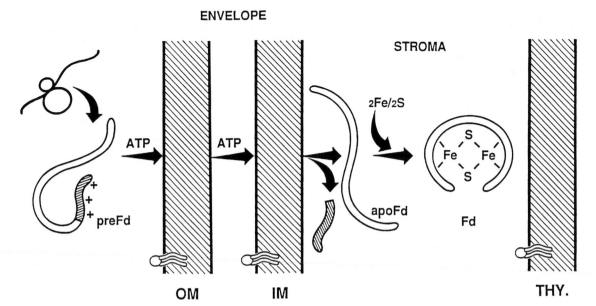

Fig. 2. The general chloroplast import pathway illustrated for ferredoxin.

involved. Several of these proteins were recently identified mainly via cross linking and affinity purification approaches (Hirsch et al, 1994; Kessler et al., 1994; Perry and Keegstra, 1994; Schnell et al., 1994). Reactive sulfhydryls are involved in import of preFd (Pilon et al., 1992a) and Cu^{2+} appears to be a selective import inhibitor because of its ability to oxidize these sulfhydryls (Seedorf and Soll, 1995). Once the precursor emerges from the inside of the inner membrane it becomes processed. A specific protease removes the transit sequence and liberates the mature protein into the stroma (Robinson and Ellis, 1984). The transit peptide is rapidly digested and it appears that the resulting products are efficiently exported out of the organelle by an at yet unknown mechanism (Van 't Hof and De Kruijff, 1995b). Once in the stroma the mature proteins can undergo further reactions finally resulting in the active molecule. ApoFd generated from the precursor is still an unfolded protein which matures upon coupling of a 2 iron-2 sulfur cluster onto four cysteines in the protein therefore folding the protein into the compact and functional Fd molecule (Pilon et al., 1992c). Proteins destined for the thylakoid systems also initially follow this general import route but subsequent sorting has to take place which occurs in the stroma and which is directed by an additional targeting signal located between the stromal targeting signal and the mature protein (Hageman et al., 1990).

The molecular mechanism of protein transport across the envelope membranes is virtually unknown but appears to involve specific recognition between the precursor and the target organelle and membrane insertion followed by membrane passage most likely through flexible tunnel-like structures. In analogy with other protein transport systems it is believed that chloroplast proteins move across the envelope as unfolded or loosely folded (Eilers and Schatz, 1986; Müller and Zimmermann, 1988) structures passing the envelope at sites of close contact between the outer and inner membrane (Schnell et al., 1990). It is assumed that each membrane has its own translocation machinery but that these can be coupled for efficient translocation.

The similarities between mitochondrial and chloroplast protein import are often emphasized. However, at least two obvious differences between these routes exist. Firstly, the targeting signals have a different architecture (see Section II). Secondly, there is a requirement of a membrane potential across the inner membrane for import of mitochondrial proteins (Gasser et al., 1982; Schleyer et al., 1982) and not for chloroplast proteins.

B. Chloroplast Lipids

The different membranes within a nucleated cell have in general a specific protein and lipid composition. This specificity is most apparent for membrane proteins. A given protein is restricted to a specific membrane. However, a membrane lipid is often found in different cellular membranes but in different concentrations. There are some exceptions to this general rule. For instance, cardiolipin is an abundant lipid found only in mitochondria (Hovius et al., 1990). Even more striking examples can be found for chloroplasts. Their membranes contain several lipid classes and species found nowhere else in the plant cell. These include the abundant galactolipids monogalactosyldiglyceride (MGDG) and digalactosyldiglyceride (DGDG) and the sulfolipid sulfoquinovosyldiglyceride (SQDG) (Douce et al., 1984). The chemical structures of MGDG and SQDG are shown in Fig. 3. As a consequence chloroplast membranes have a relatively low phospholipid content as compared to non-chloroplast membranes in the plant cell.

Figure 4 schematically illustrates the lipid composition of the outer and inner membrane of pea chloroplasts. In particular the inner membrane is very rich in galactolipids and has a composition which is similar to that of thylakoids. Zwitterionic phosphatidylcholine (PC) is an abundant lipid in the outer membrane. The main anionic lipids in the envelope are SQDG, phosphatidylglycerol (PG) and phosphatidylinositol (PI). Several chloroplast lipid classes have in addition a special fatty acid composition and distribution (Douce et al., 1984).

Only little is known about the topology of the membrane lipids across the envelope membranes. PC in the outer membrane is preferentially localized in the outer leaflet (Dorne et al., 1985). Galactolipid specific antibodies are able to recognize the galactolipids at the surface of the chloroplast (Billecocq, 1974). This demonstrates that these latter lipids are not only exposed but also accessible to large molecules. In this context it is important to recall that the outer membrane of both chloroplasts and mitochondria have a much higher lipid-to-protein ratio then other intracellular membranes including thylakoid membranes. Therefore, it is expected that extended lipid domains in these membranes face the

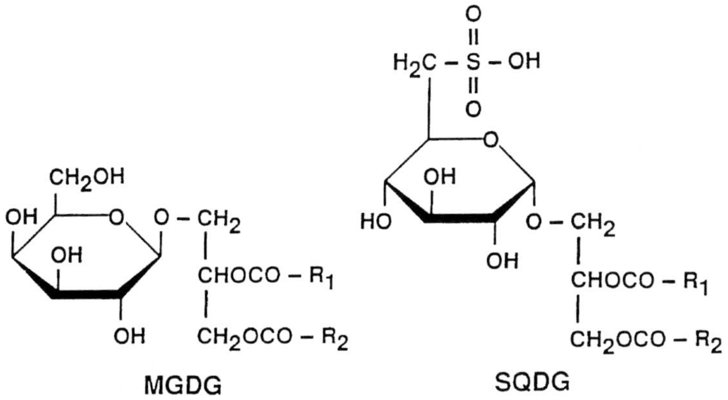

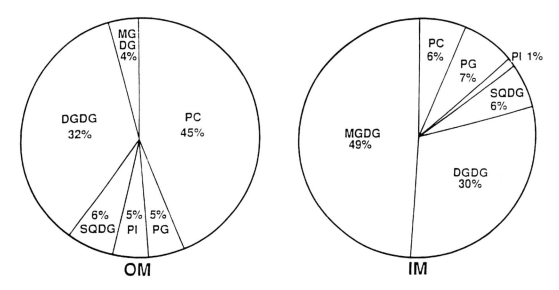

Fig. 3. The chemical structures of MGDG and SQDG.

Fig. 4. The membrane lipid composition of the outer and inner membrane of pea chloroplast (Douce et al., 1984).

cytosol. Such domains might have the ability to interact with the incoming precursor proteins which could be an important first step in the translocation process in particular if transit sequences would interact specifically with chloroplast lipids.

II. Structure and Function of Transit Sequences

Understanding chloroplast protein import will require the elucidation of the structure and function of transit sequences. From the vast number of transit sequences now available several conclusions can be drawn (Von Heijne et al., 1989). Firstly, transit sequences have a variable length ranging from 30 to 80 amino acids.

Secondly, they show little sequence similarity. Both properties are surprising because transit sequences are functionally interchangeable and they are believed to undergo specific interactions at various stages in the import process.

The characteristic features that transit sequences share are (1) a consensus sequence of the first amino acids MA which is believed to signal the removal of the initiator methionine in the cytosol; (2) the absence of negatively charged amino acids and (3) an enrichment of hydroxylated and small hydrophobic amino acids (Von Heijne et al., 1989). Three regions can be recognized: a N-terminal region of ± 15 amino acids that is uncharged; a middle region of variable length which lacks charges but which is enriched in hydroxylated amino acids; and a C-terminal region which contains positively charged amino acids. In the central region usually one or more prolines are found (Von Heijne et al., 1989).

The 47-amino acid transit sequence of preFd from *S. pratensis* is shown in Fig. 5. The hydrophobicity profile identifies four more hydrophobic regions (A,C,D,F) and two more polar regions (B and E). This amphiphilicity which is a hallmark of most targeting sequences makes the transit sequence prone to interact with membrane lipids. Theoretical analysis of secondary structure preferences of transit sequences (Von Heijne and Nishikawa, 1991) as well as circular dichroism measurements on corresponding peptides (Endo et al., 1992; Müller and Zimmermann, 1988) (transit peptides) in solution strongly suggest that transit sequences are structurally flexible polypeptides with little preference to adopt a defined structure (see also Section III.D.1).

A. Transit Sequence Domain Structure and Function

In contrast to other targeting signals such as those functional in the secretion pathway (Gierash, 1989) and in mitochondrial protein import (Roise and Schatz, 1988) no specific motifs have yet been identified in transit sequences (Von Heijne et al., 1989). Nevertheless, it can be expected that such motifs must be present in order for transit sequences to fulfill their functions. A substitution and deletion analysis of the ferredoxin transit sequence has given some insights into the importance of the various domains of the sequence for the different steps in the translocation process.

The N- and C-terminus appear to be important for targeting because small deletions in these regions strongly interfere with chloroplast binding as illustrated for some deletion mutants in Fig. 6. Large deletions in the central region are less harmful for binding but only entire transit sequence gives maximal binding. It is not surprising that impaired binding will result in impaired import. Indeed, deletions in the N- and C-terminus of the transit sequence also greatly reduce the import of the precursor (Fig. 7). This approach identified a third region from approximately residues 15-25 that is important for translocation but less for initial binding. Amino acids 26-38 constitute a region which is less essential for in vitro import. This region is rich in α-helix breaking residues and might act as a flexible linker between functional domains. Deletions in the C-terminus

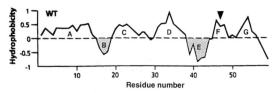

Fig. 5. The wild type (WT) transit sequence of preFd form *S. pratensis* and its hydrophobicity profile (Pilon et al., 1995). The arrow indicates the position corresponding to the processing site. The dots indicate hydroxylated residues.

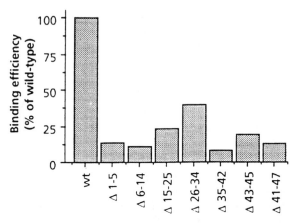

Fig. 6. Binding of wild type preFd and some transit sequence deletion mutants to chloroplasts. For further details see Pilon et al. (1995). Deletions are indicated by Δ followed by the sequence deleted.

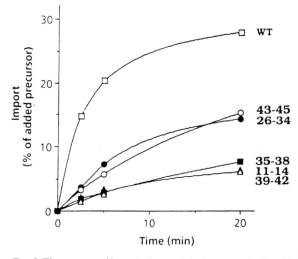

Fig. 7. Time course of import of some deletion mutants of preFd into chloroplasts. For further details see (Pilon et al., 1995).

strongly interfere with processing which is expected because the protease will have to recognize the amino acids around the cleavage site (Pilon et al., 1995). A mechanistic attractive possibility for functioning of transit sequences would be that they could open up channels through which the precursor could pass the envelope membranes. Some support for this hypothesis is obtained. Analysis of the electrical resistance of giant chloroplasts patch-clamped in the whole cell configuration revealed that the transit sequence greatly increased the envelope conductance most likely by opening translocation pores (Bulychev et al., 1994). Import of preFd does not require cytosolic factors. Therefore, the transit sequence will have to specifically recognize components of the envelope membranes. Likely candidates are protein-aceous receptors as well as the membrane lipids as will become clear in Section III.

B. Import of the Transit Peptide

One way to dissect the relative importance of the mature and transit part of a precursor protein for import is to analyze in a comparative way the import of a precursor and the corresponding transit peptide. The chemically synthesized transit peptide of ferredoxin (trFd) associates with isolated chloroplasts and becomes protease protected (Fig. 8). Protease protection is the result of its import into the organelle via the general import pathway (Van 't Hof and De Kruijff, 1995b). The rate of import of the transit peptide can be estimated to be 16.000 molecules per minute per chloroplast which is close to the maximal rate of import of the intact precursor (Pilon et al., 1992b). This suggests that for this precursor the rate limiting step is the interaction of the transit sequence with the machinery.

Like in the case of the precursor, ATP stimulates import of the transit peptide (Fig. 9). This ATP requirement for membrane translocation of a targeting signal is unique among the various cellular transport routes and supports the view that protein import into chloroplasts has very distinct features. In contrast to import of preFd, import of trFd is not inhibited by protease pretreatment of the chloroplasts (Fig. 10). The most simple explanation for this observation is that protease sensitive protein(s) on the outside of the outer membrane are involved in passage of the mature part of the precursor. Apparently no surface exposed receptor proteins are present for binding of the transit sequence. Instead it appears that the transit

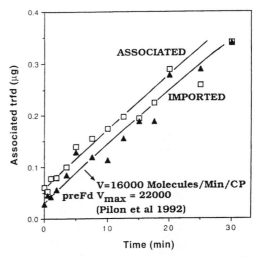

Fig. 8. Time course of association and import of trfd into chloroplasts. For details see Van 't Hof and De Kruijff (1995b).

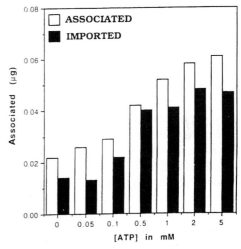

Fig. 9. ATP dependency of trfd import into chloroplasts. For details see Van 't Hof and De Kruijff (1995b).

sequence either interacts with the outer membrane lipids or with protease insensitive components of the outer membrane. The first possibility requires that the transit sequence interacts with the outer membrane lipids.

III. Transit Sequence-Lipid Interactions

A. Monolayer Insertion

Lipid extracts from biological membranes when spread at the air-water interface form monomolecular layers which are useful membrane models to analyze

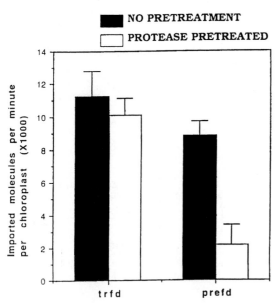

Fig. 10. Effect of thermolysin pretreatment of chloroplasts on trfd and preFd import. Mean values and the standard deviation of three individual experiments are shown. For details see Van 't Hof and De Kruijff (1995b).

the possible affinity of a protein for lipids.

When pure preFd is injected underneath such a monolayer, prepared from the total lipids of the chloroplast outer membrane, the surface pressure rapidly increases and soon reaches a stable level (Fig. 11). This demonstrates that the precursor has an interaction with its target lipids and that this results in insertion of (part of) the precursor between the lipid molecules, thereby increasing the packing density which corresponds to the change in surface pressure. The figure also demonstrates that this is not a general or aspecific effect of any protein. Injection of the unfolded apoFd or the folded holo protein does not cause a surface pressure increase at this initial surface pressure despite the fact that they have an identical polypeptide sequence as the mature part of the precursor. This strongly suggests that the lipid insertion of the precursor is due to the transit sequence. Indeed, injection of the corresponding transit peptide also causes a surface pressure increase which however is somewhat lower suggesting that the interaction of the transit sequence with lipids is enhanced by the mature part of the precursor.

B. Lipid Specificity

It could now be argued that the transit sequence itself has an aspecific mode of interaction with lipids and

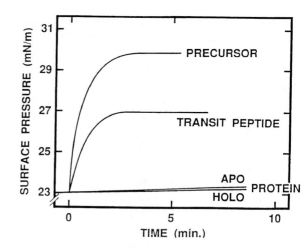

Fig. 11. Surface pressure changes induced by injection of precursor preFd, the corresponding transit peptide, apo Fd and holo Fd underneath a monolayer of the total lipid extract of the chloroplast outer envelope membrane. For details see Van 't Hof et al. (1994).

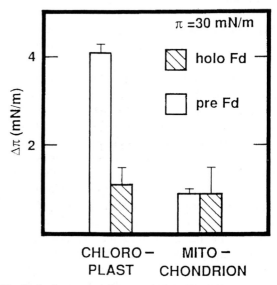

Fig. 12. Surface pressure increases induced by preFd and holo Fd when injected underneath lipid monolayers prepared from lipid extracts of chloroplast and mitochondrial outer membranes. The initial surface pressure was 30 mN/m. For details, see Van 't Hof et al. (1994).

would insert into any lipid layer. This is not the case as for instance illustrated in Fig. 12. At an initial pressure of 30 mN/m which is assumed to be the relevant pressure of biological membranes (Van 't Hof et al., 1994) an insertion dependent on the presence of the transit sequence can be observed only in the target membrane lipid extract. No transit

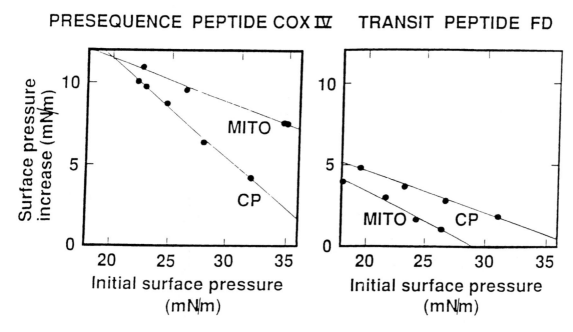

Fig. 13. Surface pressure increases induced by injection of $P_{25}L18W$ peptide (an analogue of the presequence of cytochrome oxidase subunit IV from yeast in which leu-18 is replaced by tryptophan) (left panel) and trfd (right panel) underneath monolayers of lipid extracts of chloroplast (CP) and mitochondrial outer membranes (MITO). The initial surface pressure is changed in the range indicated and the maximal polypeptide induced surface pressure increase is measured. For further details, see Van 't Hof et al. (1994).

sequence dependent insertion occurs when the precursor is injected underneath a monolayer of the lipid extract of mitochondrial membranes. The mitochondrial lipid extract was chosen for such a comparison because protein import in mitochondria also occurs posttranslationally and is mediated by the presequence, which is also an amphiphatic targeting sequence. If specific targeting sequence-lipid interactions would occur between the two systems, then these could contribute to organelle specific targeting.

A comparison of monolayer insertion of a typical presequence peptide and the transit peptide of Fd into their target lipid extracts is shown in Fig. 13 in the form of a Πi-$\Delta\Pi$ plot. The initial surface pressure (Πi) is varied and the peptide-induced surface pressure increase ($\Delta\Pi$) is measured. Increasing the initial surface pressure results in a closer lipid packing and concordingly the peptide causes a smaller increase in surface pressure. Like for the precursor, also the transit peptide preferentially inserts into its target lipids (right panel) because the largest surface pressure increases are observed for the chloroplast lipids. Extrapolation of the Πi-$\Delta\Pi$ plot to $\Delta\Pi=0$ provides the limiting insertion pressure which is around 36 mN/m for insertion of the transit peptide

into the chloroplast outer membrane extract. At the 'physiological' surface pressure of around 30 mN/m no insertion can be detected in the mitochondrial lipid extract. The presequence peptide has a much stronger capacity to insert into lipid layers because the resulting surface pressure increases are much higher (left panel). This might be due to the stronger amphiphilicity of the presequence peptide. Also the presequence peptide preferentially inserts into its target lipid extract, suggesting that indeed targeting sequence-lipid interactions could contribute to targeting of the precursor to the correct organelle. The preferential association of the targeting sequence with the lipid extract from the target membrane could be the result of either the overall properties of the lipid mixture or the presence of specific lipid classes. A way to discriminate between these possibilities is to analyze the peptide-lipid interactions for the various main lipid classes found in the outer membranes (Fig. 14). The transit peptide inserts most efficiently in monolayers of PG, SQDG and MGDG suggesting that these lipid classes are mainly responsible for insertion into the target lipid extract and that the transit sequence-lipid interaction displays electrostatic (the transit peptide is positively charged, whereas PG and SQDG are negatively charged) but

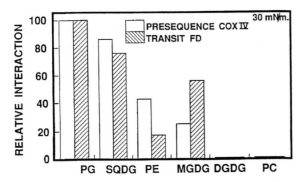

Fig. 14. The interaction of targeting peptides with monolayers prepared from several individual lipid classes. The dioleoyl species of PG, PE and PC are compared to SQDG, MGDG and DGDG from chloroplasts and the resulting surface pressure increases observed at an initial surface pressure of 30 mN/m are normalized to the value observed for PG. For further details, refer to the legend of Fig. 13 and to Van 't Hof et al. (1994).

also other contributions. The strong interaction with the neutral MGDG is particularly intriguing. The transit peptide does not show an interaction with DGDG nor with PC which is zwitterionic but also has no net charge overall. Comparing the overall pattern of lipid insertion of the transit and the presequence peptide teaches that the anionic lipids are the strongest determinants for lipid insertion and that PE (which is an abundant lipid in mitochondria) and MGDG contribute most to the specific insertion of the peptides into the target lipid extract.

Peptides corresponding to the various parts of the transit sequence of the precursor of the small subunit of ribulose-1,5-biphosphate carboxylase/oxygenase display a similar lipid specific insertion as for preFd (Van 't Hof et al., 1991) which suggests that this is a general property of transit sequences. The interaction of preFd with MGDG showed some unexpected features. MGDG from chloroplasts is highly unsaturated and model membranes made from this lipid will be in a fluid state. Yet preFd also preferentially interacts with MGDG in the gel state (Demel et al., 1995) which indicates the involvement of specific head group interactions of which hydrogen bonding between the galactose group and the hydroxylated amino acids in the transit sequence are an obvious possibility. Surprisingly, the conformation of the sugar moiety and the sugar-glycerol linkage is of little influence in the preFd-monolayer interaction. However, methylation of position 3 of the galactose largely inhibits the interaction (Demel et al, 1995).

The mode of insertion of preFd into MGDG is also

different from that into PG. In the glycolipid the penetrated domain of the precursor occupies an area of 650Å² when interaction was initiated at an initial surface pressure of the monolayer of 20 mN/m (Van 't Hof and De Kruijff, 1995a). Despite the stronger interaction with PG (compare in Fig. 14) the inserted domain only amounts to 400Å². This demonstrates that larger parts of the precursor insert into the glycolipid layer. The same specific affinity of the transit sequence for anionic lipids and MGDG was also observed for lipid bilayers in the form of large unilamellar vesicles (Van 't Hof and De Kruijff, 1995a). This study also established that the preFd-bilayer interaction did not lead to a loss of barrier function of the bilayer and that the affinity of preFd for the lipid extract is less than the affinity of a precursor for chloroplasts.

A final intriguing aspect of the preFd-MGDG interaction is shown in Fig. 15. Very low concentrations of MGDG in PC monolayers already greatly enhance insertion of the transit peptide. The outer membrane of the chloroplast envelope contains such a low concentration of this lipid that it could be sufficient for efficient and specific insertion of the transit sequence between the lipids. This result suggests that the head group of MGDG could function as an insertion site for the transit sequence. In Section III.B.2 the specific packing properties of mixed PG-

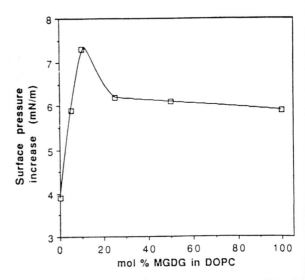

Fig. 15. Effect of the MGDG concentration in DOPC (dioleoyl phosphatidylcholine) monolayers on the interaction with trfd. The initial surface pressure was 20 mN/m. Reproduced with permission from Van 't Hof et al. (1994). See that reference for further details.

MGDG systems will be described. The precise nature of the transit sequence-MGDG interaction remains to be determined.

C. Localization of Insertion Domains

In a search to localize the domains within the transit sequence which are involved in lipid insertion a selection of transit sequence deletion mutants of preFd were expressed in *E. coli* and purified (Pilon et al., 1995). The functionality of the resulting proteins was tested via their ability to compete for import of in vitro-synthesized wild type precursor into chloroplasts. The results obtained were fully consistent with the import studies of the in vitro synthesized deletion mutants (Pilon et al., 1995) as they were summarized in section II.A. The precursors were analyzed for their ability to insert into monolayers of either a total lipid extract of the outer envelope membrane of chloroplasts, MGDG or DOPG (Fig. 16). Most deletions cause a decrease in insertion ability into the lipid extract. This is not due to just a shortening of the transit sequence because the largest deletion tested (Δ15-25) inserts even more

efficiently than the wild type protein. This already suggests that specific domains of the transit sequence are responsible for insertion into the target lipid extract. The limiting surface pressure for insertion of WT preFd is around 35 mN/m (Van 't Hof et al., 1994) which is close to the value observed for the transit peptide (see Fig. 13). This is increased to 38 mN/m for Δ15-25 but is decreased to around 30 mN/m for the N-terminal (Δ6-14) and C-terminal (Δ41-47) deletion mutants (data not shown). This implies that these mutants would be unable to penetrate a biological membrane. Because these mutants also have almost completely lost the ability to bind to chloroplasts (Fig. 6) the obvious suggestion is that these domains are involved in initial binding of the precursor to the surface of the chloroplast.

Interestingly Δ15-25 which is completely unable to import (Pilon et al., 1995) but which still has considerable activity to bind to chloroplasts also interacts strongly with lipids suggesting that the 15-25 domain is involved in translocation but less in binding. Analysis of the insertion of the deletion mutants into the MGDG and PG monolayers allowed to identify the regions responsible within the transit

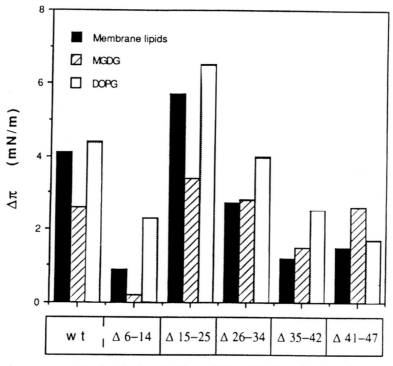

Fig. 16. Surface pressure increases caused by the injection of wild type (WT) preFd and a series of transit sequence deletion mutants of this precursor underneath monolayers of the indicated composition. Total membrane lipids were extracted from chloroplast outer membranes, MGDG was isolated from chloroplasts. The surface pressure increases at an initial surface pressure of 28 mN/m were determined from $\Pi_i/\Delta\Pi$ diagrams. For further details, see Pilon et al. (1995).

sequence for the preferential interaction with these lipids (Fig. 16). The wild type precursor inserts both in the PG and the MGDG monolayer with a preference for the former lipid as already discussed in previous sections. This pattern of specificity is also seen for the deletion mutants Δ15-25, Δ26-34 and Δ35-42. The N- and most C-terminal deletion mutants are clearly exceptional. Removal of amino acid sequence 6-14 nearly completely eliminates the interaction with MGDG which identifies this region as a putative MGDG recognition domain. In contrast, removal of amino acid residues 41 to 47 shifts the insertion preference mainly to MGDG which identifies this C-terminal region, which contains one lysine and two arginine residues, as the putative domain interacting with the negatively charged lipids.

D. Structural Consequences

Up till now the nature and the extend of the transit-sequence-lipid interaction has been described. It can be expected that as a result of the interaction the structure of the interacting species changes. Insight into such structural changes is likely to provide insight into the function of these interactions for the import process. Both aspects of the structure of the transit sequence and the structure of the lipids will be considered.

1. Transit Sequence

The transit peptide of ferredoxin is like the complete precursor a relatively unstructured polypeptide as can be inferred from a circular dichroism analysis of solutions of these polypeptides (Pilon et al., 1992). Upon interaction with detergent micelles and membrane lipids in small unilamellar vesicles the transit sequence undergoes a two-state random coil-helix transition (Horniak et al., 1993). The CD spectra can be deconvoluted to estimate the α-helix content. Figure 17 illustrates some typical results for a selection of detergents studied. All detergents had an identical acyl chain but differed in their head group. The phosphoglycol (Pglycol) head group mimics the head group of PG, C_{12}-PN has like PC a phosphocholine head group. C_{12}-maltose carries a sugar head group which models the MGDG head group. The figure shows that α-helix formation is induced by C_{12}-Pglycol and to a lesser extent also by C_{12}-PN. This random coil-helix transition thus requires the presence of a lipid-water interface. C_{12}-maltose

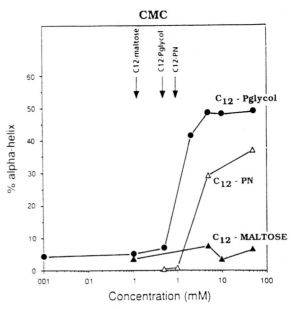

Fig. 17. Effect of different detergents on the α-helix content of trFd in buffer. The arrows indicate the critical micelle concentrations (cmc) of the detergents used and the α-helix is estimated from circular dichroism spectra. For details, see Horniak et al. (1993).

micelles do not increase the α-helix content. Anionic detergents like C_{12}-Pglycol or lipids like PG and SQDG are the strongest α-helix promoters. They can convert up to 50% of the polypeptide into a helical conformation. The location of the induced α-helical parts within the sequence is not known. Speculations on its location are hampered by lack of clear predictions on the secondary structure. Secondary structure prediction programs fail to identify regions with high propensity to form defined secondary structures in the transit sequence of ferredoxin (Pilon et al., 1992c) as well as in other transit sequences (Von Heijne and Nishikawa, 1991). The only precise information on transit sequence structure comes from a 2D-^{1}H NMR analysis of the transit peptide of ferredoxin from *Chlamydomonas reinhardtii* (Lancelin et al., 1994). In water this transit peptide is also largely in a random coil conformation but in a more apolar environment the α-helix content is increased. The α-helix was found to be localized in the N-terminus. Transit sequences from this primitive organism are small and have characteristics of both mitochondrial presequences and higher plant transit sequences. Therefore, it can be questioned whether this result can be extrapolated to the transit sequence of ferredoxin from *S. pratensis*. But if this is the case an interesting property of the transit sequence is revealed (Fig. 18). A helical conformation of the first 12 amino acids of the transit sequence of ferredoxin would generate an amphipatic helix of special design. One side of the helix would be hydrophobic whereas the other would be formed by the hydroxyl group carrying serine and threonine residues. Such a helix would have a strong potential for interaction with glycolipids because the hydroxyl rich site of the helix could participate in the hydrogen binding network of the glycolipid head groups. Such a putative amphipathic helix can also be identified in the N-terminus (residues 6-13) of the small subunit of ribulose-1,5-biphosphate carboxylase/oxygenase. Whether this is a more general property of transit sequences is unclear. The observed structural flexibility of transit sequences is consistent with the view that transit sequences function in import by acquiring different structures in various steps of the envelope translocation process. Future research will have to give a more precise picture of the conformation of transit sequences.

2. Lipids

Given the specific interaction between the transit sequence and MGDG it is useful to consider in more detail the properties of this lipid and then to discuss the consequences of the transit sequence-MGDG interaction for the structure of the lipid. MGDG from plants is highly enriched in linolenic (18:3) acid. This polyunsaturation makes that MGDG/water systems will be in a disordered (fluid) state at all temperatures. The galactose head group is small and appears to be oriented in such a way that it is extended away from the membrane surface (Howard and Prestigard, 1995). Moreover, the head groups have a strong tendency to interact intermolecularly via hydrogen bonding. These properties are responsible for the phase properties of this lipid. In aqueous dispersion MGDG from plants does not organize in bilayers but instead forms an inverse hexagonal phase, H_{II} phase (Shipley et al., 1973). This phase consists of hexagonally arranged tubes in which the lipid head groups face inward and line the narrow aqueous channels present in the tubes. The geometry of the phase is such that the headgroups can be in close interaction whereas the unsaturated acyl chain can occupy a much more extended area.

MGDG can be considered a typical non-bilayer lipid. Its presence in a membrane can be expected to

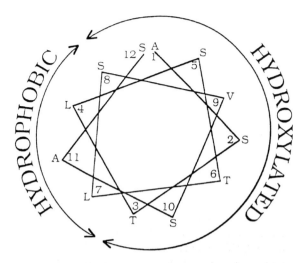

Fig. 18. Helical wheel representations of the 1-12 sequence of the transit sequence of ferredoxin.

destabilize the bilayer. Bilayer preferring lipids like DGDG, PC and PG stabilize a bilayer structure of the MGDG. Membranes very rich in MGDG like the inner chloroplast envelope membrane and the thylakoid membrane system will form relatively unstable structures because they are close to a phase transition (bilayer $\rightarrow$ non-bilayer). For thylakoid systems it has been described that a slight increase in temperature triggers the induction of the H_{II} phase.

There have been many suggestions on the functional role of non-bilayer lipids in membranes (Cullis and De Kruijff, 1979) including their involvement in membrane passage of proteins. Recently, the first experimental evidence is obtained that non-bilayer lipids are important for efficient protein transport across the *E. coli* inner membrane (Rietveld et al., 1995). The general belief is that the lipid packing properties arising from the presence of non-bilayer lipids in membranes are important for proper integration of proteins into membranes.

^{2}H NMR studies on mixed model membranes of ^{2}H acyl chain labeled synthetic MGDG and PC revealed another interesting property of MGDG (Chupin et al., 1994). ^{2}H NMR can provide information on the order of specific segments of a lipid molecule in an aqueous dispersion of that lipid. MGDG and PC were synthesized with oleoyl chains in which on the 11-position the protons were replaced by deuterons. The peak separation in the characteristic doublet type of the ^{2}H NMR spectrum is the residual quadrupolar splitting which is a measure of the local order. For pure $^2H_4$-PC bilayers a quadrupolar splitting of ~6.2 kHz is observed (see Fig. 19). Pure $^2H_4$-MGDG in water does not form bilayers but is in a H_{II} phase. The quadrupolar splitting of the ^2H NMR spectrum of the lipid in that phase is ~2kHz (Chupin et al., 1994). This small value reflects both the rapid diffusion of the MGDG molecules around the tubes of the H_{II} phase and the geometry of the phase allowing more disordered acyl chain conformations.

MGDG can be incorporated in PC in bilayers up to 60 mol%. By preparing two samples in which each lipid is alternatively labeled, insight into the individual behavior of each lipid can be obtained (Fig. 19). The quadrupolar splitting of $^2H_4$-MGDG increases with the MGDG concentration. Extrapolation predicts that the quadrupolar splitting of the acyl chains in a hypothetical pure MGDG bilayer would be around 8 kHz which is higher than the quadrupolar splitting of pure PC bilayer and reflects a tight acyl chain packing in a MGDG bilayer. This is exactly according to expectation. Because of the strong inter head group interactions and the unsaturated acyl chains the MGDG/water interface will try to acquire a convex (inverted) curvature but this is counteracted by the opposed monolayer resulting in a 'frustrated' bilayer in which the acyl chains are more squeezed towards each other resulting in a more ordered system with larger quadrupolar splittings. The surprise lies in the fact that the quadrupolar splitting of $^2H_4$-MGDG in the mixture with PC is *smaller* than the value observed for PC in the mirror sample. If the mixture would behave in an ideal way then an *identical* quadrupolar splitting would be expected which would increase in a linear way with the molar fraction of MGDG. Alternatively, if the two molecules in the mixture would maintain more their individual behavior the quadrupolar splitting of MGDG would be expected to be *larger* than that of PC.

How can this unexpected behavior be explained? It was proposed (Chupin et al., 1994) that the two lipid molecules in the mixed bilayer are vertically displaced such that the PC head group would extend more into the aqueous phase thereby displacing the acyl chain between the two lipid classes. The 11-position of the acyl chain on MGDG would be located more towards the methyl end of the oleoyl chain on PC and thereby sense a more disordered environment causing a decrease in quadrupolar splitting. Figure 20 illustrates this model. An intriguing possibility provided by the irregular hydrocarbon-polar interface is that it might provide

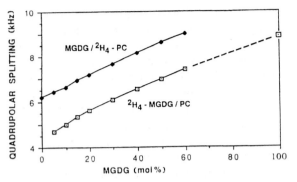

Fig. 19. Effect of the MGDG concentration on the quadrupolar splitting of mixed $^2H_4$-MGDG/PC (□) and MGDG/$^2H_4$-PC (◆) dispersions. The values were obtained from 46.1 MHz ^2H NMR spectra of aqueous dispersions of these lipids. Reproduced from Chupin et al. (1994).

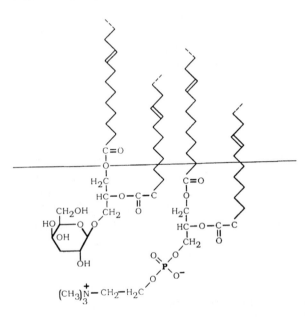

Fig. 20. A model for the organization of mixed PC-MGDG bilayers. The line is to indicate the direction of the surface of the bilayer. Both lipids are depicted with part of two oleoyl chains including the double bond between the 9-10 carbon position. The figure merely illustrates the vertical displacement of the two molecules. The precise conformation of the molecules is not known.

potential insertion sites for the transit sequence. When the precursor of ferredoxin interacts with such mixed MGDG/PC bilayers a transit sequence dependent reorganization of both lipid molecules was revealed by ^{31}P and ^2H NMR spectroscopy (Chupin et al., 1994). The precursor induces a bilayer → isotropic transition for part of the lipid molecules. The isotropic structure was interpreted to be of an inverted nature. Isotropic structures are commonly encountered as intermediates between the bilayer and the H_{II} phase. This observation demonstrates that the transit sequence promotes the formation of type II non-bilayer lipid structures in this system. Extrapolation of this observation to chloroplast protein import leads to the view that upon interaction of the transit sequence with the membrane lipids, the lipid packing equilibrium is shifted towards a non-bilayer situation which could be important for the translocation process.

IV. A Model for Import

The knowledge of transit sequence-lipid interactions summarized in this overview can be integrated in a model for import of precursor proteins into chloroplasts. The model pictured in Fig. 21 solely concentrates on transit sequence-lipid interactions in the outer chloroplast envelope membrane. Also, in the inner membrane such interactions might occur which similarly could contribute to translocation and processing.

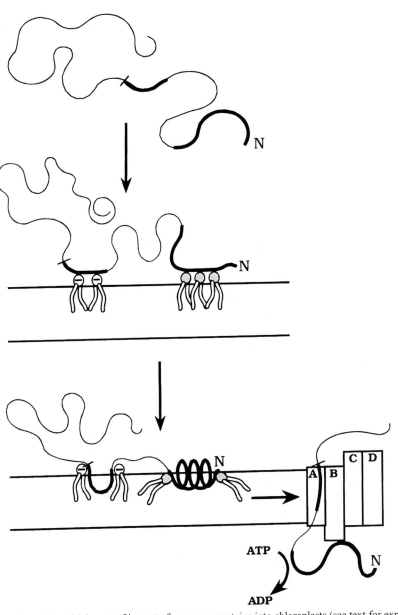

Fig. 21. A model of the initial stage of import of precursor proteins into chloroplasts (see text for explanation).

The model proposes that the newly synthesized precursor initially interacts via its transit sequence with cytosol-exposed lipid domains in the outer membrane. The interaction is twofold. The C-terminus of the transit sequence is assumed to interact with anionic lipids whereas the N-terminus interacts preferentially with chloroplast specific MGDG. The interaction is followed by insertion of both domains into the lipid layer. This is accompanied by the induction of a specific folding pattern and membrane topology of the transit sequence possibly involving an N-terminal helix. The local lipid structure is changed towards a non-bilayer organization. The membrane-inserted transit sequence encounters via two-dimensional diffusion the proteins of the transport apparatus. Here the transit sequence docks on the translocator complex with high affinity resulting in a productive interaction. The transit sequence translocates across the outer membrane in an ATP dependent manner. It is proposed that for these steps two 75 kDa protein (A or B) are responsible. These proteins were identified via cross linking and affinity purification approaches (Perry and Keegstra, 1994; Schnell et al., 1994) and do not expose domains to the cytosol. The B component is an integral protein and a member of the heatshock protein 70 family (Schnell et al., 1994) it exposes a domain to the inter membrane space and might be a protein responsible for the ATP dependency of protein translocation. The precursor is shown to interact with two other components of the translocation complex i.e. the 34 and 86 kDa proteins (Kessler et al., 1994). Both are GTP binding proteins and expose domains to the cytosol. They might act as unfoldases and reductases. The translocation complex might act as a pore to allow passage of the precursor across the outer membrane. In the inner membrane a separate translocation complex might be present which facilitates translocation of the precursor across that membrane. The main attractions of the model are that it gives some insight into the way transit sequences function in the early steps of the pathway. The model explains organelle specific targeting of the precursor as achieved by a combination of specific protein-lipid and protein-protein interactions and provides a mechanism for efficient delivery of the incoming precursor to the translocation complex. Two-dimensional diffusion over the membrane surface can deliver the precursor three orders of magnitude faster to a receptor then three-dimensional diffusion through the aqueous phase.

V. Concluding Remarks

Much attention has been paid in recent years to the lipid composition of chloroplasts in relation to chloroplast functioning and in particular to photosynthesis. It has been clearly established as described in other chapters of this book (Chapter 2, Joyard et al.; Chapter 6, Williams; Chapter 8, Siegenthaler and Trémolières; Chapter 9, Trémolières and Siegenthaler) that a specific and well controlled composition is vital to optimal functioning of the chloroplasts.

Research from our group over the past years has allowed to identify another possible important role for chloroplast lipids. They interact specifically with the targeting sequences of proteins to be imported into chloroplasts. This suggests that they play an important role in the biogenesis of the organelle. The present view on the translocation apparatus and the transit sequence structure and function is still very limited. However, recent advances such as isolation of translocation complexes (Kessler et al., 1994; Perry and Keegstra, 1994; Schnell et al., 1994), introduction of multi-dimensional NMR techniques to determine the structure of lipid-associated amphipathic peptides (Lancelin et al., 1994), the further developments in plant genetics and the possibility to analyze the chloroplast envelope directly via electrophysiological methods (Bulychev et al., 1994) offer the promise that in the near future a much more detailed picture of the process and the lipid-protein interactions involved will emerge.

Acknowledgments

Most of the research described in this review is the result of the collaboration between the groups of Peter Weisbeek and Ben de Kruijff on chloroplast protein import. The research was supported by grants from the University of Utrecht, the Institute of Biomembranes, the Life Science Foundation and the Netherlands Organization for Research. Monique Regter is thanked for preparing the manuscript and Nico van Galen for several of the drawings.

References

Billecocq A (1974) Structures of biological membranes: Localization of galactosyldiglycerides in chloroplasts by means of specific antibodies. Biochim Biophys Acta 352: 245–251

Bulychev A, Pilon M, Dassen H, Van 't Hof R, Vredenberg W and De Kruijff B (1994) Precursor-mediated opening of translocation pores in chloroplast envelopes. FEBS Lett 356: 204–206

Chupin V, Van 't Hof R and De Kruijff B (1994) The transit sequence-dependent binding of the chloroplast precursor protein ferredoxin to lipid vesicles and its implications for membrane stability. FEBS Lett 350: 104–108

Cornwell KL and Keegstra K (1987) Evidence that a chloroplast surface protein is associated with a specific binding site for the precursor to the small subunit of ribulose-1,5-biphosphate carboxylase. Plant Physiol 85: 780–785

Cullis PR and De Kruijff B (1979) Lipid polymorphism and the functional roles of lipids in biological membranes. Biochim Biophys Acta 559: 399–420

De Boer AD and Weisbeek PJ (1991) Chloroplast topogenesis: Protein import, sorting and assembly. Biochim Biophys Acta 1071: 221–253

Demel RA, De Swaaf ME, Mannock D, Van 't Hof R and De Kruijff B (1995) The specificity of glycolipid-preferredoxin interaction. Mol Membr Biol 12: 255–261

Dorne A-J, Joyard J, Block MA and Douce R (1985) Do thylakoids really contain phosphatidylcholine. Proc Natl Acad Sci USA 87: 71–74

Douce R and Joyard J (1990) Biochemistry and function of the plastid envelope. Annu Rev Cell Biol 6: 173–216

Douce R, Block MA, Dorne, A-J and Joyard, J (1984) The plastid envelope membranes: their structure, composition and role in chloroplast biogenesis. Subcell Biochem 10: 1–84

Douglas MG, McCammon MT and Vassaroti A (1986) Targeting proteins into mitochondria. Microbiol Rev 50: 166–178

Eilers M and Schatz G (1986) Binding of a specific ligand inhibits import of a purified precursor protein into mitochondria. Nature 322: 228–232

Ellis RJ (1981) Chloroplast proteins: synthesis, transport and assembly. Annu Rev Plant Physiol 32: 111–137

Endo T, Kawamura M and Nakai M (1992) The chloroplast-targeting domain of a plastocyanin transit peptide can form a helical structure but does not have a high affinity for lipid bilayers. Eur J Biochem 207: 671–675

Friedman AL and Keegstra K (1989) Quantitative analysis of precursor binding. Plant Physiol 89: 993–999

Gasser SM, Daum G and Schatz G (1982) Import of proteins into mitochondria. Energy-dependent uptake of precursors by isolated mitochondria. J Biol Chem 257: 13034–13041

Gierash LM (1989) Signal sequences. Biochemistry 28: 923–930

Hageman J, Baecke C, Ebskamp M, Pilon M, Smeekens S and Weisbeek PJ (1990) Protein import into and sorting inside the chloroplast are independent processes. Plant Cell 2: 479–494

Hall PO and Rao KK (1977) Ferredoxin. In: Pirson A and Zimmerman MH (eds.) Encyclopedia of Plant Physiology 5 (pp. 206–215) Springer Verlag Berlin, Heidelberg, New York

Hall DQ and Rao KK (1987) Photosynthesis. New studies in biology, 4th Ed. Edward Arnold Ltd, London

Hay R, Bahni P and Gasser S (1984) How mitochondria import proteins. Biochim Biophys Acta 779: 65–87

Hirsch S, Michael E, Heemeyer F, Von Heijne G and Soll J (1994) A receptor component of the chloroplast protein translocation machinery. Science 266: 1989–1992

Horniak L, Pilon M, Van 't Hof R and De Kruijff B (1993) The secondary structure of the ferredoxin transit sequence is modulated by its interaction with negatively charged lipids. FEBS Lett 334: 241–246

Hovius R, Lambrechts H, Nicolay K and De Kruijff B (1990) Improved methods to isolate and subfractionate rat liver mitochondria. Lipid composition of the inner and outer membrane. BBA 1021: 217–226

Howard KP and Prestigard JH (1995) Membrane and solution conformations of monogalactosyldiacylglycerol using NMR/molecular modeling methods. J Am Chem Soc 117: 5031–5040

Joyard J, Maréchal E, Miège C, Block MA, Dorne AJ and Douce R (1997) Structure, distribution and biosynthesis of glycerolipids. In: Siegenthaler PA and Murata N (eds) Lipids in Photosynthesis: Structure, Function and Genetics, pp 21–52. Kluwer Academic Publishers, Dordrecht

Keegstra K (1989) Transport and routing of proteins into chloroplasts. Cell 56: 247–253

Kessler F, Blobel G, Patel HA and Schnell DJ (1994) Identification of two GTP-binding proteins in the chloroplast protein import machinery. Science 266: 1035–1039

Lancelin J-M, Bally I, Arland GJ, Blackedge M, Gans P, Stein M and Jacquot J-P (1994) NMR structure of ferredoxin chloroplastic transit peptide from Chlamydomonas reinhardtii promoted by trifluoroethanol in aqueous solution. FEBS Lett 350: 104–108

Müller G and Zimmermann R (1988) Import of honeybee prepromelittin into the endoplasmic reticulum: energy requirements for membrane insertion. EMBO J 7: 639–648

Neupert W and Lill R (1992) Membrane biogenesis and protein targeting. Elsevier, Amsterdam

Olsen LJ, Theg SM, Selman BR and Keegstra K (1989) ATP is required for the binding of precursor proteins to chloroplasts. J Biol Chem 264: 6724–6729

Perry SE and Keegstra K (1994) Envelope membrane proteins that interact with chloroplast precursor proteins. The Plant Cell 6: 93–105

Pilon M, De Kruijff B and Weisbeek PJ (1992a) New insights into the import mechanism of the ferredoxin precursor into chloroplasts. J Biol Chem 267: 2548–2556

Pilon M, Weisbeek PJ and De Kruijff B (1992b) Kinetic analysis of translocation into isolated chloroplasts of the purified ferredoxin precursor. FEBS Lett 302: 65–68

Pilon M, Rietveld AG, Weisbeek PJ and de Kruijff B (1992c) Secondary structure and folding of a functional chloroplast precursor protein. J Biol Chem 267: 19407–19413

Pilon M, Wienk H, Sips W, De Swaaf ME, Talboom F, Van 't Hof R, De Korte-Kool G, Weisbeek PJ and De Kruijff B (1995) Functional domains of the ferredoxin transit sequence involved in chloroplast import. J Biol Chem 270: 3882–3893

Rapoport TA (1992) Transport of proteins across the endoplasmic reticulum membrane. Science 258: 931–936

Rietveld AG, Koorengevel MC and De Kruijff B (1995) Non-bilayer lipids are required for efficient protein transport across the plasma membrane of Escherichia coli. EMBO J 14: 5506–5513

Robinson G and Ellis RJ (1984) Transport of proteins into chloroplasts. Partial purification of a chloroplast protease involved in the processing of imported precursor polypeptides. Eur J Biochem 142: 337–342

Roise D and Schatz G (1988) Mitochondrial presequences. J Biol

Chem 263: 4509–4511

Sanders SL and Schekman R (1992) Polypeptide translocation across the endoplasmic reticulum membrane. J Biol Chem 267, 13791–13794

Schleyer M, Schmidt B and Neupert W (1982) Requirement of a membrane potential for the posttranslational transfer of proteins into mitochondria. Eur J Biochem 125: 109–116

Schnell DJ, Blobel G and Pain D (1990) The chloroplast import receptor is an integral membrane protein of chloroplast envelope contact sites. J Cell Biol 111: 1825–1838

Schnell DJ, Kessler F and Blobel G (1994) Isolation of components of the chloroplast protein import machinery. Science 266: 1007–1012

Seedorf M and Soll J (1995) Copper chloride, an inhibitor of protein import into chloroplasts. FEBS Lett 367: 19–22

Shipley GG, Green JP and Nichols BW (1973) The phase behavior of monogalactosyl, digalactosyl and sulphoquinovosyl. Biochim Biophys Acta 311: 531–544

Siegenthaler PA and Trémolières A (1997) Role of acyl lipids in the function of higher plant photosynthetic membranes. In: Siegenthaler PA and Murata N (eds) Lipids in Photosynthesis: Structure, Function and Genetics, pp 145–173. Kluwer Academic Publishers, Dordrecht

Smeekens S, van Binsbergen J and Weisbeek PJ (1985) The plant ferredoxin precursor: nucleotide sequence of a full length cDNA clone. Nucleic Acid Res 13: 3179–3194

Smeekens S, van Steeg H, Bauerle C, Bettenbroek H, Keegstra K and Weisbeek PJ (1987) Import into chloroplasts of a yeast mitochondrial protein directed by ferredoxin and plastocyanin transit peptides. Plant Mol Biol 9: 377–388

Sugiura M (1992) The chloroplast genome. Plant Mol Biol 19: 149–168

Theg SM, Bauerle C, Olsen LJ, Selman BR and Keegstra K (1989) Internal ATP is the only energy requirement for the translocation of precursor proteins across chloroplastic membranes. J Biol Chem 264: 6730–6736

Trémolières A and Siegenthaler PA (1997) Reconstitution of photosynthetic structures and activities with lipids. In: Siegenthaler PA and Murata N (eds) Lipids in Photosynthesis: Structure, Function and Genetics, pp 175–189. Kluwer Academic Publishers, Dordrecht

Umesono K and Ozeki H (1987) Chloroplast gene organization in plants. Trends Genet 3: 281–287

Van 't Hof R and De Kruijff B (1995a) Transit sequence-dependent binding of the chloroplast precursor protein ferredoxin to lipid vesicles and its implications for membrane stability. FEBS Lett 356: 204–206

Van 't Hof R and De Kruijff B (1995b) Characterization of the import process of a transit peptide into chloroplasts. J Biol Chem 270: 22368–22373

Van 't Hof R, Demel RA, Keegstra K and De Kruijff B (1991) Lipid-peptide interactions between fragments of the transit peptide of ribulose-1,5-bisphosphate carboxylase/oxygenase and chloroplast membrane lipids. FEBS Lett 291: 350–354

Van 't Hof R, Van Klompenburg W, Pilon M, Kozubek A, De Korte-Kool G, Demel RA, Weisbeek PJ and De Kruijff B (1994) The transit sequence mediates the specific interaction of the precursor of ferredoxin with chloroplast envelope membrane lipids. J Biol Chem 268: 4037–4042

Von Heijne G and Nishikawa K (1991) Chloroplast transit peptides. The perfect random coil? FEBS Lett 278: 1–3

Von Heijne G, Steppuhn J and Herrmann RG (1989) Domain structure of mitochondrial and chloroplast targeting peptides. Eur J Biochem 180: 535–545

Waegemann K, Paulsen H and Soll J (1990) Translocation of proteins into isolated chloroplasts requires cytosolic factors to obtain import competence. FEBS Lett 261: 89–92

Williams WP (1997) The physical properties of thylakoid membrane lipids and their relation to photosynthesis. In: Siegenthaler PA and Murata N (eds) Lipids in Photosynthesis: Structure, Function and Genetics, pp 103–118. Kluwer Academic Publishers, Dordrecht

Chapter 11

Development of Thylakoid Membranes with Respect to Lipids

Eva Selstam

Department of Plant Physiology, Umeå University , S-901 87 Umeå, Sweden

Summary

Chloroplast membranes contain four highly unsaturated lipids: two galactolipids, one sulfolipid and one phospholipid. These are also present in proplastid and etioplast membranes. During chloroplast differentiation thylakoid membranes are formed by vesicles which bud off from the inner envelope membrane and plastids multiply by division. Monogalactosyldiacylglycerol, which has a reversed hexagonal phase structure, facilitates this fusion process. Monogalactosyldiacylglycerol, together with protochlorophyllide oxidoreductase, also mediates the formation of the prolamellar body, a membrane with a cubic phase structure. The chloroplast membranes also contain negatively charged lipids and the fraction of the charged lipids is very constant. The maintenance of a stable fraction of negatively charged lipids is probably due to the role that these lipids play in regulating the synthesis of monogalactosyldiacylglycerol. Current data from studies with mutants and transgenic plants suggests that the high level of thylakoid lipid unsaturation probably is more important for plastid differentiation than for photosynthetic function at low temperature. Furthermore, the repair of Photosystem II following photoinhibition at low temperature has been shown to be sensitive to the level of

P.-A. Siegenthaler and N. Murata (eds): Lipids in Photosynthesis: Structure, Function and Genetics, pp. 209–224.
© 1998 Kluwer Academic Publishers. Printed in The Netherlands.

unsaturation suggesting a function for lipid unsaturation in membrane protein assembly and processing. Specific functions for the different chloroplast lipids have been difficult to assign. However, there is increasing evidence that phosphatidyl glycerol is tightly bound to the LHC polypeptide and facilitates trimer formation of the LHC complex.

I. Introduction

Chloroplasts are the green plastids characteristic of photosynthetic tissues. In higher plants the number of chloroplasts per cell varies from 50 to 200 (Possingham, 1980). Chloroplasts contain a protein rich stroma and a large flattened membrane sack, the thylakoid, and are enclosed in two envelope membranes. The thylakoid is divided into stroma and grana regions, consisting of single and stacked membranes respectively. Chloroplasts are not made de novo, but are inherited across generations as proplastids (Kirk and Tilney-Basset, 1978; Possingham, 1980). In most angiosperms the plastids are inherited either maternally or biparentally, while in most gymnosperms the plastid has a paternal inheritance (Birky, 1995). All meristematic cells contain proplastids. These multiply by division and during cell division the proplastids distribute themselves between the daughter cells. During cell differentiation, proplastids develop into different types of plastids depending on their location in the plant, the physiological state of the plant and environmental factors (Kirk and Tilney-Basset, 1978). Chloroplast development is light dependent and best known from leaf tissue, but chloroplasts also form in herbaceous stems, fruits and seeds. The development of photosynthetic function during plastid differentiation has been summarized by Bradbeer (1981), Baker (1984), Hoober (1987) and Hachtel and Friemann (1993). The lipid composition of thylakoids from cyanobacteria and from algal chloroplasts is very similar to that of higher plants [refer to Chapters 3 (Harwood) and 4 (Wada and Murata)]. However, it is not known whether the specific lipid composition of the chloroplast is needed for proper function of the

photosynthetic membrane or whether it reflects evolutionary conservation in oxygen evolving photosynthetic organisms (refer to Chapter 8, Siegenthaler and Trémolières). Studies of thylakoid lipids during chloroplast differentiation is one approach to investigate the function of the different lipids in relation to membrane structure and photosynthesis. Mutants deficient in specific desaturases have given additional insight into the role of the fatty acids esterified to the chloroplast lipids for the development of normal structure and function of the chloroplast (Somerville and Browse, 1991; Ohlrogge et al., 1991). In this chapter I will begin with the physico-chemical properties of the different chloroplast lipids and give examples of how these contribute to function in the chloroplast.

II. Lipid Composition and Biosynthesis

Chloroplast membranes are enriched in the highly unsaturated lipids; monogalactosyldiacylglycerol (MGDG), digalactosyldiacylglycerol (DGDG), sulfoquinovosyldiacylglycerol (SQDG), phosphatidylglycerol (PG) and phosphatidylinositol (PI) (Harwood, 1980; Joyard et al., 1997). Phosphatidylcholine (PC) is probably not present in the thylakoid but is an important part in the outer envelope membrane (Block et al., 1983; Dorne et al., 1990). Biosynthetic intermediates such as diacylglycerol and phosphatidic acid are also minor components in the chloroplast membranes. The galactolipids account for 80–85% of the thylakoid lipids, the remaining 15–20% being SQDG and PG in approximately equal proportion.

The acyl groups esterified to the galactolipids are mainly the trienoic fatty acids, 16:3 and 18:3. SQDG contains mainly 16:0 and 18:3, and PG also contains mainly 16:0 and 18:3 and also some 16:1 trans-Δ3. The fatty acid composition of the chloroplast lipids is regulated by the different biosynthetic pathways, and the specificity of the different desaturases (Browse and Somerville, 1991; Ohlrogge and Browse, 1995; Murata and Wada, 1995; refer also to Chapter 2, Joyard et al.).

Abbreviations: CP43 and CP47 – chlorophyll proteins of the core light-antenna of Photosystem II; DGDG – digalactosyldiacylglycerol; D1 – core subunit of the Photosystem II reaction center; LHC – light-harvesting complex; L_α – lamellar liquid crystalline; L_β – lamellar gel; H_{II} – reversed hexagonal; I_{II} – reversed cubic; MGDG – monogalactosyldiacylglycerol; PC – phosphatidylcholine; PCOR – protochlorophyllide oxidoreductase; PG – phosphatidylglycerol; PI – phosphatidylinositol; P700 – Photosystem I reaction center protein; SQDG – sulfoquinovosyldiacylglycerol; 16:1(3t) – trans-Δ3-hexadecenoic acid

Biosynthesis of chloroplast lipids begins in the chloroplast stroma where glycerol-3-phosphate, 16:0, 18:0 and 18:1 are synthesized. Glycerolipids can be synthesized by two separate pathways; one prokaryotic taking place in the inner chloroplast envelope, and one eukaryotic via phosphatidylcholine (PC) synthesis in the endoplasmic reticulum. PC is an intermediate lipid that is desaturated in the endoplasmic reticulum and then delivered back to the chloroplast envelope, where the diacylglycerol backbone is incorporated into MGDG, DGDG and SQDG. In the prokaryotic pathway the diacylglycerol formed in the envelope is used directly for MGDG, DGDG, SQDG and PG synthesis. Insertion of the double bond into the fatty acids of the complex lipids is performed by membrane bound desaturases.

III. Properties of Chloroplast Lipids

A. Phase Behavior

The phase behavior of chloroplast lipids has been reviewed by Quinn and Williams (1983), Murphy (1986), Webb and Green (1991) and Williams (1997). The major structures formed by chloroplast lipids are the lamellar liquid crystalline (L_α) phase, lamellar gel (L_β) phase, reversed hexagonal (H_{II}) phase and cubic phase (Fig. 1). DGDG, SQDG and PG mainly form L_α phases. The H_{II} phase is formed by MGDG, diacylglycerol and phosphatidic acid. Below $-2\ °C$ MGDG forms highly disordered L_β and crystal lamellar phase structures (Sanderson and Williams, 1992). MGDG does not have a distinct phase transition temperature (Sanderson and Williams, 1992). L_β phases are usually found only at subzero temperatures, except in PG isolated from some chilling sensitive plants (Murata and Yamaya, 1984). PG forms an L_β phase because PG contains molecular species with two saturated acyl chains or with one saturated and one 16:1(3t) acyl chain. The influence of high melting point PG species is discussed further below.

The formation of different phase structures can be explained by the concept of lipid molecular shape (Israelachvili et al., 1980, Israelachvili, 1992). According to this concept, an H_{II} phase is formed by lipids that have a conical shape, while more cylindrically shaped lipids form a bilayer or an L_α phase (Fig. 2). The shape of the lipid is determined by the relation between the area of the polar headgroup

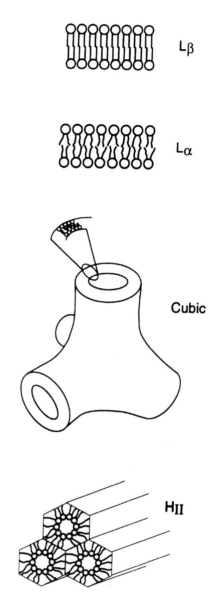

Fig. 1. The phase structures of chloroplast lipids. L_β, lamellar gel phase; L_α, lamellar liquid crystalline phase; H_{II}, reversed hexagonal phase. The cubic phase is demonstrated as a tetrapodel unit, where the position of the bilayer is shown in an enlargement. The internal proportions in the tetrapodel unit are not shown correctly. The cylinders in the H_{II} phase are shown with hexagonal cross sections to show that the acyl chains fill the space between the water channels.

and the length and volume of the acyl chains. This 'shape' concept explains why MGDG with its small polar headgroup and long polyunsaturated acyl chains, producing a conically shaped lipid, will form an H_{II} phase. DGDG on the other hand has a much larger polar headgroup, producing a more cylindrical shaped lipid and forms an L_α phase. The shape of the

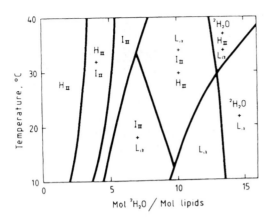

Fig. 2. A temperature-composition phase equilibrium of heavy water and a 2:1 mixture of MGDG and DGDG. H_{II} reversed hexagonal; L_{α}, lamellar, liquid crystalline; I_{II} reversed cubic Ia3d phase. This figure demonstrates the relation between the molecular shape and the phase structure of the galactolipids. Two factors influence the molecular shape of the lipids, the water content and temperature. The lower the water content the smaller the area of the polar head group. The higher the temperature the larger the hydrophobic volume of the acyl chains and the more conical the shape of the lipids. MGDG and DGDG alone form an H_{II} phase and an L_{α} respectively at all water concentrations in this temperature range (Brentel et al., 1985)

molecule is also influenced by the length and unsaturation of the acyl chain, by the hydration of the polar headgroup, by ions and by temperature (Rilfors et al., 1984). The phase preference can also be explained by the 'monolayer intrinsic curvature' concept developed by Helfrich (1973) and Gruner (1985) (see also Lindblom and Rilfors, 1989). This model suggests that the tendency of a membrane monolayer to curve induces packing strains in the hydrocarbon region. In lipid phases there is a frustration between two opposing forces, the monolayer intrinsic curvature energy and the stretching free energy of the hydrocarbon chains. The lipids will form a structure that has the lowest free energy. This concept made it possible to calculate when cubic phase stuctures are formed, namely when the lipid or lipid mixture can not form either a stable L_{α} phase or a stable H_{II} phase (Anderson et al., 1988).

Cubic phase structures are formed in intermediate mixtures of MGDG and DGDG, and in mixtures of all thylakoid lipids (Fig. 2; Brentel et al., 1985; Selstam and Widell-Wigge, 1993). The formation of cubic phases in these mixtures depends on the influence of the H_{II} phase lipid MGDG (Lindblom and Rilfors, 1989). The three dimensional structure

of the lipids in the cubic phase is difficult to illustrate. In freeze-fracture electron microscopy, a cubic phase appears as rounded particles organized in a uniform symmetrical lattice (Sen et al., 1982; Quinn and Williams, 1983). In Fig. 1, the structure of a 4-armed repeating unit from a cubic structure is shown. Several different types of cubic structures have been identified (Lindblom and Rilfors, 1989; Tate et al., 1991; Caffrey and Cheng, 1995). The type formed by thylakoid galacto-lipids has a reversed Ia3d symmetry (Brentel et al., 1985). This is a bicontinuous structure where both water and lipids are continuous in the structure. The water forms two independent and interwoven branched channels. At each branching point, three branches meet. A monolayer of lipids surrounds each water channel. The hydrocarbon chains from one monolayer meet the hydrocarbon chains from the other, forming a branched bilayer between the water channels. Thus, MGDG forms an H_{II} phase by itself and together with DGDG it will form either an H_{II} cubic or L_{α} phase depending on the proportion and the two lipids and the water content and temperature of the lipid mixture (Fig. 2).

B. Significance of Reversed Hexagonal Lipid

Approximately 50% of the thylakoid lipid component is MGDG. This high concentration of an H_{II} phase lipid has been correlated with the high protein content in this membrane. The more protein in the chloroplast membrane the higher the MGDG to DGDG ratio (Block et al., 1983; Fig. 3). The MGDG to DGDG ratio in the thylakoid is usually between 1 and 3. According to the fluid mosaic membrane model, the function of the membrane lipids is to form a continuous and closed bilayer, and to make a barrier between the different compartments in the cell. This role is fulfilled by lipids forming a lamellar phase with water. Despite this, 20 to 50% of the lipids in most membranes form an H_{II} phase, not a lamellar phase, with water. Thus, the function of a high concentration of the H_{II} phase lipid MGDG has been difficult to define.

Biological membranes also participate in dynamic events such as cell and organelle division and membrane trafficking. These processes all involve membrane fusion. Fusion involves a local breaking of membrane continuity and the membrane lipids transiently form nonbilayer structures (Siegel, 1993; Chernomordik et al., 1995). The various fusion reactions taking place in different cells have common

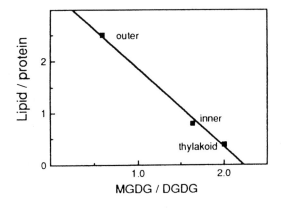

Fig. 3. Lipid to protein ratio plotted against the MGDG to DGDG ratio in outer, inner envelope membranes and in thylakoids isolated from spinach. The lipids are expressed as mg fatty acids per mg protein. The membrane preparation was made after inhibition of the galactolipid:galactolipid galactosyltransferase to prevent changes in galactolipid composition (after Block et al., 1983)

motifs. Initiation is dependent on fusion proteins and membrane fusion is dependent on the lipid composition (Chernomordik and Zimmerberg, 1995). Micelle forming lipids (hexagonal I phase) such as lysolipids prevent membrane fusion, while H_{II} lipids facilitate fusion (Chernomordik et al., 1995). Fusion can also be stimulated by phospholipase C, which converts L_α forming phospholipids to diacylglycerol, an H_{II} lipid (Goñi et al., 1994) or by incorporation of a hydrocarbon (Walter et al., 1994), which facilitates H_{II} formation by L_α lipids (Sjölund et al., 1987). Studies with plant galactolipids show that vesicles of only DGDG form nonfusing aggregates (Webb et al., 1988; Menikh and Fragata, 1993), while in mixtures of MGDG and DGDG lipidic particles or inverted micelles and fusion pores are formed (Sen et al., 1981; Sprague and Staehelin, 1984). The formation of these structures in mixtures of MGDG and DGDG are indications of MGDG dependent fusion between the bilayers of adjacent vesicles. Thus, one important function of MGDG in the chloroplast membrane is to facilitate fusion, necessary for plastid division and trafficking of vesicles from the inner envelope membrane to the thylakoid during plastid differentiation. The thylakoid lipids are synthesized primarily in the chloroplast envelope. Transportation of these lipids to the thylakoid could be performed by vesicles budding off from the envelope. Direct membrane bridges have also been visualized by

electron microscopy (Bradbeer et al., 1974; Whatley et al., 1982; Morré et al., 1991b) and by studies of the transport of lipids from the envelope to the thylakoids (Morré et al., 1991a; Rawyler et al., 1995). Other functions of MGDG are to facilitate the insertion of the transit peptide of nuclear encoded chloroplast proteins into the envelope (refer to Chapter 10, de Kruijff et al.) and to facilitate the formation of prolamellar bodies (see below).

C. Negatively Charged Lipids

Both SQDG and PG have a net negative charge at physiological pH. In the chloroplast membranes the total concentration of these negatively charge lipids is 15–20%. The same is found in most membranes containing charged phospholipids. PC and phosphatidylethanolamine are zwitterionic. PG, phosphatidylserine, and phosphatidylinositol have a net negative charge and phosphatidic acid and diphosphatidylglycerol have two negative charges. In the *act1* mutant of *Arabidopsis thaliana* the envelope glycerol-3-phosphate acyltransferase activity was reduced to less than 4% of the wild type level (Kunst et al., 1988). This enzyme is crucial for PG synthesis, since PG is only synthesized by the prokaryotic pathway. However, the concentration of PG in the mutant was not reduced more than by 25% (Kunst et al., 1988). Similarly, in SQDG mutants of *Rhodobacter sphaeroides*, *Chlamydomonas reinhardtii* and *Synechococcus* sp. PCC7942 and in *Gloeobacter violaceus* deficient in SQDG, the amount of negatively charged lipids was also kept constant by an increase in PG (Benning et al., 1993; Sato et al., 1995; Güler et al., 1996; Selstam and Campbell, 1996). These examples indicate that a relatively constant concentration of negatively charged lipids in the membrane is of importance.

The balance between the charged SQDG and PG and the neutral galactolipids is probably controlled by the regulation of MGDG synthesis, since the activity of UDP-galactose:diacylglycerol galactosyltransferase, the MGDG synthase, is dependent on these negatively charge lipids (Covés et al., 1988). UDP-galactose:diacylglycerol galactosyltransferase activity is especially dependent on PG content (Covés et al., 1988). Thus, in the *act1* mutant of *Arabidopsis thaliana* the rate of PG synthesis is regulating MGDG synthesis, and thereby keeping the balance between the charged and neutral lipids fairly constant. This regulation mechanism is also shown in the *dgd1* mutant of *Arabidopsis thaliana* where the concentration of negatively charged lipids was almost the same as in the wild type, even thought the mutant had a greatly reduced amount of DGDG (Dörmann et al., 1995). The low DGDG content resulted in an increase in the MGDG concentration and not in the other negatively charged L_α lipids.

IV. Chloroplast Differentiation from Proplastids

A. Chloroplast Differentiation

In grasses, the leaf grows from an intercalary meristem at the base of the leaf. Thus, the oldest part of the leaf is at the tip and the youngest at the base. This makes it possible to isolate plastids at different developmental stages and to use them for compositional and functional studies (Leech and Baker, 1983). In developing grass leaves, the proplastids in the youngest part of the leaf are 1–2 μm in diameter. The plastid has two envelope membranes and a more or less structureless interior. Vesicles bud off from the inner envelope membrane and form short perforated membranes in the organelle (Whatley et al., 1982; Leech and Baker, 1983). Within 4 cm of the basal part of the growing wheat leaf, the proplastids gradually develop to mature chloroplasts. In wheat this differentiation takes 30 h and involves a gradual increase in chloroplast size and in the length of the thylakoid membrane, and a differentiation of the thylakoid into stroma and grana membranes. Finally, there is an increase in the number of membranes in the grana stacks (Leech and Walton, 1983). The differentiation of a proplastid to a chloroplast in maize takes 3 h and involves, on a plastid basis, a 10-fold increase in volume, a 10-fold increase in chlorophyll and membrane lipid content and a 50-fold increase in the area of internal membranes (Leech and Baker, 1983). Proplastids in leaves contain small amounts of chlorophyll and can evolve oxygen. The oxygen evolving capacity increases during plastid development concomitant with the increase in thylakoid and grana formation (Baker and Leech, 1977). The number of plastids per cell also increases during leaf development. In the youngest dividing cells of a wheat leaf there are approximately 50 proplastids. After the termination of cell division the young chloroplasts continue to divide. In the mature wheat leaf there is around 150 chloroplasts per cell (Leech, 1984). The number of plastids per cell is dependent on cell size (Pyke and Leech, 1987).

B. Membrane Lipids and Chloroplast Differentiation

All typical thylakoid membrane lipids are present in proplastids although the relative content of galactolipids is lower and that of phospholipids is higher than in the mature chloroplast (Leese and Leech, 1976). The acyl chains of the membrane lipids are also less unsaturated, and 16:1(3t) specifically esterified to PG is not found in proplastids (Leech et al., 1973). During normal development of proplastids to chloroplasts the relative content of galactolipids increases. This is especially true for MGDG. The degree of unsaturation increases in all thylakoid lipids by a decrease in 18:2 and an increase in 18:3. There is also a decrease in the relative amount of 16:0 in both galactolipids. These changes in the concentration of the various lipids, and in the degree of unsaturation, correlates with the structural and functional development of the mature chloroplast. The increase in unsaturation of the membrane lipids during chloroplast development correlates with the increase in grana formation. However, thylakoids with large grana also developed when the amount of 18:3 was reduced in barley seedlings by treatment with San 9785 (Leech and Walton, 1983). Despite a decrease from 76 to 28% of 18:3 and an increase from 7 to 49% of 18:2, both thylakoid structure and photosynthesis were almost normal. Thus, the development of a large grana and normal photosynthesis was not dependent on the synthesis of highly unsaturated thylakoid lipids, at least not when grown at 25 °C. However, the San treatment also resulted in a 30% increase in the lipid to chlorophyll ratio. A similar increase in the lipid to protein ratio was found in the *fad6* mutant of *Arabidopsis thaliana* deficient in the synthesis of both di- and tri-enic fatty acids (Hugly et al., 1989). Thus, the loss in unsaturation might have been compensated for by an increase in the total amount of lipids in the membrane.

C. Role of Unsaturation During Chloroplast Differentiation

All plant lipids have a relatively high degree of unsaturation. This is important for plant survival in fluctuating temperatures, since highly unsaturated membrane lipids form a liquid crystalline phase at physiological temperatures. The highest degree of unsaturation is found in the thylakoid membrane lipids. This is surprising as oxygen is released into the thylakoid lumen during photosynthesis. The possibility for free radical formation during oxygen passage through the photosynthetic membrane should make the system sensitive to degradation of the acyl chains. However, the high degree of unsaturation is also necessary for chloroplast differentiation and the plant cell has developed advanced defense systems to avoid damage by oxygen radicals (Asada and Takahashi, 1987; Asada, 1996).

1. Unsaturation and Chloroplast Differentiation

A number of mutants of *Arabidopsis thaliana* have been identified that are deficient in desaturases or glycerol-3-acyltransferases involved in the biosynthesis of acyl lipids. These mutants have been an invaluable tool for understanding the function of unsaturation of the acyl lipids in the chloroplast (Somerville and Browse, 1991; refer also to Chapter 14, Vijayan et al.). In the *fad6* mutant, which is unable to synthesize 18:2, 18:3 and 16:3 in the prokaryotic pathway, the amount of these polyunsaturated fatty acids esterified to the galactolipids decreased considerably. A minor decrease in the amount 18:2 and 18:3 was also found in the cytoplasmic phospholipids (Browse et al., 1989). Despite this, the mutant was indistinguishable in appearance from the wild type when grown at 22 °C (Hugly and Somerville, 1992). Development at 5 °C resulted in chlorotic leaves and slower growth relative to the wild type. However, no chlorotic lesions appeared in mutant leaves that had developed at 22 °C prior to being shifted to 5 °C (Hugly and Somerville, 1992). This indicated that membrane function was maintained at the low temperature despite the lower level of unsaturation of the thylakoid lipids but that assembly of the thylakoid in developing chloroplasts was partly inhibited. In the *fad2* mutant of *Arabidopsis,* which is deficient in the microsomal 18:1-PC desaturase, a considerable decrease in 18:2 and 18:3 was mostly localized in the cytoplasmic phospholipids, with only a minor decrease in chloroplast lipids (Miquel and Browse, 1992). This mutant was also very similar to the wild type when grown at 22 °C and it could complete its life cycle at 12 °C, although stem elongation failed at this temperature. When shifted to 6 °C the *fad2* mutant

slowly died (Miquel et al., 1993). This lack of viability at 6 °C was correlated with an inability to increase the amount of 18:2 and 18:3 in PC at this low temperature. Thus, in both the *fad2* and *fad6* mutants there is a requirement for polyunsaturated membrane lipids during development at low temperature and this is more important for the cytoplasmic membranes. The importance of the degree of unsaturation has also been demonstrated in several mutants where a deficiency in a desaturase is compensated for by a shift in biosynthetic pathway. For example in the *fad6* mutant of *Arabidopsis*, MGDG synthesis by the prokaryotic pathway is reduced by one third, but this is compensated for by a corresponding increase in the eukaryotic pathway (Browse et al., 1989). This shift in pathways decreased the effect of the *fad6* mutation and may reflect the existence of a regulatory mechanism that responds to the physical properties of the chloroplast membrane (Browse et al., 1989).

Transformation of *Arabidopsis* with glycerol-3-phoshpate acyl-transferase from *Escherichia coli* resulted in the synthesis of high temperature melting point species of PG and chilling sensitivity, indicating that one reason for chilling sensitivity is due to the presence of saturated species of PG (Wolter et al., 1992). However, the *fab1* mutant of *Arabidopsis* that also had enhanced levels of high melting point PG in the chloroplast membrane did not show chilling sensitivity (Wu and Browse, 1995). An increase in saturated PG was also shown in tobacco plants transformed with the glycerol-3-phoshpate acyltransferase gene from squash, and this increase in saturation correlated with increased sensitivity to photoinhibition at low temperature (Moon et al., 1995). Photoinhibition is caused by overexcitation of Photosystem II, leading to the degradation of the reaction center D1 protein. Tolerance against photoinhibition is dependent on synthesis and assembly of a new D1 polypeptide in the reaction center (Gombos et al., 1994; Aro et al., 1993). At low temperature, either the assembly of new D1 protein into active Photosystem II reaction centers or the processing of the pre-D1 protein appears to be dependent on the degree of unsaturation of the thylakoid lipids (Murata and Wada, 1995; Nishida and Murata, 1996). It is not known whether the limiting factor at low temperature for the 'unsaturation' dependent differentiation of the chloroplasts in the *fad6* mutants of *Arabidopsis* (Hugly and Sommerville, 1992) and maize (Kodama et al., 1994) also involves the assembly and processing of

Photosystem II. However, assembly of thylakoid proteins in *fad5* and *fad6* mutants of *Arabidopsis* at 22 °C resulted in smaller protein particles than in the wild type, when analyzed by freeze-fracture electron microscopy. This indicates that lower unsaturation of thylakoid lipids results in incomplete assembly of the oligomeric protein complexes (Tsvetkova et al., 1994).

2. Unsaturation and Chloroplast Morphology

Some 'lipid' mutants also show changes in leaf and chloroplast morphology (Ohlrogge et al., 1991). For example, differentiation of chloroplasts in an *Arabidopsis* mutant deficient in chloroplastic ω3 desaturase resulted in reduced amounts of 18:3 and 16:3 fatty acids in the chloroplast lipids (Browse et al., 1986; Iba et al., 1993). The decrease caused a reduction in the size of the chloroplast and an increase in the number of chloroplasts per cell but had no effect on photosynthetic electron transport or CO_2 fixation (McCourt et al., 1987). Similarly, differentiation of chloroplasts in an *Arabidopsis* mutant deficient in the chloroplastic sn-glycerol-3-phosphate acyltransferase resulted in a decrease in 16:3 and an increase in 18:3, mostly in MGDG (Kunst et al., 1988). This change in the type of trienoic fatty acid resulted in an increase in the number of grana per chloroplast and a decrease in the average number of thylakoid membranes in the grana stacks but had only minor effects on photosynthesis (Kunst et al., 1989). In the *fab2* mutant of *Arabidopsis,* an increase in the concentration of 18:0 in all leaf membrane lipids did not change the size or anatomy of the chloroplast but decreased the number of chloroplast per cell (Lightner et al., 1994a, b). The decrease in the number of chloroplasts per cell was correlated with a reduction in cell and plant size (Lightner et al., 1994b; Pyke and Leech, 1987). These three examples show that the fatty acid composition of the membrane lipids may influence the number of chloroplasts and their morphology more than the photosynthetic functions of the chloroplast.

D. Trans-Δ3-Hexadecenoic Acid and Plastid Differentiation

All double bonds in the highly unsaturated chloroplast lipids are in a *cis*-configuration except one, the *trans*-Δ3-hexadecenoic acid (16:1(3t)). This fatty acid is only found esterified to PG since the Δ3 desaturase is specific for 18:0(18:1)/16:0-PG (Browse and Somerville, 1991). PG-16:1(3t) is found in nearly all chloroplasts (Murata, 1983; Roughan, 1986; Harwood, 1997; Tasaka et al., 1990). Although it was not found in chloroplast envelopes from *Vicia faba* (Mackender and Leech, 1974), it is abundant in chloroplast envelopes from spinach (Block et al., 1983) and wheat (Bahl et al., 1976). A *trans* double bond shortens the chain 0.2 Å and gives a *trans*-unsaturated fatty acid physical properties similar to a saturated fatty acid (Chapman et al., 1966). The phase transition temperature of 16:0/16:1(3t)-PG is 10 °C lower than that of 16:0/16:0-PG (Bishop and Kenrick, 1987).

In both proplastids and etioplasts there is no, or a very low amount, of 16:1(3t) esterified to PG (Leese and Leech, 1976; Roughan and Boardman, 1972; Selldén and Selstam, 1976; Bahl et al., 1976; Ohnishi and Yamada, 1980). The amount of 16:1(3t) esterified to PG in etioplasts of wheat increases with the age of the seedling (Bahl et al., 1976). Both *Picea abies* and *Pinus sylvestris* synthesize chlorophyll protein complexes and form thylakoids in darkness (Walles and Hudák, 1975). However, while PG-16:1(3t) is not found in dark grown seedlings of *Picea abies* (Lemoine et al., 1982), it is present in dark grown *Pinus sylvestris* (Selstam and Widell, 1986). During chloroplast development from either proplastids or etioplasts, there is a correlation between grana formation and an increase in the concentration of 16:1(3t) in PG (Selldén and Selstam, 1976; Tuquet et al., 1977).

In rye leaves a decrease in 16:1(3t) esterified to PG was correlated to a decrease in the oligomeric form of the light-harvesting complex of Photosystem II (LHCII) during development in low temperature, (Huner et al., 1984; 1987; 1989; Krol et al., 1988). In maize leaves the relative content of 16:1(3t) in PG in bundle sheath chloroplasts, stroma and grana thylakoids was 8, 18 and 36% respectively. These correlations have led to the suggestion that 16:1(3t)-PG is associated with the synthesis of the oligomeric form of LHCII (Trémolières, 1991) that is concentrated in the grana membrane (Andersson and Anderson, 1980). It has also been demonstrated that PG is firmly bound to LHCII trimers and that PG is needed for crystallization of LHCII trimers (Nussberger et al., 1993). The binding site of PG is in the N-terminus of LHCII (Nussberger et al., 1993; Hobe et al., 1995; Trémolières and Siegenthaler, 1997). However, the function of PG in LHCII trimer formation can not be specific to 16:1(3t)-PG, since 16:1(3t) is not found in the *fad4* mutant of *Arabidopsis* (McCourt et al., 1985) or in some orchids that have normal levels of LHCII oligomers (Selstam and Krol, 1995). It is possible that oligomers of LHCII are more stable in a complex with 16:1(3t)-PG than with 16:0-PG. LHCII oligomers from both the *Arabidopsis* mutant and the 16:1(3t) less orchids are more sensitive to detergent solubilization than wild type *Arabidopsis* and orchids containing 16:1(3t)-PG (McCourt et al., 1985; Selstam and Krol, 1995).

V. Chloroplast Differentiation from Etioplasts

A. Differentiation of Etioplasts

Chloroplast development has often been studied during greening of etiolated grass leaves, instead of during more natural plastid development in leaves grown under light-dark cycles. This use of etiolated grass leaves is practical since grasses germinate and develop large leaves in darkness. There are 33–65 etioplasts per cell in etiolated leaves of barley, and the number of etioplasts per cell increased with cell age (Robertson and Laetsch, 1974). During greening of oat leaves the number of plastids per cell increases from 80 to 100 (Ohnishi and Yamada, 1980). Etioplasts are yellow due to a high content of carotenoids and a low content of protochlorophyllide. Etioplasts are oval in shape and 1-5 μm long and contain inner membranes differentiated into prothylakoids and one or two prolamellar bodies (Fig. 4; Wellburn, 1984; 1987). The inner membranes are formed by vesicles budding off from the envelope. The vesicles fuse and form a perforated membrane that develops into prothylakoids and prolamellar bodies (Bradbeer et al., 1974). In the light, the prolamellar body looses its regular structure (transforms) and disperses into planar membranes. During the dispersal, the membrane is first perforated and later the holes close and thylakoids with grana are formed (Fig. 4; Gunning and Jagoe, 1967).

B. Membrane Lipids During Etiochloroplast Development

Changes in membrane lipid composition during chloroplast development from etioplasts in grasses, peas and beans have been analyzed either in continuous (Trémolières and Lepage, 1971; Roughan

and Boardman, 1972; Bahl et al., 1974; Selldén and Selstam, 1976; Mackender, 1979; Ohnishi and Yamada, 1980), intermittent (Bahl et al., 1975; Tuquet et al., 1976; Guillot-Salomon et al., 1977; Krol and Huner, 1989; Krol et al., 1989) or monochromatic light (Trémolières et al., 1979; Tevini, 1977). For a summary see Tevini (1977). In the etioplast, the same lipids are present as in the chloroplast but the relative amount of MGDG is usually less than in the chloroplast. During development from etioplast to chloroplast the amount of lipids per plastid increases 3 times in maize, wheat and barley (Guillot-Salomon et al., 1973; Bahl et al., 1975; Tuquet et al., 1976). The increase in lipid content has a lag phase of 6–8 h from the beginning of the light treatment (Roughan and Boardman, 1972; Selldén and Selstam, 1976). The lag phase for lipids is longer than the lag phase for chlorophyll biosynthesis (Selldén and Selstam, 1976). This indicates that the membrane lipids in the prothylakoids and the prolamellar body are reorganized into the developing thylakoids. The fatty acid composition of the etioplast membrane lipids show some variation, depending on the age of the plastid. The older the plastid the more the fatty acid composition is similar to that of the chloroplasts (Bahl et al., 1974). Linolenic acid is synthesized in the dark. Etioplasts lack or contain very low amounts of 16:1(3t) esterified to PG. Thus, in the dark, i.e. in the absence of a functioning photosynthetic electron transport, the plant can synthesize large amounts of highly unsaturated galactolipids and phospholipids but not the amounts of 16:1(3t) found in mature chloroplasts.

C. The Prolamellar Body

1. Prolamellar Body Structure

The fine structure of the prolamellar body has been extensively studied by electron microscopy (Gunning and Steer, 1975; Henningsen et al., 1993). The prolamellar body is formed by a branching tubular membrane. The tubes have an outer diameter of 14–31 nm and a lumen with a diameter of 4.7–10 nm (Gunning, 1965; Murakami et al., 1985; Henningsen et al., 1993). These dimensions indicate that the tubes are formed by a bilayer (5–6 nm thick) with a water channel inside and not by reversed hexagonal cylinders, which have an outer diameter of 6 nm when saturated with water (Shipley et al., 1973). In most prolamellar body structures, four tubes meet at

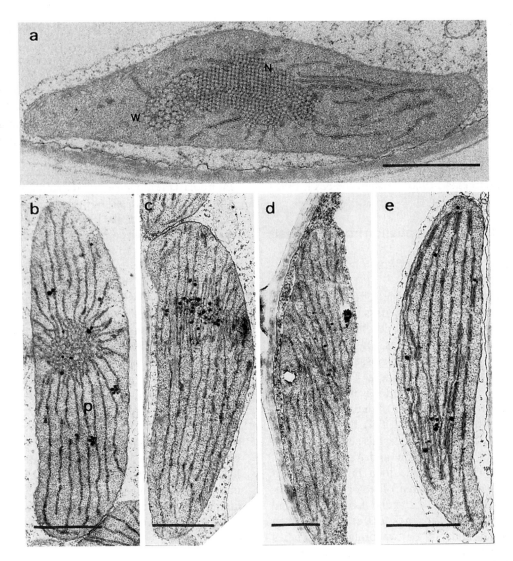

Fig. 4. Electron micrographs of plastids from dark grown barley seedlings. The plants were 5 days old and greened for a, 0 min; b, 30 min; c, 3 h; d, 9 h; e, and 12 h. N = narrow spacing; P = perforated membranes; W = wide spacing; Scale bar = 1μm.

an equal angle to make a tetrapodel unit. The tetrapodel units combine into different three dimensional structures with narrow or wide spacing. A prolamellar body structure formed by 6 armed units has also been found (Gunning, 1965; Henningsen et al., 1993; Gunning and Steer, 1975; Murakami et al., 1985). In the narrow spaced prolamellar body type, 6 tetrapodel units form a hexagon and the hexagons join together with a hexagonal symmetry similar to that found in zincblende and diamonds (Murakami et al., 1985; Selstam et al., 1995). It has been suggested that the

prolamellar body is a membrane with a bicontinuous cubic phase structure (Lindblom and Rilfors, 1989, Selstam and Widell-Wigge, 1993; Selstam et al., 1995). This suggestion is based on the similarities in the characteristics of the bicontinuous cubic phase structure on the one hand and that of the prolamellar bodies on the other. Both are formed by 4-, or 6-armed units, both have two independent water channels lined by a monolayer of lipids and both can be described by the principles of infinite periodic minimal surfaces (Lindblom and Rilfors, 1989; Lindstedt and Liljenberg, 1990; see above).

2. Components Involved in Prolamellar Body Formation

The lipid composition of the prolamellar body and the prothylakoids is similar to that of thylakoids (Selstam and Widell-Wigge, 1993). The MGDG to DGDG ratio is 1.6–1.8 and 1.1–1.4 in the prolamellar body and prothylakoids respectively. There is 30% more lipid per protein in the prolamellar body than in the prothylakoid (Ryberg et al., 1983; Selstam and Sandelius, 1984; Selstam and Widell-Wigge, 1993). Thus, the prolamellar body membrane is rich in lipids and in the H_{II} lipid MGDG, both factors that facilitate the formation of a cubic phase structure. However, the lipid composition is not the critical factor since the same composition is found in the thylakoid. Rather, prolamellar body formation is correlated with the dark accumulation of the major pigment-protein complex in this membrane, protochlorophyllide oxidoreductase (PCOR) (Ikeuchi and Murakami, 1983, Lindsten et al., 1988, Ryberg and Dehesh, 1986; Selstam and Widell, 1986). The correlation between the accumulation of PCOR and the formation of prolamellar bodies raises the question of how this protein can induce the formation of a membrane with a cubic phase structure. It is likely that this is determined by two properties of PCOR; the ability of the PCOR pigment complex to form an oligomer, and the as yet unknown anchoring of PCOR into the membrane.

The amino acid sequence of PCOR has been deduced from cDNA clones of several species (Suzuki and Bauer, 1995). A large proportion of the amino acids are hydrophobic, and the enzyme exhibits amphipathic properties when partitioning with Triton X-114 (Selstam and Widell-Wigge, 1989). Treatment with salt does not remove PCOR from the membrane (Grevby et al., 1989; Widell-Wigge and Selstam, 1990). However, neither the hydrophobicity plot (Benli et al., 1991) nor the secondary structure prediction (Birve et al., 1996) could identify a hydrophobic α-helix capable of anchoring PCOR in the membrane. Thus, in the prolamellar body the dominating protein is an extrinsic protein where the lipid protein interaction is in the polar region of the membrane.

The spectral properties of protochlorophyllide in complex with PCOR indicate that the pigment molecules are in an aggregated form in the prolamellar body (Böddi et al., 1989; 1990). Cross linking studies of the proteins in the prolamellar body also indicate that PCOR forms oligomers or higher homologs (Wiktorsson et al., 1993). PCOR oligomers are probably phosphorylated, and dephosphorylation of PCOR may regulate monomer formation of PCOR and transformation of the prolamellar body (Wiktorsson et al., 1996). The transformation and dispersal of the prolamellar body is correlated with a series of reactions that starts in the light. Protochlorophyllide is immediately reduced to chlorophyllide and within 10 min chlorophyll a and the apoproteins P700, CP43, CP47 and D1 are formed (Klein et al., 1988; Franck, 1993). In contrast to PCOR these apoproteins are intrincic membrane proteins. Thus, when the prolamellar body disappears and the thylakoid is formed, the dominating protein in the membrane changes from extrinsic to intrinsic suggesting that the intrinsic membrane protein stabilize the lipids to a bilayer.

Acknowledgments

I am grateful to Vaughan Hurry for valuable discussions and his critical reading of the manuscript. Reprinted figure is from Brentel et al., (1985) with kind permission from Elsevier Science–NL, Sara Burgerhartstraat 25, 1055 KV Amsterdam, The Netherlands. This work was supported by the Swedish Natural Science Research Council.

References

Anderson DM, Gruner SM and Leibler S (1988) Geometrical aspects of the frustration in the cubic phases of lyotropic liquid crystals. Proc Natl Acad Sci USA 85: 5364-5368

Andersson B and Anderson JM (1980) Lateral heterogeneity in the distribution of chlorophyll-protein complexes of the thylakoid membranes of spinach chloroplasts. Biochim Biophys Acta 593: 427-440

Aro EM, Virgin I and Andersson B (1993) Photoinhibition of Photosystem II. Inactivation, protein damage and turnover. Biochim Biophys Acta 1143:113–134

Asada K (1997) Radical production and scavenging in the chloroplasts. In: Baker NR (ed) Photosynthesis and the Environment, pp 123–150. Kluwer Academic Publishers, Dordrecht

Asada K and Takahashi M (1987) Production and scavenging of active oxygen in photosynthesis. In Kyle DI, Osmond CB and Arntzen CJ (eds) Photoinhibition, pp 227–287. Elsevier, Amsterdam

Bahl J, Lechevallier D and Monéger R (1974) Etude comparée de l'évolution à l'obscurité et à la lumière des lipides plastidiaux de feuilles étiolées de Blé. Physiol Vég 12: 229–249

Bahl J, Phung-Nhu-Hung S, Lechevallier D and Monéger R (1975) Effets de la lumière intermittente suivie ou non d'éclairement continu sur les lipides plastidiaux de feuilles étiolées de Blé. Comportement particulier du phosphatidylglycérol. Physiol Vég 13: 115–124

Bahl J, Francke B and Monéger R (1976) Lipid composition of envelopes, prolamellar bodies and other plastid membranes in etiolated, green and greening wheat leaves. Planta 129: 193–201

Baker NR (1984) Development of chloroplast photochemical functions. In: Baker NR and Barber J (eds) Chloroplast Biogenesis, pp 207–252. Elsevier Science Publishers, Amsterdam

Baker NR and Leech RM (1977) Development of Photosystem I and Photosystem II activities in leaves of light-grown maize (*Zea mays*). Plant Physiol 60: 640-644

Benli M, Schulz R and Apel K (1991) Effect of light on the NADPH-protochlorophyllide oxidoreductase of *Arabidopsis thaliana*. Plant Mol Biol 16: 615-625

Benning C, Beatty JT, Prince RC and Somerville CR (1993) The sulfolipid sulfoquinovosyldiacylglycerol is not required for photosynthetic electron transport in *Rhodobacter sphaeroides* but enhances growth under phosphate limitation. Proc Natl Acad Sci USA 90: 1561–1565

Birky Jr. CW (1995) Uniparental inheritance of mitochondrial and chloroplast genes: Mechanisms and evolution. Proc Natl Acad Sci USA 92: 11331–11338

Birve SJ, Selstam E and Johansson LB-Å (1996) Secondary structure of NADPH:protochlorophyllide oxidoreductase examined by circular dichroism and prediction methods. Biochem J 315:549-555

Bishop DG and Kenrick JR (1987) Thermal properties of 1-hexadecanoyl-2-trans-3-hexadecenoyl phosphatidylglycerol. Phytochem 26: 3065–3067

Block MA, Dorne A-J, Joyard J and Douce R (1983) Preparation and characterization of membrane fractions enriched in outer and inner envelope membranes from spinach chloroplasts. II. Biochemical characterization. J Biol Chem 258: 13281–13286

Böddi B, Lindsten A, Ryberg M and Sundqvist C (1989) On the aggregational states of protochlorophyllide and its protein complexes in wheat etioplasts. Physiol Plant 76: 135–143

Böddi B, Lindsten A, Ryberg M and Sundqvist C (1990) Phototransformation of aggregated forms of protochlorophyllide in isolated etioplast inner membranes. Photochem Photobiol 52: 83-87

Bradbeer JW (1981) Development of photosynthetic function during chloroplast biogenesis. In: Hatch MD and Boardman NK (eds) The Biochemistry of Plants, Vol 8, pp 423-472. Academic Press, San Diego

Bradbeer JW, Ireland HMM, Smith JW, Rest J and Edge HJW (1974) Plastid development in primary leaves of *Phaseolus vulgaris*. VII. Development during growth in continuous darkness. New Phytol 73: 263–270

Brentel I, Selstam E and Lindblom G (1985) Phase equilibria of mixtures of plant galactolipids. The formation of a bicontinuous cubic phase. Biochim Biophys Acta 812: 816-826

Browse J and Somerville C (1991) Glycerolipid synthesis: Biochemistry and regulation. Annu Rev Plant Physiol Plant Mol Biol 42: 467-506

Browse J, McCourt P and Somerville C (1986) A mutant of *Arabidopsis* deficient in $C_{18:3}$ and $C_{16:3}$ leaf lipids. Plant Physiol 81: 859-864

Browse J, Kunst L, Anderson S, Hugly S and Somerville C (1989) A mutant of *Arabidopsis* deficient in the chloroplast 16:1/18:1 desaturase. Plant Physiol 90: 522–529

Caffrey M and Cheng A (1995) Kinetics of lipid phase changes. Curr Opin Struct Biol 5: 548-555

Chapman D, Owens NF and Walker DA (1966) Physical studies of phospholipids. II. Monolayer studies of some synthetic 2.3-diacyl-DL-phosphatidylethanolamines and phosphatidylcholines containing trans double bonds. Biochim Biophys Acta 120: 148–155

Chernomordik LV and Zimmerberg J (1995) Bending membranes to the task: structural intermediates in bilayer fusion. Curr Opin Struct Biol 5:541-547

Chernomordik L, Kozlov MM and Zimmerberg J (1995) Lipids in biological membrane fusion. J Membrane Biol 146: 1–14

Covès J, Joyard J and Douce R (1988) Lipid requirement and kinetic studies of solubilized UDP-galactose:diacylglycerol galactosyltransferase activity from spinach chloroplast envelope membranes. Proc Natl Acad Sci USA 85: 4966-4970

De Kruijff B, Pilon R, van't Hof R and Demel R (1998) Lipid-protein interactions in chloroplast protein import. In: Siegenthaler PA and Murata N (eds) Lipids in Photosynthesis: Structure, Function and Genetics, pp 191–208. Kluwer Academic Publishers, Dordrecht

Dorne A-J, Joyard J and Douce R (1990) Do thylakoids really contain phosphatidylcholine? Proc Natl Acad Sci USA 87: 71-74

Dörmann P, Hoffmann-Benning S, Balbo I and Benning C (1995) Isolation and characterization of an *Arabidopsis* mutant deficient in the thylakoid lipid digalactosyl diacylglycerol. Plant Cell 7: 1801–1810

Franck F (1993) Photosynthetic activities during early assembly of thylakoid membranes. In: Sundqvist C and Ryberg M (eds) Pigment-Protein complexes in Plastids. Synthesis and Assembly, pp 365–381. Academic Press, San Diego

Gombos Z, Wada H and Murata N (1994) The recovery of photosynthesis from low-temperature photoinhibition is accelerated by the unsaturation of membrane lipids: A mechanism of chilling tolerance. Proc Natl Acad Sci USA 91: 8787-8791

Goñi FM, Nieva JL, Basañez G, Fidelio G and Alonso A (1994) Phospholipase-C-promoted liposome fusion. Biochem Soc Trans 22:839-844

Grevby C, Engdahl S, Ryberg M and Sundqvist C (1989) Binding properties of NADPH-protochlorophyllide oxidoreductase as revealed by detergent and ion treatments of isolated and immobilized prolamellar bodies. Physiol Plant 77:493-503

Gruner SM (1985) Intrinsic curvature hypothesis for biomembrane lipid composition: A role for nonbilayer lipids. Proc Natl Acad Sci USA 82: 3665–3669

Guillot-Salomon T, Douce R and Signol M (1973) Rapport entre l'évolution ultrastructurale des plastes de feuilles de plantules étiolées de Maïs soumises à l'action de la lumière et la synthèse de nouvelles molécules de phosphatidylglycérol. Plant Sci Lett 1: 463-471

Guillot-Salomon T, Tuquet C, Hallais M-F and Signol M (1977) Effets d'un éclairement continu sur l'ultrastructure et la composition lipidique des plastes de feuilles étiolées d'orge verdies en éclairement intermittent. Biol Cellulaire 28: 169–178

Güler S, Seeliger A, Härtel H, Renger G and Benning C (1996) A null mutant of *Synechococcus* sp. PC7942 deficient in the sulfolipid sulfoquinovosyl diacylglycerol. J Biol Chem 271:7501-7507

Gunning BES (1965) The greening process in plastids. 1. The structure of the prolamellar body. Protoplasma 60: 111–130

Gunning BES and Jagoe MP (1967) The prolamellar Body. In: Goodwin TW (ed) Biochemistry of Chloroplasts, pp 655-676. Academic Press, London

Gunning BES and Steer MW (eds) (1975) Plant Cell Biology: An ultrastructural approach, Edward Arnold, London

Hachtel W and Friemann A (1993) Regulation, synthesis and integration of chloroplast- and nuclear-encoded proteins. In: Sundqvist C and Ryberg M (eds) Pigment-Protein complexes in Plastids. Synthesis and Assembly, pp 279–310. Academic Press Inc, San Diego

Harwood JL (1980) Plant acyl lipids: Structure, distribution, and analysis. In: Stumpf PK (ed) The Biochemistry of Plants, vol. 4. Lipids: structure and function, pp 1-55. Academic Press, New York

Harwood JL (1998) Membrane lipids in algae. In: Siegenthaler PA and Murata N (eds) Lipids in Photosynthesis: Structure, Function and Genetics, pp 53–64. Kluwer Academic Publishers, Dordrecht

Helfrich W (1973) Elastic properties of lipid bilayers: Theory and possible experiments. Z Naturforsch 28 C: 693-703

Henningsen KW, Boynton JE and von Wettstein D (eds) (1993) Mutants at *xantha* and *albina* Loci in Relation to Chloroplast Biogenesis in Barley (*Hordeum vulgare* L.). Biologiske Skrifter 42, Munksgaard, Copenhagen

Hobe S, Kuttkat A, Förster R and Paulsen H (1995) Assembly of trimeric light-harvesting chlorophyll *a/b* complex in vitro. In: Mathis P (ed) Photosynthesis: From Light to Biosphere, Vol 1, pp 47-52. Kluwer Academic Publishers, Dordrecht

Hoober JK (1987) The molecular basis of chloroplast development. In: Hatch MD and Boardman NK (eds) The Biochemistry of Plants, Vol 10 Photosynthesis, pp 1-64. Academic Press Inc, San Diego

Hugly S and Somerville C (1992) A role for membrane lipid polyunsaturation in chloroplast biogenesis at low temperature. Plant Physiol 99: 197–202

Hugly S, Kunst L, Browse J and Somerville C (1989) Enhanced thermal tolerance of photosynthesis and altered chloroplast ultrastructure in a mutant of *Arabidopsis* deficient in lipid desaturation. Plant Physiol 90: 1134–1142

Huner NPA, Elfman B, Krol M and McIntosh A (1984) Growth and development at cold-hardening temperatures. Chloroplast ultrastructure, pigment content and composition. Can J Bot 62: 53-60

Huner NPA, Krol M, Williams JP, Maissan E, Low PS, Roberts D and Thompson JE (1987) Low temperature development induces a specific decrease in *trans*-Δ^3-hexadecenoic acid content which influences LHCII organization. Plant Physiol 84: 12–18

Huner NPA, Williams JP, Maissan EE, Myscich EG, Krol M, Laroche A and Singh J (1989) Low temperature-induced decrease in *trans*-Δ^3-hexadecenoic acid content is correlated with freezing tolerance in cereals. Plant Physiol 89: 144–150

Iba K, Gibson S, Nishiuchi T, Fuse T, Nishimura M, Arondel V,

Hugly S and Somerville C (1993) A gene encoding a chloroplast ω-3 fatty acid desaturase complements alterations in fatty acid desaturation and chloroplast copy number of the *fad7* mutant of *Arabidopsis thaliana*. J Biol Chem 268: 24099–24105

Ikeuchi M and Murakami S (1983) Separation and characterization of prolamellar bodies and prothylakoids from squash etioplasts. Plant Cell Physiol 24: 71-80

Israelachvili JN, Marcelja S and Horn RG (1980) Physical principles of membrane organization. Q Rev Biophys 13:121–200

Israelachvili JN (1992) Intermolecular and Surface Forces, Academic Press, San Diego

Joyard J, Maréchal E, Miège C, Block MA and Douce R (1998) Structure, distribution and biosynthesis of glycerolipids from higher plant chloroplasts. In: Siegenthaler P.A. and Murata N. (eds) Lipids in Photosynthesis: Structure, Function and Genetics, pp 21–52. Kluwer Academic Publishers, Dordrecht

Kirk JTO and Tilney-Bassett RAE (1978) Proplastids, etioplasts, amyloplasts, chromoplasts and other plastids. In: Kirk JTO and Tilney-Bassett RAE (eds) The Plastids Their Chemistry, Structure, Growth and Inheritance, pp 219–241. Elsevier North-Holland Biomedical Press, Amsterdam

Klein RR, Gamble PE and Mullet JE (1988) Light-dependent accumulation of radiolabeled plastid-encoded chlorophyll *a*-apoproteins requires chlorophyll *a*. Plant Physiol 88: 1246–1256

Kodama H, Hamada T, Horiguchi G, Nishimura M and Iba K (1994) Genetic enhancement of cold tolerance by expression of a gene for chloroplast ω-3 fatty acid desaturase in transgenic tobacco. Plant Physiol 105: 601-605

Krol M and Huner NPA (1989) Low temperature development under intermittent light conditions results in the formation of etiochloroplasts. J Plant Physiol 134: 623-628

Krol M, Huner NPA, Williams JP and Maissan E (1988) Chloroplast biogenesis at cold-hardening temperatures. Kinetics of trans-Δ^3-hexadecenoic acid accumulation and the assembly of LHCII. Photosynth Res 15: 115–132

Krol M, Huner NPA, Williams JP and Maissan EE (1989) Prior accumulation of phosphatidylglycerol high in trans-Δ^3-hexadecenoic acid enhances the in vitro stability of oligomeric light harvesting complex II. J Plant Physiol 135: 75-80

Kunst L, Browse J and Somerville C (1988) Altered regulation of lipid biosynthesis in a mutant of *Arabidopsis* deficient in chloroplast glycerol-3-phosphate acyltransferase activity. Proc Natl Acad Sci USA 85: 4143-4147

Kunst L, Browse J and Somerville C (1989) Altered chloroplast structure and function in a mutant of *Arabidopsis* deficient in plastid glycerol-3-phosphate acyltransferase activity. Plant Physiol 90: 846-853

Leech RM (1984) Chloroplast development in angiosperms: Current knowledge and future prospects. In: Baker NR and Barber J (eds) Chloroplast Biogenesis, pp 1–21. Elsevier Science Publishers, Amsterdam

Leech RM and Baker NR (1983) The development of photosynthetic capacity in leaves. In: Dale JE and Milthorpe FL (eds), Growth and Functioning of Leaves, pp 271–307. Cambridge University Press, Cambridge

Leech RM and Walton CA (1983) Modification of fatty acid composition during chloroplasts ontogeny and the effects on thylakoid appression and primary photochemistry. In: Thomson WW, Mudd JB and Gibbs M (eds) Biosynthesis and Function of Plant Lipids, pp 56-80. American Society of Plant Physiologists

Leech RM, Rumsby MG and Thomson WW (1973) Plastid differentiation, acyl lipid, and fatty acid changes in developing green maize leaves. Plant Physiol 52: 240–245

Leese BM and Leech RM (1976) Sequential changes in the lipids of developing proplastids isolated from green maize leaves. Plant Physiol 57: 789-794

Lemoine Y, Dubacq J-P and Zabulon G (1982) Changes in light-harvesting capacities and Δ3-*trans* hexadecenoic acid content in dark- and light-grown *Picea abies*. Physiol Vég 20: 487-503

Lightner J, Wu J and Browse J (1994a) A mutant of *Arabidopsis* with increased levels of stearic acid. Plant Physiol 106: 1443–1451

Lightner J, James Jr. DW, Dooner HK and Browse J (1994b) Altered body morphology is caused by increased stearate levels in a mutant of *Arabidopsis*. Plant J 6: 401-412

Lindblom G and Rilfors L (1989) Cubic phases and isotropic structures formed by membrane lipidspossible biological relevance. Biochim Biophys Acta 988: 221–256

Lindstedt I and Liljenberg C (1990) On the periodic minimal surface structure of the plant prolamellar body. Physiol Plant 80: 1-4

Lindsten A, Ryberg M and Sundqvist C (1988) The polypeptide composition of highly purified prolamellar bodies and prothylakoidsfrom wheat (*Triticum aestivum*) as revealed by silver staining. Physiol Plant 72: 167–176

Mackender RO (1979) Galactolipid and chlorophyll synthesis and changes in fatty acid composition during the greening of etiolated maize leaf segments of different ages. Plant Sci Lett 16: 101–109

Mackender RO and Leech RM (1974) The galactolipid, phospholipid, and fatty acid composition of the chloroplast envelope membranes of *Vicia faba* L. Plant Physiol 53: 496-502

Mascia PN and Robertson DS (1978) Studies of chloroplast development in four maize mutants defective in chlorophyll biosynthesis. Planta 143: 207–211

McCourt P, Browse J, Watson J, Arntzen CJ and Somerville CR (1985) Analysis of photosynthetic antenna function in a mutant of *Arabidopsis thaliana* (L.) lacking *trans*-hexadecenoic acid. Plant Physiol 78: 853-858

McCourt P, Kunst L, Browse J and Somerville CR (1987) The effects of reduced amounts of lipid unsaturation on chloroplast ultrastructure and photosynthesis in a mutant of *Arabidopsis*. Plant Physiol 84: 353–360

Menikh A and Fragata M (1993) Fourier transform infrared spectroscopic study of ion binding and intramolecular interactions in the polar head of digalactosyldiacylglycerol. Eur Biophys J 22: 249–258

Miquel M and Browse J (1992) *Arabidopsis* mutants deficient in polyunsaturated fatty acid synthesis. Biochemical and genetic characterization of a plant oleoyl-phosphatidylcholine desaturase. J Biol Chem 267: 1502–1509

Miquel M, James D, Dooner H and Browse J (1993) *Arabidopsis* requires polyunsaturated lipids for low-temperature survival. Proc Natl Acad Sci USA 90: 6208-6212

Moon BY, Higashi S-I, Gombos Z and Murata N (1995) Unsaturation of the membrane lipids of chloroplasts stabilizes the photosynthetic machinery against low-temperature photoinhibition in transgenic tobacco plants. Proc Natl Acad Sci USA 92: 6219-6223

Morré DJ, Morré JT, Morré SR, Sundqvist C and Sandelius AS (1991a) Chloroplast biogenesis, cell-free transfer of envelope monogalactosyl glycerides to thylakoids. Biochim Biophys Acta 1070: 437-445

Morré DJ, Selldén G, Sundqvist C and Sandelius AS (1991b) Stromal low temperature compartment derived from the inner membrane of the chloroplast envelope. Plant Physiol 97: 1558–1564

Murakami S, Yamada N, Nagano M and Osumi M (1985) Three-dimensional structure of the prolamellar body in squash etioplasts. Protoplasma 128: 147–156

Murata N (1983) Molecular species composition of phosphatidylglycerols from chilling-sensitive and chilling-resistant plants. Plant Cell Physiol 24: 81-86

Murata N and Wada H (1995) Acyl-lipid desaturases and their importance in the tolerance and acclimatization to cold of cyanobacteria. Biochem J 308: 1-8

Murata N and Yamaya J (1984) Temperature-dependent phase behavior of phosphatidylglycerols from chilling-sensitive and chilling-resistant plants. Plant Physiol 74: 1016–1024

Murphy DJ (1986) The molecular organisation of the photosynthetic membranes of higher plants. Biochim Biophys Acta 864: 33-94

Nishida I and Murata N (1996) Chilling sensitivity in plants and cyanobacteria: The crucial contribution of Membrane lipids. Annu Rev Plant Physiol Plant Mol Biol 47:541-568

Nussberger S, Dörr K, Wang DN and Kühlbrandt W (1993) Lipid-protein interactions in crystals of plant light-harvesting complex. J Mol Biol 234: 347–356

Ohlrogge J and Browse J (1995) Lipid biosynthesis. Plant Cell 7: 957-970

Ohlrogge JB, Browse J and Somerville CR (1991) The genetics of plant lipids. Biochim Biophys Acta 1082: 1–26

Ohnishi J and Yamada M (1980) Glycerolipid synthesis in *Avena* leaves during greening of etiolated seedlings I. Lipid changes in leaves. Plant Cell Physiol. 21:1595–1606

Possingham JV (1980) Plastid replication and development in the life cycle of higher plants. Ann Rev Plant Physiol 31: 113–129

Pyke KA and Leech RM (1987) The control of chloroplast number in wheat mesophyll cells. Planta 170: 416-420

Quinn PJ and Williams WP (1983) The structural role of lipids in photosynthetic membranes. Biochim Biophys Acta 737: 223–266

Rawyler A, Meylan-Bettex M and Siegenthaler PA (1995) (Galacto)lipid export from envelope to thylakoid membranes in intact chloroplasts. II. A general process with a key role for the envelope in the establishment of lipid asymmetry in thylakoid membranes. Biochim Biophys Acta 1233: 123–133

Rilfors L, Lindblom G, Wieslander Å and Christiansson A (1984) Lipid bilayer stability in biological membranes. In: Kates M and Manson LA (eds) Membrane Fluidity, pp 205–245, Plenum, New York

Robertson D and Laetsch WM (1974) Structure and function of developing barley plastids. Plant Physiol 54: 148–159

Roughan G (1986) A simplified isolation of phosphatidylglycerol. Plant Sci 43: 57-62

Roughan PG and Boardman NK (1972) Lipid composition of pea and bean leaves during chloroplast development. Plant Physiol 50: 31–34

Ryberg M and Dehesh K (1986) Localization of NADPH-protochlorophyllide oxidoreductase in dark-grown wheat (*Triticum aestivum*) by immuno-electron microscopy before and after transformation of the prolamellar bodies. Physiol Plant 66: 616-624

Ryberg M, Sandelius AS and Selstam E (1983) Lipid composition of prolamellar bodies and prothylakoids of wheat etioplasts. Physiol Plant 57: 555-560

Sanderson PW and Williams WP (1992) Low-temperature phase behaviour of the major plant leaf lipid monogalactosyldiacyl-glycerol. Biochim Biophys Acta 1107: 77-85

Sato N, Tsuzuki M, Matsuda Y, Ehara T, Osafune T and Kawaguchi A (1995) Isolation and characterization of mutants affected in lipid metabolism of *Chlamydomonas reinhardtii*. Eur J Biochem 230: 987-993

Selldén G and Selstam E (1976) Changes in chloroplast lipids during the development of photosynthetic activity in barley etio-chloroplasts. Physiol Plant 37: 35-41

Selstam E and Campbell D (1996) Membrane lipid composition of the unusual cyanobacterium *Gloeobacter violaceus,* which lacks sulfoquinovosyl diacylglycerol. Arch Microbiol 166: 132–135

Selstam E and Krol M (1995) Trimeric LHC II in plants lacking 16:1t in phosphatidyl-glycerol. In: Mathis P (ed) Photosynthesis: From Light to Biosphere, Vol 1, pp 95-98. Kluwer Academic Publishers, Dordrecht

Selstam E and Sandelius AS (1984) A comparison between prolamellar bodies and prothylakoid membranes of etioplasts of dark-grown wheat concerning lipid and polypeptide composition. Plant Physiol 76: 1036–1040

Selstam E and Widell A (1986) Characterization of prolamellar bodies, from dark-grown seedlings of Scots pine, containing light- and NADPH-dependent protochlorophyllide oxido-reductase. Physiol Plant 67: 345–352

Selstam E and Widell-Wigge A (1989) Hydrophobicity of protochlorophyllide oxidoreductase, characterized by means of Triton X-114 partitioning of isolated etioplast membrane fractions. Physiol Plant 77: 401-406

Selstam E and Widell Wigge A (1993) Chloroplast lipids and the assembly of membranes. In: Sundqvist C and Ryberg M (eds) Pigment Protein Complexes in Plastids: Synthesis and assembly, pp 241–277. Academic Press, San Diego

Selstam E, Williams P, Brain T and Bras W (1995) Electron microscopy and X-ray scattering studies of the structural organisation of prolamellar bodies isolated from *Zea mays*. In: Kader JC, Mazliak P (eds) Plant lipid metabolism, pp 141–143. Kluwer Academic Publishers, Dordrecht

Sen A, Williams WP, Brain APR, Dickens MJ and Quinn PJ (1981) Formation of inverted micelles in dispersions of mixed galactolipids. Nature 293: 488-490

Sen A, Brain APR, Quinn PJ and Williams WP (1982) Formation of inverted lipid micelles in aqueous dispersions of mixed sn-3-galactosyldiacylglycerols induced by heat and ethylene glycol. Biochim Biophys Acta 686: 215–224

Siegel DP (1993) Energetics of intermediates in membrane fusion: comparison of stalk and inverted micellar intermediate mechanisms. Biophys J 65: 2124–2140

Siegenthaler PA and Trémolières A (1998) Role of acyl lipids in the function of higher plant photosynthetic membranes. In: Siegenthaler PA and Murata N (eds) Lipids in Photosynthesis: Structure, Function and Genetics, pp 145–173. Kluwer

Academic Publishers, Dordrecht

Sjölund M, Lindblom G, Rilfors L and Arvidson G (1987) Hydrophobic molecules in lecithin-water systems. Biophys J 52: 145–153

Shipley GG, Green JP and Nichols BW (1973) The phase behavior of monogalactosyl, digalactosyl, and sulphoquinovosyl diglycerides. Biochim Biophys Acta 311: 531-544

Somerville C and Browse J (1991) Plant lipids: metabolism, mutants, and membranes. Science 252: 80-87

Sprague SG and Staehelin LA (1984) Effects of reconstitution method on the structural organization of isolated chloroplast membrane lipids. Biochim Biophys Acta 777: 306–322

Suzuki JY and Bauer CE (1995) A prokaryotic origin for light-dependent chlorophyll biosynthesis of plants. Proc Natl Acad Sci USA 92: 3749–3753

Tasaka Y, Nishida I, Higashi S, Beppu T and Murata N (1990) Fatty acid composition of phosphatidylglycerols in relation to chilling sensitivity of woody plants. Plant Cell Physiol 31: 545-550

Tate MW, Eikenberry EF, Turner DC, Shyamsunder E and Gruner SM (1991) Nonbilayer phases of membrane lipids. Chem Phys Lipids 57: 147–164

Tevini M (1977) Light, function, and lipids during plastid development. In: Tevini M and Lichtenthaler HK (eds) Lipids and Lipid Polymers in Higher Plants, pp 121–145. Springer-Verlag, Berlin

Thompson WF and White MJ (1991) Physiological and molecular studies of light-regulated nuclear genes in higher plants. Annu Rev Plant Physiol Plant Mol Biol 42:423-466

Trémolières A (1991) Lipid-protein interactions in relation to light energy distribution in photosynthetic membrane of eukaryotic organisms. Role of trans-Δ3-hexadecenoic acid-containing phosphatidylglycerol. Trends Photochem Photobiol 2: 13–32

Trémolières A and Lepage M (1971) Changes in lipid composition during greening of etiolated pea seedlings. Plant Physiol 47: 329–334

Trémolières A and Siegenthaler PA (1997) Reconstitution of Photosynthetic Structures and Activities with Lipids. In: Siegenthaler PA and Murata N (eds) Lipids in Photosynthesis: Structure, Function and Genetics, pp 175–189. Kluwer Academic Publishers, Dordrecht

Trémoliéres A, Guillot-Salomon T, Dubacq J-P, Jacques R, Mazliak P and Signol M (1979) The effect of monochromatic light on α-linolenic and trans-3-hexadecenoic acids biosynthesis, and its correlation to the development of the plastid lamellar system. Physiol Plant 45: 429-436

Tsvetkova NM, Brain APR and Quinn PJ (1994) Structural characteristics of thylakoid membranes of *Arabidopsis* mutants deficient in lipid fatty acid desaturation. Biochim Biophys Acta 1192: 263–271

Tuquet C, Guillot-Salomon T, Farineau J and Signol M (1976) Biogenèse des membranes plastidiales dans les feuilles étiolées d'Orge soumises à des éclairs répétés: développement d'accolements de thylacoïdes, synthèse de phosphatidylglycérol et apparition de cytochrome b_{559} (forme haut potentiel). Physiol Vég 14: 11–30

Tuquet C, Guillot-Salomon T, de Lubac M and Signol M (1977) Granum formation and the presence of phosphatidyl-glycerol containing *trans*-Δ3-hexadecenoic acid. Plant Sci Lett 8: 59-64

Vijayan P, Routaboul J-M and Browse J (1998) A genetic approach to investigating membrane lipid structure and photosynthetic function. In: Siegenthaler P.A. and Murata N. (eds) Lipids in Photosynthesis: Structure, Function and Genetics, pp 263–285. Kluwer Academic Publishers, Dordrecht

Wada H and Murata N (1998) Membrane lipids in cyanobacteria. In: Siegenthaler P.A. and Murata N. (eds) Lipids in Photosynthesis: Structure, Function and Genetics, pp 65–81. Kluwer Academic Publishers, Dordrecht

Walles B and Hudák J (1975) A comparative study of chloroplast morphogenesis in seedlings of some conifers (*Larix decidua, Pinus sylvestris* and *Picea abies*). Studia Forestalia Suecica 127: 2–22

Walter A, Yeagle PL and Siegel DP (1994) Diacylglycerol and hexadecane increase divalent cation-induced lipid mixing rates between phosphatidylserine large unilamellar vesicles. Biophys J 66: 366–376

Webb MS and Green BR (1991) Biochemical and biophysical properties of thylakoid acyl lipids. Biochim Biophys Acta 1060: 133–158

Webb MS, Tilcock CPS and Green BR (1988) Salt-mediated interactions between vesicles of the thylakoid lipid digalactosyldiacylglycerol. Biochim Biophys Acta 938: 323–333

Wellburn AR (1984) Ultrastructural, respiratory and metabolic changes associated with chloroplast development. In:. Baker NR and Barber J (eds) Chloroplast Biogenesis, pp 253–303. Elsevier Science Publishers, Amsterdam

Wellburn AR (1987) Plastids. Int Rev Cytol Suppl 17: 149–210

Whatley JM, Hawes CR, Horne JC and Kerr JDA (1982) The establishment of the plastid thylakoid system. New Phytol 90: 619-629

Widell-Wigge A and Selstam E (1990) Effects of salt wash on the structure of the prolamellar body membrane and the membrane binding of NADPH-protochlorophyllide oxidoreductase. Physiol Plant 78: 315–323

Wiktorsson B, Ryberg M and Sundqvist C (1996) Aggregation of NADPH-protochlorophyllide oxidoreductase-pigment complexes is favoured by protein phosphorylation. Plant Physiol Biochem 34: 23–34

Wiktorsson B, Engdahl S, Zhong LB, Böddi B, Ryberg M and Sundqvist C (1993) The effect of cross-linking of the subunits of NADPH-protochlorophyllide oxidoreductase on the aggregational state of protochlorophyllide. Photosynthetica 29:205–218

Williams WP (1998) The physical properties of thylakoid membrane lipids and their relation to photosynthesis. In: Siegenthaler PA and Murata N (eds) Lipids in Photosynthesis: Structure, Function and Genetics, pp 103–118. Kluwer Academic Publishers, Dordrecht

Wolter FP, Schmidt R and Heinz E (1992) Chilling sensitivity of *Arabidopsis thaliana* with genetically engineered membrane lipids. EMBO J 11: 4685-4692

Wu J and Browse J (1995) Elevated levels of high-melting-point phosphatidylglycerols do not induce chilling sensitivity in an *Arabidopsis* mutant. Plant Cell 7: 17–27

Chapter 12

Triglycerides As Products of Photosynthesis. Genetic Engineering, Fatty Acid Composition and Structure of Triglycerides

Daniel Facciotti and Vic Knauf
Calgene Inc., 1920 Fifth Street, Davis, CA 95616, U.S.A.

P.-A. Siegenthaler and N. Murata (eds): Lipids in Photosynthesis: Structure, Function and Genetics, pp. 225–248.

Summary

In this chapter we will review the contributions of recombinant DNA technology to the study of triglyceride synthesis. The success of this technology will firstly be illustrated by research concerned with fatty acid chain length. Until recently, there was no explanation for the preferential accumulation of fatty acids of a given chain length by certain species. Only through gene cloning and gene transfer technologies was it possible to prove unequivocally that specific thioesterases controlled the elongation of fatty acids. So, the diversity of fatty acids accumulated by different species can now be explained by the diversity of thioesterases active in these species.

Research on other enzymes involved in fatty acid synthesis and lipid assemblage is also progressing rapidly, following the path established for the study of thioesterases. In all likelihood, cloning 'through homology' will soon deliver the entire ketoacyl-ACP synthases gene family and provide detailed characteristics on the role of each KAS in fatty acid elongation. As for lipid assemblage, recent experiments with a coconut lysophosphatidylacyltransferase expressed in canola seeds support the expectations that future manipulations of specific acyltransferases will enable the production of a variety of new structured triglycerides.

With respect to fatty acid desaturation, the genetic manipulation of desaturase genes has already created potentially new oil products. The isolation of membrane-bound desaturase genes, however, remains hindered by their relative lack of homology. Instead of strict homology, the different desaturases share common motifs such as the 'histidine box' which was used to identify and isolate a delta-6 desaturase gene from borage. Since the sequences of entire plant genomes are now available, motif recognition could become a general approach to the cloning of new genes.

I. Introduction

It may be surprising to find a chapter on triglycerides in a volume dedicated to lipids in photosynthesis. Indeed, the lipids described throughout this book are structural lipids, some with well defined roles in the maintenance of specific membrane functions, while triglycerides are storage lipids. If we consider their syntheses, however, structural lipids and triglycerides share mostly a common pathway. The main differences may reside in the activity of a diacyl-glycerol acyltransferase, which in storage tissues esterifies the third fatty acid on the diacylglycerol molecule, and the yet unexplained mechanisms by which number of lipids, accepted as storage molecules, are excluded from cell membranes. Therefore studies enabling us to compare the similarities and the differences existing between

synthesis of triglycerides and structural lipids seemed, at least to us, relevant and complementary to each other. We believe this will be demonstrated in this chapter and the following Chapters 13 (Gombos and Murata 1997) and 14 (Vijayan et al., 1997).

There are excellent reviews which present classical biochemical approaches to the study of triglyceride synthesis (Stymne and Stobart, 1987; Somerville and Browse, 1991; Frentzen, 1993). In this chapter we focus on a more recent approach based on the use of recombinant DNA and gene transfer technologies. These technologies have made it possible to manipulate directly, in vivo, content and activity of enzymes involved in fatty acid and triglyceride syntheses (Knauf, 1993; Töpfer et al., 1995a). Yet, in a field where experimental activities are very intense there is no way to be exhaustive on the subject. While summarizing triglyceride synthesis, we organized this review around the following fatty acid characteristics: chain length, position on the glycerol backbone and degree of desaturation.

II. Generalities

Triglycerides are triesters of fatty acids and glycerol (Fig. 1). Since there is rotational asymmetry in a glycerol molecule, the position of each carbon is distinct, and so is the position of the fatty acids (R1,

---

Abbreviations: ACC – acetyl-CoA carboxylase; ACP – acyl carrier protein; BC – biotin carboxylase; CoA – coenzyme A; CPT – choline phosphotransferase; DAG – diacylglycerol; DAGAT – diacylglycerol acyltransferase; ER – endoplasmic reticulum; GPAT – glycerol-3-P acyltransferase; KAS – ketoacyl synthase; KCS – ketoacyl-CoA synthase; LPA – lysophosphatidic acid; LPAT – lysophosphatidic acid acyltransferase; LPC – lysophosphatidylcholine; LPA – lysophosphatidic acid; LPCAT – lysophosphatidic acid acyltransferase; MCFA – medium chain fatty acid; PA – phosphatidic acid; PC – phosphatidylcholine; PCR – polymerase chain reaction; TAG – triacylglycerol

C - R1 *(sn - 1)*

C - R2 *(sn - 2)*

C - R3 *(sn - 3)*

Fig. 1. Triglyceride structure. R1, R2 and R3: Fatty acids = CH_3-$(CH_2)_n$COOH.

R2, R3) esterified onto it. A *stereochemical numbering* system: *sn*-1, *sn*-2 and *sn*-3 designates these positions. Triglycerides are the main form of storage lipids in the majority of plant species. They are almost exclusively accumulated in the cotyledons and/or the endosperm of developing embryos. Besides seeds, the mesocarp tissues of oil-palm and avocado fruit are the only known tissues capable of storing significant amounts of triglycerides. Within the storage cell, triglycerides accumulate as discrete cytoplasmic oil bodies that are bordered by a single phospholipid layer with proteins comprising a half-unit membrane (Murphy et al., 1989; Huang 1993; Murphy, 1993). This proteinaceous membrane may prevent these oil bodies from fusing into a single, larger oil drop. A fragmented system of small oil bodies may offer a larger, more accessible surface to catalytic enzymes resulting in more efficient digestion during seedling development. Huang (1993) has suggested that the half-unit membrane bordering the oil bodies derived directly from the endoplasmic reticulum (ER), where triglycerides are initially synthesized and trapped between the double layer of proteins and phospholipids constituting the membrane (Huang, 1993).

The most common fatty acids found in triglycerides of oil crops grown under temperate latitudes are 16 and 18 carbon long molecules (C16, C18). These fatty acids can be saturated (C16:0, C18:0) or unsaturated, with up to three, usually *cis*, double bonds as in linolenate (C18:3). Some tropical crops, such as coconut and palm produce oils mostly composed of saturated, shorter C8:0–C14:0 fatty acids, while other crops such as rapeseed, grown under more northern latitudes, may deposit very long, monounsaturated fatty acids (C22:1). Regarding degree of desaturation, black currant oil contains small amounts (up to 4%) of stearidonate (C18:4), one of the most unsaturated fatty acids of terrestrial plants. Some algae, on the other hand, can produce fatty acids with 5 (C20:5) and 6 (C22:6) double bonds. The double bonds are usually three carbons apart from one another (divinyl methane blocks), as in linolenate, where they occupy positions C9, C12, C15 counted from the carboxyl (COOH⁻) terminus.

Chain length and degree of desaturation give fatty acids their chemical and physical properties, as presented in more detail in Chapter 6 (Williams). Usually, saturated fatty acids are chemically stable which is the reason why palm and coconut oils are the preferred oils for frying. The viscosity of a saturated fatty acid increases with its chain length. Oils rich in stearate (C18:0), usually prepared by artificial hydrogenation of unsaturated fatty acids, are solid at refrigerator temperature and are suitable for margarine preparation. In contrast, desaturation reduces chemical stability and oils rich in linolenate rapidly polymerize in the presence of oxygen and are used as varnishes. The off flavor of some oils has often been attributed to the chemical instability of linolenate (C18:3). Desaturation also lowers fatty acid melting points: 70 °C for stearate (C18:0), 13 °C for oleate (C18:1), and –57 °C for stearidonate (C18:4) (Heinz, 1993). When nutritional and health values are considered, even the position of a fatty acid on the triglycerol molecule is relevant. Indeed, the digestion of triglycerides in the upper intestine often leaves one fatty acid in its original *sn*-2 position. This determines how this fatty acid is transported, metabolized and how it may finally promote coronary disease (Small, 1991).

Plant oils are a mixture of triglycerides whose characteristics are genetically inherited. The variation of these genetic characteristics between crops creates the diversity of composition and properties that make oils suitable for a wide range of applications in the food, lubricant, polymer and cosmetic industries. Even greater diversity in oil compositions can be found in wild species than in domesticated crops (Stymne, 1993; van de Loo et al., 1993). The seeds of some wild species, such as bay laurel, accumulate oils essentially composed of C12 fatty acids. Oils from other species are characterized by even shorter (C8, C10) fatty acids, by hydroxy-fatty acids, such as ricinoleic acid in castor bean, by odd carbon chain lengths, by the presence of double bonds in unusual positions, or by epoxy, cyclic or branching groups. Unfortunately, the species producing these unusual oils such as *Vernonia* or *Euphorbia*, are not manageable, domesticated plants. As a consequence, there is only a marginal supply of these unusual oils and limited experimentation devoted to the discovery of their use in new applications and new products. The recent development of recombinant DNA and gene transfer technologies offers the prospect of isolating genes for specific oil characteristics from any plant or other living source, and introduce them into well domesticated crops such as canola or soybean. While attempting to create new products, this approach generates valuable information on triglyceride synthesis and accumulation.

III. The Building Blocks of Triglyceride Synthesis

A. Photosynthesis and Triglycerides

Photosynthates produced in leaves, tendrils or pods in the case of *Brassica* species, are transported as sucrose via the phloem and delivered to the embryo through the seed coat. The role of the seed coat on embryo development has been recently investigated. The specific expression of invertases in this organ may have particular repercussions on the size and thereby the storage capacity of the embryos. According to recent studies (Weber et al., 1996), high seed coat invertase activity at early stages of development would maintain the embryos in high hexose conditions, which correlates with extended mitotic activity in the cotyledons. This, in turn, would determine the number of cotyledonary cells and, finally, the storage capacity of the embryo. Lower invertase activity following the mitotic phase would then provide high sucrose condition, which correlates with the deposition of storage molecules in the cotyledons.

In non-chlorophyllous embryos such as sunflower, imported sucrose provides the embryos with both precursors and reducing power necessary for the synthesis of triglycerides. In green, chlorophyll-rich cotyledons of immature embryos it has not yet been established if photosynthesis plays any direct role in the accumulation of storage molecules, either by producing some of the photosynthates and/or some of the reducing molecules. Gametic embryo cultures of canola accumulated more lipids (ca. 80%) when grown in the light than in the dark (D. Facciotti, unpublished). Recent experiments with in vitro grown zygotic *Brassica* embryos showed that the rate of ^{14}C acetate incorporated into oils was also higher in the light than in the dark, suggesting some active role of light (J. B. Ohlrogge, personal communication). This

role on lipid accumulation, however, may not be through photosynthetic activities. Since the experiments with gametic and zygotic embryos were both conducted in media containing sucrose, light may have influenced, perhaps, some aspect of sucrose uptake.

B. Glycerol Backbone

Triglycerides are formed in storage tissues by the esterification of three fatty acids to the three hydroxyls of a glycerol molecule. Fatty acid chains are themselves the products of the condensation of acetate and malonate. We will briefly review the origin of the building blocks, glycerol, acetate and malonate.

The immediate precursors of glycerol-3-P, the active form of glycerol, are the products of the sucrose glycolysis in the cytoplasm (Fig. 2). Glycerol-3-P could derive directly from the reduction of dihydroxyacetone-P by glycerophosphate dehydrogenase (Gee et al., 1988) as it is the case in animal cells. Groundnut seeds, however, do not display glycerophosphate dehydrogenase activity (Ghosh and Sastry, 1988). In these seeds a second source has been proposed: glyceraldehyde-3-P could be dephosphorylated, reduced to glycerol and rephosphorylated by glycerol kinase when needed. This alternative path was suggested by experiments in which glycerol fed to tissue slices readily incorporated into triglycerides (Ghosh and Sastry, 1988; Sangwan et al., 1992). The activity of the glycerol kinase, however, is relatively low in oil seed tissues of other crops and does not account for the full accumulation of triglycerides (Barron and Stumpf, 1962; Gee et al., 1988). It is possible that both routes are used according to plant species and plant tissues. Usually glycerol synthesis is not considered a limiting step in triglyceride synthesis. Competition between the deposition of proteins and triglycerides observed in oil-seeds (Canvin, 1964; Kumar and Tsunoda, 1980) suggests that the pool of acetyl-CoA precursors may become a limiting factor in fatty acid synthesis when diverted towards the synthesis of amino acids and proteins (Fig. 2). This should not affect the production of glycerol that branches out earlier from the glycolytic pathway. Intuitively, one would expect some coordination between triglycerides and glycerol synthesis. A tight negative feed-back by glycerol may control the dephosphorylation of glyceraldehyde-3-P. The use of glycerol-3-P in lipid synthesis would pull the reaction towards dephosphorylation and the

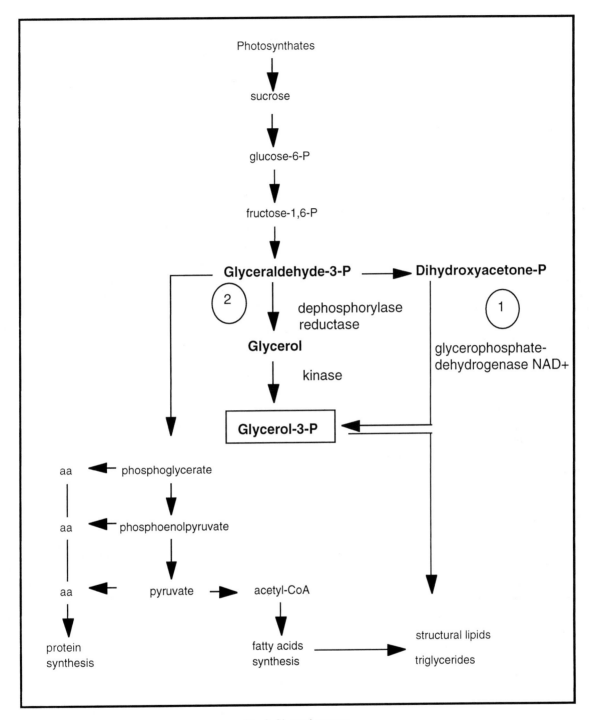

Fig. 2. Glycerol sources.

synthesis of glycerol would occur only when needed. Since some plastids contain the enzymes necessary for glycolysis (Dennis and Miernyk, 1982; Myernik and Dennis, 1982), it is not clear whether they are self-sufficient in glycerol-3-P or if they also use imports from the cytoplasm when they build up their structural lipids.

C. Acetate

Fatty acids, the other triglyceride components, result from the repeated condensation of two carbon molecules. Starting with acetate as an initiator, repeated condensations of malonate, the 'elongator' molecule, bring the chain to its full length. The malonate, as we will see it below, is itself derived from acetate. Several acetate sources have been proposed (Fig. 3). One recent suggestion is that malate produced by cytoplasmic glycolysis is the precursor of acetate (Smith et al., 1992). This suggestion is based on results from feeding experiments in which malate given to leucoplasts of castor endosperm appears to affect the fatty acid synthesis rate more efficiently than acetate or pyruvate precursors. Alternatively, in leucoplasts from the same tissues, the acetate moiety might also derive from pyruvate via glycolysis occurring within the leucoplasts themselves (Dennis and Miernyk, 1982; Myernik and Dennis, 1982). Indeed, the activity of glycolytic enzymes measured in leucoplast extracts could account for all the structural lipids and triglycerides deposited by these seeds. Mitochondrial acetyl-CoA could be a third possible source for acetate. Under this hypothesis, acetyl-CoA formed in the mitochondria would be released as free acetate in the cytoplasm, it would enter the plastid compartment and be esterified to CoA (Murphy and Stumpf, 1981; Liedvogel and Stumpf, 1982). The possible role of carnitine as acetate carrier into the plastids has also been discussed in the case of pea seed (Masterson et al., 1990). No evidence, however, exists to confirm this route in other species (Roughan et al., 1993). A fourth cytoplasmic source of acetyl-CoA is now being investigated in plants. Acetyl-CoA might originate from citrate by the action of an ATP:citrate lyase as this is the case in animal and yeast cells. The activity of this enzyme has been detected in various plant tissues (Fritsch and Beevers, 1979; Kaetner and Rees, 1985). Finally, in chloroplasts a portion of the acetyl-CoA could also be generated either as the products of ribulose-bisphosphate carboxylase (Andrews and Kane, 1991) or from the activity of pyruvate dehydrogenase on pyruvate. The pyruvate dehydrogenase activity measured in spinach chloroplasts, however, appears insufficient to account for the actual synthesis of lipids (Roughan et al., 1979). To our knowledge such activity has not been reported for storage tissues of 'green' cotyledons. All the studies we have just reported suggest that there may be several routes to provide cells with acetate for fatty acid synthesis. Each plant cell appears to be completely independent for the synthesis of its lipids. Fatty acids are synthesized in the plastid compartment which varies considerably from tissue to tissue in morphology, function and metabolism: proplastids in meristems and young cells, amyloplasts in roots and tuber cells, leucoplasts in the endosperm of certain species,

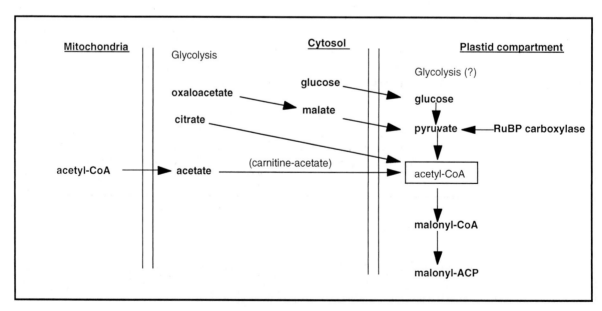

Fig. 3. Acetyl-CoA sources.

chromoplasts in fruit cells, chloroplasts in photosynthetic cells. It is possible that the different sources of acetate that have been reported fit in with the metabolism and the type of plastid present in each cell. This flexibility would assure that fatty acids can be synthesized in the cells of every tissue.

D. Malonate

A large portion of the acetyl-CoA, either produced or imported into the plastid, is then converted into malonyl-CoA through the activity of the enzyme acetyl-CoA carboxylase (ACC). This is often considered the first step committing acetate towards the synthesis of fatty acids. Although malonyl-CoA is found in the cytoplasm as a precursor of flavonoids or stilbenes (Ebel and Hahlbrock, 1977) the plastidial malonate may be reserved almost exclusively to the synthesis of fatty acids. The formation of malonyl-CoA corresponds to the fixation of CO_2 onto acetyl-CoA (Fig. 4). The reaction, catalyzed by an acetyl-CoA carboxylase, proceeds in two steps: an ATP mediated fixation of CO_2 on biotin, followed by the transfer of CO_2 onto acetyl-CoA. A possible regulatory role in lipid synthesis has been attributed to ACC because its activity in leaves is modulated by light and parallels fatty acid synthesis, with a higher activity during exposure to light (Eastwell and Stupmf, 1983; Nikolau and Hawke, 1984; Post-Beittenmiller et al., 1992). Acetyl-CoA carboxylases have been studied intensely in the past years in both animals and bacteria. A type II ACC complex has been purified from *E. coli*. It is composed of a biotin carboxyl carrier protein (BCCP), a biotin carboxylase (BC) and two dissimilar subunits of the carboxyl-transferase (CT). The complex can be dissociated and the activities of its components tested separately (Guchheit et al., 1974). Each protein is coded by a different gene (Li and Cronan, 1992). In animals, a type I ACCase has been isolated: a single polypeptide which carries the three functional domains of the bacterial ACC complex (Luo et al., 1989). In plants the two types of ACCases may coexist within the same cells (Nikolau et al., 1984). Large single polypeptides (type I) have been characterized in dicots and monocots with sizes over 200 kDa (Egin-Buhler and Ebel, 1983; Slabas and Hellyer, 1983). In dicots it is believed that they represent the cytoplasmic form of ACCase which catalyzes the malonyl-CoA used for the synthesis of stilbenes, flavonoids or for the extraplastidial elongation of very long fatty acids such as erucate (C22:1). Smaller individual proteins and genes, corresponding to type II bacterial ACC components, have also been isolated from plastids of dicots cells (Shorrosh et al., 1995; Konishi et al., 1996; Roesler et al., 1996). In monocots, on the other hand, a type I ACC, ca. 2300 amino acids long and terminated by a transit peptide for transfer into the chloroplast has found (M. Egli, personal communication).

The genetic engineering of ACC has begun. A gene for the cytoplasmic type I ACCase, isolated from *Arabidopsis* has been overexpressed behind a seed specific promoter in canola. The gene product was directed either to the cytoplasm or to the plastid by addition to the coding region of a sequence encoding a transit peptide. In the latter case the transgenic seeds were reported to display some (5%) increase in oil content (Ohlrogge and Jaworsky, 1997). It is fair to add, however, that the data collected so far

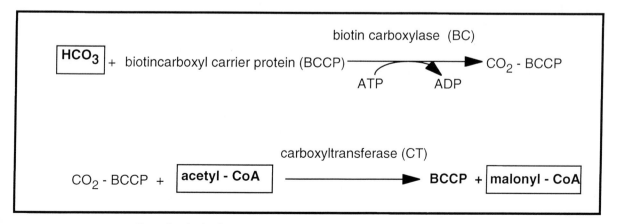

Fig. 4. Synthesis of malonyl-CoA by acetyl-CoA carboxylase (ACC) complex.

from transgenic ACC plants are preliminary and it is still too early to reach any conclusion on the regulatory role of ACC in lipid and triglyceride synthesis.

Finally, before entering fatty acid synthesis the malonate is transferred from malonyl-CoA to an acyl carrier protein (ACP) by a malonyl-CoA:ACP transacylase. The transfer of malonyl to ACP is not considered a limiting step in fatty acid synthesis. The malonyl-CoA:ACP transacylase has been purified to homogeneity from avocado tissues (Caughey and Kekwick, 1982). We are not aware, however, of further attempts to clone the avocado gene or to produce transgenic plants with modified levels of the plant enzyme. A bacterial malonyl-CoA:ACP transacylase, was cloned from *E. coli* and expressed in plants under the control of the seed specific promoter napin (Stuitje et al., 1993). The gene product was targeted to the chloroplast and the presence of an active bacterial transacylase was found in in vitro assays. There was no difference in oil content or composition in the transgenic seeds.

IV. Fatty Acid Synthesis and Fatty Acid Chain Length

A. Fatty Acid Synthesis

With acetyl-CoA and malonyl-ACP as initiator and elongator molecules respectively, the fatty acid synthesis begins in the plastid stroma. The fatty acid chain elongates by repeated increments of two carbons provided by malonate until the chain reaches its final length (Fig. 5). It starts with the condensation of acetyl-CoA and malonyl-ACP. The condensing enzyme, a ketoacyl-ACP synthetase III (KASIII), mediates the formation of acetoacetyl-ACP. Acetoacetyl-ACP is then reduced to hydroxybutyryl-ACP by a ketoacyl-ACP reductase. A dehydration step follows and yields crotonyl-ACP which is finally converted to butyryl-ACP by the action of an enoyl-ACP reductase. A second elongation cycle including the same four steps can then start. A different condensing enzyme, however, KASI, catalyzes the formation of C6 acyl-ACP. With repeated elongation cycles, and KASI as condensing enzyme the chain of acyl-ACP is elongated to up to 16 carbons. An additional elongation cycle usually occurs involving yet another condensing enzyme, KASII. In the majority of plant species, most of the C18:0-ACP

(stearoyl-ACP) molecules formed are then desaturated (see Section VI.B1) by a delta 9 desaturase and converted to oleoyl-ACP. At this point the ACP group of C16:0-ACP, C18:0-ACP, or C18:1-ACP is hydrolyzed by a thioesterase. The free fatty acids are then thought to diffuse into the plastid membranes. An acyl CoA synthase, located presumably in the outer envelope, further catalyzes the formation of soluble acyl-CoAs and directs them to the cytosol.

In some crucifer and in jojoba seed tissues, additional elongation cycles, mediated by endoplasmic reticulum bound enzymes, will elongate oleoyl-CoA (C18:1-CoA) mainly to erucyl-CoA (C22:1-CoA). In this case malonyl-CoA is used as elongator molecule rather than malonyl-ACP.

Some of the C16:0-ACP, C18:0-ACP, and C18:1-ACP molecules formed in the plastid are diverted towards the synthesis of plastidial lipids. In this case the acyl groups are directly transferred by acyltransferases to glycerol-3-P or lysophosphatidic acid. The plastidial acyltransferases place C16:0 specifically in the *sn*-2 position (Frentzen et al., 1983; Cronan and Roughan, 1987) and C18:1 preferentially in *sn*-1 (Murata et al., 1982; Murata, 1983). The glycerol-3-phosphate-acyltransferase (GPAT) specificity for C16:0 and C18:1 may vary considerably between species (Frentzen, 1993) and may determine some important adaptive plant features. GPAT from chilling insensitive *Arabidopsis* appears to be more selective for C18:1 than GPAT from chilling sensitive squash. Plastidial GPAT, cloned from squash and *Arabidopsis* and an *E. coli* C16:0-selective GPATs were overexpressed in tobacco leaves (Murata et al., 1992; Wolter et al., 1992). The transgenic tobacco carrying the *E. coli* or the squash GPAT were more sensitive to chilling than control plants. Chilling sensitivity was somewhat reduced in tobacco expressing the *Arabidopsis* GPAT.

The fatty acid elongation pattern described above is common to all plants. This is well reflected in the similarities displayed by their structural lipids. In contrast, there can be remarkable variability between the seed oil compositions of different species, especially with respect to fatty acid chain length. As mentioned earlier, some crops are rich in triglycerides with chains shorter than C18 or C16. Palm and coconut, in particular, are crops in which oil fatty acids are mostly C12, C14 and C16 long. In other non-domesticated plants such as the California bay laurel or elm, the seed oil is composed essentially of

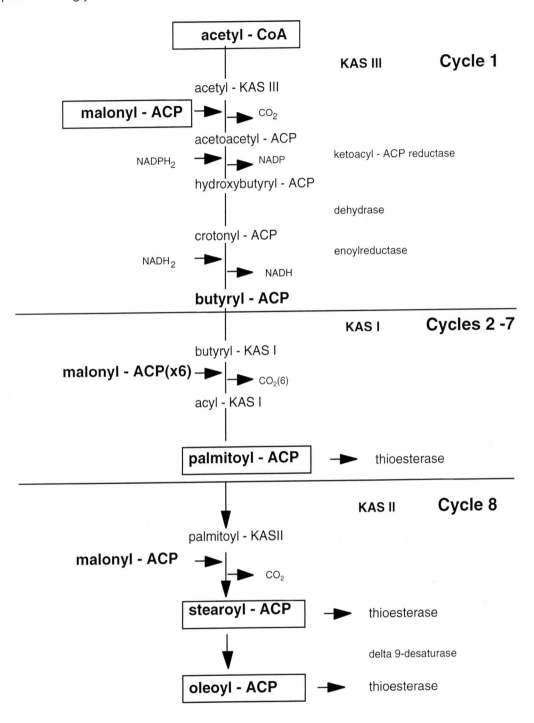

Fig. 5. Fatty acid synthesis.

C12:0 or C10:0, respectively (Pollard et al., 1991; TA Voelker, personal communication). Some *Cuphea* varieties produce mainly shorter fatty acids C8:0 and C10:0 while others accumulate preponderantly C10 and C14:0 rich triglycerides. Since these differences are genetic in nature and therefore potential targets for modifications, we will review the recent studies on the mechanisms which determine fatty acid chain lengths. Some of the enzymes participating in elongation cycles such as ketoacyl reductases, dehydrases, and enoyl reductases have not been considered here because they seem unlikely to play a role in determining chain length.

B. Acyl Carrier Proteins (ACP)

Early in vitro experiments comparing two leaf ACP isomers of spinach pointed to their possible role in determining substrate specificity: one isomer would make oleoyl-ACP the substrate for a thioesterase while the other isomer would direct it to the plastidial glycerol-3-P acyltransferase (Guerra et al., 1986). It was not established whether this potential role in partitioning fatty acids between plastid and cytoplasm had any relevance in vivo. Since the isolation of the first ACP gene from spinach leaves (Scherer and Knauf, 1987), numerous ACP isomers have been cloned. All contain highly conserved domains (90% homology) around the point of attachment of the prosthetic group phosphopantetheine as well as differences, with an overall homology averaging approximately 70% (Ohlrogge et al., 1993). Some of the ACP isomers appear to be expressed specifically in leaf tissues (Ohlrogge and Kuo, 1985; Battey and Ohlrogge, 1990). Others may be specific to seeds as reported for *Brassica napus* (de Silva et al., 1990) and *Arabidopsis* (Hlousek-Radojcic et al., 1992). Because of these structural differences and the putative role in fatty acid partitioning there have been speculations that different ACP isomers may be specific to different fatty acids and may confer specific affinities to either ketoacyl-synthases (KAS) or thioesterases. None of these hypotheses has yet been confirmed.

There may be as many as 35 different ACP genes in *Brassica napus* (de Silva et al., 1990). This diversity would suggest that ACPs form a relatively large and degenerated gene family with little or no functional differences between isomers.

C. The Thioesterases

In early experiments thioesterase-enriched fractions from avocado, coconut and jojoba were characterized in vitro for their specific activities on different acyl-ACP substrates (Shine, 1976; Ohlrogge et al., 1978). The fractions from the three species preferentially hydrolyzed C18:1-ACP (oleoyl-ACP) over C18:0-ACP and C16:0-ACP. This specificity pattern failed to explain the fatty acid profile of the oils from coconut, which is composed predominantly of shorter, C8:0–C14:0 fatty acids (medium chain fatty acids = MCFA). The failure to find thioesterases with enhanced activity on medium chain acyl-ACPs in coconut (Oo and Stumpf, 1979) led to other hypothetical chain termination mechanisms (Harwood, 1988). The problem was re-examined using seeds from the California bay laurel (Pollard et al., 1991). This material was chosen because bay laurel seeds accumulate approximately 60% of fatty acids as C12:0 (laurate) and 30% as C10:0 (caprate). This oil composition had a simpler profile than that of coconut, where a mixture of MCFA are present, perhaps as the result of a complex interaction between different thioesterases or other chain termination factors. Using the California bay seeds, Pollard et al. (1991) found preferential hydrolytic activity for C12:ACP (lauryl-ACP). The activity could be attributed to a soluble enzyme. Following further purification to near homogeneity, a native enzyme with MW of approximately 42 kDa was isolated (Davies et al., 1991). Amino acid sequences of the enzyme were obtained and used to synthesize DNA probes. The screening of maturing bay seed cDNA libraries soon delivered a cDNA clone of the expected size. Sequences encoding an apparent transit peptide, presumably necessary for the transfer of the enzyme into the plastid, were present. This gene was expressed in *E. coli* as a *lacZ* fusion product and yielded a protein with thioesterase activity specific for C12:0-ACP. Consequently the gene was transferred into *Arabidopsis* and canola under the control of a napin promoter for specific expression in the seed. The laurate content of the oil from transgenic seeds exceeded 40% of the total fatty acids extracted (Voelker et al., 1992). This experiment clearly demonstrated that in bay laurel, chain elongation/termination depended mainly on the activity of this specific thioesterase.

The next research steps consisted of generalizing the role of thioesterases as the main factor determining fatty acid chain length in other species. Based on the comparisons between the primary structure of the safflower oleoyl-ACP thioesterase (Knutzon et al., 1992b) and that of the California bay lauryl-ACP thioesterase, it seemed that acyl-ACP thioesterases were related and formed a large homologous gene family. Thus for the first time it became possible to design PCR primers on the basis of potential homology and to use synthetic probes to screen the various cDNA libraries prepared from the developing seeds of various plant species. The advantage of this approach was that it bypassed the relatively long and costly protein purification steps that had been necessary to isolate the bay laurel thioesterase. Although the cDNA libraries of many species were screened, *Cuphea* was perhaps the most investigated of all genuses (Voelker et al., 1995; Toepfer et al., 1995b). Indeed, there is a variety of medium chain fatty acids stored by the seeds of different *Cuphea* species. Individual species accumulate mostly C8:0, C10 or C12, whereas others have a mixture of C10 and C12 and some even a mixture of unsaturated C18 fatty acids resembling that of soybean oil. A number of *Cuphea* thioesterase genes have been isolated, expressed in *E. coli* for quick specific activity tests, and finally transferred into canola, a species which does not naturally accumulate MCFA. The oils from the transgenic seeds contained significant amounts of the MCFA that had been specified by the *Cuphea* thioesterase gene (Töpfer et al., 1995; Dehesh et al., 1995).

Thioesterases with specificity for longer chain acyl-ACPs have also been cloned from crops producing C18:1. The expression of a safflower C18:1-ACP thioesterase in transgenic canola increased, unexpectedly, the content of palmitate (J. Kridl, personal communication). This confirmed, however, in vitro activity profiles where, besides its marked preference for oleoyl-ACP (C18:1), the enzyme displayed some affinity for palmitoyl-ACP (C16:0). Conversely, C18:1 producing crops may also possess thioesterases with affinity for shorter chains, as did C16:0-ACP thioesterase found in canola (Jones et al., 1995). The overexpression of this gene in canola seeds increased their palmitate content. These results are a strong indication that in vivo, the combination of several thioesterase activities may be responsible for the relative ratios of C16:0, C18:0 and the unsaturated C18 fatty acids. More recently, a thioesterase with some specificity for C18:0-ACP has been cloned from mangosteen seeds, a species which accumulate large amounts of C18:0. When expressed in canola seeds, the mangosteen thioesterase affected their fatty acid composition by increasing, as expected, the levels of stearate (C18:0) up to 20% (J. Kridl, personal communication). All the thioesterase genes with specificity for the saturated fatty acids from C8:0-ACP to C16:0-ACP share considerable DNA and protein homology. They seem to be more closely related to each other than to the

genes from several species coding for the stearoyl-ACP and oleoyl-ACP thioesterases (Jones et al., 1995). Thus, two families of thioesterases have been designated on the basis of their DNA homology. The thioesterases with specificity for oleoyl-ACP and/or stearoyl-ACP belong to a group A, all the others to a group B.

D. Ketoacyl-ACP-Synthases (KAS)

There is no doubt that thioesterases play the major role in determining the length of a fatty acid. It is interesting to note, however, that discrepancies in thioesterase specificities seem to exist, in vivo, between native and transgenic plants. In California bay seeds, for example, considerable C10:0 can be found in addition to C12:0. In transgenic canola expressing the cloned bay thioesterase gene, however, C10:0 (caprate) is absent. This discrepancy might be due to differences in specific activities between canola and bay KASs. Presumably these condensing enzymes compete with thioesterases for the same pool of acyl-ACPs (Davies, 1993; Voelker and Davies, 1994). Therefore it is possible that in canola the KAS elongating short chains are more efficient than those of bay seeds, particularly with C10:0-ACP. Species-specific differences in KAS activities might exist among all MCFA producing plants and be genetically determined. KAS genes, especially those transcribed in triglyceride storing tissues, may form a large gene family encoding isomers with different specificities for fatty acids of given chain lengths and also with different efficiencies as condensing enzymes. The size of the different acyl-ACP pools available to thioesterases would depend on the specificity and activity of each different KAS. Some support for this hypothesis can be found with bacterial KAS. Different KAS genes have been isolated from *E. coli*. In terms of specificity, it was believed for quite a long time that only two KAS enzymes were responsible for the elongation of fatty acids up to C18:0: KASI using acyl-ACP as initiator was thought to elongate the molecule up to C16:0 and KASII would then complete the elongation to C18:0 (Fig 5). Since then KASIII was discovered which has high specificity for acetyl-CoA (Tai and Jaworsky, 1993). A KASIV with preference for C6:0-ACP and C8:0-ACP was also reported (Siggaard-Andersen et al., 1994). The KASIV specificity, however, remains somewhat controversial (Edwards and Dehesh, 1997). Such diversity in specificity may also exist for plant KASs, especially when triglyceride synthesis is concerned.

KASI and KASII proteins from castor bean endosperm have been purified (Genez et al., 1991). The KASI appears to be a dimer of two 50 KDa subunits. This agrees well with the KASI gene identified from barley (Siggaard-Andersen et al., 1991). The castor bean KASII preparation contained two polypeptides of respectively 50 KDa and 46 KDa. Monoclonal antibodies raised against the 50 KDa subunit of KASII cross-reacted with KASI but not with the 46 KDa subunit. The KASI 50KDa polypeptide and the KASII 50 KDa subunit may be identical. The cDNAs for both 50 KDa and the 46 KDa polypeptides have been cloned from castor bean and rapeseed (Genez et al., 1991). There is clear DNA and amino acid homology between the genes of both species and between the 50 KDa and 46 KDa genes, and with KASI and KASII of *E. coli*. The two castor bean genes coding for 46 KDa and 50 KDa subunit of KASII were transferred into *Arabidopsis* to be specifically overexpressed in seed tissues (G. A. Thompson, personal communication). KASII proteins were detected with antibodies raised against the 50 KDa in preparations of transgenic *Arabidopsis*. In the case, where the highest KASII expression was detected, the content of seed palmitate (C16:0) decreased by approximately 40%, but in most other transgenic lines more modest reductions were observed. Similar results were recently reported for transgenic *Arabidopsis* seeds expressing a putative Cuphea KASII (J. M. Leonard, personal communication). The moderate effects of extra KASII on the C16:0 content suggest that in *Arabidopsis* seeds KASII may already be present at levels adequate to support larger flux of C16:0-ACP. Higher level of expression of KASII in seed tissues may be necessary to compete with both thioesterases and the plastidial transacylases, and to reduce more effectively the content of C16:0. At the present time, the role of KAS in determining the length of fatty acids plants remains mostly hypothetical due to the limited number of plant KAS genes cloned. There is now a considerable effort in cloning KAS genes by taking advantage of their possible homology, as this was so successfully done with thioesterases.

E. Ketoacyl-CoA Synthases (KCS)

As described above (Section IV.A), the elongation of fatty acids takes place in the plastid stroma. In some plants such as rapeseed, jojoba and meadow foam, which accumulate very long fatty acids (C20, C22 and C24), additional elongation cycles are completed at the ER level, as it was demonstrated with microsomal preparations of jojoba. Recently, a membrane-bound ketoacyl-CoA synthase (KCS) was purified from jojoba (Lassner et al., 1996). A probe was derived from a partial amino acid sequence of the enzyme and used to clone the KCS gene from a jojoba cDNA library. The gene was transferred into canola to be expressed specifically in seed tissues. Canola was chosen because it is essentially a rapeseed mutant in which erucic acid (C22:1) synthesis has been selected out. The jojoba ketoacyl-CoA synthase complemented the canola 'lesion' and generated transgenic seeds containing more than 50% of very long chains fatty acids (C20, C22 and C24) (Lassner et al., 1996). These experiments confirmed the hypothesis that in canola the 'lesion' directly affected ketoacyl-CoA synthase activity.

V. Assemblage of Triglycerides. Acyltransferases and Structured Lipids

A. Assemblage of Triglycerides

The assemblage of fatty acids with glycerol-3-P to form structural lipids and triglycerides takes place at the ER level and involves membrane-bound acyltransferases (Fig. 6, and for review: Frentzen, 1993). A first acyltransferase (glycerol-3-P acyltransferase = GPAT) esterifies the fatty acid to the *sn*-1 of the glycerol-3-P to form lysophosphatidic acid (LPA). A second fatty acid is esterified in *sn*-2 by a lysophosphatidate-acyltransferase (LPAT) to form phosphatidate (PA). A phosphatase then restores the hydroxyl group on *sn*-3 of the PA to form a diacylglycerol (DAG). At this point there are two possible fates for the diacylglycerol molecule. A third esterification may take place, mediated by a diacylglycerol-acyltransferase (DAGAT) to form a triglyceride molecule (TAG). Alternatively DAG may be converted to phosphatidylcholine (PC) in a fully reversible reaction catalyzed by a cholinephosphotransferase (CPT). The PC itself can generate two isomers of lysophosphatidic acid (LPC) with free *sn*-1 or *sn*-2, and liberate an acyl-CoA by a reversible reaction catalyzed by a lysophosphatidylacyltransferase (LPCAT). The reversibility makes

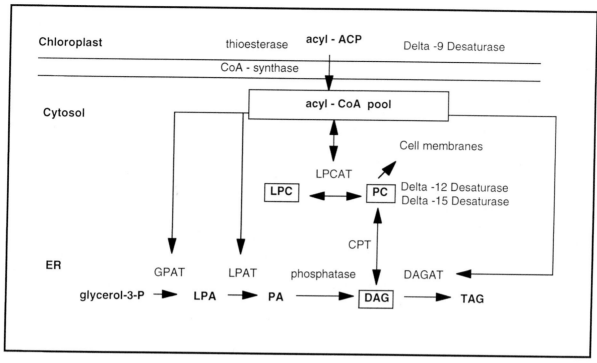

Fig. 6. Assemblage of triglycerides and formation of polyunsaturated fatty acids. CoA – coenzyme A; CPT – cholinephosphotransferase; DAG – diacylglycerol; DAGAT – diacylglycerol acyltransferase; ER – endoplasmic reticulum; GPAT – glycerol-P acyltransferase; LPA – lysophosphatidic acid; LPAT – lysophosphatidylacyltransferase; LPC – lysophosphatidic acid; LPCAT – lysophosphatidic acid acyltransferase; PA – phosphatidic acid; PC – phosphatidylcholine; TAG – triacylglycerol

LPC a receptor of acyl-CoA and PC a donor to the pool of acyl-CoA. The acyl-CoAs from this pool can then be (re)esterified by any of the acyltransferases. As we will see below (Section VI.C) the polyun-saturated fatty acids are produced at the PC level. The reversible CPT reaction, by reconverting PC into DAG, makes possible the incorporation of poly-unsaturated fatty acids into triglycerides. LPCAT makes polyunsaturated fatty acids available to DAGAT for esterification in *sn*-3.

B. Acyltransferases and Structured Lipids

In crops such as safflower, soybean, rapeseed and canola, the position of the fatty acids on the glycerol backbone of triglycerides is not random (Stymne and Stobart, 1987). Saturated fatty acids C16:0 and C18:0, ordinarily present in small amounts, are only found in *sn*-1 and *sn*-3. The *sn*-2 position is normally occupied by unsaturated fatty acids. In the last decade, studies on microsome acyltransferases from the seeds of several species have indicated that the position of a fatty acid on triglycerides is determined by the

selectivity of the acyltransferases (Griffiths et al., 1985; Ichihara et al., 1987; Bafor et al, 1990). This selectivity can vary from species to species. In a crop like canola GPAT selects for saturated fatty acids and LPAT selects exclusively for unsaturated ones, while DAGAT accepts both saturated and unsaturated fatty acids (Oo and Huang, 1989). This was confirmed in transgenic canolas expressing the California bay C12:0-ACP thioesterase, where C12:0 was found almost exclusively in *sn*-1 and *sn*-3 (Voelker et al., 1996). By contrast, laurate (C12:0) is also present on *sn*-2 both in bay and coconut triglycerides. The canola LPAT appears to exclude saturated fatty acids and thereby may limit the total laurate accumulation to ²/₃ of its potential on a mole/mole basis. In the attempt to bypass this limitation, an LPAT accepting lauryl-CoA as a substrate was identified in coconut endosperm (Davies et al., 1995). The membrane-bound enzyme was solubilized and purified, so as to provide the primary amino acid sequences necessary to the synthesis of a DNA probe. A putative LPAT gene was then cloned and expressed in *E. coli* where its product confirmed the expected LPAT medium-chain activity. The same activity was later confirmed in preparations from transgenic canola seeds expressing the same gene (Knutzon et al., 1995). Finally the coconut-LPAT canola lines were crossed with canola lines expressing the bay laurel thioesterase. Preliminary results indicate that the amount of laurate (C12:0) in the sn-2 position increased significantly in the hybrid seeds (M. Davies, personal communication). It remains to be established whether the esterification in sn-2 will lead to an overall increase in laurate content or if it will only result in the redistribution of laurate among all the sn- positions. LPAT selectivity has also been identified as the major limitation to the accumulation of erucate in rapeseed. Some rapeseed lines contain up to 50% erucate (22:1), that is found exclusively in sn-1 and sn-3. In some plants such as meadow foam (Limnanthes alba), erucate is present in sn-2 (Cao et al., 1990). PCR primers, designed on the basis of the coconut LPAT gene template, helped the cloning of a meadow foam LPAT. The oil from transgenic rapeseed expressing the meadow foam LPAT contains a portion of the expected trierucin molecules (Lassner et al., 1995).

Although the results obtained with coconut and meadow foam LPATs are still preliminary, they confirm, in vivo, the results obtained earlier, in vitro with microsome fractions. They also prove that lipid structure can be engineered. Additional studies will determine to what extent acyltransferases can be manipulated to develop new structured lipid products.

VI. Fatty Acid Desaturation

A. Desaturation in Relation to Low Temperature

Oleate, linoleate and linolenate, with respectively one, two, and three cis- double bonds are the most common unsaturated fatty acids found in plants. Stearidonic acid (C18:4) with four double bonds is an exception present in limited proportions (3–4%) in black currant and borage oil. When compared to numerous algae, fungi and most animals, the degree of desaturation in the fatty acids of terrestrial plants is unusually low. In the oils of some algae, fungi, and animals, fatty acids with four (C20:4, arachidonate), five (C20:5, eicosapentaenoic), and six (C22:6, docosahexaenoic) double bonds are relatively common (Ratledge, 1993). The degree of desatura-

tion, as we have already presented above (Section II), gives a fatty acid some of its physical and chemical properties (see Chapter 6, Williams). A higher degree of desaturation confers a lower viscosity at low temperature. This property has been discussed for structural lipids in relation to adaptation to cold environment (Murata et al., 1982; Murata, 1983). Desaturation would assure survival by maintaining membrane fluidity at low temperature [see also Chapters 13 (Gombos and Murata) and 14 (Vijayan et al.)]. Experiments with blue-green algae have provided direct support to this hypothesis (Wada et al., 1990). Anacystis nidulans synthesizes mainly monounsaturated fatty acids and is sensitive to low temperatures. Synechocystis PCC 6803, in contrast, synthesizes linolenic acid in response to low temperatures and survives. The delta-12 desaturase gene from Synechocystis was cloned and expressed in Anacystis. Not only did the transformed Anacystis produce linolenic acid, but it was also more resistant to low temperatures. The need for polyunsaturated fatty acids under low temperature conditions was also observed with an Arabidopsis mutant (Kunst et al., 1989a,b). The mutation affected a chloroplast monogalactosyldiacylglycerol desaturase and the formation of C16:1 (delta-9 hexadecenoic acid) and led to membranes rich in its precursor, C16:0. The plants grew faster at high temperature but, contrary to expectations, no sensitivity to low temperature was observed. The chloroplast had compensated the lack of C16:1 by incorporating polyunsaturated fatty acids synthesized at the ER level. This also demonstrated the existence of a flow of fatty acids from the cytoplasm to the organelles. Other Arabidopsis mutants suggest a flow of polyunsaturated fatty acids in the opposite direction (Ha and Thompson, 1992; Miquel and Browse, 1992). It has not been established if such a flow could modify the fatty acid composition of triglycerides in some species.

When canola is grown at relatively low temperatures (<18 °C) the seeds produce oils which are richer in polyunsaturated fatty acids (Canvin, 1964). This could reflect the higher degree of desaturation affecting structural lipids and would suggest that in canola a single mechanism underlies the synthesis and desaturation of both structural and storage lipids in response to low temperatures. This, however, may not be the case in other plant species.

B. Plastid Desaturases

1. Delta-9 Desaturase: Low and High Stearate Canolas

The first desaturation of a fatty acid takes place in the plastid and converts stearoyl-ACP into oleoyl-ACP. The plastid stearoyl desaturase is a soluble stroma enzyme using ferredoxin as electron donor. Genes coding for this enzyme have been cloned from several plant species including safflower, castor bean, and rapeseed (Knutzon et al., 1991; Shanklin and Somerville, 1991; Shanklin et al., 1991; Thompson et al., 1991; Knutzon et al., 1992a; Sato et al., 1992). A transit sequence for transfer into the plastid was present in the genes from the three species. The gene from safflower was transferred into canola and overexpressed in seeds. In most of the transgenic seeds the levels of stearate (C18:0) dropped significantly (ca. 40%) from ca. 1.5%–2% to ca. 1% (G. A. Thompson, personal communication). At the same time the contents of C20:0 and C22:0 which totaled ca. 2% in control seeds, were further reduced indicating that C18:0 may be the direct precursor of these fatty acids.

When a delta-9 desaturase (stearoyl-ACP desaturase) gene from *Brassica* was specifically expressed in canola seeds in an antisense orientation, it caused the seed stearate content to increase substantially, in some cases to up to 40% (w/w) (Knutzon et al., 1992a). Along with C18:0 the level of C20:0 also increased, sometimes to up to 10%. It is not known if the elongation to C20:0 takes place in the plastid as a result of non specific activity of KASII, driven by increased stearate concentration in the organelle. Alternatively, increased pools of stearoyl-CoA in the cytoplasm, could force the elongation by ER bound enzymes with some affinity for C18:0 (stearate). These elongases, if they exist, differ from those participating in the synthesis of erucic acid since their activity has been eliminated in canola. Unexpectedly, high stearate seeds also contained large amounts (up to 20%) of linolenate (C18:3). A plausible explanation for this increase in C18:3 relies on the specific activities of both acyltransferases and desaturases for their substrates. The high content of stearate and the preference of GPAT for saturated fatty acids will direct more stearate (C18:0) onto *sn*-1 and more oleate (C18:1) onto *sn*-2. In this position C18:1 may be a better substrate for the delta-12

desaturase and, in turn, the resulting linoleate (C18:2) may be a better substrate for the delta-15 desaturases (see below Section V.C). In addition, a low LPCAT activity on stearate PC could also favor desaturation by maintaining C18:1 and C18:2 on *sn*-2. The linolenate (C18:3) ratio on the *sn*-3 position could provide information on the LPCAT activity. Should this ratio be low, it would support the idea that in high stearate seeds, LPCAT does not have much affinity for stearate-PC, and that the triglycerides containing polyunsaturated fatty acids are formed essentially through DAG route by reversion from PC to DAG. At the moment, however, all these hypotheses remain unsupported.

The creation of high stearate canola has more than scientific implications. Their stearate is synthesized without the generation of *trans*-fatty acids which accompanies artificial saturation (hydrogenation). The presence of *trans*-fatty acids in margarine is, at the least, controversial and has been linked by some authors to coronary disease (reviewed by Hunter, 1992). High stearate canola could substitute for hydrogenated vegetable oils in the preparation of margarine.

2. Delta-6 Desaturase

Seeds from *Apiaceae*, such as coriander, synthesize petroselinic acid, a monounsaturated fatty acid with the double bond in the delta-6 position instead of the delta-9 position of oleic acid. The delta-6 stearoyl-ACP desaturase is also a soluble enzyme and cross reacts with antibodies raised against the delta-9 desaturase from avocado (Cahoon et al., 1992). This homology has allowed the direct cloning of the delta-6 desaturase gene. Its expression in tobacco callus led to the synthesis of petroselinic acid. Note that hexadecenoic C16:1 with a double bond in delta-4 was detected along with petroselinic acid suggesting that the enzyme may have some activity on palmitate. If this was the case, petroselinic acid could result from the elongation of delta-4-C16:1 and the enzyme would qualify as a delta-4 desaturase.

C. Endoplasmic Reticulum Desaturases: High Oleic Soybean and Canola

The desaturations responsible for the formation of linoleic (C18:2) and linolenic (C18:3) acids take place at the ER level and are mediated by membrane-bound delta-12 and delta-15 desaturases, respectively. Membrane-bound delta-12 and delta-15 desaturases are also found in plastids (Somerville and Browse, 1991) but are not considered here, as they do not appear to contribute significantly to seed triglyceride composition. In borage, evening primrose, and black currant, linoleate (C18:2) is converted to gamma-linolenate by a membrane-bound delta-6 desaturase. All of these desaturation reactions involve a complex electron acceptor/donor system including at least the desaturase itself, a cytochrome (Cyt) b_5 as the main electron donor to the desaturase, and two flavo-proteins: NADH-Cyt b_5 oxidoreductase and NADPH-Cyt-P-450 oxidoreductases which in turn provide electrons to the cytochrome b_5 (Heinz, 1993). The total capacity of the electron donor system is several orders of magnitude beyond the activity of the desaturase itself and therefore is non limiting. The last electron acceptor is O_2, which is presumably reduced to water. No molecule substituting for O_2 has yet been found.

The substrates for the desaturases are phos-phatidylcholine (PC) molecules (Stymne and Appelqvist, 1978). After reaching the endoplasmic reticulum, the oleoyl-CoA molecules are esterified on the glycerol backbone by a glycerol-3-P acyltransferase (GPAT) (sn-1) and LPAT (sn-2) to form, respectively, lysophosphatidic acid (LPA) and phosphatidic acid (PA) eventually converted to DAG by a phosphatase. A portion of DAG can then enter the PC pool (Fig. 6) under the reversible action of a cholinephosphotransferase (CTP). Alternatively, the oleoyl-CoA can also enter the PC pool by direct esterification to LPC mediated by LPCAT. Only as PC components are oleates further desaturated. Those esterified on the sn-2 are desaturated preferentially to those on sn-1 position. The linoleate formed on the PC can be further desaturated by the delta-15 desaturase. The preference displayed by delta-15 desaturase for the sn-2 linoleate over the sn-1 is more marked than that of delta-12 desaturase (Stymne et al., 1987). The reversible action of choline-phosphotransferase makes DAG with unsaturated fatty acids available for the synthesis of triglycerides (Slack et al., 1983). The reversible activity of LPCAT, on the other hand, recycles the unsaturated fatty acids into the acyl-CoA pool and make them available to the other acyltransferases and especially to DAGAT (Hares and Frentzen, 1987; Frentzen, 1990). This explains the presence of polyunsaturated fatty acids

in the TAG's sn-3. Both delta-12 and delta-15 desaturases are very specific with respect to the fatty acid substrate. In the 'petroselinic' transgenic tobacco tissues (Cahoon et al., 1992) no fatty acid was found with both delta-6 and delta-12 double bonds. This supports in vitro experiments suggesting that plant delta-12 desaturases somehow recognize both the presence of the delta-9 desaturation and the distance from the delta-12 desaturation site, counting from the carboxyl group.

The gene coding for delta-15 desaturase has been cloned from *Arabidopsis* by a T-DNA tagging approach (Arondel et al., 1992; Yadav et al., 1993a,b). Briefly, large numbers of plants were transformed with *Agrobacterium* with the purpose of inserting the T-DNA into a desaturase gene (insertion mutant), and thereby creating a dysfunctional enzyme. Transgenic seeds containing the T-DNA were then screened for linolenate content. Genomic DNA digests (alternatively cDNA libraries) of seeds with low linolenate content were screened with T-DNA specific probes. The DNA portions adjacent to the T-DNA insert, that were presumably part of the desaturase genes, were sequenced. In turn the new sequences were used to probe c-DNA libraries of a non transgenic control seeds where the desaturase gene was finally fished out. The gene was down-regulated specifically in soybean and canola seeds by expressing it in antisense orientation (Kinney, 1994; Fader, 1995). This reduced the content of linolenic from 8–10% to 1–2%. Both linoleate and oleate contents increased and accounted for the decrease in linolenate. The delta-12 desaturase gene has also been cloned (Okule et al., 1994) and specifically downregulated in canola and soybean seeds. As expected there was a reduction of polyunsaturated fatty acids (64% to 10% in soybean and 26% to 10% in canola) exactly compensated for by gains in oleate (Kinney, 1994; Fader et al., 1995). Besides delta-12 and delta-15 desaturases, it may be possible to modify the content of polyunsaturated fatty acids by engineering other genes. Low linolenic mutants have been isolated from several species including linseed, *Arabidopsis* and canola. Independent mutations could be combined to generate canola lines producing an oil with less than 1% linolenate (D. Facciotti, unpublished). The analysis of these mutations suggests that, although several desaturases may contribute additively to the synthesis of C18:3, other genes may also be involved in the accumulation of this fatty acid. Such genes may include cytochrome b_5 (Kearns et al., 1992) and associated redox systems, which could become obvious targets of genetic engineering. The downregulation of cytochrome b_5 could lead to low polyunsaturated plant lines. To be successful, however, the reduction of cytochrome b_5 should be drastic since the desaturases use only a small portion of the electrons that the cytochrome complex can supply. Such a drastic reduction may cause unwanted side effects. Other targets may include DAGAT, CPT and LPCAT genes. The effectiveness with which engineering for higher DAGAT and lower CPT and LPCAT activities would modify the content of polyunsaturated fatty acids has not yet been determined.

The two delta-12 and delta-15 desaturases do not display much homology, although certain motifs, the so-called 'histidine boxes' that are common to many desaturases, are present in both enzymes (Shanklin et al., 1995). These motifs may make it possible to identify desaturase or related genes from randomly sequenced cDNAs. This is precisely how a gene coding for a microsomal delta-6 desaturase was identified and cloned from a cDNA library of growing borage seeds (Sayanova et al., 1997). Once expressed in tobacco leaves, this gene promoted the synthesis of gamma linolenic acid. Homology with delta-12 and delta-15 desaturases was largely limited to the histidine motifs. The substrate of this desaturase appears to be linoleoyl-PC. No molecules with delta-6 and delta-9 double bonds were found in the transgenic tobacco. The enzyme substrate appears to require double bonds in C9 and C12.

VII. Present Success and Future Challenges

A. Significant Achievements: Summary

Recombinant DNA technology has been used successfully to modify the fatty acid composition of triglycerides in oleaginous crops. It has helped by generating both basic information on triglyceride synthesis as well as opportunities for new products.

One of its main contributions to the understanding of fatty acid synthesis has been in describing how specific thioesterases control fatty acid chain-lengths. In the first step, a gene encoding a thioesterase with specific activity for laurate, was isolated and proven to be mainly responsible for the accumulation of this

fatty acid in bay laurel seeds. Following this breakthrough, DNA homology between thioesterases was exploited to isolate the thioesterase genes from other species, each known to accumulate preferentially fatty acids of different chain lengths. These genes were then expressed in transgenic plants where they promoted the accumulation of the same fatty acids as in their plants of origin (Voelker, 1996). A commercial variety of high laurate canola has been developed and the oil produced is already marketed to the detergent and food industries. The development of other canola varieties for the production of oils rich in saturated fatty acids of different chain lengths is in progress.

Research on KAS and KCS proceeds according to the model established with thioesterases: following the isolation of the first KAS, gene homology is then exploited for the cloning of other KASs. In all likelihood, this research will soon deliver the entire family of KAS genes and provide detailed characteristics of each KAS and its role in determining fatty acid chain length. KCS genes responsible for the synthesis of erucic acid have been cloned from rapeseed on the basis of their homology with a membrane-bound KCS from jojoba.

As for acyltransferases, the techniques necessary to purify, solubilize, reconstitute membrane-bound enzymes such as the jojoba KCS have also progressed considerably. The improvements have led to the solubilization of the coconut LPAT (Davies et al., 1995) and finally to the cloning of the gene encoding it (Knutzon et al., 1995). As described above (Section V.B), the gene was expressed in canola seeds where it promoted the insertion of medium chain fatty acids in *sn*-2. Another LPAT, believed to be responsible for the esterification of erucic acid on *sn*-2 in meadow foam has also been cloned through homology (Lassner et al., 1995). Future manipulation of specific acyltransferases may lead to the production of a variety of new structured triglycerides.

Overall, the cloning of thioesterases, KAS and acyltransferase genes is progressing rapidly. Now that the DNA sequences of many isozymes are available, it is also possible to compare these sequences and link structural differences to differences in substrate specificity and activity existing between the isozymes. The practical follow up of such studies has already led to the creation of new thioesterases in which functional domains of two different enzymes were combined into a new hybrid protein with new activity profiles (Yuan et al., 1995).

With respect to desaturation of fatty acids the three most common desaturases: delta-9, delta-12, and delta-15, have been cloned. The manipulation of these genes has potentially created new food-oil products (Kinney, 1994). The cloning of other desaturases, however, has been hindered by the relative lack of homology between desaturases. Instead of strict homology, different desaturases share common motifs such as the 'histidine box'. This motif was used to identify and clone a putative microsomal delta-6 desaturase from a cDNA library of maturing borage seeds (Sayanova et al., 1997). The gene was effectively expressed in transgenic tobacco which synthesized gamma linolenic acid (30%), a fatty acid usually absent in these species (Sayanova et al., 1997). Since entire plant genomes are now being sequenced, motif recognition could become a general approach to cloning other desaturase genes.

It is worth adding that most of the triglyceride engineering reported here was achieved by a single gene modification of the plant genome. The esterification of laurate on the *sn*-2 position was obtained by the independent transfers into canola of two genes, a bay laurel thioesterase and a coconut LPAT, later combined in hybrid progenies. At the moment, only two to three genes, including a selective marker, can be transferred and expressed simultaneously into a plant. One can anticipate that some desirable oil modifications will require the engineering of numerous genes. The present limitations in gene transfer technology would generate long and costly breeding programs necessary to achieve complex gene combinations. Therefore a 'one-step' transfer of entire gene clusters and the control of their expression remain on the list of future challenges. Other challenges are presented below.

B. New Fatty Acids in Common Crops

Unusual and rare fatty acids will soon be new oil-products of common crops. Ricinoleate, for example, an hydroxy fatty acid used in the polymer industry, is presently extracted from castor bean. Unfortunately, along with ricinoleate castor beans accumulate ricin, a storage protein so toxic that the personnel involved in the oil extraction are required to wear special protective equipment. Therefore it would be safer to produce ricinoleate in *Brassica rapa* or *B. juncea*. In addition to safety, some commercial advantage could also be derived from the sales of *B. rapa* or *B. juncea* storage proteins to the feed industry. Recently, the delta-12 hydroxylase genes from castor bean and *Lesquerella fendleri* were expressed in tobacco and *Arabidopsis* (Broun et al., 1996). The transgenic plants contained significant amounts of hydroxylated fatty acids (20%). The focus is now on expressing the hydroxylase genes in oleaginous crops.

Other unusual and rare oils with potential for industrial applications have been mentioned elsewhere (Stymne, 1993; van de Loo et al., 1993) and among them are oils rich in branched fatty acids, and oils rich in epoxy fatty acids such as vernolic acid. To our knowledge, only limited resources have yet been allocated to understanding the synthesis of these rare fatty acids.

C. Limits of Triglyceride Engineering

The future of new oil products may reside not only in the ability to synthesize unusual fatty acids in a well domesticated crop, but also to produce them in proportions responding to industrial specifications. Will it be possible, for example, to generate canola synthesizing mainly ricinoleate, or laurate? Will it be possible to produce triglycerides with cocoa butter structures and characteristics? One can anticipate many obstacles to the achievement of such goals. In some cases, success may depend on the ability with which genes from distantly related genera can be efficiently expressed in a crop, assuming the resulting enzymes are stable in the new crop environment. In other cases, such as an enzyme participating in a complex reaction, success may depend on the presence of other enzymes and cofactors involved in the reaction and on the degree to which they complement the enzyme activity in the host crop. Surprising results may also occur, due to our ignorance of the regulation of fatty acid and triglyceride synthesis. This was illustrated in transgenic 'high stearate' canola (see Section VI. B.1) that also produced unexpectedly high levels of linolenate.

In addition, when the goal is to deposit unusual fatty acids or to accumulate unusually large amounts of any fatty acid, the seeds may have to overcome physiological challenges, particularly at germination. Deleterious effects related to modifications in the fatty acid composition of cell membranes are the main concern. It is known that the fatty acid composition of storage and structural lipids may differ considerably within the same plant. Bay laurel seeds accumulate triglycerides rich in laurate while leaves do not contain laurate. Recently, a bay laurel seed thioesterase gene with specificity for lauryl-ACP was transferred into canola under the control of a constitutive CaMV 35S promoter (Eccleston et al., 1996). The gene was expressed in leaves, roots and also in seeds. Seeds accumulated a small but significant amount of laurate (7%). In the leaves, although the protein was present and its activity measurable in vitro, no laurate could be detected. It is not known whether or not the acyltransferases (Fig. 6) present in leaves are so specific as to discriminate against laurate. In addition, the absence of laurate from the polar lipids of mature seeds producing laurate was also reported. Since acyltransferases capable of esterifying laurate are present in seeds, it is not known at this time to what degree LPCAT or CPT limit the inclusion of laurate into structural lipids. Recent analyses of immature 'laurate canola' seeds (moderate laurate producers) indicated that in fact, some laurate was present in the PC of growing embryos but that it disappeared at full maturity (S. Stymne, personal communication). This supports the hypothesis that an active mechanism may also exclude laurate from structural lipids of seed cells. This mechanism could involve, for example, specific lipases capable of degrading unusual phospholipids containing laurate or other unwanted fatty acids. When higher contents of laurate are reached, however, the exclusion mechanism, whatever this may be, is insufficient to prevent the incorporation of the fatty acid into the cell membranes. This was found to be the case in high 'laurate canola' lines (>55% laurate w/w) which retained a small portion of laurate in the PC of mature seeds (S. Stymne, personal commun-ication). There were no negative effects associated with the presence of laurate in the polar lipids of these lines. Though this may not remain true for higher levels of laurate or with other fatty acids. Low germination rates were recorded in seeds of a high 'stearate canola' line (> 40% of the lipids present as stearate). The polar lipids of these seeds were found to be unusually rich in stearate (Thompson and Li, 1996) and the poor germination was tentatively attributed to the crossing of a threshold percentage of stearate in membrane lipids. These germination

problems, although isolated, should be an incentive for further study, which may lead to understanding and then utilizing these mechanisms of exclusion to maintain the integrity of cell membranes.

D. Yield

The improvement of oil yield in any given crop is always part of future challenges. When the oil products are commodities with low benefit margins, improving yields even by 1% gives a competitive advantage.

1. Partitioning Photosynthates Between Oils and Proteins

The ratio between oil and protein accumulation in a seed is often a function of the environment. When canola plants are grown in the presence of nitrogen-rich fertilizers, proteins accumulate at the expense of oil. Under cold temperature, perhaps because of low nitrogen uptake and assimilation, the situation is reversed and more oil is produced (Canvin, 1964). Varietal differences, on the other hand, suggest the existence of genetic components to photosynthate partitioning and therefore the possibility of modifying them in favor of either protein or oil deposition.

Present approaches to divert more photosynthates towards TAG deposition focus on increasing the content and the activity of the enzymes believed to limit fatty acid or triglyceride synthesis. The conversion of acetate to malonyl-CoA and malonyl-ACP (Section III.D), the first committed reaction towards fatty acid synthesis in the plastid, has often been considered to be a limiting step to TAG synthesis. The preliminary results observed after overexpressing acetyl-CoA carboxylase (ACC) in canola (Ohlrogge and Jaworsky, 1997) may support this hypothesis. Should malate be the main source of acetyl-CoA (Fig. 3), the increase of malic enzyme content would divert more photosynthates to the plastid and towards oil accumulation. To our knowledge, this has not yet been attempted. There are speculations that DAGAT, the last key enzyme of TAG synthesis, may be a limiting factor and that by increasing its activity one may speed up oil deposition or even increase the import of photosynthates into the embryo. At the present time we are not aware of any report on the cloning and manipulation of the DAGAT gene. In our opinion, however, fatty acid synthase (FAS) enzymes, ACC or DAGAT, compete at a disadvantage by mediating reactions which occur only after the photosynthates may have already been diverted towards amino acid and protein synthesis (Fig. 2).

It is possible that in 'high oil' canola varieties the 'protein route' is less efficient than the 'oil route'. Following this hypothesis one could attempt to increase oil contents by limiting protein deposition. The genes coding for storage proteins of canola, napin and cruciferin, constitute relatively large gene families (10–15 genes) (Josefsson et al., 1987). Recently a napin gene was reintroduced in canola and expressed in antisense orientation. In most of the transgenic plants the level of napin was significantly reduced, sometimes to zero. The losses in napin were compensated, however, by gains in cruciferin, but no effect on oil content was noted (Kohno-Murase et al., 1994). Curiously enough, the content of unsaturated fatty acids in the 'low napin' seeds also increased and it was hypothesized that more desaturases were synthesized in place of napin. This would suggest that in maturing embryos, the pools of some active enzymes may be regulated by the deposition of storage protein. We are not aware of any further attempt to lower the content of both cruciferin and napin in the transgenic 'low napin' canola.

2. Source/Sink

As an alternative to the partitioning of photosynthates between oils and proteins, oil yield may be improved by providing embryos with more photosynthates. The production of storage molecules in the embryo may be limited mainly by the supply of photosynthates. The effectiveness with which yield may be improved by increasing the flux of photosynthates to the embryos at the expense of other photosynthate sinks, is illustrated by canola 'low fiber coat' mutants. The seeds of these mutants are protected by a very thin, yellow coat which lacks pigments and is low in fiber content (lignin and non digestible cell wall components). They display significantly higher contents (1%–5%) of oil and protein than black coated seeds. It is believed that the photosynthates no longer used for the synthesis of coat fibers accumulate in the embryo as oils and proteins. This supports the hypothesis that the pool of the enzymes acting on storage molecules can handle more photosynthates than it actually does. The reduction of the seed coat 'sink', by making more photo-synthates available to the embryo, determines the amount of reserves stored. In canola and *Arabidopsis* 'low fiber coat' seeds are usually obtained by combining several recessive mutations which likely affect the synthesis of phenols and lignin. Recently a canola plant with a single recessive mutation has been selected and the trait transferred into elite varieties (D. Facciotti, unpublished). In the mutant, the alteration of a single gene seemed sufficient to reduce the fiber (lignin) content of the coat. When the mutated gene was isolated it could be engineered to reduce the lignin sink in the seed coat of other species or in other plant organs, including in the tissues of the embryo itself. The modification could make more photosynthates available for the synthesis of storage molecules. Even modest reductions in lignification, at levels that would not affect other plant features, may increase oil yields competitively. Alternatively, other genes encoding enzymes involved in lignification are now available. We know, however, that the engineered downregulation of an enzyme such as cinnamyl alcohol dehydrogenase alone, does not reduce lignification in transgenic tobacco (Halpin et al., 1994). The downregulation of other enzymes along with cinnamyl alcohol dehydrogenase may be necessary to achieve this reduction.

Finally, instead of reducing the size of other sinks one could attempt to improve the efficiency of the embryo to drain photosynthates. The overexpression of specific enzymes such as sucrose synthase in the embryo comes to mind and should be relatively easy to attempt. One may also consider achieving larger seed sinks. We have seen earlier (Section III.A) the possible effects of seed coat invertases on the number of cotyledonary cells and their storage capacity. Also, by increasing the ratio of the number of seeds/plant, canola breeders have effectively selected for high yielding varieties. To our knowledge, however, the way to boost seed production via genetic engineering has not yet been defined.

Acknowledgments

This review is the updated version of two lectures presented at the 'IIIe Cycle Romand en Sciences Biologiques' held at the Université de Neuchâtel (Switzerland), in March 1995. We thank Professor Paul-André Siegenthaler, the organizer, for his invitation to Neuchâtel, his superb organization and his warm welcome. We also thank J. Metz, D. Scherer, M. Facciotti, E. Carlson and G. Thompson for reviewing this manuscript and Professor Norio Murata and Professor Paul-André Siegenthaler, the editors, for their patience.

References

Andrews TJ and Kane HJ (1991) Pyruvate is a by product of catalysis by ribulose-bisphosphate carboxylase/oxygenase. J Biol Chem 266: 9447–9452

Arondel V, Lemieux B, Hwang I, Gibson S, Goodman HM and Somerville CR (1992) Map-based cloning of a gene controlling omega-3 fatty acid desaturation in *Arabidopsis*. Science 258: 1353–1355

Bafor M, Stobart AK and Stymne S (1990) Properties of the glycerol acylating enzymes in microsomal preparations from the developing seeds of safflower (*Carthamus tinctorius*) and turnip rape and their ability to assemble cocoa-butter type fats. J Am Oil Chem Soc 67: 217–225

Barron EJ and Stumpf PK (1962) Fat metabolism in higher plants. The biosynthesis of triglycerides by avocado mesocarp enzymes. Biochim Biophys Acta 60: 329–337

Battey JF and Ohlrogge JB (1990) Evolutionary and tissue-specific control of expression of multiple acyl carrier protein isoforms in plants and bacteria. Planta 180: 352–360

Broun P, Hawker N, and Somerville C (1996) Expression of castor and *Lesquerella Fendleri* oleate-12 hydroxylases in transgenic plants: Effects on lipid metabolism and inferences on structure-function relationship in fatty acid hydroxylases. In: Williams JP, Khan MU, Wan Lem N (eds) Physiology, Biochemistry and Molecular Biology of Plant Lipids, pp 342–344. Kluwer Academic Publishers, Dordrecht

Cahoon EB, Shanklin J and Ohlrogge JB (1992) Expression of a coriander desaturase results in petroselinic acid production in transgenic tobacco. Proc Natl Acad Sci USA 89: 11184–11188

Canvin DT (1965) The effects of temperature on the oil content and fatty acid composition of the oils from several oilseed crops. Can J Bot 43: 63–69

Cao Y Z, Oo KC and Huang AHC (1990) Lysophosphatidate acyltransferase in the microsomes from maturing seeds of meadowfoam (*Limnanthes alba*). Plant Physiol 94: 1199–1206

Caughey I and Kekwich RGO (1982) The characteristics of some components of the fatty acid synthetase system in the plastids from the mesocarp of avocado fruit. Eur J Biochem 123: 553–561

Cronan JE and Roughan PG (1987) Fatty acid specificity and selectivity of the chloroplast *sn*-glycerol-3-phosphate acyltransferase of the chilling sensitive plant *Amaranthus lividus*. Plant Physiol 83: 676–680

Davies HM (1993) Medium Chain acyl-ACP hydrolysis activities of developing oilseeds. Phytochemistry 33: 1353–1356

Davies HM, Anderson L, Fan C and Hawkins DJ (1991) Developmental induction, purification, and further characterization of 12:0-ACP thioesterase from immature cotyledons of *Umbellularia californica*. Arch Biochem Biophys 290: 37–45

Davies HM, Hawkins DJ and Nelsen JS (1995) Lysophosphatidic acid acyltransferase from immature coconut endosperm having medium chain-length substrate specificity. Phytochemistry 39: 989–996

Dehesh K, Jones A, Knutzon DS, and Voelker TA (1996) Production of high levels of 8:0 and 10:0 fatty acids in transgenic canola by over-expression of Ch FatB2, a thioesterase cDNA from *Cuphea hookeriana*. Plant J 9: 167–172

Dennis D and Miernyk J (1982) Compartmentation of non synthetic carbohydrate metabolism. Annu Rev Plant Physiol 33: 27–50

DeSilva J, Loader N, Jarman C, Windust J, Hughes S and Safford R (1990) The isolation and sequence analysis of two seed-expressed acyl carrier protein genes from *Brassica napus*. Plant Mol Biol, 14: 537–548

Eastwell K and Stumpf PK (1983) Regulation of plant acetyl-CoA carboxylase by adenylate nucleotides. Plant Physiol 72: 50–55

Ebel J and Hahlbrock K (1977) Enzymes of flavone and flavonol glycoside biosynthesis: Coordinated and selective induction in cell suspension cultures of petroselinum hortense. Eur J Biochem 75: 201–209

Eccleston V, Voelker TA and Ohlrogge J (1996) Medium-chain fatty acid biosynthesis and utilization in *Brassica napus* plants expressing lauroyl-acyl carrier protein thioesterase. Planta 198: 46–53

Edwards J, Nelsen JS, Metz JG and Dehesh K (1997) Cloning of the *fabF* gene in an expression vector and in vitro characterization of recombinant *fabF* and *fabB* encoded enzymes from *Escherichia coli*. FEBS Lett 402: 62–66

Egin-Buhler B and Ebel J (1983) Improved purification and further characterization of acetyl-CoA carboxylase from cultured cells of parsley. Eur J Biochem 133: 335–339

Frentzen M (1990) Comparison of certain properties of membrane bound and solubilized acyltransferase activities of plant microsomes. Plant Sci 69: 39–48

Frentzen M (1993) Acyltransferases and triacylglycerol. In: Moore TS (ed) Lipid Metabolism in Plants, pp 195–230. CRC Press, Boca Raton

Frentzen M, Heinz E, Mc Keon TA, and Stumpf PK (1983) Specificities and selectivities of glycerol-3-phosphate acyltransferases and monoacyl-glycerol-3-phosphate acyltransferase from pea and spinach chloroplasts. Eur J Biochem 129: 629–636

Fritsch H and Beevers H (1979) ATP citrate lyase from germinating castor bean endosperm. Plant Physiol 63: 687–691

Gee RW, Byerrum RU, Gerber DW, and Tolbert NE (1988) Dihydroxyacetone phosphate reductase in plants. Plant Physiol 86: 98–103

Genez A, McCarter D, Nelsen J, Bleibaum J, Thomson GA, Knauf V and Stalker D (1991) Molecular cloning and characterization of beta-keto-acyl-ACP synthase genes from *Ricinus communis*. In: RB Hallick (ed) Program and Abstracts, Third International Congress of plant Molecular Biology. Molecular Biology of Plant Growth and Development, October 6–11,1991. Abstact nr 719. Tucson

Ghosh S and Sastry PS (1988) Triacylglycerol synthesis in developing seeds of groundnut (*Arachys hypogea*): Pathway and property of enzymes of sn-glycerol 3-phosphate formation.

Arch Biochem Biophys 262: 508–516

Gombos Z and Murata N (1998) Genetically engineered modulation of the unsaturation of glycerolipids and its consequences in tolerance of photosynthesis to temperature stress. In: Siegenthaler PA and Murata N (eds) Lipids in Photosynthesis: Structure, Function and Genetics, pp. 249–262. Kluwer Academic Publishers, Dordrecht

Griffiths G, Stobart AK, and Stymne S (1985) The acylation of sn-glycerol-3-phosphate and the metabolism of phosphatidate in microsomal preparations from the developing cotyledons of safflower (Carthamus tinctorius) seed. Biochem J 230: 379–388

Guchhait RB, Polakis SE, Dimroth P, Stall E, Moss J and Lane MD (1974) Acetyl-CoA carboxylase system of E coli. J Biol Chem 249: 6633–6645

Guerra DJ, Ohlrogge JB, and Frentzen M (1986) Activity of acyl carrier protein isoforms in reactions of plant fatty acid metabolism. Plant Physiol 82: 448–453

Ha KS and Thompson GA (1992) Biphasic change in the level and composition of Dunaliella salina plasma membrane diacylglycerols following hypoosmotic shock. Biochemistry 31: 596–603

Halpin C, Knight ME, Foxon GA, Campbell MM, Boudet AM, Boon JJ, Chabbert B, Tollier MT, and Schuch W (1994) Manipulation of lignin quality by downregulation of cinnamoyl alcohol dehydrogenase. The Plant J 6: 339–350

Hares W and Frentzen M (1987) Properties of the microsomal acyl-CoA:sn-1-acyl glycerol 3-phosphate acyltransferase from spinach (Spinacia oleracea) leaves. J Plant Physiol 131: 49–59

Harwood JL (1988) Fatty acid metabolism. Annu Rev Plant Physiol Plant Mol Biol 39: 101–138

Heinz E (1993) Biosynthesis of polyunsaturated fatty acids. In: Moore TS (ed) Lipid Metabolism in Plants, pp 33–89. CRC Press, Boca Raton

Hlousek-Radojcic A, Post-Beittenmiller D and Ohlrogge JB (1992) Expression of constitutive and tissue specific acyl carrier protein isoforms in Arabidopsis. Plant Physiol 98: 206–214

Huang AHC (1993) Oil bodies in maize and other species. In: Murata N and Somerville C (eds) Biochemistry and Molecular Biology of Membranes and Storage Lipids of Plants, Current Topics in Plant Physiology, Vol 9, pp 215–227. ASPP Series, Rockville

Hunter JE (1992) Safety and health effects of isomeric fatty acids. In: Chow CK (ed) Fatty Acids in Foods and Their Health Implications, pp 857–868. Marcel Dekker, New York

Ichihara K, Asahi T and Fujii S (1987) 1-Acyl-sn-glycero-3-phosphate acyltransferase in maturing safflower seeds and its contribution to the non-random fatty acid distribution of triacylglycerol. Eur J Biochem 167: 339–347

Jones A, Davies HM and Voelker TA (1995) Palmitoyl-acyl carrier protein (ACP) thioesterase and the evolutionary origin of plant acyl-ACP thioesterases. Plant Cell 7: 359–371

Josefsson LG, Linman M, Erickson L and Lask L (1987) Structure of a gene encoding the 1.7S storage protein, napin, from Brassica napus. J Biol Chem 262: 12196–12201

Kaetner TM and Rees AT (1985) Intracellular location of ATP citrate lyase in leaves of Pisum sativum L. Planta 163: 290–294

Kearns EV, Keck P and Somerville CR (1992) Primary structure of cytochrome b_5 from cauliflower (Brassica oleracea) deduced

from peptides and cDNA sequences. Plant Physiol 99: 1254–1257

Kinney AJ (1994) Genetic modification of the storage lipids of plants. Current Opinion in Biotechnology 5: 144–147

Knauf V (1993) Progress in the cloning of genes for plant storage lipid biosynthesis. Genetic Engineering 15: 149–164

Knutzon DS, Scherer DE and Schreckengost WE (1991). Nucleotide sequence of a complementary DNA clone encoding stearoyl-acyl carrier protein desaturase from castor bean, Ricinus communis. Plant Physiol 96: 344–345

Knutzon DS, Thompson GT, Radke SE, Johnson WB, Knauf V and Kridl JC (1992a) Modification of Brassica seed oil by antisense expression of a stearoyl-acyl carrier protein desaturase gene. Proc Natl Acad Sci USA 89: 2624–2628

Knutzon DS, Bleibaum JL, Nelsen JS, Kridl JC and Thompson GA (1992b) Isolation and characterization of two safflower oleoyl-acyl carrier protein thioesterase cDNA clones. Plant Physiol 100: 1751–1758

Knutzon DS, Lardizabal KD, Nelsen JS, Bleibaum JL, Davies HM and Metz JG (1995) Cloning of a coconut endosperm cDNA encoding a 1-acyl sn-glycerol-3-phosphate acyltransferase which accepts medium-chain substrates. Plant Physiol 109: 999–1006

Kohno-Murase J, Murase M, Ichikawa H and Imamura J (1994) Effects of an antisense napin gene on seed storage compounds in transgenic Brassica napus seeds. Plant Mol Biol 26: 1115–1124

Konishi T, Shinohara K, Yamada K and Sasaki Y (1996) Acetyl-CoA carboxylase in higher plants: Most plants other than Graminaceae have both the prokaryotic and the eukaryotic forms of these enzymes. Plant Cell Physiol 37: 117–122

Kumar PR and Tsunoda S (1980) Variation in oil content and fatty acid composition among seeds from the Cruciferae. In: Brassica Crops and Wild Allies, pp 235–252. Japan Scientific Society Press, Tokyo

Kunst L, Browse J and Somerville C (1989a) A mutant of Arabidosis deficient in desaturation of palmitic acid in leaf lipids. Plant Physiol 90: 943–947

Kunst L, Browse J and Somerville C (1989b) Enhanced thermal tolerance in a mutant of Arabidopsis deficient in palmitic acid unsaturation. Plant Physiol 91: 401–408

Lassner MW, Lardizabal K and Metz JG (1996) A jojoba beta-ketoacyl-CoA synthase cDNA complements the canola fatty acid elongation mutation in transgenic plants. The Plant Cell 8: 281–292

Lassner MW, Levering CK, Davies HM and Knutzon DS (1995) Lysophosphatidic acid acyltransferase from Limnanthes alba mediates insertion of erucic acid at the sn-2 position of triacylglycerol in transgenic rapeseed oil. Plant Physiol 109: 1389–1394

Li SJ and Cronan JE (1992) The genes encoding the two carboxyltransferase subunits of Escherichia coli acetylCoA carboxylase. J Biol Chem 267: 16841–16847

Liedvogel B and Stumpf PK (1982) Origin of acetate in spinach leaf cells. Plant Physiol 69: 897–903

Luo X, Park K, Lopez-Casillos F and Kim KH (1989) Structural features of acetyl-CoA carboxylase gene. Mechanism for the generation of mRNA with 5' end heterogeneity. Proc Acad Sci USA 86: 4042–4046

Masterson C, Wood C, and Thomas DR (1990) L-acetylcarnitine, a substrate for fatty acid synthesis by isolated chloroplasts. Plant Cell Environ 13: 755–765

Miernyk J and Dennis D (1982). Isozyme of the glycolytic enzymes in endosperm from developing castor oil seeds. Plant Physiol 69: 825–828

Miquel M and Browse J (1992) *Arabidopsis* mutant deficient in polyunsaturated fatty acids synthesis. J Biol Chem 267: 1502–1509

Murata N (1983) Molecular species composition of phosphatidylglycerol from chilling-sensitive and chilling-resistant plants. Plant Cell Physiol 24: 81–86

Murata N, Sato N, Takahashi N and Hamazaki Y (1982) Composition and positional distributions of fatty acids in phospholipids from leaves of chilling-sensitive and chilling-resistant plants. Plant Cell Physiol 23: 1071–1079

Murata N, Ishizaki-Nishizawa O, Higashi S, Hayashi H, Tasaka Y and Nishida I (1992) Genetically engineered alteration in the chilling sensitivity of plants. Nature 356: 710–713

Murphy DJ (1993) Biochemical and molecular regulation of storage product formation in oilseeds. In: Murata N and Somerville C (eds) Biochemistry and Molecular Biology of Membranes and Storage Lipids in Plants, Current Topics in Plant Physiology, Vol 9, pp 228–237. ASPP Series, Rockville

Murphy DJ and Stumpf PK (1981) The origin of chloroplast acetyl-CoA. Arch Biochem Biophys 212: 730–739

Murphy DJ, Cummins I and Kang AS (1989) Synthesis of the major oil-body membrane protein in developing rapeseed (*Brassica napus*) embryos. Biochem J 258: 285–293

Nawrath C, Poirier Y and Somerville C (1994) Targeting of the polihydroxybutyrate biosynthetic pathway to the plastids of *Arabidopsis thaliana* results in high levels of polymer accumulation. Proc Natl Acad Sci USA 91: 12760–12764

Nikolau BJ and Hawke JC (1984) Purification and characterization of maize leaf acetyl-coenzyme A carboxylase. Arch Biochem Biophys 228: 86–96

Nikolau BJ, Wurtele ES and Stumpf PK (1984) Tissue distribution of acetyl-CoA carboxylase in leaves. Plant Physiol 75: 895–901

Ohlrogge JB and Jaworski JG (1997) Regulation of fatty acid synthesis. Annu Rev Plant Physiol Plant Mol Biol 48: 109–136

Ohlrogge JB and Kuo TM (1985) Plants have isoforms of acyl carrier proteins that are expressed differently in different tissues. J Biol Chem 260: 8032–8037

Ohlrogge JB, Shine WE and Stumpf PK (1978) Purification and characterization of plant acyl-ACP and acyl-CoA hydrolases. Arch Biochem Biophys 189: 382–391

Ohlrogge JB, Jaworski JG and Post-Beittenmiller D (1993) De novo fatty acid biosynthesis. In: Moore TS (ed) Lipid Metabolism in Plants, pp 3–32. CRC Press, Boca Raton

Okuley J, Lightner J, Feldmann K, Yadav NS and Browse J (1994) The *Arabidopsis* Fad 2 gene encodes the enzyme that is essential for polyunsaturated lipid synthesis. The Plant Cell 6: 147- 158

Oo KC and Stumpf PK (1979) Fatty acid synthesis in the developing endosperm of *Cocus nucifera*. Lipids 14: 132–143

Pollard MR, Anderson L, Fan C, Hawkins DJ and Davies HM (1991) A specific acyl-ACP thioesterase implicated in medium-chain fatty acid production in immature cotyledons of *Umbellularia californica*. Arch Biochem Biophys 284: 306–312

Post-Beittenmiller D, Roughan G, and Ohlrogge J (1992) Regulation of plant fatty acid biosynthesis. Plant Physiol 100: 923–930

Ratledge C (1993) Single cell oils—have they a biotechnological future? Tibtech 11: 278–284

Roesler K, Savage L Shintani D, Shorrosh B and Ohlrogge J (1996) Co-purification, co-immunoprecipitation and co-ordinate expression of acetyl-coenzyme A, biotin carboxylase, and biotin carboxyl carrier protein of higher plants. Planta 198: 517–525

Roughan PG, Holland R and Slack CR (1979) On the control of long-chain-fatty acid synthesis in isolated intact spinach (*Spinacia oleracea*) chloroplasts. Biochem J 184: 193–202

Roughan G, Post-Beittenmiller D, Ohlrogge J and Browse J (1993) Acetylcarnitine and fatty acid synthesis by isolated chloroplasts. Plant Physiol 101: 1157–1162

Sato A, Becker C and Knauf V (1992) Nucleotide sequence of a complementary DNA clone encoding stearoyl-acyl carrier protein desaturase from *Simmondsia chinensis*. Plant Physiol 99: 362–364

Sayonava O, Smith MA, Lapinskas P, Stobart AK, Dobson G, Christie WW, Shewry PR and Napier JA (1997) Expression of a borage desaturase cDNA containing an N-terminal cytochrome b_5 domain results in the accumulation of high levels of Δ^6-desaturated fatty acids in transgenic tobacco. Proc Natl Acad Sci USA 94: 4211–4216

Shanklin J and Somerville C (1991) Stearoyl-acyl-carrier-protein desaturase from higher plants is structurally unrelated to the animal and fungal homologs. Proc Natl Acad Sci USA 88: 2510–2514

Shanklin J, Whittle EJ and Fox B (1995) Membrane bound desaturases and hydroxylases: structure function studies. In: Kader JC and Mazliak P (eds) Plant Lipid Metabolism, pp 18–20. Kluwer Academic Publishers, Dordrecht

Scherer DE and Knauf VC (1987) Isolation of a cDNA clone for the acyl carrier protein-I of spinach. Plant Mol Biol 9: 127–134

Shine WE, Mancha M, and stumpf PK (1976) The function of acyl thioesterases in the metabolism of acyl-CoAs and acyl-acyl carrier proteins. Arch Biochem Biophys 172: 110–116

Shorrosh BS, Roesler KR, Shintani D, van de Loo FJ and Ohlrogge J (1995) Structural analysis, plastid localization, and expression of the biotin carboxylase subunit of acetyl-CoA carboxylase from tobacco. Plant Physiol 108: 805–812

Siggaard-Andersen M, Kauppinen S and von Wettstein-Knowles P (1991) Primary structure of a cerulenin-binding beta-ketoacyl-(acyl carrier protein) synthase from barley chloroplasts. Proc Natl Acad Sci USA 88: 4114–4118

Siggaard-Andersen M, Wissenbach M, Chuck JA, Svendsen I, Olsen JG and von Wettstein-Knowles P (1994) The fabJ-encoded beta-ketoacyl-(acyl carrier protein) synthase IV from *Escherischia coli* is sensitive to cerulenin and specific for short-chain substrates. Proc Natl Acad Sci USA 91: 11027–11031

Slabas AR and Hellyer A (1985) Rapid purification of a high molecular weight subunit polypeptide form of rape seed acetyl-CoA carboxylase. Plant Sci 39: 177–182

Slack CR, Campbell LC, Browse JA and Roughan PG (1983) Some evidence for the reversibility of the choline-phosphotransferase-catalyzed reaction in developing linseed cotyledons in vivo. Biochim Biophys Acta 754: 10–20

Small DM (1991) The effects of glyceride structure on absorbtion and metabolism. Annu Rev Nutr 11: 413–434

Smith RG, Gauthier DA, Dennis DT and Turpin DH (1992) Malate- and pyruvate-dependent fatty acid synthesis in leucoplasts from developing castor endosperm. Plant Physiol 98: 1233–1238

Somerville C and Browse J A (1991) Plant Lipids: Metabolism, Mutants, and Membranes. Science 252: 80–87

Stuitje AR, Kater M, Verwoert IIGS, Fawcett T, Slabas AR and Nijkamp HJJ (1993) Molecular genetic studies of plant enoyl-ACP reductase and bacterial malonyl CoA-ACP transacylase genes. In: Murata N and Somerville C (eds) Biochemistry and Molecular Biology of Membranes and Storage Lipids in Plants, Current Topics in Plant Physiology, Vol 9, pp 121–132. ASPP Series, Rockville

Stymne S (1993) Biosynthesis of 'uncommon' fatty acids and their incorporation into triacylglycerols. In: Murata N and Somerville C (eds) Biochemistry and Molecular Biology of Membranes and Storage Lipids in Plants, Current Topics in Plant Physiology, Vol 9, pp 150–158. ASPP Series, Rockville

Stymne S and Appelqvist LA (1978) The biosynthesis of linoleate from oleoyl-CoA via oleoyl-phosphatidylcholine in microsomes of developing safflower seeds. Eur J Biochem 90: 223–229

Stymne S and Stobart AK (1987) Triacylglycerol Biosynthesis. In: Stumpf PK (ed) The Biochemistry of Plants, Vol 9, pp175–214. Academic Press Inc, New York

Stymne S, Tonnet ML, and Green AG (1992) Biosynthesis of linolenate in developing embryos and cell-free preparations of high linolenate linseed (Linum usitatissimum) and low-linolenate mutants. Arch Biochem Biophys 294: 557–563

Tai H and Jaworski JG (1993) 3-Ketoacyl-acyl carrier protein synthase III from spinach (Spinacia oleracea) is not similar to other condensing enzymes of fatty acid synthase. Plant Physiol 103: 1361–1367

Thompson GA and Li C (1996) Altered fatty acid composition of membrane lipids in seeds and seedling tissues of high saturated canolas. In: Williams JP, Khan MU, Wan Lem N (eds) Physiology, Biochemistry and Molecular Biology of Plant Lipids, pp 313–315. Kluwer Academic Publishers, Dordrecht

Thompson GA, Scherer DE, Foxall-Van Aken S, Kenney JW, Young HL, Shintani DK, Kridl JC and Knauf VC (1991) Primary structures of the precursor and mature forms of stearoyl-acyl carrier protein desaturase from safflower embryos and requirement of ferredoxin for enzyme activity. Proc Natl Acad Sci USA 88: 2578–2582

Töpfer R, Martini N, and Shell J (1995a) Modification of plant lipid synthesis. Science 268: 681–686

Töpfer R, Martini N, and Shell J (1995b) International Patent Application No. WO 95/06740. Published March 9, 1995

van de Loo FJ, Fox BJ and Somerville C (1993) Unusual Fatty Acids. In: Moore TS (ed) Lipid Metabolism in Plants, pp 91–126. CRC Press, Boca Raton

Vijayan P, Routaboul JM and Browse J (1998) A genetic approach to investigating membrane lipid structure and photosynthetic function. In: Siegenthaler PA and Murata N (eds) Lipids in Photosynthesis: Structure, Function and Genetics, pp 263–285. Kluwer Academic Publishers, Dordrecht

Voelker TA (1996) Plant acyl-ACP thioesterases: Chain length determining enzymes in plant fatty acid biosynthesis. In: Setlow JK (ed) Genetic Engineering, Vol 18, pp 111–131. Plenum Press, New York

Voelker TA, Worrell AC, Anderson L, Bleibaum J, Fan C, Hawkins DJ, Radke SE and Davies HM (1992) Fatty acid biosynthesis redirected to medium chains in transgenic oilseed plants. Science 257: 72–74

Voelker TA, Yuan L, Kridl J, and Hawkins D (1995) International Patent Application No. W-O-95-13390. Published May 18, 1995.

Voelker TA, Hayes TR, Cranmer AM, Turner JC and Davies HM (1996) Genetic engineering of a quantitative trait: Metabolic and genetic parameters influencing the accumulation of laurate in rapeseed. The Plant J 9: 229–241

Wada H, Gombos Z and Murata N (1990) Enhancement of chilling tolerance of a cyanobacterium by genetic manipulation of fatty acid desaturation. Nature 347: 200–203

Weber H, Borisjuk L and Wobus U (1996) Controlling seed development and seed size in Vicia faba: A role for seed coat-associated invertases and carbohydrate state. Plant J 10: 823–834

Williams WP (1998) The physical properties of thylakoid membrane lipids and their relation to photosynthesis. In: Siegenthaler PA and Murata N (eds) Lipids in Photosynthesis: Structure, Function and Genetics, pp 103–118. Kluwer Academic Publishers, Dordrecht

Wolter FP, Schmidt R and Heinz E (1992) Chilling sensitivity of Arabidopsis thaliana with genetically engineered membrane lipids. EMBO J 11: 4685–4692

Yadav N, Wierzbicky A, Knowlton S, Pierce J, Ripp K, Hitz W, Aegerter M and Browse J (1993a) Genetic manipulation to alter fatty acid profiles of oilseed crops. In: Murata N and Somerville C (ed) Biochemistry and Molecular Biology of Membranes and Storage Lipids in Plants, Current Topics in Plant Physiology, Vol 9, pp 60–66. ASPP Series, Rockville

Yadav NS, Wierzbicki A, Aegerter M, Caster CS, Perez-Grau L, Kinney AJ, Hitz WD, Booth R Jr, Schweiger B, Stecca KL, Allen SM, Blackwell M, Reiter RS, Carlson TJ, Russell SH, Feldmann KA, Pierce J and Browse J (1993b) Cloning of higher plant ω-3 fatty acid desaturases. Plant Physiol 103: 467–476

Yuan L, Voelker TA and Hawkins DJ (1995) Modification of the substrate specificity of an acyl-acyl carrier protein thioesterase by protein engineering. Proc Natl Acad Sci USA 92: 10639–10643

Chapter 13

Genetic Engineering of the Unsaturation of Membrane Glycerolipid: Effects on the Ability of the Photosynthetic Machinery to Tolerate Temperature Stress

Zoltan Gombos[1,2] and Norio Murata[1]

[1]*Department of Regulation Biology, National Institute for Basic Biology, Myodaiji, Okazaki 444, Japan*

[2]*Institute of Plant Biology, Biological Research Center of Hungarian Academy of Sciences, H-6701 Szeged, P.O. Box 521, Hungary*

Summary

Most of the photosynthetic machinery is embedded in thylakoid membranes, which are composed of proteins, lipids and pigments, in chloroplasts and in cyanobacterial cells. The specific lipid composition of thylakoid membranes is regarded as being essential for the functions of these membranes, and alterations in the fatty acids of the membrane glycerolipids can affect the physical characteristics of the membranes and, consequently, the activities of the photosynthetic machinery. Inactivation of individual genes for fatty-acid desaturases in *Synechocystis* sp. PCC 6803 has allowed us to decrease the level of unsaturation of membrane glycerolipids in a step-wise manner. Moreover, introduction of a gene for the Δ12 desaturase from *Synechocystis* sp. PCC 6803

P.-A. Siegenthaler and N. Murata (eds): Lipids in Photosynthesis: Structure, Function and Genetics, pp. 249–262.
© *1998 Kluwer Academic Publishers. Printed in The Netherlands.*

into *Synechococcus* sp. PCC 7942 endowed the latter strain with the ability to synthesize diunsaturated fatty acids that are not found in the wild-type strain. In higher plants, genetic manipulation of glycerol-3-phosphate acyltransferase and cyanobacterial $\Delta 9$ desaturase has allowed modulation of the level of unsaturation of phosphatidylglycerol in thylakoid membranes. Characterization of the photosynthetic functions in such transformed strains of cyanobacteria and higher plants has revealed that the polyunsaturated fatty acids are essential for the protection of the photosynthetic machinery against photoinhibition at low temperatures. In their presence, the recovery of the photosynthetic machinery from the photoinhibited state is accelerated. However, the unsaturation of fatty acids does not affect the ability of the photosynthetic machinery to resist heat-induced inactivation.

I. Introduction

A. The Role of Membrane Lipids in the Chilling Sensitivity of Plants

A relationship between the phase transition of membrane lipids and the chilling sensitivity of plants was first postulated in the early, 1970's by Raison (1973) and Lyons (1973). They proposed that the formation of a gel (or solid) phase by lipids in biological membranes at non-freezing chilling temperatures induces damage to some plant tissues that can lead to death of the plant. It was argued that the membrane lipids of chilling-sensitive plants enter the gel phase at chilling temperatures, whereas those of chilling-tolerant plants remain in the liquid-crystalline phase. Since the physical phase of each membrane glycerolipid depends on the type of polar head group, on the length of the component fatty acids and, in particular, on the level of unsaturation of fatty acids (Chapter 6, Williams), it has been proposed that the chilling sensitivity of plants should be affected by modifications of membrane glycerolipids. This hypothesis was tested in the cyanobacterium *Anacystis nidulans* (Murata, 1989), in which the cell membrane has a structure similar to that of chloroplasts of eukaryotic plants, with a thylakoid membrane, a plasma membrane and an outer membrane. Temperatures for the phase transition in *A. nidulans*, as determined by various

methods, indicated that the plasma membrane enters a phase-separated state at a temperature 10 °C lower than that at which the thylakoid membrane does so (Murata et al., 1984). In the phase-separated state at low temperature, the plasma membrane becomes leaky with respect to electrolytes and small molecules (Ono and Murata, 1981; Wada et al., 1984), and this leakiness leads to damage to the cells at low temperature. Changes in the physical phase of the thylakoid membrane lipids induce changes in the efficiency of photosynthesis, but they are not critical for the survival of the cells (Murata, 1989).

When cells of *A. nidulans* are grown at different temperatures, the phase-transition temperature of the plasma membrane and the temperature critical for the death of the cells shift in parallel with the growth temperature (Murata, 1989). This phenomenon is explained by growth temperature-dependent alterations in the unsaturation of membrane lipids in the cyanobacterium (Sato et al., 1979).

In higher plants, the phase transition of membrane lipids, as observed in cyanobacteria, has not been demonstrated. In early studies, several research groups attempted to correlate the chilling sensitivity of plants to the extent of unsaturation of the fatty acids of their membrane lipids. However, these efforts were unsuccessful, and success was only achieved when unsaturation of fatty acids was analyzed in individual lipid classes. We found a clear correlation between the chilling sensitivity of herbaceous plants and the level of saturated and *trans*-monounsaturated molecular species (known as 'high-melting-point' molecular species) of phosphatidylglycerol (PG) in chloroplast membranes (Murata et al., 1982; Murata, 1983). These molecular species undergo the transition from the liquid-crystalline phase to the gel phase at room temperature and, therefore, they might be expected to induce a phase transition in thylakoid membranes.

Similar correlations have been observed in a wide

Abbreviations: ACP – acyl-carrier protein; Cm^r – gene encoding resistance to chloramphenicol; DGDG – digalactosyldiacylglycerol; GPAT – glycerol-3-phosphate acyltransferase; Km^r – gene encoding resistance to kanamycin; MGDG – monogalactosyldiacyglycerol; PG – phosphatidylglycerol; PS II – Photosystem II; SQDG – sulfoquinovosyldiacylglycerol; X:Y(Z$_1$,Z$_2$,...) – fatty acid in which X and Y indicate numbers of carbon atoms and double bonds, respectively, and Z$_1$, Z$_2$, etc., in parenthesis, indicate positions of double bonds, as counted from the carboxyl (Δ) terminus of the fatty-acyl chain variety of higher plants (Roughan, 1985; Bishop, 1986; Dorne et al., 1986; Li et al., 1987; Tasaka et al., 1990). However, the phase transition of thylakoid-membrane lipids at low temperatures has not been detected in either chilling-sensitive or chilling-tolerant plants. This failure might be explained by the fact that the amount of high-melting-point PG as a percentage of the total glycerolipids in the thylakoid membrane is only about 5% and also by the possibility that the methods used to detect the phase transition are insufficiently sensitive.

Raison et al. (1982) proposed another hypothesis, namely, that the saturation of membrane glycerolipids stabilizes the photosynthetic machinery against inactivation by heat. Moreover, Pearcy (1978) and Raison et al. (1982) also speculated that photosynthetic electron transport might be affected by the fluidity of the thylakoid membrane lipids. These hypotheses have been tested with transgenic systems, as described below.

B. Glycerolipids in Photosynthetic Membranes

The most abundant constituents, in addition to proteins, of thylakoid membranes are glycerolipids, which form bilayers that are necessary for the functioning of the membrane proteins (Trémolières et al., 1981; Doyle and Yu, 1985). The thylakoid membranes of chloroplasts and cyanobacterial cells contain four major glycerolipids: monogalactosyldiacylglycerol (MGDG); digalactosyldiacylglycerol (DGDG); sulfoquinovosyldiacylglycerol (SQDG); and phosphatidylglycerol (PG). These glycerolipids are individually associated with specific proteins (Chapter 7, Siegenthaler) and they are assumed to play specific roles in the functions of the individual proteins in the photosynthetic machinery (Chapter 8, Siegenthaler and Trémolières; Chapter 9, Trémolières and Siegenthaler).

The physical characteristics, such as the molecular motion of glycerolipids, depend on the extent of unsaturation of the fatty acids that are esterified to the glycerol backbone of the lipids (Chapman, 1975; Coolbear et al., 1983; Quinn, 1988; Chapter 6, Williams). Consequently, changes in unsaturation should affect various functions of membrane-bound proteins, such as the photochemical and electron-transport reactions that occur in thylakoid membranes.

The unsaturation of the fatty acids in the glycerolipids of biological membranes can be modified by various environmental factors (Chap-

ter 15, Harwood). Growth temperature is the major factor that influences the unsaturation of fatty acids in thylakoid membranes. In higher plants and cyanobacteria, a decrease in growth temperature induces the desaturation of the fatty acids of membrane lipids. Such temperature-induced changes in the extent of unsaturation can be explained in terms of the regulation of membrane fluidity, which is decreased at low temperatures and is increased by the desaturation of the fatty acids of the membrane lipids. It has been suggested that such a regulatory mechanism is necessary to maintain the optimal functioning of biological membranes (Cossins, 1994).

Decreases in ambient temperature induce not only the unsaturation of membrane lipids but also alterations in a large number of cellular constituents. In order to study the direct relationship between a change in the unsaturation of membrane lipids and cellular functions, it is essential to manipulate the unsaturation of membrane lipids genetically without any concomitant changes in environmental conditions, such as temperature. Techniques for such genetic manipulation are now available for studies in higher plants and cyanobacteria.

Key enzymes that have been genetically engineered to date are the acyl-lipid desaturases of cyanobacteria and glycerol-3-phosphate acyltransferase (GPAT) of higher-plant chloroplasts (Nishida and Murata, 1996). Acyl-lipid desaturases have been inactivated in *Synechocystis* sp. PCC 6803 (Tasaka et al., 1996), and an acyl-lipid desaturase has been introduced into another strain, namely, *Synechococcus* sp. PCC 7942 (Wada et al., 1990;, 1994) and also into tobacco (*Nicotiana tabacum)* (Ishizaki-Nishizawa et al., 1996). cDNAs for chloroplast-specific GPAT from squash (*Cucurbita moschata*) and from *Arabidopsis thaliana* have been introduced into tobacco (Murata et al., 1992a) and the gene for GPAT from *E. coli* has been introduced into *A. thaliana* (Wolter et al., 1992) to modify the level of fatty acid unsaturation of PG. In such transformed cells and plants, the unsaturation of membrane lipids was modified without any significant modification of other cellular constituents.

C. Acyl-Lipid Desaturases in Cyanobacteria

In cyanobacteria, only saturated fatty acids such as 18:0 and 16:0 are synthesized in the ACP-bound form, and they are esterified to the *sn*-1 and *sn*-2 positions of glycerol 3-phosphate (Chapter 4, Wada and Murata). The resultant phosphatidic acid is converted to MGDG, DGDG, SQDG and PG (Murata and Nishida, 1987). The saturated fatty acids bound to these glycerolipids are desaturated to unsaturated fatty acids with various numbers of double bonds (unsaturated bonds) in reactions catalyzed by fatty-acid desaturases (Murata and Wada, 1995).

Fatty-acid desaturases are the enzymes that catalyze the conversion of single bonds to double bonds in fatty-acyl chains. They are classified into three groups: the acyl-lipid desaturases, the acyl-CoA desaturases, and the acyl-ACP desaturases (Murata and Wada, 1995). In cyanobacteria, the acyl-lipid desaturases are the only type of desaturase that has been found (Murata and Wada, 1995). These enzymes introduce double bonds into fatty acids that have been esterified to glycerolipids.

The cyanobacterium *Synechocystis* sp. PCC 6803 contains four acyl-lipid desaturases; namely, the $\Delta 9$, $\Delta 12$, $\Delta 15$ ($\omega 3$) and $\Delta 6$ acyl-lipid desaturases. Therefore, the most desaturated fatty acid in this strain is 18:4(6,9,12,15) (Wada and Murata, 1989). The *desA* gene for the $\Delta 12$ desaturase was the first to be isolated (Wada et al., 1990). Subsequently, the *desD* gene for the $\Delta 6$ desaturase (Reddy et al., 1993), the *desB* gene for the $\omega 3$ ($\Delta 15$) desaturase (Sakamoto et al., 1994a) and the *desC* gene for the $\Delta 9$ desaturase (Sakamoto et al., 1994b) were isolated from the same strain.

The cyanobacterium *Synechococcus* sp. PCC 7942 contains only the $\Delta 9$ desaturase. Thus, this strain has saturated and monounsaturated fatty acids in its membrane lipids. The *desC* gene for the $\Delta 9$ desaturase has been isolated from *Synechococcus* sp. PCC 7942 (Ishizaki-Nishizawa et al., 1996).

To examine the effects of the unsaturation of the fatty acids of membrane lipids on the ability to tolerate temperature stress, it is necessary to alter the level of unsaturation of glycerolipids exclusively by manipulation of the genes that encode fatty-acid desaturases. Since some strains of cyanobacteria, such as *Synechocystis* sp. PCC 6803 and *Synechococcus* sp. PCC 7942, are autotransformable (Grigoreva and Shestakov, 1982; Williams and Szalay, 1983), it is possible to alter the level of unsaturation of membrane glycerolipids by genetic manipulation of the genes for the fatty-acid desaturases.

D. Glycerol-3-Phosphate Acyltransferase in the Chloroplasts of Higher Plants

Biosynthesis of glycerolipids in the chloroplasts of higher plants differs from that in cyanobacterial cells. Both 16:0 and 18:1 (9) are synthesized in the ACP-bound form and then they are esterified to the *sn*-1 and *sn*-2 positions of glycerol 3-phosphate. The resultant phosphatidic acid is converted to the glycerolipids in the chloroplasts, although some of the chloroplast glycerolipids are synthesized via the cytoplasmic pathway (Chapter 2, Joyard and Douce). However, PG is synthesized exclusively in the chloroplasts.

Glycerol-3-phosphate acyltransferase (GPAT) catalyzes the first step in the synthesis of glycerolipids, namely, the esterification of the acyl group to the *sn*-1 position of glycerol-3-phosphate (Roughan and Slack, 1982). In plant cells, there are two types of GPAT; one is present in the cytoplasm, and the other is found in chloroplasts (McKeon and Stumpf, 1982). These enzymes differ in the use of acyl donors. The cytoplasmic enzyme uses acyl-CoA as the acyl donor (Holloway, 1983), whereas the enzyme in the chloroplasts uses acyl-(acyl-carrier protein), abbreviated as acyl-ACP, as the acyl donor (McKeon and Stumpf, 1982). The substrate specificity of acyl-ACP: glycerol-3-phosphate acyltransferase in chloroplasts for 16:0 or 18:1 determines the synthesis of saturated molecular species of PG in chloroplasts (Murata et al., 1982; Murata, 1983).

The chloroplast-specific GPAT was purified first from the plastids of squash cotyledons by Nishida et al. (1987) and the cDNA for the precursor to the enzyme was isolated by Ishizaki et al. (1988). Subsequently, cDNAs for GPAT were isolated from cucumber (Johnson et al., 1992), pea (Weber et al., 1991) and spinach (Ishizaki-Nishizawa et al., 1995). Overexpression of cDNAs for GPAT from various sources in tobacco and *Arabidopsis* resulted in changes in levels of saturated molecular species of PG (Murata et al., 1992a; Wolter et al., 1992).

II. Genetic Dissection of Fatty-Acid Desaturation in *Synechocystis*

A. Inactivation of Genes for Acyl-Lipid Desaturases

Synechocystis sp. PCC 6803 belongs to group 4 of cyanobacteria, according to the classification that is based on the desaturation of fatty acids (Murata et al., 1992b; Murata and Wada, 1995; Chapter 4, Wada and Murata). In this strain, fatty acids are desaturated at the Δ6, Δ9, Δ12 and Δ15 (ω3) positions. The *desA*, *desB*, *desC* and *desD* genes that encode the Δ12, Δ15 (ω3), Δ9 and Δ6 acyl-lipid desaturases, respectively, have been cloned from *Synechocystis* sp. PCC 6803 (Murata and Wada, 1995). An examination of the sequence of the entire genome (Kaneko et al., 1996) reveals that there are no other genes that are homologous to genes for fatty-acid desaturases. Insertional disruption of the *desA*, *desB* and *desD* genes provided a useful experimental system in which we were able to manipulate the number of double

bonds in glycerolipid molecules in *Synechocystis* sp. PCC 6803 (Wada et al., 1990, 1992; Tasaka et al., 1996).

The *desA* and *desB* genes were disrupted by insertion of a kanamycin-resistance gene cartridge, and the *desD* gene was disrupted by insertion of a chloramphenicol-resistance gene cartridge (Fig. 1). The inactivation of the *desD* gene completely eliminated the activity of the ω3 desaturase and, consequently, the transformed cells did not contain tetra-unsaturated fatty acids (Fig. 2). Inactivation of

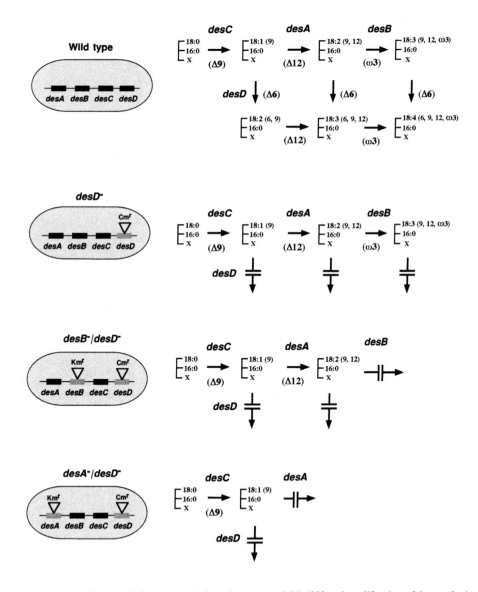

Fig. 1. Insertional mutagenesis of fatty-acid desaturases in *Synechocystis* sp. PCC 6803 and modification of the synthetic pathways to various molecular species of glycerolipids. X represents a polar group.

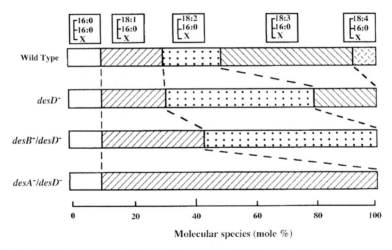

Fig. 2. Changes in molecular species composition of glycerolipids after manipulation of fatty-acid desaturases by insertional mutagenesis. 18:3 includes 18:3(6,9,12) and 18:3(9,12,15). Data were calculated from the results of Tasaka et al. (1996). X represents a polar head group.

the *desB* and *desA* genes in the *desD⁻* cells further eliminated the triunsaturated and diunsaturated fatty acids, respectively (Fig. 2; Tasaka et al., 1996). These studies demonstrated that the extent of unsaturation of glycerolipids in membranes could be manipulated solely by the inactivation of genes for individual desaturases. The properties of the transformed cells were analyzed in physiological studies in an attempt to identify direct relationships between the unsaturation of glycerolipids in photosynthetic membranes and the temperature-dependent processes of photosynthesis.

B. Tolerance to Low Temperature and Low-Temperature Photoinhibition

The sensitivity of the photosynthetic machinery to high-intensity light is considerably enhanced when photosynthetic organisms are exposed to low temperature (Öquist et al., 1987; Greer et al., 1991). This phenomenon results in so-called low-temperature photoinhibition. The primary target of this process is the D1 protein of the PS II complex (Nanba and Sato, 1987; Aro et al., 1990). Light-induced inactivation of the PS II complex is caused by damage to the D1 protein. Restoration of the PS II complex from the photoinhibition by repairing the D1 protein depends on a series of reactions, such as degradation of the damaged D1 protein, synthesis of the precursor to the D1 protein, incorporation of the precursor into the PS II complex, and processing of the precursor to yield the mature D1 protein (Aro et

al., 1993). We examined the effects of the unsaturation of membrane lipids on photoinhibition in cells in which genes for various desaturases had been inactivated.

Wild-type, *desD⁻* and *desA⁻/desD⁻* cells of *Synechocystis* sp. PCC 6803 (Figs. 1 and 2), which had been grown under isothermal conditions, were exposed to illumination for various lengths of time at low temperatures. The extent of photoinhibition of photosynthesis depended on the strain and the temperature of treatment. The *desA⁻/desD⁻* cells, which contained monounsaturated but no polyunsaturated lipid molecules, were the most sensitive to low-temperature photoinhibition (Fig. 3A). The *desD⁻* cells, which contained di- and tri unsaturated lipid molecules, were indistinguishable from the wild-type cells with respect to the sensitivity to low-temperature photoinhibition.

The extent of photoinhibition depends on the balance between the process of photoinduced inactivation and the recovery from the photo-inactivated state of the photosynthetic machinery. Therefore, an attempt was made to separate the two processes by a biochemical method. Chloramphenicol, an inhibitor of protein synthesis in prokaryotes, was used to demonstrate that the inactivation of the PS II complex was unaffected by alterations in the extent of unsaturation of fatty acids (Fig. 3B). This conclusion was reinforced by results of measurements of the inactivation of electron transport in isolated thylakoid membranes, in which the recovery process does not occur (Fig. 3C). Direct

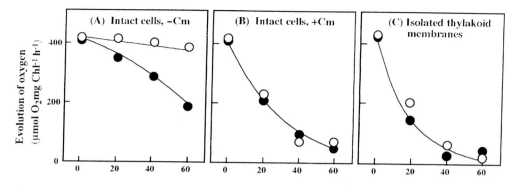

Fig. 3. The photoinhibition of photosynthesis in intact cells of wild-type (○) and desA⁻/desD⁻ (●) strains of *Synechocystis* sp. PCC 6803 and in thylakoid membranes isolated from them. The photoinhibitory treatment was performed at 20 °C and at a light intensity of 1.5 mmol m⁻² sec⁻¹. (A) Intact cells in the absence of chloramphenicol (Cm). (B) Intact cells in the presence of chloramphenicol at 0.2 mg ml⁻¹. (C) Isolated thylakoid membranes in the absence of chloramphenicol. Taken from Tasaka et al. (1996).

measurements of the recovery from the photoinhibited state of the PS II complex in intact cells demonstrated that the unsaturation of membrane lipids accelerated the recovery process (Fig. 4; Gombos et al., 1994b; Tasaka et al., 1996).

The turnover of the D1 protein was studied in wild-type and *desA⁻/desD⁻* cells. Immunoblotting analysis indicated that changes in the extent of unsaturation of glycerolipids in the thylakoid membranes did not affect the rate of removal of the damaged D1 protein from the membrane, nor did it affect the rate of transcription of the *psbA* gene, which encodes the D1 protein. The results suggested that polyunsaturated fatty acids might affect protein synthesis at the translational or the post-translational level (Kanervo et al., 1995). It seems likely that the processing of the precursor into the D1 protein requires polyunsaturated glycerolipids (Kanervo et al., 1997).

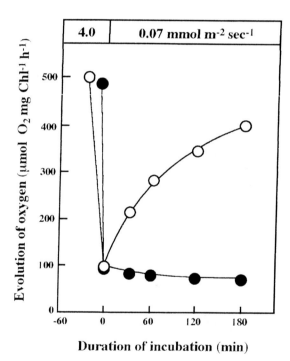

Fig. 4. The recovery of the PS II complex from the photoinhibited state in wild-type (○) and *desA⁻/desD⁻* (●) strains of *Synechocystis* sp. PCC 6803. The activity of the PS II complexes was monitored at 34 °C by measuring of the evolution of oxygen in the presence of 1,4-benzoquinone as the electron acceptor. Photoinhibition was induced by incubating cells at a light intensity of 4.0 mmol m⁻² sec⁻¹ and then the recovery was allowed to proceed at 0.07 mmol m⁻² sec⁻¹, both at 20 °C. Taken from Tasaka et al. (1996).

C. Heat Tolerance

Cyanobacterial cells respond to an increase in temperature by a reduction in the level of their polyunsaturated fatty acids (Sato and Murata, 1981; Sato et al., 1979; Wada and Murata, 1990). To confirm a direct relationship between high-temperature tolerance and the unsaturation of glycerolipids, we performed studies with wild-type and Fad6/*desA⁻* mutant cells of *Synechocystis* sp. PCC 6803. To obtain the Fad6/*desA⁻* mutant strain, we first generated the Fad6 mutant of *Synechocystis* sp. PCC 6803 by

mutation of the gene for the Δ6 desaturase with ethylmethane sulfonate (Wada and Murata, 1989). The resultant Fad6 mutant cells were further mutated by insertional disruption of the *desA* gene to generate the Fad6/*desA*⁻ strain. In this strain polyunsaturated fatty acids were completely eliminated, as they were in the *desA*⁻/*desD*⁻ cells when the Fad6/*desA*⁻ cells were grown at a high temperature, such as 35 °C (Wada et al., 1992).

We observed no significant effect of the complete absence of polyunsaturated fatty acids from Fad6/*desA*⁻ cells on the ability of the photosynthetic machinery to tolerate high temperatures (Gombos et al., 1994a). These results conflicted with the earlier hypothesis, which was based on physiological studies (Raison et al., 1982), that the saturation of glycerolipids stabilizes the photosynthetic machinery against inactivation by heat. An examination of the effects of light on heat stability revealed that, in neither wild-type nor Fad6/*desA*⁻ cells, heat and light had any synergistic effects on the inactivation of the photosynthetic machinery (Gombos et al., 1994a).

These findings seem to indicate that the acclimation of photosynthetic organisms to high temperature might involve factors other than saturated glycerolipids. Moreover, it has been demonstrated, for example, that some protein factors in thylakoid membranes are required for the acclimation of the photosynthetic machinery to high temperature (Nishiyama et al., 1993, 1994).

D. Photosynthetic Activity and Cell Growth

When the photosynthetic activities of *Synechocystis* sp. PCC 6803 that had been grown at various temperatures were compared under isothermal conditions, those of cells grown at lower temperatures were found to be higher than those of cells grown at elevated temperatures (Gombos et al., 1992). Such changes in photosynthetic activity with growth temperature were regarded as a result of alterations in the extent of unsaturation of glycerolipids (Pearcy, 1978).

However, our recent studies (Tasaka et al., 1996) demonstrated clearly that changes in the extent of unsaturation of glycerolipids in photosynthetic membranes in wild-type and *desA*⁻/*desD*⁻ cells affect neither photosynthesis nor the transport of electrons from H_2O to p-benzoquinone, when cells are grown under isothermal conditions. Consequently, we can conclude that variations in photosynthetic activities with growth temperature are unrelated to the extent of lipid unsaturation.

A comparison of growth rates of wild-type

Synechocystis sp. PCC 6803 and transformed cells demonstrated that elimination of 18:3 fatty acids from the membranes of *desD*⁻ cells did not affect the growth rate at all temperatures tested (20–35 °C). However, the *desA*⁻/*desD*⁻ cells grew at 35 °C at a rate similar to that of the wild-type and other mutant cells, while the growth rate of *desA*⁻/*desD*⁻ cells at 25 °C was lower than that of wild-type, *desD*⁻ and *desB*⁻/*desD*⁻ cells. The *desA*⁻/*desD*⁻ cells were unable to grow at 20 °C (Tasaka et al., 1996). These results suggest that, in order to retain the ability to grow at low temperatures, cells require polyunsaturated fatty acids.

The discrepancy between the effects of unsaturation of membrane lipids on photosynthetic activities and the growth rates at low temperatures may be explained by a mechanism whereby the growth of cells requires the stability of the photosynthetic machinery. Thus, the presence of polyunsaturated fatty acids might be essential for the growth of cells at low temperatures because of their role in protecting the photosynthetic machinery against low-temperature photoinhibition.

III. Genetic Enhancement of Fatty-Acid Desaturation in *Synechococcus*

A. Introduction of a Novel Gene for a Δ12 Acyl-Lipid Desaturase

One strategy that was used to modify the extent of desaturation of glycerolipids in membranes involved the introduction of a fatty-acid desaturase that is not found in wild-type cells of *Synechococcus* sp. PCC 7942 (*Anacystis nidulans* R2). This strain contains only saturated and monounsaturated fatty acids. *Synechococcus* cells were transformed with the *desA* gene from *Synechocystis* sp. PCC 6803 either by use of a shuttle vector, namely, plasmid pUC303 (Wada et al., 1990), or by insertion into the chromosome (Sakamoto et al., 1994a). The latter method was better because the *desA*⁺-transformed cells stably retained the introduced gene. By contrast, when the shuttle vector was used, the transformed cells were unstable (Gombos et al., 1997). Transformation by either method allowed the *Synechococcus* cells to synthesize diunsaturated fatty acids such as 16:2(9,12) and 18:2(9,12). As a result, the transformed cells contained almost equal amounts of monounsaturated and diunsaturated glycerolipid molecules as major components. By contrast, in the wild-type cells, 90%

of the total lipid molecules were monounsaturated molecules with only minor levels of saturated molecules.

B. Low-Temperature Tolerance

In *Anacystis nidulans* (*Synechococcus* sp. PCC 6301, a strain very similar to *Synechococcus* sp. PCC 7942), the thermotrophic transition of the physical phase of the plasma membrane is closely correlated with the susceptibility to chilling (Murata, 1989). Since the physical phase of membrane lipids depends on the extent of their unsaturation, modulation of the unsaturation of glycerolipids in response to changes in ambient temperature alters the temperature for the phase transition as well as that for the chilling-induced damage to cells (Murata, 1989).

To obtain direct evidence for a relationship between chilling temperature and the extent of unsaturation of glycerolipids, we transformed *Synechococcus* sp. PCC 7942 cells with the novel *desA* gene, using a shuttle vector, as mentioned above (Wada et al., 1990). The responses to low temperatures from 0 to 10 °C differed between the wild-type cells and the *desA*+-transformed cells. The *desA*+-transformed cells survived and remained photosynthetically active even after a one-hour incubation at 5 °C. By contrast, wild-type cells irreversibly lost 50% of their photosynthetic activity and failed to regain full photosynthetic activity even after incubation at a normal growth temperature (Wada et al., 1990). The irreversible inactivation of photosynthesis in darkness is related to the phase transition of plasma membranes (Murata, 1989). Consequently, it appears that the presence of diunsaturated glycerolipids in the membranes of photosynthetic organisms enhances the ability of cells to protect themselves against dramatic decreases in the temperature of the natural environment.

C. Low-Temperature Photoinhibition

When cells of the wild-type strain and of the *desA*+-transformed strain of *Synechococcus* sp. PCC 7942, in which the *desA* gene had been integrated into the chromosome, were grown under isothermal conditions and then exposed to strong light at temperatures below the growth temperature, photosynthesis was markedly inhibited in both cases. However, the *desA*+-transformed cells that contained diunsaturated glycerolipids were more resistant to low-temperature

photoinhibition than were the wild-type cells that contained only monounsaturated fatty acids. Treatment with an inhibitor of protein synthesis during the photoinhibitory treatment showed clearly that there was no difference in the kinetics of photoinhibition between the *desA*+-transformed and wild-type cells. Direct measurement of the recovery of the PS II complex from the photoinhibited state confirmed that diunsaturated glycerolipids accelerated the recovery process (Gombos et al., 1997). Therefore, diunsaturated fatty acids appear to play a critical role in the recovery from photoinhibition.

D. Heat Tolerance and Photosynthetic Activity

Synechococcus sp. PCC 7942 exhibits moderate tolerance to heat. Transformation with the *desA* gene altered the unsaturation of glycerolipids but it did not affect the profile of heat inactivation of the PS II complex. By contrast, an increase in growth temperature increased the temperature for the heat-induced inactivation of the PS II complex in both the wild-type and the *desA*+-transformed cells (Wada et al., 1994). These results indicate that 16:2(9,12) and 18:2(9,12) do not affect the ability of photosynthetic organisms to tolerate high-temperature stress. Furthermore, they support the hypothesis that acclimation to high temperatures involves factors other than the extent of unsaturation of glycerolipids.

Despite the difference in the extent of fatty-acid unsaturation between the wild-type and *desA*+-transformed cells, we observed no significant differences in photosynthetic activities and PS II-mediated electron transport when these cells were grown under isothermal conditions (Wada et al., 1994; Gombos et al., 1997). These results serve to emphasize the fact that the introduction of a second double bond into glycerolipids in membranes does not affect the activity of the PS II complex when it is measured over a short time.

IV. Genetic Engineering of Phosphatidylglycerol in Tobacco

A. Transformation with cDNA for Glycerol-3-phosphate Acyltransferase

Glycerol-3-phosphate acyltransferase (GPAT) in chloroplasts (or plastids) catalyzes the transfer of the acyl group of acyl-ACP to the *sn*-1 position of glycerol

3-phosphate (Nishida and Murata, 1996). This enzyme in chilling-tolerant plants, such as pea, spinach and *Arabidopsis thaliana*, specifically uses 18:1-ACP as its substrate (Frentzen et al., 1983, 1987), while the enzyme in chilling-sensitive plants, such as squash, uses both 18:1-ACP and 16:0-ACP.

The chilling sensitivity and the level of high-melting-point species (saturated and *trans*-mono-unsaturated species) of PG in tobacco are intermediate among those of higher plants. Moreover, squash is more sensitive and *Arabidopsis* is less sensitive to chilling than tobacco. cDNAs for the chloroplast-specific GPAT from squash and from *Arabidopsis* were incorporated into the binary plasmid pBI-121 under the control of the 35S constitutive promoter of cauliflower mosaic virus and the resultant plasmids were designated pSQ and pARA, respectively. Tobacco plants were transformed with these plasmids by *Agrobacterium*-mediated transformation. As a result, the level of high-melting-point PG in tobacco plants was altered significantly, but the relative levels of lipid classes and the fatty-acid composition of the other glycerolipids were hardly affected (Murata et al., 1992a). The level of high-melting-point molecular species of PG in tobacco leaves that expressed the cDNA for squash GPAT increased from 36% to 76%, whereas the plants that expressed the cDNA for *Arabidopsis* GPAT contained a lower level of high-melting-point PG (28%) than that in the wild-type plants. These results provide direct evidence that GPAT plays a critical role in the regulation of the level of high-melting-point molecular species of PG.

B. Low-Temperature Tolerance

During incubation at low temperatures under strong light, leaves of pSQ-transformed tobacco plants lost their capacity for photosynthesis at a higher rate than those of wild-type plants. This result indicates that plants with a higher level of high-melting-point PG were more sensitive to chilling. When transgenic plants were exposed to continuous illumination at a normal light intensity but at a low temperature, such as 1 °C, for 10 days, leaves from wild-type, control (pBI-121-transformed) and pSQ-transformed plants exhibited symptoms of chlorosis but leaves of pARA-transformed tobacco plants did not (Murata et al., 1992a). These results emphasize the importance in the chilling sensitivity of higher plants of *cis*-unsaturated molecular species of PG in chloroplasts, and they provide evidence that the chilling tolerance

of higher plants can be manipulated by trans-formation. Thus, transgenic plants can be produced with decreased levels of high-melting-point PG and increased tolerance to chilling stress.

C. Low-Temperature Photoinhibition

Photoinhibition is observed in many plant species upon their exposure to strong light (Powles, 1984). Photoinhibition appears even under moderate illumination when plants are subjected to various types of environmental stress, such as chilling (Berenyi and Krause, 1985; Boese and Huner, 1990; Falk et al., 1990; Somersalo and Krause, 1990b) and freezing (Strand and Öquist, 1985; Somersalo and Krause, 1990a; Greer et al., 1991).

To examine the correlation between low-temperature photoinhibition and unsaturated glycerolipids, we compared transgenic tobacco plants that had been transformed with pSQ with wild-type plants. The former plants were more sensitive to low-temperature photoinhibition than the latter (Moon et al., 1995).

Experiments with intact leaves that had been treated with lincomycin, an inhibitor of translation in chloroplasts, revealed that photoinduced damage to the PS II complex was unaffected by alterations in levels of high-melting-point PG. No significant difference in the rate of photoinactivation of the PS II complex was detected between thylakoid membranes from leaves of wild-type and transformed plants in which the recovery process had been totally blocked. A study of the recovery of the photosynthetic machinery from the photoinhibited state suggested that the unsaturation of PG might be important in the protection against low-temperature photoinhibition via acceleration of the recovery process.

D. Heat Tolerance and Photosynthetic Activity

The relationship between the unsaturation of membrane glycerolipids and the tolerance of higher plants to heat has been studied since the 1970s (Pearcy, 1978). The high-temperature-induced inactivation of the oxygen-evolving PS II complex is caused by the release of Mn^{2+} ions from the complex (Nash et al., 1985). Pearcy (1978) and Raison et al. (1982) postulated that an increase in the growth temperature might increase the extent of saturation of fatty acids and, as a consequence, it might enhance the stability of the photosynthetic machinery at higher temperatures. In contrast to the above results, Santarius and Müller (1979) observed that, in spinach, an increase in the heat tolerance of the photosynthetic machinery during the acclimation to a high temperature was not accompanied by changes in the saturation of membrane lipids.

We compared the profiles of the temperature-dependent disappearance of photosynthetic activities in thylakoid membranes that had been isolated from wild-type and the pSQ-transformed tobacco plants (Moon et al., 1995). The inactivation of photosynthesis was observed after a 20-min incubation at various temperatures above 30 °C. The stability of the photosynthetic machinery at high temperature was unaffected by changes in the level of high-melting-point PG in transformed plants, demonstrating that the extent of saturation of PG had no effect on the stability at high temperatures of the oxygen-evolving PS II complex of the plants. This observation is in accordance with the results of experiments with transformed cyanobacterial cells.

We studied the effects of alterations in the level of unsaturated PG on the activity of the PS II complex in the transgenic tobacco plants. The activity of the PS II-mediated transport of electrons in thylakoid membranes was unaffected by the level of unsaturation of PG when the activity of the PS II complex in various transformed and wild-type tobacco plants was examined under isothermal growth conditions (Moon et al., 1995).

E. Transformation with a Gene for a Cyanobacterial Desaturase

Tobacco plants have also been transformed with the *desC* gene for the Δ9 desaturase from *Synechococcus* sp. PCC 7942 (Ishizaki-Nishizawa et al., 1996). This gene encodes a fatty acid desaturase that introduces a *cis*-double bond at the Δ9 position of fatty acids that are linked to the *sn*-1 position of all membrane lipids (Murata et al., 1992b). In the transformed tobacco plants the level of high-melting-point PG relative to that of the total PG was reduced from 44% to 24%. The reduced level of saturation of membrane glycerolipids enhanced the tolerance of the plants to chilling temperatures. The transformed tobacco plants showed no symptoms after plants had been exposed to a chilling temperature of 4 °C for 11 days. Moreover, seeds germinated and seedlings were grown successfully at 10 °C for 70 days, an indication that the transgenic plants were able to grow as chilling-

tolerant plants. These results provide a further demonstration of the fact that the high-melting-point PG is responsible for the chilling-induced damage to chilling-sensitive plants.

V. Genetic Engineering of Phosphatidylglycerol in *Arabidopsis thaliana*

Wolter et al. (1992) increased the level of high-melting-point PG in *Arabidopsis thaliana* by transformation with the *plsB* gene, which encodes a membrane-bound glycerol-3-phosphate acyltransferase in *E. coli*. The product of the *plsB* gene was targeted to the chloroplasts in transformed plants. In the transformed plants, the relative level of high-melting-point PG increased from 10% in the wild-type plants to about 50%.

Arabidopsis plants that had been transformed with the *plsB* gene from *E. coli* and contained PG with elevated levels of high-melting-point molecular species were more susceptible to chilling than the wild-type plants. When transformed plants were incubated in darkness for 7 days at 4 °C and then returned to 20 °C for 2 days, the older leaves wilted and became brown and necrotic (Wolter et al., 1992). The chilling susceptibility of *plsB*-transformed plants was enhanced by light, and *plsB*-transformed plants died after a 3- to 4-day incubation at 4 °C under normal illumination.

Wu and Browse (1995) reported the isolation of a mutant (*fab1*) of *Arabidopsis* that was defective in palmitoleyl-ACP elongase. The cellular lipids of this mutant had an elevated level of 16:0 and the relative level of high-melting-point molecular species of PG was increased to 43% from a value of only 9% in the wild type. The growth of the *fab1* mutant plants was similar to that of wild-type plants when plants were incubated at 2 °C for 7 days under continuous, moderately strong illumination (Wu and Browse, 1995). These results seem to contradict the results obtained with transformed *A. thaliana* (Wolter et al., 1992) and transformed tobacco (Murata et al., 1992a) plants. However, no data have yet been reported on the sensitivity to low-temperature photoinhibition of the mutated and transformed plants of *A. thaliana*.

VI. Conclusions

From studies that were performed with genetically

manipulated strains of cyanobacteria and higher plants we can draw the following conclusions:

(1) The genetic manipulation of fatty-acid desaturases in cyanobacteria provides unique experimental systems to analyze the importance of the unsaturation of fatty acids of membrane lipids in the tolerance to low- and high-temperature stress.

(2) The transformation of tobacco and *Arabidopsis* with glycerol-3-phosphate acyltransferase produces plants with altered levels of the high-melting-point molecular species (saturated and *trans*-monounsaturated molecular species) of PG. Using these systems, we can examine directly the roles of the saturation of PG in the tolerance of higher plants to temperature stress.

(3) The unsaturation of fatty acids in membranes is clearly essential for the ability of the photosynthetic machinery to tolerate low-temperature stress. In particular, the unsaturation of fatty acids accelerates the recovery of the PS II complex from low-temperature photoinhibition. This phenomenon is related to the accelerated maturation of the D1 protein in the PS II complex.

(4) The tolerance of the photosynthetic machinery to high temperature is not affected by the unsaturation of membrane lipids.

Acknowledgments

The preparation of this review was supported, in part, by a Grant-in-Aid for Specially Promoted Research (no. 08102011) to N. Murata and by grants from the Hungarian Science Foundation (OTKA; nos. T 020293 and F 023794) to Z. Gombos. It was also supported, in part, by the NIBB Cooperative Research Program on the Stress Tolerance of Plants.

References

Aro E-M, Hundal T, Carlberg I and Andersson B (1990) In vitro studies on light-induced inhibition of Photosystem II and D1-protein degradation at low temperature. Biochim Biophys Acta 1019: 269–275

Aro E-M, Virgin I and Andersson B (1993) Photoinhibition of photosystem II, protein damage and turnover. Biochim Biophys

Acta 1143: 113–134

Berenyi B and Krause GH (1985) Inhibition of photosynthesis by light. Planta 163: 218–226

Bishop DG (1986) Chilling sensitivity in higher plants: The role of phosphatidylglycerol. Plant Cell Environ 9: 613–16

Boese SR and Huner NPA (1990) Effect of growth temperature and temperature shifts on spinach leaf morphology and photosynthesis. Plant Physiol 94: 1830–1836

Chapman D (1975) Phase transition and fluidity characteristics of lipids and cell membranes. Quart Rev Biophys 8: 185–235

Coolbear KP, Berde CB and Keough KMW (1983) Gel to liquid-crystalline phase transition of aqueous dispersions of polyunsaturated mixed-acid phosphatidylcholines. Biochemistry 22: 1466–1473

Cossins AR (1994) Homeoviscous adaptation of biological membranes and its functional significance. In: Cossins AR (ed) Temperature Adaptation of Biological Membranes, pp 63–76. Portland Press, London

Dorne A-J, Cadel G and Douce R (1986) Polar lipid composition of leaves from nine typical alpine species. Phytochemistry 25: 65–68

Doyle MF and Yu C-A (1985) Preparation and reconstitution of phospholipid-deficient cytochrome b_6-f complex from spinach chloroplasts. Biochem Biophys Res Commun 131: 700–706

Falk S, Samuelsson G and Öquist G (1990) Temperature-dependent photoinhibition and recovery of photosynthesis in the green alga *Chlamydomonas reinhardtii* acclimated to 12 and 27 °C. Physiol Plant 78: 173–180

Frentzen M, Heinz E, McKeon TA and Stumpf PK (1983) Specificities and selectivities of glycerol-3-phosphate acyltransferase from pea and spinach chloroplasts. Eur J Biochem 129: 629–636

Frentzen M, Nishida I and Murata N (1987) Properties of the plastidial acyl-(acyl-carrier-protein): Glycerol-3-phosphate acyltransferase from the chilling-sensitive plant squash (*Cucurbita moschata*). Plant Cell Physiol 28: 1195–1201

Gombos Z, Wada H and Murata N (1992) Unsaturation of fatty acids in membrane lipids enhances tolerance of the cyanobacterium *Synechocystis* PCC 6803 to low-temperature photoinhibition. Proc Natl Acad Sci USA 89: 9959–9963

Gombos Z, Wada H, Hideg E and Murata N (1994a) The unsaturation of membrane lipids stabilizes photosynthesis against heat stress. Plant Physiol 104: 563–567

Gombos Z, Wada H and Murata N (1994b) The recovery of photosynthesis from low-temperature photoinhibition is accelerated by the unsaturation of membrane lipids: A mechanism of chilling tolerance. Proc Natl Acad Sci USA 91: 8787–8791

Gombos Z, Kanervo E, Tsvetkova N, Sakamoto T, Aro E-M and Murata N (1997) Genetic enhancement of the ability to tolerate photoinhibition by introduction of unsaturated bonds into membrane glycerolipids. Plant Physiol 115: 551–559

Greer DH, Ottander C and Öquist G (1991) Photoinhibition and recovery of photosynthesis in intact barley leaves at 5 and 20 °C. Physiol Plant 81: 203–210

Grigoreva G and Shestakov S (1982) Transformation in the cyanobacterium *Synechocystis* sp. PCC 6803. FEMS Microbiol Lett 13: 367–370

Holloway PW (1983) Fatty acid desaturation. In: Boyer PD (ed) The Enzymes, Vol 16, pp 63–83. Academic Press, Orlando

Ishizaki O, Nishida I, Agata K, Eguchi G and Murata N (1988)

Cloning and nucleotide sequence of cDNA for the plastid glycerol-3-phosphate acyltransferase from squash. FEBS Lett 238: 424–430

Ishizaki-Nishizawa O, Azuma M, Ohtani T, Murata N and Toguri T (1995) Nucleotide sequence of cDNA from *Spinacia oleracea* encoding plastid glycerol-3-phosphate acyltransferase. Plant Physiol 108: 1342

Ishizaki-Nishizawa O, Fujii T, Azuma M, Sekiguchi K, Murata N, Ohtani T and Toguri T (1996) Low-temperature resistance of higher plants is significantly enhanced by a nonspecific cyanobacterial desaturase. Nature Biotech 14: 1003–1006

Johnson TC, Schneider JC and Somerville C (1992) Nucleotide sequence of acyl-acyl carrier protein: Glycerol-3-phosphate acyltransferase from cucumber. Plant Physiol 99: 771–772

Kaneko T, Sato S, Kotani H, Tanaka A, Asamizu E, Nakamura Y, Miyajima N, Hirosawa M, Sugiura M, Sasamoto S, Kimura T, Hosouchi T, Matsuno A, Muraki A, Nakazaki C, Naruo K, Okumura S, Shimpo S, Takeuchi C, Wada T, Watanabe A, Yamada M, Yasuda M and Tabata S (1996) Sequence analysis of the genome of the unicellular cyanobacterium *Synechocystis* sp. strain PCC 6803. II. Sequence determination of the entire genome and assignment of potential protein-coding regions. DNA Research 3: 109–136

Kanervo E, Aro E-M and Murata N (1995) Low unsaturation level of thylakoid membrane lipids limits turnover of the D1 protein of Photosystem II at high irradiance. FEBS Lett 364: 239–242

Kanervo E, Tasaka Y, Murata N and Aro E-M (1997) Membrane lipid unsaturation-modulated processing of Photosystem II reaction-center protein D1 at low temperatures. Plant Physiol 114: 841–849

Li T, Lynch DV and Steponkus PL (1987) Molecular species composition of phosphatidylglycerols from rice varieties differing in chilling sensitivity. Cryo-Lett 8: 314–321

Lyons J M (1973) Chilling injury in plants. Annu Rev Plant Physiol 24: 445–466

McKeon TA and Stumpf PK (1982) Purification of the stearyl-acyl carrier protein desaturase and the acyl-acyl carrier protein thioesterase from maturing seeds of safflower. J Biol Chem 257: 12141–12147

Moon BY, Higashi S, Gombos Z and Murata N (1995) Unsaturation of the membrane lipids of chloroplasts stabilizes the photosynthetic machinery against low-temperature photoinhibition in transgenic tobacco plants. Proc Natl Acad Sci USA 92: 6219–6223

Murata N (1983) Molecular species composition of phosphatidylglycerols from chilling-sensitive and chilling-resistant plants. Plant Cell Physiol 24: 81–86

Murata N (1989) Low-temperature effects on cyanobacterial membranes. J Bioenerg Biomembr 21: 61–75

Murata N and Nishida I (1987) Lipids of blue-green algae (cyanobacteria). In: Stumpf PK (ed) The Biochemistry of Plants, Vol 9, Structure and Functions of Lipids, pp 314–347. Academic Press, Orlando

Murata N and Wada H (1995) Acyl-lipid desaturases and their importance in the tolerance and acclimatization to cold of cyanobacteria. Biochem J 308: 1–8

Murata N, Sato N, Takahashi N and Hamazaki Y (1982) Compositions and positional distributions of fatty acids in phospholipids from leaves of chilling-sensitive and chilling-resistant plants. Plant Cell Physiol 23: 1071–1079

Murata N, Wada H and Hirasawa R (1984) Reversible and irreversible inactivation of photosynthesis in relation to the lipid phases of membrane in the blue-green algae (cyanobacteria) *Anacystis nidulans* and *Anabaena variabilis*. Plant Cell Physiol 25: 1027–1032

Murata N, Ishizaki-Nishizawa O, Higashi S, Hayashi H, Tasaka Y and Nishida I (1992a) Genetically engineered alteration in the chilling sensitivity of plants. Nature 356: 710–713

Murata N, Wada H and Gombos Z (1992b) Modes of fatty-acid desaturation in cyanobacteria. Plant Cell Physiol 33: 933–941

Nanba O and Satoh K (1987) Isolation of a photosystem II reaction center consisting of D-1 and D-2 polypeptides and cytochrome *b*-559. Proc Natl Acad Sci USA 84: 109–112

Nash D, Miyao M and Murata N (1985) Heat inactivation of oxygen evolution inPhotosystem II particles and its acceleration by chloride depletion and exogenous manganese. Biochim Biophys Acta 807: 127–133

Nishida I and Murata N (1996) Chilling sensitivity in plants and cyanobacteria: The crucial contribution of membrane lipids. Annu Rev Plant Physiol Plant Mol Biol 47: 541–568

Nishida I, Frentzen M, Ishizaki O and Murata N (1987) Purification of isomeric forms of acyl-[acyl-carrier-protein]: Glycerol-3-phosphate acyltransferase from greening squash cotyledons. Plant Cell Physiol 28: 1071–1079

Nishiyama Y, Kovács E, Lee CB, Hayashi H, Watanabe T and Murata N (1993) Photosynthetic adaptation to high temperature associated with thylakoid membranes of *Synechococcus* PCC 7002. Plant Cell Physiol 34: 337–343

Nishiyama Y, Hayashi H, Watanabe T and Murata N (1994) Photosynthetic oxygen evolution is stabilized by cytochrome c_{550} against heat inactivation in *Synechococcus* sp. PCC 7002. Plant Physiol 105: 1313–1319

Ono T-A and Murata N (1981) Chilling susceptibility of the blue-green alga *Anacystis nidulans*. II. Stimulation of the passive permeability of cytoplasmic membrane at chilling temperatures. Plant Physiol 67: 182–187

Öquist G, Greer DH and Ögren E (1987) Light stress at low temperature. In: Kyle DJ, Osmond CB and Arntzen CJ (eds), Photoinhibition, pp 67–87. Elsevier, Amsterdam

Pearcy RW (1978) Effect of growth temperature on the fatty acid composition of the leaf lipids in *Atriplex lentiformis* (Torr.) Wats. Plant Physiol 61: 484–486

Powles SB (1984) Photoinhibition of photosynthesis induced by visible light. Annu Rev Plant Physiol 35: 15–44

Quinn PJ (1988) Regulation of membrane fluidity in plants. In: Aloia RC, Curtain CC and Gordon LM (eds) Physiological Regulation of Membrane Fluidity, Vol 3, pp 293–321. Alan R Liss, Inc., New York

Raison JK (1973) The influence of temperature-induced phase changes on kinetics of respiratory and other membrane-associated enzymes. J Bioenerg 4: 258–309

Raison JK, Roberts JKM and Berry JA (1982) Correlations between the thermal stability of chloroplast (thylakoid) membranes and the composition and fluidity of their polar lipids upon acclimation of the higher plant, *Nerium oleander*, to growth temperature. Biochim Biophys Acta 688: 218–228

Reddy AS, Nuccio ML, Gross LM and Thomas TL (1993) Isolation of Δ6-desaturase gene from *Synechocystis* sp. strain PCC 6803 by gain-of-function expression in *Anabaena* sp. strain PCC 7120. Plant Mol Biol 22: 293–300

Roughan PG (1985) Phosphatidylglycerol and chilling sensitivity in plants. Plant Physiol 77: 740–746

Roughan PG and Slack CR (1982) Cellular organization of glycerolipid metabolism. Annu Rev Plant Physiol 33: 97–132

Sakamoto T, Los DA, Higashi S, Wada H, Nishida I, Ohmori M and Murata N (1994a) Cloning of ω3 desaturase from cyanobacteria and its use in altering the degree of membrane-lipid unsaturation. Plant Mol Biol 26: 249–264

Sakamoto T, Wada H, Nishida I, Ohmori M and Murata N (1994b) Δ9 acyl-lipid desaturases of cyanobacteria: molecular cloning and substrate specificities in term of fatty acids, sn-positions, and polar head groups. J Biol Chem 269: 25576–25580

Santarius U and Müller M (1979) Investigation on heat resistance of spinach leaves. Planta 146: 529–538

Sato N and Murata N (1981) Studies on the temperature shift-induced desaturation of fatty acids in monogalactosyl-diacylglycerol in the blue-green alga (cyanobacterium), Anabaena variabilis. Plant Cell Physiol 22: 1043–1050

Sato N, Murata N, Miura Y and Ueta N (1979) Effect of growth temperature on lipid and fatty acid compositions in the blue-green algae, Anabaena variabilis and Anacystis nidulans. Biochim Biophys Acta 572: 19–28

Somersalo S and Krause GH (1990a) Effects of freezing and subsequent light stress on photosynthesis of spinach leaves. Plant Physiol Biochem 28: 467–475

Somersalo S and Krause GH (1990b) Reversible photoinhibition of unhardened and cold-acclimated spinach leaves at chilling temperatures. Planta 180: 181–187

Stanier RY and Cohen-Bazire G (1977) Phototrophic prokaryotes: The cyanobacteria. Annu Rev Microbiol 31: 225–274

Strand M and Öquist G (1985) Inhibition of photosynthesis by freezing temperatures and high light levels in cold-acclimated seedlings of Scots pine (Pinus silvestris). I. Effects on the light-limited and light-saturated rates of CO_2 assimilation. Physiol Plant 64: 425–430.

Tasaka Y, Nishida I, Higashi S, Beppu T and Murata N (1990) Fatty acid composition of phosphatidylglycerols in relation to chilling sensitivity of woody plants. Plant Cell Physiol 31: 545–550

Tasaka Y, Gombos Z, Nishiyama Y, Mohanty P, Ohba T, Ohki K and Murata N (1996) Targeted mutagenesis of acyl-lipid desaturases in Synechocystis: Evidence for the important roles of polyunsaturated membrane lipids in growth, respiration and photosynthesis. EMBO J 15: 6416–6425

Trémolières A, Dubacq J-P, Ambard-Bretteville F and Remy R (1981) Lipid composition of chlorophyll-protein complexes. Specific enrichment in trans-hexadecenoic acid of an oligomeric form of light-harvesting chlorophyll a/b protein. FEBS Lett 130: 27–31

Wada H and Murata N (1989) Synechocystis PCC 6803 mutants defective in desaturation of fatty acids. Plant Cell Physiol 30: 971–978

Wada H and Murata N (1990) Temperature-induced changes in the fatty acid composition of the cyanobacterium, Synechocystis PCC 6803. Plant Physiol 92: 1062–1069

Wada H, Hirasawa R, Omata T and Murata N (1984) The lipid phase of thylakoid and cytoplamic membranes of the blue-green algae (cyanobacteria), Anacystis nidulans and Anabaena variabilis. Plant Cell Physiol 25: 907–911

Wada H, Gombos Z and Murata N (1990) Enhancement of chilling tolerance of a cyanobacterium by genetic manipulation of fatty acid desaturation. Nature 347: 200–203

Wada H, Gombos Z, Sakamoto T and Murata N (1992) Genetic manipulation of the extent of desaturation of fatty acids in membrane lipids in the cyanobacterium Synechocystis PCC 6803. Plant Cell Physiol 33: 535–540

Wada H, Gombos Z and Murata N (1994) Contribution of membrane lipids to the ability of the photosynthetic machinery to tolerate temperature stress. Proc Natl Acad Sci USA 91: 4273–4277

Weber S, Wolter F-P, Buck F, Frentzen M and Heinz E (1991) Purification and cDNA sequencing of an oleate-selective acyl-ACP: sn-glycerol-3-phosphate acyltransferase from pea chloroplasts. Plant Mol Biol 17: 1067–1076

Williams JGK and Szalay AA (1983) Stable integration of foreign DNA into the chromosome of the cyanobacterium Synechococcus R2. Gene 24: 37–51

Wolter FP, Schmidt R and Heinz E (1992) Chilling sensitivity of Arabidopsis thaliana with genetically engineered membrane lipids. EMBO J 11: 4685–4692

Wu J and Browse J (1995) Elevated levels of high-melting-point phosphatidylglycerols do not induce chilling sensitivity in an Arabidopsis mutant. Plant Cell 7: 17–27

Chapter 14

A Genetic Approach to Investigating Membrane Lipid Structure and Photosynthetic Function

Perumal Vijayan, Jean-Marc Routaboul and John Browse
Institute of Biological Chemistry, Washington State University, Pullman, WA 99164-6340, U.S.A.

Summary

Chloroplast membranes have a unique lipid composition, and its importance for normal photosynthetic function has been emphasized by biochemical and biophysical studies. In the last ten years, a new and rewarding way to approach this structure-function relationship has been through the investigation of a series of *Arabidopsis* mutants with specific alterations in lipid composition. The majority of lipid mutations that have been characterized have little or no impact on the growth and photosynthesis of plants grown at 22 °C and 150 μmol quanta $m^{-2}s^{-1}$. In general, reduced membrane unsaturation in a mutant line is often (though not always) associated with a slight increase in thermostability of photosynthesis. Conversely, growth and photosynthesis of these mutant plants is often adversely affected at low temperatures. By contrast, two particular mutations, *dgd1* and *fab2*, have shown wide-ranging effects on growth and photosynthesis at all temperatures.

Since there are two distinct pathways of membrane lipid synthesis, a single mutation rarely blocks the

P.-A. Siegenthaler and N. Murata (eds): Lipids in Photosynthesis: Structure, Function and Genetics, pp. 263–285.
© *1998 Kluwer Academic Publishers. Printed in The Netherlands.*

synthesis of a particular lipid completely. More recently double and triple mutants have become available and these have provided more clear-cut indications about the membrane fatty acid composition required for photosynthesis. Trienoic fatty acids (18:3 and 16:3) comprise more than 60% of the acyl groups that make up the central portion of the thylakoid bilayer. However, a triple mutant *fad3 fad7-2 fad8* that completely lacks 18:3 and 16:3 has growth and photosynthetic characteristics at 25 °C that are indistinguishable from wild type. Although photosynthesis is affected at lower and higher temperatures, the consequences of removing trienoic fatty acids from the thylakoid membrane are clearly more subtle than expected. In contrast, the *fad2-2 fad6* double mutant which is deficient in dienoic as well as trienoic fatty acids is incapable of autotrophic growth. Studies of this mutant indicate that photosynthesis is the only cellular function that absolutely requires polyunsaturated fatty acids during vegetative growth.

I. Introduction

A. Arabidopsis *as a Model System*

Although mutational genetics has been used to study photosynthesis for several decades (Levine, 1969; von Wettstein and Kristiansen, 1973; Miles, 1982; Somerville, 1986), it is only in the last ten years that the genetic approach has been adopted to investigate the relationship between thylakoid glycerolipid structure and photosynthetic function. As in many other fields of plant biology, such investigations have benefited considerably from the use of the small crucifer *Arabidopsis thaliana* as a model. *Arabidopsis* has a number of traits that render it particularly well-suited for physiological genetics; chief among these are its variable size and generation time. Plants may be grown at densities as high as 10 plants.cm^{-2}, so the large number of individuals required for most genetic experiments can be handled relatively easily. Because plants produce their first seed in less than six weeks in continuous light, generations can be rapidly

advanced. Alternatively the plants can be grown in short day cycles for more than 2 months without bolting, and individual plants may attain up to 13 cm in diameter with up to 5 g fresh leaf weight (Fig. 1).

The plants are self-fertile and outcross at a very low frequency (<0.1%). However, the flowers can easily be emasculated to permit crosses between different lines, and male sterile lines are also available. Despite its relatively small size, *Arabidopsis* is a typical dicotyledonous plant and has been the subject of many physiological and biochemical studies (Somerville and Meyerowitz, 1994). In addition, *Arabidopsis* has several attributes that make it a suitable subject for molecular genetics (Meyerowitz, 1994). These include a small nuclear genome (100,000 kb), a very small amount of dispersed repetitive DNA and a high percentage of single copy sequences. A large number of well documented phenotypic and molecular markers are available. Facile transformation protocols exist for this species, whereas this is not true of better studied higher plant photosynthetic model systems. These characteristics simplify many procedures including library screening, chromosome walking and gene tagging (Bell and Ecker, 1994). Additionally, seeds of various ecotypes and a variety of mutants are available from academic and commercial sources in the US and Europe.

B. The Mutational Approach as a Tool in Biology

In contrast to many alternative approaches, mutant analysis offers the potential to provide clear and unequivocal information about how lipid composition affects plant function. Once a single mutation has been established in an otherwise uniform wild-type genetic background, it then follows that all the differences between the mutant and wild-type must

---

Abbreviations: ACP – acyl carrier protein; Chl – chlorophyll; DCPIP – 2,6-dichlorophenol-indophenol; DCIPH$_2$ – reduced 2,6-dichlorophenol-indophenol; DGD – digalactosyldiacylglycerol; DPH – diphenylhexatriene; EFs – exoplasmic fracture face of stacked thylakoids; PFs – protoplasmic fracture face of unstacked thylakoids; Fv/Fm – maximum quantum yield of PS II photochemistry; ΦII – quantum yield of non-cyclic electron transfer; LHC – light harvesting chlorophyll-protein complex; M2 – second generation progeny of mutagenized seeds; MGD – monogalactosyldiacylglycerol; MV – methylviologen; PBQ – p-benzoquinone; PC – phosphatidylcholine; PE – phosphatidylethanolamine; PG – phosphatidylglycerol; PQ – plastoquinone; PS I – Photosystem I; PS II – Photosystem II; SL – sulphoquinovosyldiacylglycerol; TMPD – N, N, N′,N′- tetramethyl-p-phenylenediamine; X:Y – a fatty acyl group containing X carbon atoms and Y double bonds (*cis* unless specified); double bond positions are indicated relative to the carboxyl end of the chain (e.g., Δ12).

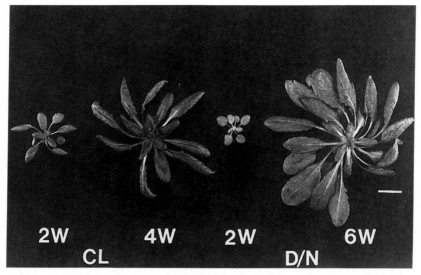

Fig. 1. Rosettes of *Arabidopsis thaliana* var Columbia grown in continuous light (CL) for 2 weeks (2W) and 4 weeks (4W); and 10 h photoperiod (D/N) for 2 weeks (2W) and 6 weeks (6W). (bar = 2 cm)

be related directly or indirectly to the mutation. Of course, pleiotropic or compensatory effects may occur in the mutants, but once identified, these effects will also contribute to our understanding of lipid metabolism and membrane biology. Mutational analysis is useful in three main contexts. First, it provides an alternative approach to elucidating the mechanisms regulating synthesis and desaturation of membrane glycerolipids. Second, the availability of a series of mutants with specific alterations in leaf lipid composition offers a novel method for studying the role of lipids in the structure and function of plant membranes. Finally, because of the advantages of *Arabidopsis* as a model for molecular biology, the mutations can be used as markers to facilitate the cloning of the desaturase genes by chromosome walking or gene tagging. The cloned structural genes and their promoters can also be used as powerful tools to understand the biochemistry and regulation of the enzymes in lipid biosynthetic pathways. Equally important is the potential use of these genes to transform diverse plant systems and modify their lipid composition in a directed fashion (Murata et al., 1992; Wolter et al., 1992).

The difficult task of screening tens of thousands of mutagenized plants is often facilitated by the availability of easily detectable visual markers like altered morphology or pigmentation. However, there is no obvious way to identify mutants with altered lipid composition on the basis of gross phenotype. Therefore, fatty acid desaturase mutants have been screened by direct assay of fatty acid composition of leaf tissue by gas chromatography (Browse et al., 1986a). Using this method, it is possible to obtain quantitative information on the fatty acid composition of the lipids from as little as 5 mg of leaf tissue within several hours. This procedure allows one to analyze the fatty acid composition of total lipids from single leaves from a population of ethylmethane sulfonate mutagenized *Arabidopsis* plants. Seven mutants with major changes in fatty acyl composition were identified from among the first 2,000 M2 plants examined in this way (Somerville and Browse, 1991) and, in subsequent searches, several additional mutant lines have been identified (Browse et al., 1994, Browse and Somerville, 1994). Table 1 shows the leaf fatty acid compositions of nine classes of mutants deficient in (nuclear encoded) chloroplast enzymes involved in lipid synthesis. Because chloroplast membranes account for 75% of the total leaf glycerolipids, the data in Table 1 are sufficient to illustrate the overall change in chloroplast membrane fatty acid composition in each mutant line. The phenotypes of the mutants are summarized in Tables 2 and 3.

C. Two Pathways for Glycerolipid Synthesis

Biochemical and genetic studies have contributed to our current understanding of the two major pathways by which the acyl-ACP products of plastid fatty acid synthesis are utilized in plant cells for the biosynthesis of glycerolipids and the associated production of

Table 1. The overall fatty acid compositions of leaf lipids from *Arabidopsis* mutants. Data are mole% (Browse and Somerivlle, 1994)

Mutant Line	Fatty Acid								
	16:0	16:1c	16:1t	16:2	16:3	18:0	18:1	18:2	18:3
Wild type	15	tr	3	tr	14	1	3	14	48
fab1	23	1	4	tr	17	1	3	11	39
fab2	14	tr	2	tr	6	14	3	18	42
act1	10	tr	2	tr	1	1	8	23	54
fad4	18	tr	0	tr	12	1	3	19	47
fad5	24	1	3	tr	tr	1	3	17	50
fad6	14	11	4	tr	tr	1	16	17	37
fad7[1]	17	3	3	6	2	1	9	39	19
fad2	12	1	3	tr	17	1	21	4	41
fad3	15	tr	3	tr	16	1	3	21	41
Multiple mutants:									
fad7,fad8[2]	15	1	2	10	0	1	4	49	17
fad2,fad6	11	16	4	1	0	1	60	5[3]	0
fad3,fad7,fad8	15	1	2	9	0	1	6	65	0

[1] Plants grown at 28°C.

[2] The fatty acid composition of *fad8* alone is indistinguishable from wild type.

[3] Probably the Δ9, Δ15 isomer.

Table 2. Comparison of chloroplast ultrastructure and excitation energy distribution changes in lipid mutants of *Arabidopsis* and SAN 9785-treated plants

Mutant/Plant	Chl *a/b*		Ratio of appressed to unappressed thylakoids	FII/FI ratio[1]		References
	Wild type/ untreated	Mutant/treated		Low Salt	High Salt	
fad5	2.9	3.2	–	Higher	Higher	Kunst et al., 1989a
fad6	2.9	3.3	Lower	Higher	Higher	Hugly et al., 1989
fad7	3.4	3.7	Higher	No change	No change	McCourt et al., 1987
act1	2.9	3.1	Lower	Higher	Higher	Bose et al., 1992
Peas (50 μM SAN 9785)	2.7	2.2	Higher	Higher	Higher	Bose et al., 1992
Barley (100 μM SAN 9785)	3.5	2.7	Higher	Higher	Higher	Laskay and Lehoczki, 1986 Leech et al., 1985

[1] Ratio of PS II fluorescence yield to PS I fluorescence yield at 77°K. Results given are mutant relative to wild type or chemically-treated plants relative to untreated controls.

polyunsaturated fatty acids. The evidence for the two pathway model has been summarized elsewhere (Browse and Somerville, 1991 and 1994). In brief, the model proposes that in all plant tissues, fatty acids are synthesized de novo in plastids and either 1) used directly for production of chloroplast lipids by a pathway in the plastid (the 'prokaryotic pathway') or 2) exported to the cytoplasm as CoA ester and then incorporated into lipids in the endoplasmic reticulum by an independent set of acyltransferases (the 'eukaryotic pathway'). The major glycerolipids are first synthesized using only 16:0 and 18:1 acyl groups; subsequent desaturation of the lipids to the highly unsaturated forms typical of the membranes of plant cells is carried out by membrane-bound desaturases of the chloroplast and the endoplasmic reticulum (Browse and Somerville, 1991; Heinz, 1993). The essential features of this model are depicted in Fig. 2, which also shows the proposed or confirmed sites of the various mutations discussed in this chapter.

In many species of higher plants, phosphatidyl-

Table 3. High and low temperature phenotypes of *Arabidopsis* lipid mutant compared to wild-type plants

| Mutant lines | Former name(s) | Mutant phenotypes compared to wild-type | | | | | | | References |
| | | 22 °C | Low temperatures lesions | | | Thermotolerance | | | |
		Phenotype Chl deficiency	Growth rate	Phenotype	Quantum yield of photosynthesis	Growth rate	Ti [1]	Photosynthetic electron transport	
fab1	– [2]	None detected	Lower	Death after prolonged exposure to 2 °C; Severe chlorosis	Inhibition after prolonged exposure	–	–	–	Wu et al., 1994; Wu and Browse, 1995; Wu et al., 1997
fab2	–	Dwarf (ameliorated over 35°C)	–	–	–	–	–	–	Lightner et al., 1994a,b
fad2	–	None detected	Lower	Reduced stem elongation at 12 °C; Death at 6 °C; Severe chlorosis	No change	–	–	–	Miquel and Browse, 1992; Miquel et al., 1993
fad3	–	None detected	–	None detected	No change	–	–	–	Browse et al., 1993
fad4	fadA	None detected	Lower	None detected	–	–	–	–	Browse et al., 1985; Mc Court et al., 1985
fad5	fadB, JB67	Slight chlorosis –18%	Lower	Slight leaf chlorosis	No change	Higher	Higher	Higher	Kunst et al., 1989 a,c
fad6	fadC, LK3	Slight chlorosis –19%	Lower	Leaf chlorosis	No change	Lower	Higher	Higher	Hugly et al., 1989; Browse et al., 1989
fad7	fadD, JB1, JB101	Slight chlorosis –14%	–	None detected	–	–	–	–	Browse et al., 1986c; McCourt et al., 1987
act1	JB25, LK8	Slight chlorosis –8%	No change	None detected	–	Higher	Higher	Higher	Kunst et al., 1988; Kunst et al., 1989b
fad3–fad7–fad8	–	Slight chlorosis –19%	Lower	Severe leaf chlorosis	Inhibition after short or prolonged exposure	–	–	–	Mc Conn and Browse, unpublished; Routaboul and Browse, 1995; Routaboul and Browse, unpublished
fad2–fad6	–	Severe chlorosis	–	–	–	–	–	–	Mc Conn and Browse, unpublished

(1) Threshold temperature for heat induced increase in Fo.
(2) No data available

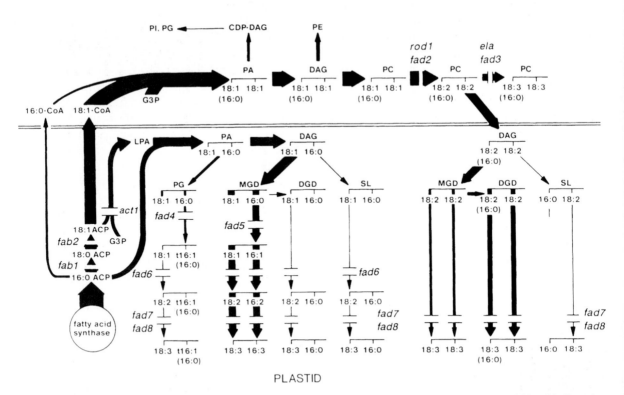

Fig. 2. The two-pathway scheme of membrane glycerolipid synthesis in *Arabidopsis* leaves (See text for details). Widths of the lines show the relative fluxes through different reactions. The breaks indicate the putative enzyme deficiencies in various mutants (see Table 1 also). No attempt has been made to accurately represent the various lipid species that are transferred between the endoplasmic reticulum and the plastid (adapted from Browse and Somerville, 1994.)

glycerol (PG) is the only product of the prokaryotic pathway, and the remaining chloroplast lipids are synthesized entirely by the eukaryotic pathway (Heinz and Roughan, 1983). In other species, including *Arabidopsis*, in which both pathways contribute about equally to the synthesis of monogalactosyldiacylglycerol (MGD), digalactosyldiacylglycerol (DGD), and sulphoquinovosyldiacylglycerol (SL) (Browse and Somerville, 1994), the leaf lipids characteristically contain substantial amounts of hexadecatrienoic acid (16:3), which is found only in MGD and DGD molecules produced by the prokaryotic pathway. These plants have been termed '16:3 plants' to distinguish them from the other angiosperms (18:3 plants) whose galactolipids contain predominantly linolenate. A detailed quantitative analysis of the relative contributions of the two pathways was carried out in *Arabidopsis* (Browse et al., 1986b). These studies indicated that approximately 38% of newly synthesized fatty acids enter the prokaryotic pathway

of lipid biosynthesis. Of the 62% that is exported as acyl-CoA species to enter the eukaryotic pathway, 56% (34% of the total) is ultimately re-imported into the chloroplast. Thus, chloroplast lipids of *Arabidopsis* are about equally derived from the two pathways. The contribution of the eukaryotic pathway to MGD, DGD, and SL synthesis is reduced in lower plants, and in many green algae the chloroplast is almost entirely autonomous with respect to membrane lipid synthesis.

D. The Significance of Chloroplast Glycerolipid Composition

The photosynthetic characteristics of all lipid mutants of *Arabidopsis* and other plants have to be considered in the context of the characteristic and unusual lipid composition of chloroplast membranes. They contain a high percentage of uncharged galactolipids, especially the nonbilayer forming MGD. Typically,

18:3 or a combination of 18:3 and 16:3 fatty acids account for approximately two-thirds of all the thylakoid membrane fatty acids and over 90% of the fatty acids of MGD — the most abundant chloroplast lipid. The atypical fatty acid $\Delta3$ *trans*-hexadecenoate (*trans* 16:1) is present as a component of the major thylakoid phospholipid, PG. The fact that these and other characteristics of chloroplast lipids are common to most or all higher plant species suggests that the lipid fatty acid composition is important for maintaining photosynthetic function. The occurrence of trienoic fatty acids as a major component of the thylakoid membrane lipids is especially remarkable since these fatty acids form highly reactive targets for active oxygen species and free radicals, which are often the byproducts of oxygenic photosynthesis (Asada, 1994). Hence, a high degree of unsaturation of thylakoid lipids is expected to offer great selective advantage to the photosynthetic system of the higher plant to be of such universal occurrence. It has been argued that this advantage is the high fluidity it confers on the membrane matrix (Webb and Green, 1991; Laskay and Lehoczky, 1986).

E. Nine Classes of Arabidopsis *Lipid Mutants*

The characterization of nine different classes of *Arabidopsis* mutants (all involving nuclear genes) has contributed to the understanding of both the regulation of lipid synthesis in plants and also to the role of some of these fatty acids in the normal physiology of the plant (Somerville and Browse, 1991; Browse and Somerville, 1994). A brief description of these well documented mutants is provided in this section as a background to the discussion of photosynthetic characteristics of these mutants in the following sections.

Four different classes of mutants, each one defective in one of the chloroplast desaturases in the prokaryotic pathway, were originally called *fadA*, *fadB*, *fadC* and *fadD*; but these have now been renamed *fad4*, *fad5*, *fad6* and *fad7*, respectively. Two of the chloroplast desaturases are highly substrate specific. The *FAD4* (*FADA*) gene product controls a $\Delta3$ desaturase that inserts a *trans* double bond into 16:0 esterified to position *sn*-2 of PG while the *FAD5* (*FADB*) gene product is responsible for the synthesis of $\Delta7$ 16:1 on MGD and possibly DGD (Browse et al., 1985; Kunst et al., 1989a; 1989c). In contrast, the other two chloroplast desaturases act on acyl chains with no apparent specificity for the length of the fatty acid

chain (16- or 18-carbon), its point of attachment to the glycerol backbone (*sn*-1 or *sn*-2) or to the nature of the lipid head group. The 16:1/18:1 desaturase is encoded by the *FAD6* (*FADC*) gene (Browse et al., 1989), whereas two 16:2/18:2 isozymes are encoded by *FAD7* (*FADD*) and *FAD8* (*FADE*) (Browse et al., 1986c; McConn et al., 1994). The endoplasmic reticulum 18:1 (*FAD2*) and 18:2 (*FAD3*) desaturases also act on fatty acids at both the *sn*-1 and *sn*-2 positions of the molecule (Miquel and Browse, 1992; Browse et al., 1993). They have been characterized as phosphatidylcholine desaturases, but it is possible that they also act on other phospholipids.

The *act1* mutation reduces the activity of the first enzyme of the prokaryotic pathway, glycerol-3-phosphate acyltransferase (Kunst et al., 1988 and 1989b). This mutant has greatly reduced levels of 16:3 fatty acids and correspondingly higher levels of C_{18} fatty acids in MGD. In addition, there is a 20% decrease in the content of PG and minor decrease in the plastid lipid classes MGD, DGD and SL accompanied by minor increase in the extraplastidic lipids PC and PE. Thus the act1 mutation effectively converts *Arabidopsis* from a 16:3 plant to an 18:3 plant.

The *fab1* and *fab2* mutants of *Arabidopsis* are characterized by increased levels of 16:0 and 18:0, respectively, in seed and leaf tissues (James and Dooner, 1990; Wu et al., 1994; Lightner et al., 1994a). In the *fab1* mutant, the biochemical defect appears to be a reduction in the activity of the condensing enzyme (3-ketoacyl-ACP synthase II) responsible for the elongation of 16:0 to 18:0. In *fab2*, it is assumed that 18:0-ACP desaturase activity is reduced. In both mutants the saturated fatty acids are incorporated into all the major membrane glycero-lipids (although MGD contains relatively low proportions). Both the mutants appear to be leaky so that the changes in overall membrane fatty acid composition are moderately small. Nevertheless, the changes do have profound effects on the biology of these plants (Lightner et al., 1994b; Wu and Browse, 1995, 1997).

F. More Extreme Phenotypes: Generation of Multiple-Mutant Lines

With only one exception, all the *Arabidopsis* mutants deficient in lipid biosynthesis have been isolated as healthy plants, which are not readily distinguished from the wild-type when grown at 22 °C. In line with this observation, relatively small changes have been found in the photosynthetic characteristics, growth or development of these plants, except at extreme temperatures. However, in all of these mutants, the level of polyunsaturated fatty acids is at least 60% to 80% of that found in the wild-type (Table 1) because each pathway of lipid synthesis has an associated set of desaturases (Fig. 2). In order to expand the range of mutant phenotypes available, it was necessary to generate multiple mutants with more extensive alterations in the fatty acid composition of their thylakoid membranes. Some of the double mutants produced show only a small additional change in chloroplast lipid composition. However, in some cases, it has been possible to produce very substantial alterations.

The most extreme fatty acid composition obtained to date is that for the *fad2 fad6* double mutant. The *fad6* mutant is deficient in the synthesis of polyunsaturated fatty acids by the prokaryotic pathway, but leaves of *fad6* plants still contain 52% 18:2 + 18:3 + 16:3 (Table 1) compared with 77% in the wild-type (Browse et al., 1989). The *fad2* mutant is deficient in activity of the endoplasmic reticulum 18:1 desaturase, but chloroplast lipids derived from the eukaryotic pathway in this mutant are desaturated by the *FAD6* gene product. This results in *fad2* plants containing 63% polyunsaturated fatty acids in their leaf lipids (Miquel and Browse, 1992). Analysis of 146 F_2 plants derived from a cross between the two mutants revealed a large deviation from the expected Mendelian ratio and no double mutants were recovered. Instead, plants that were homozygous for *fad6* and heterozygous for *fad2* were kept. Individual seeds from these plants were analyzed to show that they were segregating 1:2:1 at the *fad2* locus. Subsequent germination tests revealed that only 40% of the homozygous *fad2 fad6* seeds germinated and that the seedlings that were produced were not capable of autotrophic growth (McConn, 1994).

There are three gene products in *Arabidopsis* that mediate the synthesis of trienoic fatty acids. The *fad7* and *fad8* genes encode chloroplast isozymes, and the *fad3* gene product is the endoplasmic reticulum desaturase (see Fig. 2). We have recently succeeded in generating triple mutants *fad3 fad7 fad8* that contain no detectable 18:3 in their leaf lipids. These multiple desaturase mutants have provided us the means to explore the limits of variation in thylakoid lipid unsaturation within which the autotrophic mode can operate in higher plants. In

addition, the different alleles of the individual *Arabidopsis* desaturase mutants and their combinations can provide a gradient of fatty acid desaturation levels in the thylakoids against which relevant photosynthetic and other physiological functions can be correlated (see Section III.B).

G. DGD-Deficient and Chlorophyll Tail Mutants

As will be evident from the following sections, the study of mutants has begun to play a very useful role in dissecting relationships between fatty acid unsaturation and thylakoid membrane function. But most of what we know about the role of the different lipid classes in photosynthesis has been the result of painstaking biochemical experiments on isolated thylakoids (Siegenthaler et al., 1989). Although SL-deficient mutants of the photosynthetic bacterium *Rhodobacter sphaeroides* (Benning and Somerville, 1992; Benning et al., 1993) and the unicellular alga *Chlamydomonas reinhardtii* (Sato et al., 1995a,b) have been described, no higher plant mutant with a deficiency in any thylakoid lipid class was available until very recently. The isolation of a DGD-deficient *Arabidopsis* mutant, *dgd1*, by Benning and collaborators has shed new light on the role of DGD in determining the organization, ultrastructure and function of the photosynthetic apparatus (Dörmann et al., 1995). Another important new addition to the set of photosynthetically relevant lipid mutants is a rice mutant that accumulates chlorophyll *a* and *b* esterified to incompletely reduced C-20 alcohol chains instead of the fully reduced phytol (Nishimoto et al., 1992).

H. The Nature of Linkage Between Mutation and Phenotypes

One of the most powerful applications of the genetic approach toward understanding cellular biochemistry and physiology is the ability to select secondary mutations that suppress the phenotype. Characterization of these suppressor mutations frequently provides unanticipated insights about cell metabolism and function. Co-segregation of the phenotypes in crosses and co-reversion of the phenotypes in true revertants are two of the commonly used tools to confirm the linkage of a biochemical phenotype (e.g., lipid deficiency) to a physiological or morphological one like a photosynthetic function or thylakoid structure. In a few instances, chemical complementation of a fatty acid defect has also served the purpose (Garnier et al., 1990; McConn and Browse, 1996; see Section II.D). A unique method that has worked effectively for the fatty acid desaturase mutant *fad7* exploited the fact that lowering growth temperature restored the wild-type fatty acid phenotype in the mutant, and the concomitant reversion of the photosynthetic phenotypes confirmed the linkage between lipid composition and photosynthetic function (McCourt et al., 1987).

In the following sections, we have tried to present a survey of the photosynthetic parameters that are affected by various lipid mutations and attempted to draw some broad conclusions on the nature of relationships that may exist between the lipid composition of thylakoids and the structure and function of the photosynthetic system. Since fatty acid desaturation has traditionally been considered an important factor in low temperature adaptation of temperate plants, and since a number of fatty acid desaturase mutants of *Arabidopsis* have indeed shown low temperature photosynthetic phenotypes, we have tried to discussed this aspect in a different section. Two multiple desaturase mutants deficient in trienoic fatty acids and polyenoic fatty acids respectively, mutants deficient in DGD and SL, and two rice mutants with less saturated C-20 alcohol tail groups have been described separately because they enable us to pose qualitatively new types of questions about lipids and photosynthesis.

II. Structure and Function of the Photosynthetic Apparatus

A. Chloroplast Size and Number

Fatty acid unsaturation and the nature of the head groups determine the phase and fluidity properties of membrane glycerolipids and their interaction with specific membrane proteins (Murphy 1986; Webb and Green, 1991). In mutants where the lipid composition has been altered, one would expect to see ultrastructural changes related directly to the physical properties of the modified lipid component as well as secondary adaptive changes in structure induced by the primary alteration of lipid composition. Several fatty acid desaturase mutants of *Arabidopsis* exhibit alterations in chloroplast morphology and ultrastructure.

The *fad7* mutant of *Arabidopsis* has decreased

trienoic fatty acid content in the thylakoid lipids. Although it has about 14% less chlorophyll than the wild-type (Table 3), the Chl *a/b* ratio is higher than those of the wild-type (Table 2; McCourt et al., 1987). The average cross sectional area of the chloroplasts in this mutant is only 45% of the wild-type, and the total length of thylakoid membranes per plastid is also decreased comparably. However, the reduction in chloroplast size is compensated by an increase in their number, from an average of 40 per cell in the wild type to 55 in the mutant. These differences between the wild-type and the mutant in chloroplast number, size and chlorophyll content disappeared when the normal 18:3 and 16:3 content of the thylakoids was restored in the mutant by growing it at 18 °C. Apart from revealing that the fatty acid unsaturation level could still be modulated in the absence of the *FAD7* gene product (probably by *FAD8* gene product), this observation confirmed that the chloroplast structural changes are indeed caused by the altered lipid composition.

Another mutant that exhibits altered chloroplast size and number is *fab2*, in which the desaturation of 18:0 to 18:1 is inhibited. As a result, the 18:0 content of total leaf lipids increases from 0.7% to 17%. The increase in 18:0 content is mainly at the expense of 18:3 in all the lipid classes. Additionally, the total contribution of the chloroplast lipid classes DGD, PG and MGD to total leaf lipids is decreased from about 62% in wild-type *Arabidopsis* to 52% in *fab2* (Lightner et al., 1994a). This mutant has a dramatic morphological phenotype in that it is a true miniature with reduced organ and cell size. The mesophyll of the mutant leaves has a 'brick-wall' appearance with small cells packed tightly without air spaces in between them. Electron microscopy reveals that the chloroplasts in these cells are smaller than those of the wild-type. The number of chloroplasts per cell are also reduced in the mutants as revealed in transverse sections of leaves. The reduced size of the chloroplasts is at least partially accounted for by the low levels of starch accumulated in them. The thylakoid membranes of the mutant are also affected and fail to form stacked granal structures in vivo. Although normal gross morphology of the plant can be restored if the mutants are grown at an elevated temperature (above 35 °C), the plastid size and thylakoid organization in the high-temperature-grown plants have not been studied. However, the high temperature restoration of the morphological phenotype is consistent with existing knowledge of the negative physical effects of saturated fatty acids on membrane structure and function (Lightner et al., 1994b).

B. LHC Content and Structural Organization of Thylakoids

A significant role of lipids in thylakoid organization is evident from the variation in the total content and aggregation of LHC into trimers in the thylakoids of lipid mutants. A number of *Arabidopsis* mutants altered in fatty acid composition (e.g., *fad5*, *fad6*, *fad7*, *act1* and the *fad3 fad7 fad8* triple mutant) exhibit modest decreases in chlorophyll content (see Table 3). In all these instances the Chl *a/b* ratio also increases (see Table 2), indicating a net decrease in the accumulation of Chl *b* containing LHC. The fact that this decrease in chlorophyll content occurs in mutants exhibiting a range of deficiencies (*fad5* is deficient in 16:3 only, *fad7* and *fad3 fad7 fad8* are deficient in all trienoic fatty acids, and *fad6* in dienoic and trienoic acids. *act1* is deficient in 16:3 fatty acids and also in PG and other prokaryotic galactolipid species) is indicative of the requirement of an optimum level of polyunsaturated fatty acids (16:3 and 18:3) for maintaining normal stoichiometry of LHC to the photosystems.

In the *Arabidopsis* mutant *dgd1*, DGD content of the thylakoids is reduced from 16% to 1% of the total lipids, and the 16:3 fatty acid content also decreases from 17% to 6%. The total chlorophyll content of the leaves is 1.1 mg g^{-1}FW and the Chl *a/b* ratio is 2.3, compared to the corresponding values of 1.25 mg g^{-1} FW and 2.8, respectively, in the wild-type plants when both are grown at the low-light intensity of $90 \, \mu E \, m^{-2} \, sec^{-1}$. Increasing the intensity of the growth light results in only a marginal increase in the Chl *a/b* ratio of the mutant (2.7 at $1000 \, \mu E \, m^{-2} \, sec^{-1}$), whereas the wild-type plants can adapt to the high growth-light intensity by increasing the Chl *a/b* ratio to 4.3. This illustrates a loss in the plasticity of the photosynthetic system in *dgd1* mutant, which, unlike the wild-type *Arabidopsis*, appears to be unable to modulate its LHC content in response to high irradiance conditions. The severe damage to the leaves of the mutant at high light intensities is probably a result of this lack of adaptive capacity (Dörmann et al., 1995).

A role for phosphatidylglycerol and especially PG containing Δ3-*trans*-hexadecenoic (*trans*-16:1) acid has been postulated in the organization of LHC

(Dubacq and Tremolieres, 1983; see Chapter 9). Plants subjected to chilling treatment exhibit decreased *trans*-16:1 fatty acid content of their PG, which was correlated with decrease in the proportion of LHC oligomers in their thylakoids when separated on a partially denaturing polyacrylamide gel (Huner et al., 1989). The *Chlamydomonas* mutant *mf2* is deficient in *trans*-16:1 and also lacks LHC oligomers in thylakoids, as well as the ability to undergo state changes as compared to the wild-type. Cultures of this alga supplemented with *trans*-16:1 fatty acid restore the *trans*-16:1 content of thylakoid PG, which correlates with the reappearance of LHC oligomers and the ability to undergo state changes (Garnier et al., 1990).

In the *fad4* mutant of *Arabidopsis*, thylakoid membranes contain PG devoid of any *trans*-16:1 fatty acid. Thylakoids of this mutant do not show LHC oligomer bands in green gels in the presence of 30 mM NaCl. However, at 5 to 10 mM NaCl concentration, LHC oligomers can be observed at near wild-type levels in this mutant. The $C_{1/2}$ for LHC dissociation is 13 mM in the mutant and 37 mM in the wild-type (McCourt et al., 1985). These results indicate that the weaker interaction between LHC monomer units in *fad4* is possibly due to the disruption of electrostatic interactions that stabilize the oligomers. The *fad5* mutant of *Arabidopsis* has lower levels of 16:3 on the *sn*-2 position of the lipids MGD and DGD. The thylakoids of this mutant also exhibit a lowered LHC oligomer/monomer ratio in green gels (Kunst et al., 1985c).

Partially denaturing gels of octylglucoside-solubilized thylakoids of the *dgd1* mutant exhibit a decreased LHC oligomer to monomer ratio compared to the wild-type. However, wild-type levels of the trimeric LHC is restored in the gels when the thylakoids are solubilized in the presence of decylmaltoside, which is a milder detergent containing a dihexosyl head group similar to that of DGD. Although these results point to a role for DGD in stabilization of LHC trimers, inferences made from these results have to be treated with some caution since the *dgd1* mutation also causes a decrease in the (prokaryotic) 16:3 fatty acid content and an increase in the (eukaryotic) 18:3 fatty acids of the thylakoid galactolipids (Dörmann et al., 1995). Thus, the stability of the LHC trimers is adversely affected in three *Arabidopsis* mutants lacking *trans* 16:1-PG, 16:3 galactolipids, and DGD plus 16:3 galactolipids, respectively.

Stoichiometric numbers of tightly bound PG and loosely associated DGD molecules were found to be required for the stability of trimeric LHC complexes during detergent fractionation of thylakoids (Nußberger et al., 1993). Thus, while the specific role of PG in stabilization of LHC trimers in vitro and in vivo seems to be confirmed by the *fad4* mutant, the phenotypes of *dgd1* and *fad5* imply that a variety of fatty acid and head group changes can influence the stability of LHC oligomers.

A distinct ultrastructural feature of the fatty acid desaturation deficient mutants of *Arabidopsis*, *fad5*, *fad6*, *fad7* and the tobacco transgenic, *plsB* is the presence of arrays of EFs and PFs freeze fracture particles representing PS II units (Tsvetkova et al., 1994 and 1995). The three *Arabidopsis* mutants also show an increase in the Chl *a/b* ratio indicating a partial deficiency in LHC. Moreover, the relative fluorescence emission from LHC (F_{685}) is higher in these mutants. This result was interpreted by Tsvetkova and coworkers in terms of a decreased energy transfer from LHC to PS II centers. Thus, the reduction in the number of double bonds per fatty acid in these mutants seems to weaken the interaction between PS II and LHC and concomitantly facilitate close packing of the PS II units which is reflected in the formation of freeze fracture particle arrays (Simpson, 1983).

C. Lipid Content and Adaptive Features of Thylakoids

Lateral segregation of PS I into the stromal membranes and both PS II and LHCII into appressed granal membranes is a distinctive feature of photosynthesis in all higher plants. Modulation of this heterogeneity mediates the dynamic adaptation of photosynthesis to light intensity changes in the short term (state changes) and to developmental plasticity in the long term (shade and sun adaptation) (Allen, 1992). Physical parameters like membrane fluidity and lipid/protein interactions determine the diffusion rates of intersystem redox carriers like plastoquinone (PQ), the lateral migration of pigment protein complexes, and the insertion and stabilization of integral membrane proteins. Therefore, changes in lipid composition can constrain or facilitate the dynamic structure-function changes of thylakoids during these adaptive transitions (Webb and Green, 1991).

Features of thylakoid structure, like the ratio of appressed (granal) to nonappressed (mostly stromal)

membranes, its modulation in vitro by cations, and related functional parameters like intersystem excitation energy distribution, have been studied in the context of parallel changes in fatty acid unsaturation in thylakoids of plants treated with the herbicide SAN 9785 (BASF 13.338). Since the herbicide also directly inhibits PS II electron transport (Mannan and Bose, 1985) and carotenoid biosynthesis (Bose et al., 1992), no unequivocal conclusions can be drawn from these studies.

A number of fatty acid desaturase mutants of *Arabidopsis* have also been investigated from a similar perspective. The results are summarized in a table for easy comparison (Table 2). The *fad7* mutant, in which the trienoic fatty acid levels are reduced by about 50%, also have wider grana stacks, a lower average number of thylakoids per granum, and increased appressed to nonappressed membrane ratio. These structural changes, although less pronounced, are similar to those observed in SAN 9785 treated plants. However, in contrast to the herbicide-treated plants (Laskay and Lehoczky, 1986; Bose et al., 1992), the modulation of excitation energy distribution between PS II and PS I by cations is unaffected in isolated thylakoids of *fad7* (McCourt et al., 1987). Also, in contrast to reports of lipid saturation mediated inhibition of state changes in SAN 9785-treated plants (Graf et al., 1984), the lateral mobility of phosphorylated LHC (which mediates the state changes) is not affected by the decrease in lipid unsaturation in the *fad7* mutant. Relevant to these findings is the fact that DPH fluorescence polarization measurements of the *fad7* thylakoids also shows only a marginal decrease in the fluidity of these membranes. Hence, apparent effects of SAN9785 treatment on dynamic properties of thylakoid membranes appear to be unrelated to its inhibition of fatty acid desaturation.

It is often assumed that an increase in the appressed to nonappressed membrane ratio would favor PS II in excitation energy distribution and the reverse to be true when the ratio decreases. However, a comparison of thylakoid organization and excitation energy distribution reveals that a decrease in unsaturation of different lipids may result in an increase (in *fad7*) or a decrease (in *fad6* and *act1*) in the appressed/ nonappressed thylakoid ratio. Also, an increase in excitation of PS II may actually take place in the context of an increase or decrease in the appressed/ nonappressed membrane ratio. In the *fad7* mutant an increase in appressed/nonappressed thylakoid ratio does not result in a change in excitation energy distribution at high or low salt conditions (see Table 2).

Membrane fluidity may be expected to influence the diffusion rates of the mobile electron carrier plastoquinone in the hydrophobic domain and the lateral migration of the phosphorylated LHC from the appressed to the non-appressed regions of the thylakoids (Murphy, 1986; Webb and Green, 1991). Despite the substantial reduction in the trienoic fatty acid content in *fad7* thylakoids, the apparent rate of phosphorylation-induced LHC migration at room temperature was not affected by the mutation. Whole chain electron transport measured at different temperatures showed that the rate-limiting step in linear electron transport, namely plastoquinone oxido-reduction, was not adversely affected in the mutant, even at lower temperatures. However, the membrane fluidity measured by DPH fluorescence polarization also decreased only marginally in the thylakoids of *fad7* (McCourt et al., 1987). It is clear from these experiments that even substantial decrease in trienoic fatty acid content (as observed in *fad7*) is insufficient to inhibit these physiologically relevant functions, which might be fluidity dependent.

Mobility of PQ in the thylakoids and that of respiratory quinones in the mitochondrial membrane system have also been proposed to depend on the availability of voids in the interior of the membrane for these molecules to migrate (Mathai et al., 1993). Such voids could exist inside the hydrophobic domain of the thylakoid membrane because of the low acyl chain density of the highly unsaturated fatty acids. The voids could also be due to the presence of a small percentage of shorter chain (16:0, 16:1 and 16:3) fatty acids on some lipids. Mutants like *act1*, which alter the content of 16 carbon fatty acids, and different desaturase mutants can enable the investigation of this and other potentially fluidity-dependent phenomena in relation to thylakoid function.

D. Are Polyenoic Fatty Acids Critical for Photosynthesis?

Although individual desaturase mutants generate a range of lipid compositions within which the photosynthetic process can operate with modified characteristics, none of them is useful in defining the minimum requirement in lipid composition for photosynthesis. Multiple mutants with extreme fatty acid composition have been generated which serve this purpose.

The fatty acid desaturase mutants *fad3*, *fad7* and *fad8*, each deficient in a particular 18:2 desaturase activity, can be used to generate multiple mutant combinations containing varying levels of trienoic fatty acids. On its own, the *fad3* mutation reduces the desaturation level of the thylakoid galactolipids only marginally. On the other hand, *fad7*, which is deficient in one of the plastidic $\Delta 15$-desaturases, has a reduced 18:3 and 16:3 content in thylakoid specific leaf lipids. The double mutant *fad7 fad8* has about 17% trienoic fatty acids in its membranes as compared to 62% in wild-type *Arabidopsis*. A triple mutant *fad3 fad7 fad8* has essentially no linolenic acid (18:3) or hexadecatrienoic acid (16:3) in its thylakoid lipids (see Table 1). Remarkably, this triple mutant has morphological, growth and developmental charac-teristics similar to those of wild-type *Arabidopsis* for most of its life cycle. The efficiency of PS II and steady state whole chain electron transport in leaves of the mutant grown at room temperature, as measured by noninvasive Chl *a* fluorescence techniques, were close to those of the wild-type (McConn and Browse, 1996; discussed in the following sections). The absence of trienoic fatty acids from the *fad3 fad7 fad8* triple mutant produces a pronounced physio-logical effect only when it is subjected to low and high growth temperatures or to high-light stress. This triple mutant clearly demonstrates that trienoic fatty acids of thylakoid lipids are not critical for the basic process of photosynthesis, although their absence renders the photosynthetic process highly vulnerable to stress (Routaboul and Browse, 1995; see Section III).

Interestingly, the critical lesion affecting the survival of this plant is its male sterility. Because *Arabidopsis* is a self-pollinated plant, the triple mu-tant cannot survive unless it is either artificially cross-pollinated or its male fertility restored by chemical complementation with 18:3 fatty acid or its derivative plant hormone jasmonate (McConn and Browse, 1996). This surprising finding, although not directly related to photosynthesis, underscores the potential role of specific lipids and their derivatives as mediators in signaling pathways controlling the development and modification of the photosynthetic system. Such a potentially hormone-like role of some lipids (or their derivatives) complicates the task of correlating lipid deficiencies in mutants, with membrane structure and function, especially in the light of reports that jasmonate regulates the expression of several genes including the *rbcL* gene encoding the Rubisco large subunit protein and genes involved in β-carotene synthesis (see Creelman and Mullet, 1995).

The unsaturation levels of thylakoid fatty acids can be reduced even further, if the insertion of the second, Δ12 double bond in 18:1 and 16:1 fatty acids can be blocked. This is achieved in the *Arabidopsis* mutants *fad2* and *fad6*, which are defective in the desaturation of monoenoic fatty acids in the endoplasmic reticulum (*fad2*) and chloroplast (*fad6*), respectively (see Fig. 2). The *fad2* mutation by itself does not seem to affect either the fatty acid content or the function of the photosynthetic membranes. However, the desaturation level of the nonplastidic membrane lipids (clearly evident in nonphotosynthetic tissue like roots) is reduced in the mutant, along with the ability to grow at low temperatures (Miquel et al., 1993). The mutant has normal growth rates and photosynthesis at room temperature although alterations are observed in the gross morphology of the reproductive shoots. Relative decrease in chlorophyll content and accumulation of very high levels of anthocyanins is observed in the leaves of *fad2* plants after 6 and 12 days of growth at 6 °C. However, quantum yield of PS II and whole chain electron transport measured by Chl *a* fluorescence does not appear to be seriously affected even after 12 days at 6 °C (Wu et al., 1997), although the plant stops growing and eventually dies at this temperature (Miquel et al., 1993). Therefore, the primary chilling-induced lesion in fad2 plants does not seem to be the photosynthetic process.

The *fad6* mutation on the other hand decreases the 18:2, 18:3 and 16:3 content of leaf lipids (Table 1), resulting in marked changes but no critical lesion in the structure and function of the photosynthetic machinery. The double mutant *fad2 fad6* is essentially devoid of all trienoic and dienoic fatty acids except for a small quantity (6%) of dienoic fatty acids with double bonds at positions Δ9 and Δ15 instead of the usual Δ9 and Δ12 found in the wild-type. This mutant is incapable of autotrophic existence and can grow only in media supplemented with a carbon source. It has to be maintained as segregating population of seeds generated from heterozygous parents. However, the mutant plants have near normal vegetative morphology in sucrose-supplemented medium although the leaves are pale green in color due to loss of chlorophyll. Chlorophyll fluorescence measurements reveal near wild-type levels of PS II quantum yield, but about a 50% decrease in the quantum yield of steady-state photosynthetic electron transport

compared to the wild-type plants grown under the same conditions. In thylakoids isolated from the double mutant, PS II activity (measured as O_2 evolution with p-phenylene diamine as acceptor) is equal to that of the wild-type, and PS I activity measured with reduced TMPD (N, N, N′,N′-tetramethyl-p-phenylenediamine) as donor and methyl viologen as acceptor is actually higher (by about two-fold) than that of wild-type when calculated on a chlorophyll basis. However, the effect of chlorophyll deficiency becomes evident when the rates are calculated on a fresh weight basis, and the corresponding rates of PS II activity is about 20% and PS I activity about 40%, respectively, of the wild-type (McConn, 1994).

Taken together, the Chl fluorescence data of leaves and electron transport rates in isolated thylakoids indicate that the thylakoids in the *fad2 fad6* double mutant probably have far fewer but functionally competent PS II and PS I units. This mutant is either unable to accumulate pigment protein complexes or is unable to protect them from photodynamic damage, resulting in a loss of autotrophic capacity. Thus, multiple desaturase mutants like the *fad3 fad7 fad8* and *fad2 fad6* and the head group mutants like the *dgd1* have provided us the means to explore the limits of the variation in thylakoid lipid composition within which the autotrophic mode is viable in higher plants. The results described here are in sharp contrast to the situation in cyanobacteria. Some strains of cyanobacteria are completely lacking polyunsaturated fatty acids and are photosynthetically fully competent. Even in those cyanobacteria which naturally contain polyunsaturated fatty acids, their depletion by molecular genetic means does not cause any lesion in basic photosynthetic function (Gombos et al., 1992). The significant contribution of di- and trienoic fatty acids in thylakoid lipids of these strains seems to be in conferring chilling tolerance (Murata and Wada 1995; see Chapter 13).

E. Non Bilayer Forming Lipids

The decrease in the DGD content of the *dgd1* mutant does not result in any change in the total leaf lipid content. However, the net share of the nonplastidic lipids PE and PC increase from 22% to 32% in this mutant, and MGD increases only by 3%, to offset a 15% decrease in DGD content. In the absence of data on separated membrane fractions, it can be assumed that the net share of the thylakoid lipids in the leaf tissue falls from 77.6% in the wild-type *Arabidopsis* to 66.3% in *dgd1* (Dörmann et al., 1995). Taking into account changes in the number of plastids per cell, the total length of thylakoids per cell apparently increases by about 30% (as calculated from the total thylakoid length per plastid and the number of plastids per cell).

The molar ratio of nonbilayer forming lipids (MGD) to bilayer forming lipids (DGD, PG and SL) in the thylakoids increases from 1.9 in wild-type *Arabidopsis* to 4.5 in the *dgd1* mutants. The ratio of 4.5 for the total lipids in *dgd1* compares with a similar ratio proposed for the outer leaflet of oat thylakoid membranes (Siegenthaler et al., 1989) and is far above the critical ratio of 2.5, above which transition to nonbilayer structures begins in MGD/DGD mixtures (Sprague and Stahelin, 1984). The high curvature of the thylakoids observed in this mutant may, in part, be an accommodation of this highly skewed ratio of nonbilayer-forming to bilayer-forming lipids. It has also been suggested that the balancing of the nonbilayer forming tendencies in the two leaflets of the same bilayer may lead to a 'frustrated' equilibrium in which the nonbilayer forming lipids in either of the leaflets is unable to deform the bilayer (Hazel and Williams, 1990). Thus, the existence of bilayers with such high proportions of nonbilayer-forming lipids in the *dgd1* mutant may be due to the role played by thylakoid proteins, partial deformation of the planar bilayer, the balancing of the bilayer-disrupting tendencies in the opposite leaflets of the bilayer, or a combination of these factors.

F. Translocation and Localization of Proteins

The proteins involved in the biogenesis of the chloroplast, photosynthetic energy transduction, carbon fixation and other related pathways are encoded in the nuclear as well as in the plastid genomes. They are often synthesized on membrane bound polysomes, translocated across the envelope or thylakoid membranes and localized in the membrane system. Thus, at some stage of their synthesis or function most proteins involved in photosynthesis must interact with the lipid matrix of one or more membranes. Recently, direct evidence has been reported for close interaction of the signal/anchor domains of translocated and membrane localized proteins of the mammalian endoplasmic reticulum with the lipids of the bilayer during their

translocation or insertion into the membrane (Martoglio et al., 1995). The successful translation and insertion of the D1 protein of the PS II complex depends both on its acylation and interaction with pigments and cofactors localized in the thylakoid membranes (Nickelsen and Rochaix, 1994).

Direct evidence for an important role of fatty acid unsaturation in the temperature dependence of D1 protein accumulation has been demonstrated in cyanobacteria recovering from photoinhibitory damage to PS II (Kanervo et al., 1995). A similar role for fatty acid unsaturation in higher plants was indicated in transgenic tobacco expressing glycerol-3-phosphate acyl transferase gene from squash. The transgenic plants have a higher level of saturated fatty acids in thylakoid PG and show greater susceptibility to photoinhibition at chilling temperatures. Experiments with isolated thylakoids and the chloroplast protein synthesis inhibitor lincomycin have indicated a defect in turnover of the PS II protein D1 may be the cause of the greater susceptibility of this transgene to photoinhibition at lower temperatures (Moon et al., 1995).

Results from studies with the fatty acid desaturation mutant *fad3 fad7-2 fad8* completely lacking trienoic fatty acids indicate a similar defect may also exist in this mutant of *Arabidopsis*. The rates of photo-inhibition of PS II under conditions precluding recovery (3 °C) were equal in the triple mutant and wild type. However, the rate of subsequent light dependent recovery of PS II in the mutant was similar to the wild type at 27 °C, but severely retarded at lower temperatures (Vijayan et al., 1997; Fig. 3). Since dark incubation for several hours did not result in appreciable recovery in either plants, these results indicated an important role for trienoic fatty acids in the light-dependent repair of photodamaged PS II units — probably in facilitating the accumulation of newly synthesized D1 protein (van Wijk et al., 1994). A mechanistic explanation for the specific role of fatty acid desaturation in thylakoid protein synthesis and localization may emerge from a detailed study of these and other cyanobacterial and *Arabidopsis* lipid mutants.

III. Chilling Sensitivity and Fatty Acid Composition

Many species are injured or killed by exposure to low, non-freezing temperatures in the range of 1 to

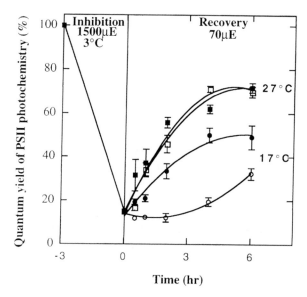

Fig. 3. Recovery of PS II from photoinhibition in leaves of wild-type and *fad3 fad7 fad8* of *Arabidopsis thaliana* at 27 °C and 17 °C. Quantum yield of PS II was measured as Fv/Fm in leaves dark adapted for 15 min after various treatments. Longer dark adaptation (up to 6 h) did not produce any substantial change in Fv/Fm (data not shown). ●, ■ wild type; ○, □ *fad3 fad7 fad8* triple mutant (from Vijayan et al., 1997).

12 °C, whereas others may tolerate temperatures below 0 °C without exhibiting any deleterious effects (Lyons, 1973; Wang, 1990). Economically important crops, such as cotton, soybean, maize and many tropical or subtropical fruits, are examples of chilling-sensitive plants, whereas most plants of temperate origin, including *Arabidopsis,* are classified as chilling-resistant plants. Functional definitions of chilling sensitivity vary to some extent (Wang, 1990), but broadly speaking, plants will be recognized as chilling sensitive if they suffer visible damage or impaired growth after one or several days exposure to temperatures in the range of 0 to 12 °C. The idea that membrane lipid unsaturation plays a role in low-temperature damage to plants originates from Lyons (1973) and Raison (1973). They linked the biochemical or biophysical changes associated with chilling injury to a single site of damage, and hypothesized that the primary event of chilling injury was a liquid-crystalline phase to gel phase (Lα to Lβ) transition in cellular membranes. Such a lipid phase transition would produce a major breakdown of membrane integrity. However, in higher plants, chloroplast membranes of even chilling sensitive plants are highly unsaturated and do not seem to

undergo a liquid-crystal to gel-phase transition in the temperature range associated with chilling injury.

A. Chilling Sensitivity and Disaturated PG

Based on a broad correlation between chilling sensitivity and the presence of disaturated high melting point PG (disaturated molecular species containing only 16:0, 16:1-*trans* or 18:0 acyl groups) in various species, Murata and coworkers proposed that chilling-sensitive species are characterized by a high content (26 to 65 %) of disaturated PG (Murata et al., 1982; Murata 1983; Murata and Yamaya, 1984; Roughan, 1985; Murata and Nishida, 1990), whereas chilling-resistant plants typically contain less than about 20% of disaturated PG. Since 16:1-*trans* fatty acid has physical properties that are similar to a saturated fatty acid, the various molecular species of PG containing only 16:0,18:0 and 16:1-*trans* fatty acids are collectively referred to as disaturated PG. This hypothesis is supported by: 1) the detection of Lα- to Lβ-phase transitions at physiological temperatures in synthetic 16:0/16:0 PG molecular species and in the PG purified from chilling-sensitive but not from chilling-resistant plants (Murata and Yamaya, 1984) and 2) the observation that the phase transition temperature of a mixture of leaf membrane polar lipids is significantly increased by addition of as little as 2% high-melting point PG (Raison and Wright, 1983).

Further support comes from modification of chilling sensitivity in transgenic plants overexpressing the glycerol-3-phosphate acyltransferase genes of the chilling sensitive squash plant and *E. coli*, as well as the chilling-resistant *Arabidopsis*. Transgenic tobacco expressing the squash and *Arabidopsis* expressing the *E. coli* acyltransferase, contained higher than normal levels of disaturated PG and exhibited increased chilling sensitivity, whereas tobacco plants overexpressing the *Arabidopsis* gene contained lower than normal levels of disaturated PG and correspondingly exhibited greater chilling tolerance than control tobacco plants (Murata et al., 1992; Wolter et al., 1992). However, one of these studies was conducted in the genetic background of a chilling-sensitive species (tobacco), and the test of chilling sensitivity was based on the appearance of lesions after short low-temperature treatments and subsequent return to room temperature.

Notwithstanding this accumulated body of evidence, an *Arabidopsis* mutant, *fab1*, demonstrates that high disaturated PG content by itself is not sufficient to cause classic chilling sensitivity. The *fab1* mutant contains higher levels of disaturated PG (47%) than approximately half of all chilling sensitive species. Nevertheless, this mutant is unaffected by a range of short-duration low-temperature treatments that led to severe necrosis and death in chilling-sensitive plants (Wu and Browse, 1995). For instance, *fab1 Arabidopsis* and two chilling-sensitive plants, cucumber and mung bean, were subjected to 10 °C and constant moderate light. After 30 days, no symptoms were visible in *fab1* whereas mung bean and cucumber exhibited wilting and necrosis after only 21 days. More severe temperatures (2 °C for 7 days or –2 °C for 24 h) led to the death of the chilling-sensitive plants, whereas *fab1* failed to show any injury. Taken together, these studies indicate that the correlation between high-disaturated PG content and susceptibility to severe short-term chilling injury is not universally applicable. Thus, studies with the *fab1* mutant of *Arabidopsis* showed that a mere increase in the content of disaturated PG was not sufficient to convert the normally chilling-tolerant *Arabidopsis* to a typically chilling-sensitive one like tobacco or cucumber. The failure of *fab1* plants to show physiologically relevant chilling damage does not necessarily disprove the general hypothesis that a change in membrane structure at low temperature is the trigger that initiates chilling; these results only indicate that the current hypothesis is inadequate. In contrast, many mutations (including *fab1*) that cause a decrease in overall membrane lipid unsaturation levels result in a loss of low-temperature tolerance in *Arabidopsis*. Experimental results of prolonged cold treatment of *fab1* show that high disaturated PG content is definitely incompatible with low-temperature adaptation in *Arabidopsis* (Wu and Browse, 1995; Wu et al., 1997).

B. Relationship Between Lipid Unsaturation and Low-Temperature Growth and Photosynthesis

Five *Arabidopsis* lipid mutants (*fab1, fad2, fad5, fad6* and the triple mutant *fad3 fad7 fad8*), which are indistinguishable from wild type when grown at 22 °C, show chlorotic lesions and decreased growth after transfer to low temperatures (Table 3). When wild-type and these chilling-intolerant mutants of *Arabidopsis* are transferred to low temperatures (in the range of 2 to 6 °C), the growth rate is significantly

reduced and the chlorophyll content drops by almost 30% in the first 7 days (Fig. 4). After this initial phase, wild-type *Arabidopsis* recovers from the effects of chilling stress, the growth rate increases and chlorophyll content is restored to near pre-treatment levels. In contrast, the mutants continue to deteriorate during this second phase (Miquel et al., 1993; Wu and Browse, 1995; Routaboul and Browse, 1996). Depending on the nature of the mutation, their leaves may become highly chlorotic (*fad2*, Miquel et al., 1993; *fab1*, Wu and Browse, 1995; *fad5* and *fad6*, Hugly and Somerville, 1992; *fad3 fad7 fad8*, Routaboul and Browse, 1995), necrosis may appear (*fad2*) and the growth rate may be reduced compared to the wild-type (*fad5, fad6, fab1* and *fad2*). In some mutants, like *fad2* and *fab1*, the severity of these symptoms may ultimately lead to the death of the plant after more than a month at low temperatures (Miquel et al., 1993; Wu et al., 1997).

In some of these mutants, noninvasive chlorophyll fluorescence techniques have been used to measure quantum yield of PS II photochemistry (measured as Fv/Fm; Kitajima and Butler, 1975) and quantum yield of steady state-linear electron transport (measured as ΦII; Genty et al., 1989) during chilling treatment. On the whole, the gradual, chronic and diverse chilling phenotypes observed in these lipid deficient mutants are consistent with specific and limited defects in membrane functions rather than a dramatic loss of membrane integrity (Miquel et al., 1993) or a single site of temperature sensing dependent on gross membrane fluidity changes. For example, an 18% decrease in ΦII occurred by 5 days after transfer to 4 °C in the trienoic deficient *fad3 fad7 fad8* triple mutant whereas no substantial inhibition of Fv/Fm was observed until after 21 days (Routaboul and Browse, 1995). In the *fab1* mutant (which has increased 16:0 in both chloroplast and extrachloroplast lipids), both Fv/Fm and ΦII were inhibited when plants were exposed to 2 °C for more than 10 days (Wu et al., 1997). Thus, the chilling lesions seem to develop at different rates and to affect different sites of the photosynthetic apparatus in these two mutant lines.

In the *fad2* mutant, which is deficient in dienoic and trienoic fatty acids of the extrachloroplast membranes, neither (dark adapted) Fv/Fm nor ΦII (at steady state under actinic illumination) appear to be affected even after 12 days at 6 °C, although the chlorophyll content of leaves drops by about 50% during this period (Wu et al., 1997). The mutant

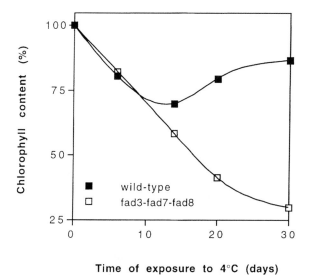

Fig. 4. Changes in chlorophyll content of wild-type and *fad3 fad7 fad8* mutant of *Arabidopsis thaliana* during low temperature treatment at 4 °C for 30 days. Results are expressed as percentage of pretreatment values. Initial chlorophyll content were 2.1 and 1.7 mg g^{-1} FW, respectively, for wild type and fad3 fad7 fad8 (from Routaboul and Browse, 1995).

plants also exhibit excessive anthocyanin accumulation and partial necrosis of leaves accompanied by striking curtailment of stem elongation, eventually leading to death of the plants (Miquel et al., 1993). This combination of chilling-sensitive photosynthetic phenotypes clearly indicates that in cold-stressed *fad2* plants, in spite of severe chlorosis, the residual chlorophylls in the leaves probably form part of a competent photosynthetic apparatus. Thus, in a mutant in which only non plastidic membrane lipids are affected, photosynthetic electron transport does not seem to be the site of primary chilling induced lesion. The loss of chlorophyll probably reflects the inability of the cells to maintain a sufficient number of active photosynthetic units rather than direct damage to their functioning by the cold treatment.

Deficiencies affecting chloroplast development were also observed in some other chilling-intolerant lipid mutants in which tissue developed during low-temperature stress was chlorotic, whereas tissues that had developed before transfer to low temperature was not visibly affected by prolonged chilling exposure. This is particularly true with the *fad5* (deficient in 16:3 fatty acids of thylakoid lipids) and *fad6* (deficient in dienoic and trienoic fatty acids of thylakoid lipids) mutants that are chlorotic when exposed to 5 °C at the cotyledon stage (7–8 days old

seedlings; Hugly and Somerville, 1992). This observation led Somerville and coworkers to propose that an effect on membrane biogenesis rather than function is the primary consequence of the lipid change.

From these examples, it is clear that the normal functioning of the photosynthetic apparatus at chilling temperatures is dependent upon a fairly specific lipid composition of both chloroplast as well as extra-chloroplast membranes. Alterations in the lipid composition of the thylakoid membranes appears to result in direct and defined lesions in the thylakoid function, whereas the effects of lipid deficiencies in the extrachloroplast membranes appear to act indirectly, perhaps through disrupting maintenance of the chloroplasts at low temperature. In either case, the nature of the effects seem to depend on the specific molecular species that is deficient rather than through the mediation of global changes in membrane fluidity or integrity in the mutants.

C. Trienoic Fatty Acids Have a Crucial Role in Tolerance of Photosynthesis to Low Temperatures

Since cold acclimation in higher plants is known to induce an increase in trienoic fatty acid content in leaves, hexadecatrienoic (16:3) and linolenic (18:3) acids are thought to protect the plants against chilling-related damage (Graham and Patterson, 1982). Recently, Kodama et al. (1994, 1995) have shown that tobacco plants over-expressing a chloroplast ω-3 fatty acid desaturase gene (the FAD7 gene) isolated from *Arabidopsis* contain increased levels of 16:3 and 18:3 fatty acids and acquire significant degree of chilling tolerance compared to the nontransformed chilling-sensitive wild-type tobacco.

The triple mutant *fad3 fad7 fad8* is deficient in the activity of the three genes responsible for the desaturation of 16:2 and 18:2 fatty acids, and this mutant is completely deficient in 16:3 and 18:3 fatty acids. When detached leaves of 25 °C grown wild-type and trienoic-deficient mutants (*fad3 fad7 fad8* triple mutant) were subjected to low-temperature treatment (20 °C to 5 °C) for 90 minutes, the quantum yield of steady-state electron transport (ΦII) was inhibited more in the mutant leaves than in the wild-type at all chilling temperatures. This inhibition was observed only if the measurement was made at the chilling temperature itself and was easily reversed if the leaves were warmed to room temperature during measurement. In contrast, the quantum yield of PS II electron transport (Fv/Fm) was unchanged during short-term cold treatment, suggesting that PS II is not inhibited by the chilling treatment (Routaboul and Browse, 1995).

Interestingly, in vivo chlorophyll fluorescence studies of low temperature (4 °C) grown plants also revealed a similar chilling-induced inhibition on the reducing side of PS II in the triple mutant as early as 5 days after transfer to 4 °C. However, this inhibition cannot be reversed by warming the leaves to 25 °C (Routaboul and Browse, 1995). In a segregating population of *fad3 fad7 fad8* heterozygotes that includes a range of individuals containing 0.3% (the triple mutant homozygote) to 62.0% (wild-type) trienoic fatty acids in their leaves, a clear relationship is observed to exist between the deficiency in trienoic fatty acid content and the susceptibility of photosynthetic quantum yield to long-term chilling inhibition (Fig. 5). Again, these results contrast with the situation in cyanobacteria in which dienoic and not trienoic fatty acids are important in conferring chilling tolerance (Wada et al., 1994; Murata and Wada, 1995; see also Chapter 13). Therefore, a specific role for trienoic fatty acids in mediating chilling tolerance is not a universal phenomenon in all oxygenic autotrophs.

IV. Effects of High Temperature on Photosynthetic and Growth Parameters

High temperature acclimation in higher plants is often accompanied by increased lipid saturation (Raison et al., 1982; Lynch and Thomson, 1984; Hugly et al., 1989; Kunst et al., 1989) which has also been shown to be associated with enhanced thermal tolerance of the photosynthetic apparatus in many plants (Thomas et al., 1986). The progressive disruption of the PS II complex by high temperature can be monitored by chlorophyll fluorescence and the threshold temperature at which Fo increase is initiated (Ti) has been widely used as a measure of thermostability of thylakoid function (Schreiber and Berry, 1977; Armond et al., 1980; Gounaris, 1984; Sundby et al., 1986). Thermostability of photosynthetic oxygen evolution in isolated thylakoids has also been used for this purpose (Hugly et al., 1989; Kunst et al., 1989a,b; Gombos et al., 1994).

Among the *Arabidopsis* lipid mutants in which

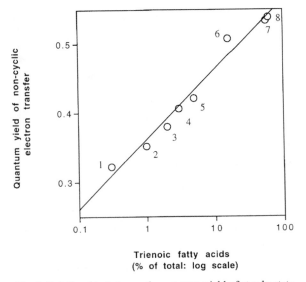

Fig. 5. Relationship between the quantum yield of steady-state photosynthesis measured after 30 days cold treatment at 4 °C and trienoic fatty acid content in leaves of different *Arabidopsis* mutants. Measurements were made at 25 °C under 100 μE m^{-2}s^{-1}. The numbered points represent (1) *fad3 fad7-2 fad8*; (2–5) individual plants of a F2 population from a cross of *fad3 fad7-1 fad8* and *fad3 fad7-2 fad8* (*fad7-1* is a leaky allele), (6) *fad7 fad8*, (7) *fad3* and (8) wild type (from Routaboul and Browse, 1995).

temperature profiles of Fo increase have been studied, the threshold for Fo increase was 2–3 °C higher than wild-type in three mutants, namely, *fad5* (Kunst et al., 1989a), *fad6* (Hugly et al., 1989) and *act1* (Kunst et al., 1989b). Isolated thylakoids of all these mutants also showed significantly higher thermostability of their photosynthetic oxygen evolution compared to the wild-type. The general picture that emerges from these studies is that a decrease in the levels of di- or tri-unsaturated fatty acids of thylakoid lipids causes an increase in thermal tolerance of thylakoid function. Moreover, thermostability of thylakoids is accompanied by better high-temperature growth rates in *fad5* (Kunst et al., 1989a) and *act1* (Kunst et al., 1989b) but not in *fad6*. When we compare the low temperature performance of these same mutants, two of them (*fad5* and *fad6*) are susceptible to chilling temperatures, whereas *act1* has no chilling phenotype. In general, these comparisons indicate that a decrease in fatty acid unsaturation results in an increase in thermal tolerance of thylakoid function whether measured by Fo increase or thermostability of electron transport. However, increased thermotolerance of thylakoids conferred by greater saturation of membrane fatty acids does not strictly correlate either with better growth performance at high temperature (e.g., *fad6*) or with increased chilling sensitivity of the plant (e.g., *act1*).

The role of lipids in stabilizing PS II at high temperature in these mutants remains difficult to explain. However, Tsvetkova et al. (1994, 1995) showed that thylakoids isolated from the *fad5* and *fad6* mutants had a propensity to form PS II freeze-fracture particle arrays similar to those observed in wild-type thylakoids rendered thermostable by the presence of tricine in the medium. Based on these results, they proposed that a decrease in the number of double bonds of membrane fatty acids facilitates greater interaction between adjacent PS II complexes, which is reflected in the formation of freeze-fracture arrays as well as greater thermostability of thylakoid function in these mutants (Tsevetkova et al., 1995). They also observed similar arrays in *fad7*, which is also deficient in fatty acid unsaturation of thylakoid lipids. However, thylakoids isolated from the mutant did not exhibit increased thermotolerance of Fo emission (McCourt et al., 1987). Array formation is also observed in the thylakoids of the *viridis* mutant of barley, which lacks functional LHC in its thylakoids (Simpson, 1983). All of the mutants that exhibit a propensity to form PS II freeze fracture arrays seem to share a deficiency in LHC. It is also true that only some treatments that confer thermotolerance to thylakoids in vitro induce the formation of PS II arrays.

In contrast to these results from *Arabidopsis* mutants is the finding of Murata and colleagues. They found that the elimination of polyenoic fatty acids does not affect thermal stability of a cyanobacterial mutant (Gombos et al., 1991; Wada et al., 1994). This is illustrative of the fact that the roles of similar species of lipids in thylakoids of evolutionarily distant taxa like cyanobacteria and angiosperms may, in fact, be very different.

The only common change in lipid composition which appears to exist in the *Arabidopsis* mutants exhibiting thermotolerance is that they contain more saturated lipids than the wild type. The three high-temperature mutants contain almost no 16:3 (<1%) and reduced amounts of 18:3. It is also notable that no clear relationship exists between chilling sensitivity and thermotolerance in them.

V. Alteration of Lipid Class Composition (Head Group Mutants)

Until recently, our understanding of the role of different lipid classes in photosynthesis was based on the correlation of thylakoid function with lipid changes brought about either by chloroplast development or by in vitro modification (see Chapter 8). Recently, a DGD-deficient mutant of *Arabidopsis* (*dgd1*; Dörmann et al., 1995) and an SL deficient mutant of *Chlamydomonas* (*hf2*; Sato et al., 1995a,b) have been isolated and promise to provide insight into the role of these lipids in photosynthesis. The DGD-deficient mutant and its photosynthetic characteristics have been discussed in an earlier section.

The SL-deficient *Chlamydomonas* mutant *hf2* shows a 17% decrease in growth rate, which is probably accounted for by an approximately 22% decrease in photosynthetic activity measured as CO_2-dependent oxygen evolution. The light-saturated rate of the PS II partial reaction (H_2O to PBQ) was 42% lower than that of the wild-type, whereas the rate of PS I electron transport ($DCIPH_2$ to MV) was relatively unaffected by the mutation. Since the spectroscopic properties and stoichiometry of pigment protein complexes in the green gels of mutant thylakoids are apparently not different from the wild-type, it has been proposed that this lipid may play a modulating role, rather than a major structural role, in PS II. Sato et al. (1995 a,b) have reported two unidentified small molecular weight polypeptides in the PS II complexes of the mutant which, when identified, may provide clues to the relationship between the changes in lipid composition and PS II function.

In contrast to the *hf2* mutant of *Chlamydomonas*, an SL-deficient mutant of the cyanobacterium *Synechocystis* PCC 7942, SY-SQDB isolated by Benning and coworkers, has photosynthetic and growth characteristics very similar to the wild-type cells. Although subtle differences were observed in the energy transfer from phycobilisomes to the PS II chlorophyll antenna, in variable chlorophyll fluorescence emission and in the oscillation pattern of flash induced oxygen evolution between the mutant and wild-type cyanobacteria, gross photosynthetic characteristics such as the stoichiometry of photosystems, and the light intensity dependence of photosynthetic oxygen evolution were nearly identical between the sulfolipid-deficient mutant and wild-type cyanobacteria (Gueler et al., 1996).

However, an important role for SL in cyanobacteria is evident under phosphate limited growth conditions where the grown of SL deficient mutant, SY-SQDB is reduced. In this respect the results are similar to those reported earlier for the anaerobic photosynthetic bacterium *Rhodobacter sphaeroides* (Benning et al., 1993). In both these systems, it appears that maintaining the net level of highly anionic lipid classes (either PG or SL) is crucial for maintaining maximum growth rates. The lack of SL only becomes important under conditions of phosphate starvation when PG availability for thylakoid biogenesis or function is restricted. Thus sulfolipid does not seem to have any crucial specific role in oxygenic photosynthesis per se. On the whole, the discrepancies between the apparent effects of SL deficiency in *Chlamydomonas* and cyanobacteria may reflect real differences in the role of sulfolipids in cyanobacterial and plastidic photosynthesis or they may reside in differences in the methodology of isolation and characterization of these mutants.

VI. Mutations in Other Lipid Components and Their Effects

Among the lipid mutants affecting photosynthesis, carotenoid and chlorophyll pigment mutants have been studied extensively using biochemical, biophysical and structural approaches (for review see Somerville, 1986). However, recently a unique and interesting set of chlorophyll mutants have been isolated by Nishimoto et al. (1992). These rice mutants accumulate chlorophyll a and b in which phytol is mostly replaced by its incompletely reduced precursors like geranylgeraniol (in M134), and tetrahydrogeranylgeraniol (in M249). Both the mutants had low Chl *a/b* ratios (2.2 and 2.3 compared to 3.0 in the wild type). However, in spite of the low Chl *a/b* ratio, the relative contents of different chlorophyll protein complexes in partially denaturing gels were found to be substantially similar to those of the wild type. The PS I complexes isolated from the mutant showed somewhat lower levels of redox-active P700 compared to wild type (Nishimoto et al., 1992). The further study of these unique mutants is likely to produce valuable information on the role of the phytyl group of chlorophyll in the organization of chlorophyll-protein complexes and photosynthetic function.

Acknowledgments

Work in the authors' laboratory reported in this chapter is supported by the U.S. National Science Foundation under grants DCB-910550 and IBN-9407902.

References

Allen JF (1992) Protein Phosphorylation in regulation of photosynthesis. Biochim Biophys Acta 1098: 275–335

Armond PA, Björkman O and Staehelin LA (1980) Dissociation of supramolecular complexes in chloroplast membranes. A manifestation of heat damage to the photosynthetic apparatus. Biochim Biophys Acta 601: 433–442

Asada K (1994) Mechanisms for scavenging reactive molecules generated in chloroplasts under light stress. In: Baker NR and Bowyer JR (eds) Photoinhibition of Photosynthesis, pp 129–142, Bios Scientific Publishers, Oxford

Bell CJ and Ecker JR (1994) Assignment of 30 microsatellite loci to the linkage map of *Arabidopsis*. Genomics 19: 137–144

Benning C and Somerville CR (1992) Isolation and genetic complementation of a sulfolipid-deficient mutant of *Rhodobacter sphaeroides*. J Bacteriol 174: 2352–2360

Benning C, Beatty JT, Prince CR and Somerville CR (1993) The sulfolipid sulfoquinovosyldiacylglycerol is not required for photosynthetic electron transport in *Rhodobacter sphaeroides* but enhances growth under phosphate limitation. Proc Natl Acad Sci USA 90: 1561–1565

Bose S, Vijayan P, Santhanam R and Kandasamy MK (1992) Inhibition of the cation-induced reversible changes in excitation energy distribution in thylakoids of BASF 13.338-grown plants. Biochim Biophys Acta 1098: 351–358

Browse J and Somerville CR (1991) Glycerolipid metabolism: Biochemistry and Regulation. Annu Rev Plant Physiol Plant Mol Biol 42: 467–506

Browse J and Somerville C (1994) Glycerolipids. In: Meyerowitz E and Somerville C (eds) *Arabidopsis*, pp 881–912. Cold Spring Harbor Press, New York

Browse J, McCourt PJ and Somerville CR (1985) A mutant of *Arabidopsis* lacking a chloroplast-specific lipid. Science 227: 763–765

Browse J, McCourt PJ and Somerville CR (1986a) Fatty acid composition of leaf lipids determined after combined digestion and fatty acid methyl ester formation from fresh tissue. Anal Biochem 152: 141–145

Browse J, Warwick N, Somerville CR and Slack CR (1986b) Fluxes through the procaryotic and eucaryotic pathway of lipid synthesis in the 16:3 plant *Arabidopsis thaliana*. Biochem J 235: 25–31

Browse J, McCourt P and Somerville CR (1986c) A mutant of *Arabidopsis* deficient in $C_{18:3}$ and $C_{16:3}$ leaf lipids. Plant Physiol 81: 751–756

Browse J, Kunst L, Anderson S, Hugly S and Somerville CR (1989) A mutant of *Arabidopsis* deficient in the chloroplast 16:1/18:1 desaturase. Plant Physiol 90: 522–529

Browse J, McConn M, James D and Miquel M (1993) Mutants of *Arabidopsis* deficient in the synthesis of α-linolenate.

Biochemical and genetic characterization of the endoplasmic reticulum linoleoyl desaturase. J Biol Chem 268: 16345–16351

Browse J, Miquel M, McConn M and Wu J (1994) *Arabidopsis* mutants and genetic approaches to the control of lipid composition. In: Cossins AR (ed) Temperature Adaptation of Biological Membranes, pp 141–154. Portland Press, London and Chapel Hill

Creelman RA and Mullet JE (1995) Jasmonic acid distribution and action in plants: Regulation during development and response to biotic and abiotic stress. Proc Natl Acad Sci USA 92: 4144–4119

Dörmann P, Hoffmann-Benning S, Balbo I and Benning C (1995) Isolation and characterization of an *Arabidopsis thaliana* mutant deficient in the thylakoid lipid digalactosyldiacylglycerol. Plant Cell 7: 1801–1810

Dubacq J-P and Tremolieres A (1983) Occurrence and function of phosphatiydylglycerol containing Δ3-*trans*-hexadecanoic acid in photosynthetic lamellae. Physiol Veg 21: 293–312

Garnier J, Wu B, Maroc J, Guyon D and Tremolieres A (1990) Restoration of both oligomeric form of the light harvesting antenna CP-II and fluorescence state II-state I transition by Δ3-*trans*-hexadecenoic acid containing phosphatidylglycerol in cell of a mutant of *Chlamydomonas reinhardtii*. Biochim Biophys Acta 1020: 153–162

Genty B, Briantais J-M and Baker N (1989) The relationship between the quantum yield of photosynthetic electron transport and quenching of chlorophyll fluorescence. Biochim Biophys Acta 990: 87–92

Gombos Z, Wada H and Murata N (1991) Direct evaluation of effects of fatty-acid unsaturation on the thermal properties of photosynthetic activities, as studied by mutation and transformation of *Synechocystis* PCC6803. Plant Cell Physiol 32: 205–211

Gombos Z, Wada H and Murata N (1992) Unsaturation of fatty acids in membrane lipids enhances tolerance of the cyanobacterium *Synechocystis* PCC6803 to low-temperature photoinhibition. Proc Natl Acad Sci 89: 9959–9963

Gombos Z, Wada H, Hideg E and Murata N (1994) The unsaturation of membrane lipids stabilizes photosynthesis against heat stress. Plant Physiol 104: 563–567

Gounaris K, Brain ARR, Quinn PJ and Williams WP (1984) Structural reorganisation of chloroplast thylakoid membranes in response to heat-stress. Biochim Biophys Acta 766: 199–208

Graf JA, Strasser RJ and Kull U (1984) State-1 State-2 transition influenced by herbicides which modify fatty acid composition in leaves. In: Sybesma C (ed) Advances in Photosynthesis Research, Vol. IV, pp 37–40. Martinus Nijhoff/Dr W Junk Publishers, The Hague

Graham D and Patterson BD (1982) Responses of plants to low, non-freezing temperatures: Proteins, metabolism, and acclimation. Annu Rev Plant Physiol 33: 347–372

Gueler S, Seeliger A, Haertel H, Renger G and Benning C (1996) A null mutant of *Synechococcus* sp. PCC7942 deficient in the sulfolipid sulfoquinovosyl diacylglycerol. J Biol Chem 271: 7501–7507

Hazel JR and Williams EE (1990) The role of alterations in membrane lipid composition in enabling physiological adaptation of organisms to their physical environment. Prog Lipid Research 29: 167–227

Heinz E (1993) Biosynthesis of polyunsaturated fatty acids. In: Moore T (ed) Lipid Metabolism in Plants, pp 33–90. CRC Press, Boca Raton

Heinz E and Roughan PG (1983) Similarities and differences in lipid metabolism of chloroplasts isolated from 18:3 and 16:3 plants. Plant Physiol 72: 273–279

Hugly S and Somerville C (1992) A role for membrane lipid polyunsaturation in chloroplast biogenesis at low temperature. Plant Physiol 99: 197–202

Hugly S, Kunst L, Browse J and Somerville C (1989) Enhanced thermal tolerance and altered chloroplast ultrastructure in a mutant of *Arabidopsis* deficient in lipid desaturation. Plant Physiol 90: 1134–1142

Huner NPA, Williams JP, Maissan EE, Myscich EG, Krol M, Laroche A and Singh J (1989) Low temperature-induced decrease in *trans*-Δ3-hexadecenoic acid content is correlated with freezing tolerance in cereals. Plant Physiol 89: 144–150

James DW and Dooner HK (1990) Isolation of EMS-induced mutants in *Arabidopsis* altered in seed fatty acid composition. Theor Appl Genet 80: 241–245

Kanervo E, Aro E-M and Murata N (1995) Low unsaturation level of thylakoid membrane lipids limits turnover of the D1 protein of Photosystem II at high irradiance. FEBS Lett 364: 239–242

Kitajima M and Butler WL (1975) Quenching of chlorophyll fluorescence and primary photochemistry in chloroplasts by dibromothymoquinone. Biochim Biophys Acta 376: 105–115

Kodama H, Hamada T, Horiguchi G, Nishimura M and Iba K (1994) Genetic enhancement of cold tolerance by expression of a gene for chloroplast ω-3 fatty acid desaturase in transgenic tobacco. Plant Physiol 105: 601–605

Kodama H, Horiguchi G, Nishiuchi T, Nishimura M and Iba K (1995) Fatty acid desaturation during chilling acclimation is one of the factors involved in conferring low-temperature tolerance to young tobacco leaves. Plant Physiol 107: 1177–1185

Kunst L, Browse J and Somerville C (1988) Altered regulation of lipid biosynthesis in a mutant of *Arabidopsis* deficient in chloroplast glycerol-3-phosphate acyltransferase activity. Proc Natl Acad Sci USA 85: 4143–4147

Kunst L, Browse J and Somerville C (1989a) Enhanced thermal tolerance in a mutant of *Arabidopsis* deficient in palmitic acid unsaturation. Plant Physiol 91: 401–408

Kunst L, Browse J and Somerville C (1989b) Altered chloroplast structure and function in a mutant of *Arabidopsis* deficient in plastid glycerol-3-phosphate acyltransferase activity. Plant Physiol 90: 846–853

Kunst L, Browse J and Somerville CR (1989c) A mutant of *Arabidopsis* deficient in desaturation of palmitic acid in leaf lipids. Plant Physiol 90: 943–947

Laskay G and Lehoczki E (1986) Correlation between linolenic-acid deficiency in chloroplast membrane lipids and decreasing photosynthetic activity in barley. Biochim Biophys Acta 849: 77–84

Leech RM, Walton CA and Baker NR (1985) Some effects of 4-chloro-5-(dimethylamine)-2-phenyl-3(2H) pyridazinone (SAN9785) on the development of chloroplast thylakoid membranes in *Hordeum vulgare* L. Planta 165: 277–283

Levine RP (1969) The analysis of photosynthesis using mutant strains of algae and higher plants. Annu Rev Plant Physiol 20: 523–540

Lightner J, Wu J and Browse J (1994a) A mutant of *Arabidopsis* with increased levels of stearic acid. Plant Physiol 106: 1443–1451

Lightner J, James Jr DW, Dooner HK and Browse J (1994b) Altered body morphology is caused by increased stearate levels in a mutant of *Arabidopsis*. Plant J 6: 401–412

Lynch D and Thompson GA (1984) Chloroplast phospholipid molecular species alterations during low temperature acclimation in *Dunaliella*. Plant Physiol 74: 198–203

Lyons J M (1973) Chilling injury in plants. Annu Rev Plant Physiol 24: 445–466

Mannan RN and Bose S (1985) Inhibition of photosynthetic electron transport in wheat chloroplast thylakoids by the herbicide BASF 13.338 (4-chloro-5-dimethylamino-2-phenyl-3(2H) pyridazinone). Indian J Biochem Biophys 22: 179–183

Martoglio B, Hofmann MW, Brunner J and Dobberstein B (1995) The protein-conducting channel in the membrane of the endoplasmic reticulum is open laterally toward the lipid bilayer. Cell 81: 207–214

Mathai JC, Sauna ZE and Sitaramam V (1993) Rate-limiting step in electron transport. Osmotically sensitive diffusion of quinones through voids in the bilayer. J Biol Chem 268: 15442–15454

McConn M (1994) Mutants of *Arabidopsis* deficient in polyunsaturated fatty acids. PhD thesis. Washington State University

McConn M and Browse J (1996) The critical requirement for linolenic acid is pollen development, not photosynthesis, in an *Arabidopsis* mutant. Plant Cell 8: 403–416

McConn M, Hugly S, Browse J and Somerville C (1994) A mutation at the *fad8* locus of *Arabidopsis* identifies a second chloroplast ω-3 desaturase. Plant Physiol 106: 1609–1614

McCourt PJ, Browse JA, Watson J, Arntzen C and Somerville CR (1985) Analysis of photosynthetic antenna function in a mutant of *Arabidopsis thaliana* (L.) lacking *trans*-hexadecenoic acid. Plant Physiol 78: 853–858

McCourt PJ, Kunst L, Browse J and Somerville CR (1987) The effects of reduced amounts of lipid unsaturation on chloroplast ultrastructure and photosynthesis in a mutant of *Arabidopsis*. Plant Physiol 84: 353–360

Meyerowitz EM (1994) Structure and organization of the *Arabidopsis thaliana* nuclear genome. In: Meyerowitz EM and Somerville CR (eds) *Arabidopsis*, pp 21–36. Cold Spring Harbor Laboratory Press, New York

Miles CD (1982) The use of mutations to probe photosynthesis in higher plants. In: Hallick R, Edelman M and Chua N (eds) Methods in Chloroplast Molecular Biology, pp 75–109. Elsevier, New York

Miquel M and Browse J (1992) *Arabidopsis* mutants deficient in polyunsaturated fatty acid synthesis: Biochemical and genetic characterization of a plant oleoyl phosphatidylcholine desaturase. J Biol Chem 267: 1502–1509

Miquel M, James D, Dooner H and Browse J (1993) *Arabidopsis* requires polyunsaturated lipids for low temperature survival. Proc Natl Acad Sci USA 90: 6208–6212

Moon BY, Higashi S-I, Gombos Z and Murata N (1995) Unsaturation of the membrane lipids of chloroplasts stabilizes the photosynthetic machinery against low-temperature photoinhibition in transgenic tobacco plants. Proc Natl Acad Sci 92: 6219–6223

Murata N (1983) Molecular species composition of phos-phatidylglycerols from chilling-sensitive and chilling-resistant plants. Plant Cell Physiol 24: 81–86

Murata N and Nishida I (1990) Lipids in relation to chilling sensitivity of plants. In: Wang CY (ed) Chilling Injury of Horticultural Crops, pp 181–199. CRC Press, Boca Raton

Murata N and Wada H (1995) Acyl lipid desaturases and their importance in the tolerance and acclimation to cold of cyanobacteria. Biochem J 308: 1–8

Murata N and Yamaya J (1984) Temperature-dependent phase behavior of phosphatidylglycerols from chilling-sensitive and chilling-resistant plants. Plant Physiol 74: 1016–1024

Murata N, Sato N, Takahashi N and Hamazaki Y (1982) Compositions and positional distributions of fatty acids in phospholipids from leaves of chilling-sensitive and chilling-resistant plants. Plant Cell Physiol 23: 1071–1079

Murata N, Ishizaki-Nishizawa O, Higashi S, Hayashi H, Tasaka Y and Nishida I (1992) Genetically engineered alteration in the chilling sensitivity of plants. Nature 356: 313–326

Murphy DJ (1986) The molecular organisation of the photosynthetic membranes of higher plants. Biochim Biophys Acta 864: 33–94

Nickelsen J and Rochaix J-D (1994) Regulation of the synthesis of the D1 and D2 proteins of Photosystem II in photoinhibition of photosynthesis. In: Baker NR and Bowyer JR (eds) Photoinhibition of Photosynthesis, pp 179–194. Bios Scientific Publishers Ltd., Oxford

Nishimoto E, Mikota T, Iwata N and Kobayashi Y (1992) Organization of pigment protein complexes in rice plants accumulating chlorophylls with incompletely reduced alcohol side chains. In: Murata N (ed) Research in Photosynthesis, Vol III, pp 59–62. Kluwer Academic Publishers, Dordrecht

Nußberger S, Dörr K, Wang DN and Kühlbrandt W (1993) Lipid-protein interactions in crystals of plant light-harvesting complex. J Mol Biol 234: 347–356

Raison JK (1973) The influence of temperature-induced phase changes on the kinetics of respiratory and other membrane associated enzyme systems. Bioenergetics 4: 285–309

Raison JK and Wright LC (1983) Thermal phase transitions in the polar lipids of plant membranes: Their induction by disaturated phospholipids and their possible relation to chilling injury. Biochim Biophys Acta 731: 69–74

Raison JK, Roberts JKM and Berry JA (1982) Correlations between the thermal stability of chloroplast (thylakoid) membranes and the composition and fluidity of their polar lipids upon acclimation of the higher plant *Nerium oleander* to growth temperature. Biochim Biophys Acta 688: 218–228

Roughan PG (1985) Phosphatidylglycerol and chilling sensitivity in plants. Plant Physiol 77: 740–746

Routaboul JM and Browse J (1995) A role for membrane lipid trienoic fatty acid in photosynthesis at low temperatures. In: Mathis P (ed) Photosynthesis: From Light to Biosphere, Vol IV, pp 861–864. Kluwer Academic Publishers, Dordrecht

Sato N, Tsuzuki M, Matsuda Y, Ehara T, Osafune T and Kawaguchi A (1995a) Isolation and characterization of mutants affected in lipid metabolism of *Chlamydomonas reinhardtii*. Eur J Biochem 230: 987–993

Sato N, Sonoike K, Tsuzsuki M and Kawaguchi A (1995b) Impaired PS II in a mutant of *Chlamydomonas reinhardtii* defective in sulfoquinovosyldiacylglycerol. Eur J Biochem 234: 16–23

Schreiber U and Berry JA (1977) Heat-induced changes in

chlorophyll fluorescence in intact leaves correlated with damage in the photosynthetic apparatus. Planta 136: 233–238

Siegenthaler PA, Rawyler A and Mayor J-P (1989) Structural and functional aspects of acyl lipids in thylakoid membranes from higher plants. In: Biacs PA, Gruiz K and Kremmer T (eds) Biological Role of Plant Lipids, pp 171–180. Plenum Press, Akadémiai Kiadó, New York, Budapest

Simpson DJ (1983) Freeze fracture studies on barley plastid membranes. VI. Location of the P700-chlorophyll *a*-protein 1. Eur J Cell Biol 31: 305–314

Somerville CR (1986) Analysis of photosynthesis with mutants of higher plants and algae. Annu Rev Plant Physiol 37: 467–507

Somerville C and Browse J (1991) Plant lipids: Metabolism mutants and membranes. Science 252: 80–87

Somerville CR and Meyerowitz EM (1994) Introduction. In: Meyerowitz EM and Somerville CR (eds) *Arabidopsis*, pp 1–6. Cold Spring Harbor Laboratory Press, New York

Sprague SG and Staehelin LA (1984) Effect of reconstitution method on the structural organization of isolated chloroplast membrane lipids. Biochim Biophys Acta 777: 306–433

Sundby C, Melis A, Mäenpää P and Andersson B (1986) Temperature-dependent changes in the antenna size of Photosystem II. Reversible conversion of Photosystem II$_\alpha$ to Photosystem II$_\beta$. Biochim Biophys Acta 851: 475–483

Thomas PG, Dominy PJ, Vigh L, Mansourian AR, Quinn PJH and Williams PW (1986) Increased thermal stability of pigment-protein complexes of pea thylakoids following catalytic hydrogeneration of membrane lipids. Biochim Biophys Acta 849: 131–140

Tsvetkova NM, Brain APR and Quinn PJ (1994) Structural characteristics of thylakoid membranes of *Arabidopsis* mutants deficient in lipid fatty acid desaturation. Biochim Biophys Acta 1192: 263–271

Tsvetkova NM, Apostolova EL, Brain APR, Williams PW and Quinn PJ (1995) Factors influencing PS II particle array formation in *Arabidopsis thaliana* chloroplasts and the relationship of such arrays to the thermostability of PS II. Biochim Biophys Acta 1228: 201–210

van Wijk KJ, Nilsson LO and Styring S (1994) Synthesis of reaction center proteins and reactivation of redox components during repair of Photosystem II after light induced inactivation. J Biol Chem 269: 28382–28392

Vijayan P, Routaboul J-M and Browse J (1997) A trienoic fatty acid deficient mutant of Arabidopsis is defective in recovery from photoinhibition at low temperatures. In: Williams JP (ed) Plant Lipid Metabolism, pp 203–205. Kluwer Academic Publishers, Dordrecht

von Wettstein D and Kristiansen K (1973) Stock list for nuclear gene mutants affecting the chloroplast. Barley Genet Newslett 3: 113–117

Wada H, Gombos Z and Murata N (1994) Contribution of membrane lipids to the ability of the photosynthetic machinery to tolerate temperature stress. Proc Natl Acad Sci USA 91: 4273–4277

Wang CY (1990) Chilling Injury of Horticultural Crops. CRC Press, Boca Raton

Webb MS and Green BR (1991) Biochemical and biophysical properties of thylakoid acyl lipids. Biochim Biophys Acta 1060: 133–158

Wolter FP, Schmidt R and Heinz E (1992) Chilling sensitivity of *Arabidopsis thaliana* with genetically engineered membrane lipids. EMBO J 11: 4685–4692

Wu J and Browse J (1995) Elevated levels of high-melting-point phosphatidylglycerols do not induce chilling sensitivity in an *Arabidopsis* mutant. Plant Cell 7: 17–27

Wu J, James DW, Dooner HK and Browse J. (1994) A mutant of *Arabidopsis* deficient in the elongation of palmitic acid. Plant Physiol 106: 143–150

Wu J, Lightner J, Warwick N and Browse J (1997) Low temperature damage and subsequent recovery of *fab1* mutant *Arabidopsis* exposed to 2 °C. Plant Physiol 113: 347–366

Chapter 15

Involvement of Chloroplast Lipids in the Reaction of Plants Submitted to Stress

John L. Harwood
School of Molecular and Medical Biosciences, University of Wales College of Cardiff,
P.O. Box 911, Cardiff, CF1 3US, United Kingdom

Summary

This chapter presents a comprehensive discussion of stresses which have been noted to affect chloroplast lipids or their metabolism. Where adaptation to the stress is possible, it is usually not clear if the alterations in lipid biochemistry are part of the adaptive response.

Light is needed for adequate rates of lipid formation and, in addition, may be absolutely required for the production of certain molecules. For temperature, most attention has focused on low temperature and chilling stresses. However, high stress temperatures may also affect lipid metabolism and function. Low temperature exposure may have a number of effects of which increased unsaturation is the most common change. This phenomenon has been relatively well studied compared to the effect of drought or salt stress.

Within the atmosphere ozone is known to increase lipid (and membrane) peroxidation. However, it also initiates a series of changes to the normal metabolism of chloroplast lipids which result in the marked disappearance of galactolipids and increase in triacylglycerol. In contrast, raised carbon dioxide (as predicted in the 'Greenhouse Effect') causes more subtle changes in lipid metabolism and cellular morphology.

Various xenobiotics have been found to alter plant lipids. Of these, two classes of herbicide (the substituted

P.-A. Siegenthaler and N. Murata (eds): Lipids in Photosynthesis: Structure, Function and Genetics, pp. 287–302.
© 1998 Kluwer Academic Publishers. Printed in The Netherlands.

pyridazinones and the Graminicides) have prominent effects. Although plants can usually survive pyridazinone exposure if only acyl lipids are affected, their ability to cope with subsequent stresses is often compromised. In contrast, the Graminicides are lethal at quite low concentrations with the susceptible *Poaceae*.

I. What is Stress?

Stress on living cells can be defined as those environmental conditions capable of causing potentially injurious effects (Levitt, 1980; Liljenberg, 1992). Clearly if the magnitude and/or duration of the external factor is sufficient, then a certain degree of strain can be developed in the organism. Resistance to stress can be inherent (genetically acquired during evolution) or may be developed following exposure to the stress. Surprisingly, it has sometimes been observed that exposure to one stress may confer subsequent tolerance to a different stress (Liljenberg, 1992).

II. Environmental Factors that Alter Chloroplast Lipids

A large number of environmental factors have been reported to produce effects—often detrimental and sometimes lethal—on plants. Many of these effects include alterations in chloroplast lipids and, by implication, in their metabolism. These factors range from 'natural' influences (such as light) over which we have little control through others (such as carbon dioxide) where Man's actions have some influence to xenobiotics which are introduced deliberately or which escape to enter the environment accidentally. As well as atmospheric carbon dioxide, Mankind may also be responsible for altering global temperatures and certainly contributes to the increase in nitrogen oxides, sulfur dioxide and ozone. With regard to the latter, it seems doubly unfortunate that the rise in atmospheric ozone at ground level to damaging concentrations is accompanied by a depletion of the protective stratospheric ozone layer— also by Mankind's industrial activities. A list of major environmental factors which can contribute to 'stress' are listed in Table 1. For general reviews on the effect of environmental factors on plant lipids and their metabolism, the reader is referred to previous publications (Harwood, 1984, 1994, 1995; Harwood et al., 1994).

III. Light

Because chloroplast membranes dominate the total membrane content of leaves and because these organelles have been shown to be major sources of lipid synthesis (see e.g. Harwood, 1989; Browse and Somerville, 1991), it is not surprising that light has been reported to produce many effects on their acyl lipids or the latter's metabolism. Some early experiments have been described in detail (Hitchcock and Nichols, 1971) and later summaries are by Harwood (1984, 1989).

Monocotyledons are a convenient system for such studies. This is for two major reasons. First, the leaves of monocotyledons expand well in the dark— thus providing plenty of experimental material. Second, because monocotyledonous leaves expand from their base, a gradient of cells in different stages of development is available up the length of the leaf. Therefore, it is possible to gain a large amount of information from single experiments and, moreover, although cell development is not in synchrony it is sufficiently close for information to be gained about the biogenesis and differentiation of, e.g. organelles which would be extremely difficult with dicotyledons. Studies with serial sections from a wide variety of monocotyledons have shown that there are increases in typical 'chloroplast' fatty acids, such as α-linolenate or *trans*-Δ3-hexadecenoate, and in the proportions of galactosylglycerides with increasing maturity of tissues (Hawke et al., 1971; Bolton and Harwood, 1978). In the same way, greening of etiolated tissue involves, following an initial degradation of (proplastid) lipids, an accumulation of thylakoid membrane components glycosyl-glycerides and phosphatidylglycerol (see Chapter 2). At the same time as chloroplast development is induced by light, the relative importance of major extrachloroplastic lipids (e.g. phosphatidylcholine, phosphatidylethanolamine) decreases markedly. These changes have been followed in a range of plants including the monocotyledons wheat, barley,

Abbreviations: ACP – acyl-carrier protein; DAG – diacylglycerol; DGDG – digalactosyl diacylglycerol; MGDG – monogalactosyl diacylglycerol; Tc – temperature for phase transition; TGA – triacylglycerol

Table 1. Important environmental stresses which can alter chloroplast lipids

Stress	Examples of Effects
Light—quantitative	High light may lead to oxidative damage of polyunsaturated membrane components. Light absolutely required for synthesis of *trans*-3-hexadecenoic acid and alters its levels once made.
Light—qualitative	Qualitative changes in lipids associated with alterations in chloroplast development.
Temperature—heat stress	Changes in membrane lipid phase structures and alterations in metabolism.
Temperature—cold stress	Induction of enzymes used for adaptation in tolerant plants or severe lipid catabolism in susceptible species.
Water deficiency (drought)	Decreases in phospho- and glycosyl-glycerides. Build up in triacylglycerol. Increased peroxidation of unsaturated fatty acids.
Soil constituents—salt stress	Increased degradation of chloroplast lipids.
Soil constituents—calcium	Changes in lipid metabolism depending on whether the plant is calcicole or calcifuge.
Atmospheric constituents—ozone	Causes oxidation of unsaturated acyl chains and degradation of galactolipids.
Atmospheric constituents—carbon dioxide	Elevated levels cause significant changes in metabolism, affecting chloroplast lipids.
Xenobiotics	Herbicides such as pyridazinones affect fatty acid desaturation while Graminicides kill grass species by preventing chloroplast fatty acid biosynthesis.

maize and oats as well as the dicotyledons pea and bean (see Harwood, 1989). Tevini (1977) has summarized a detailed series of studies comparing the effects of different light qualities on the greening process. In particular, it was noted that exposure to the light caused an initial drop in the percentage of monogalactosyldiacylglycerol because this component rose again, together with phosphatidylglycerol and digalactosyldiacylglycerol, during an 8–16 h period. These changes in lipid compositions accompanied marked differences in chloroplast morphology, as revealed by electron microscopy (Tevini, 1977). However, it was not clear from these simple experiments whether the light stress caused by unusual wavelength quality affected chloroplast development primarily and the lipid changes were merely a secondary consequence of this. To address this question, Gemmrich (1982) used cultures of *Ricinus communis* to try and differentiate light effects per se from those associated with light-induced chloroplast formation. In this case, the results showed that the light-induced changes in lipid composition were associated with thylakoid formation. In addition, light was suggested to stimulate the activities of certain enzymes such as acetyl-CoA carboxylase (see below).

Of all the chloroplast acyl constituents, *trans*-Δ3-hexadecenoate is the only one whose synthesis in

higher plants appears to be absolutely dependent on light. In contrast, some green algae have been shown to form this acid in the dark. The unusual nature of *trans*-Δ3-hexadecenoate and its ubiquitous presence in the phosphatidylglycerol of chloroplast thylakoids from all wild-type plants which have been examined strongly suggest an important function in photosynthesis. Originally, the acid was proposed to be needed for granal stacking (Tuquet et al., 1977). In fact, formation of *trans*-Δ3-hexadecenoate was correlated with exposure of etiolated maize to monochromatic wavelengths that also induced stacking (Tremolieres et al., 1979). However, experiments with different light treatments (Percival et al., 1979) or by the use of chlorophyll *b*-less mutants (Bolton et al., 1978; Selstam, 1980) have eliminated this possibility.

It is possible that *trans*-Δ3-hexadecenoate is involved in State 1/State 2 transitions which aid efficient light utilization and minimize stress damage. Following my suggestion of this (Harwood, 1984), we attempted to gain experimental evidence. We used redox-induced transitions (Telfer et al., 1983) and looked for changes in the ratio of palmitate to *trans*-Δ3-hexadecenoate. Small alterations were seen (H. Davies and J. L. Harwood, unpublished) which fitted with the hypothesis and were consistent with the apparent ability of plants to replace *trans*-Δ3-

hexadecenoate by palmitate in the dark (by reduction?) (Harwood and James, 1974) as well as with the well established desaturation of palmitate to hexadecenoate (Harwood, 1988). Tremolieres et al. (1992) have followed up my suggestions and made a detailed discussion of the possible role of trans-Δ3-hexadecenoate in influencing lipid-protein inter-actions during alterations in the distribution of light energy.

Connected with the above proposal, although originally suggested previously and independently, is the hypothesis that trans-Δ3-hexadecenoate may be involved in the oligomeric organization and stabilization of the light-harvesting chlorophyll-protein complex. This role was suggested by the presence of the acid in isolated complexes, its activity during in vitro reconstitution experiments and its theoretical ability to interact with chlorophyll (see Harwood, 1989, 1994). However, some experimental results failed to support (though do not necessarily disprove) the hypothesis. Thus, data from EPR measurements (Foley et al., 1988) and mutants of Arabidopsis which lack trans-Δ3-hexadecenoate but had normal oligomeric LHC-2 (McCourt et al., 1985) were not consistent with the proposal.

However, a series of experiments by Huner and colleagues have provided more evidence for an association of trans-Δ3-hexadecenoate with func-tional oligomeric LHC-2. They stressed rye with low temperatures and showed that cold-hardening growth was associated with a specific reduction in trans-Δ3-hexadecenoate content. This decrease was associated with a rise in monomeric LHC-2 at the expense of the oligomeric form (Huner et al., 1987). During purification of LHC-2 it was found that phosphatidyl-glycerol was tightly and specifically associated (Krol et al., 1988). Most significantly, it was observed that optimal levels of the oligomeric form were formed only at high trans-Δ3-hexadecenoate levels (Krupa et al., 1992) and it was concluded that the differences in the oligomeric form seen on cold-hardening could be fully accounted for by the difference in the trans-Δ3-hexadecenoate content of phosphatidylglycerol (Krupa et al., 1987). In order to gain further support for the idea that stabilization of the supramolecular organization of LHC-2 was specifically dependent upon molecular species of phosphatidylglycerol containing trans-Δ3-hexadecenoate, a series of lipid digestion and reconstitution experiments were carried out. Some results are shown in Table 2. It will be seen that (a) cold-hardening is associated with a relative decrease in the oligomeric form of LHC-2 (and trans-Δ3-hexadecenoate), (b) phospholipase diges-tion of phosphatidylglycerol also lowers the oligomeric form which can be restored, (c) by the addition of phosphatidylglycerol containing trans-Δ3-hexadecenoate but not by other lipids. Further developments on the function of trans-Δ3-hexa-decenoate are awaited with interest. A way forward may be provided by use of the deficient mutants of Arabidopsis (McCourt et al., 1985) which could be subjected to physiological stresses during its growth.

It has been known for many years that light markedly stimulates incorporation of ^{14}C-acetate into lipids. In mutants of Chlamydomonas, Picaud et al. (1990) found that much of this stimulation was via Photosystem I redox-driven control—presumably by NADPH (and reduced ferredoxin) supply. However, in many other situations it has been assumed that acetyl-CoA carboxylase activity was constraining the overall rate of lipid labeling and/or synthesis. In the first experiments to test this idea, Post-Beittenmiller and her colleagues looked at changes in the pattern of thioester intermediates during light-stimulated lipid synthesis in spinach leaves. Their results are summarized in Table 3 and were consistent with an important role for acetyl-CoA carboxylase in regulating carbon flux under the experimental conditions (see Post-Beittenmiller et al., 1991).

It is accepted nowadays (though perhaps still not widely known!) that it is misleading to talk of a regulatory step in a metabolic pathway. This is because regulation is shared by all the steps (and enzymes catalyzing them) as described by Kacser and Burns (1973) and Heinrich and Rapoport (1974). A measure of the importance of a given step can be gained by determining the flux control coefficient since the sum of these for each enzyme will be, from theory, equal to 1. Because of the availability of specific acetyl-CoA carboxylase inhibitors, we were able to determine flux control coefficients for this enzyme in the graminacious species, barley and maize (Table 4). The results demonstrated quantitatively the importance of acetyl-CoA carboxylase in light-stimulated leaf lipid synthesis. Thus, 45–61% of the total flux control resided in this single enzyme step despite the involvement of at least 20 other enzymes in the lipid labeling (Page et al., 1994).

Although most storage tissues are either non-photosynthetic or such that chloroplasts play a minor role in product accumulation, there are exceptions. In order to evaluate the relative supply of carbon

Table 2. Effect of lipid hydrolysis and reconstitution by phospholipids on the ratio of oligomeric to monomeric light harvesting complex II

| Hydrolysis | Addition | Ratio LHC oligomer/LHC monomer | |
		Control thylakoids	Cold-hardened thylakoids
None	None	3.51 ± 0.82	1.86 ± 0.63
PLC/PLA	None	0.81 ± 0.29	0.88 ± 0.17
PLC/PLA	18:2/16:0-PC	0.74 ± 0.19	0.77 ± 0.29
PLC/PLA	16:1-PG	1.92 ± 0.41	1.87 ± 0.30
PLC/PLA	18:1-PG	0.59 ± 0.05	0.59 ± 0.05

Data calculated from Krupa et al. (1992) where experimental details may be found.
Abbreviations : PLC, phospholipase C, PLA, phospholipase A_2; PC, phosphatidylcholine; PG, phosphatidylglycerol.
Winter rye was grown at cold-hardening (5/5 °C, day/night) and non-hardening (20/16 °C, day/night) temperatures. Light harvesting complex II (LHC II) was isolated from the two sets of plants by successive cation precipitation. Preparations of LHC II were digested with phospholipase C (from *Bacilllus cereus*) and phospholipase A_2 (from *Vipera russelli*) and then re-isolated. Suspensions of the LHC II preparations were then incubated with sonicated lipids for the reconstitution of lipid-depleted materials. In such experiments it was noted that addition of a galactolipase did not increase de-lipidation. Furthermore, galactolipid addition failed to restore the oligomer to monomer ratio.

Table 3. Changes in the levels of intermediates as the rate of fatty acid synthesis is varied in spinach chloroplasts

| Incubation condition | Fatty acid synthesis | Intermediate levels | | | |
		Acetyl-CoA	Acetyl-ACP	Malonyl-CoA	Malonyl-ACP
Dark	Low	Stable	Increased	Not detected	Not detected
Light	High	Stable	Decreased	Detected	Detected
Light and Triton	Higher	Small decrease	Decreased further	Increased	Increased

The rate of fatty acid synthesis by isolated spinach chloroplasts was varied by incubating the organelles in the dark, in the light or in the light with the addition of Triton X-100. The levels of thioester intermediates of fatty acid biosynthesis were estimated at the end of the incubations.
For further details and some actual values see Post-Beittenmiller et al. (1992).

from leaves or generated in situ for oil deposition in olive fruits, Sanchez (1994, 1995) used light stress as well as defoliation in his studies. His data showed that developing olives are able to fix atmospheric CO_2 into storage triacylglycerol at significant rates and he suggested that this process might contribute appreciably to overall oil deposition in the fruit.

IV. Temperature Effects on Photosynthetic Lipids

Changes in plant lipids during temperature adaptation (Harwood et al., 1994) and the use of *Arabidopsis* mutants in such studies (Browse et al., 1994) have been reviewed recently. Detailed discussions are made here and only the main points will be repeated here.

A number of changes in lipid content or in lipid metabolism have been noted for plants grown at

Table 4. Values for flux control coefficients for acetyl-CoA carboxylase during lipid synthesis in the light.

| Plant | Inhibitor used | |
	Fluazifop	Sethoxydim
Barley	0.61 ± 0.05	0.54 ± 0.17
Maize	0.59 ± 0.16	0.45 ± 0.13

Data show means ±S.D. for five independent experiments, each performed in triplicate. Taken from Page et al. (1994).
In the experiments, the incorporation of radioactivity from [14C]acetate into lipids was followed. This involves at least 20 enzymes. If flux control was shared equally amongst all these enzymes, then the latter would each have a flux control coefficient of 0.05. However, the data show that 50-60% of the control resides at a single step—that catalysed by acetyl-CoA carboxylase.

different temperatures. However, it is not always clear to what extent these represent true adaptations or are, merely, responses to environmental stress. Nevertheless, many of the lipid changes result in desirable alterations to the properties of membrane lipids so that normal functions can continue unimpaired. For example, fatty acid unsaturation may be increased following low temperature exposure, there may be relative alterations in the molecular species for a given lipid class and the proportions of different lipids may change. Opposite alterations may accompany high temperature exposure. These changes are consistent with functional requirements for bulk membrane lipids in terms of fluidity (see Harwood et al., 1994). Possible changes in membrane lipids observed during adaptation to temperature stress are summarized in Table 5.

A. Molecular Species Remodeling

Molecular species remodeling is the redistribution of acyl moieties within an individual lipid class. Because no new synthesis is involved, remodeling can occur very quickly and is regarded as an 'emergency' mechanism to allow survival while normal metabolism of lipids is re-adjusted. Remodeling can result in very significant changes in melting properties. For example, changing a lipid with 50% disaturated (e.g. distearoyl) and 50% di-monoenoic (e.g. dioleoyl) species to a 100% 1-sat, 2-monounsat mixture would ensure that the hydrophobic membrane phase was liquid at 3 °C instead of at 58 °C (for distearoyl-phosphatidylcholine).

Remodeling has been shown to occur following temperature stress in cyanobacteria and in the green, salt-tolerant alga *Dunaliella salina* (Lynch and Thompson 1984; Lynch et al., 1984). In the latter case, significant modifications occur rapidly in plasma membrane lipids as well as in chloroplast components such as phosphatidylglycerol.

Most studies of remodeling have focused on combinations of fatty acids in a given molecular species without determining their positional distribution. Surprisingly perhaps to those unfamiliar with membrane architecture, the latter can make a significant difference (see Harwood et al., 1994). However, I am not aware of any plant system which has been characterized in sufficient detail to date to determine whether, within a given lipid molecular species, the exchange of acyl positions takes place to a significant degree in vivo.

In plants the susceptibility to chilling damage has been proposed to be correlated with the percentage of molecular species of phosphatidylglycerol which

Table 5. Changes in lipids caused by low temperature stress

1. Molecular species remodelling.
2. Changes in fatty acid chain length.
3. Changes in the degree of unsaturation.
4. Alterations in the proportions of different lipid classes.
5. Increases in the ratio of membrane lipid to protein.

have a high Tc (i.e. 16:0/16:0- or 16:0/$t\Delta$3–16:1-species) (Murata, 1983). Because the specificity of the chloroplast acyl-ACP acyltransferases determine the combination of acyl groups on the glycerol backbone (Browse and Somerville, 1991), then it should be possible to alter molecular species by transferring genes for acyltransferases of different substrate selectivities. Thus, plants which were resistant to chilling and, hence, had unsaturated fatty acids at the *sn*-1 position of the backbone glycerol, possessed a glycerol 3-phosphate:acyl-ACP acyltransferase that showed good selectivity towards oleoyl-ACP. In contrast, the same enzyme from chilling-sensitive plants, such as squash, used palmitoyl-ACP well (Frentzen et al., 1987). Murata and his group have been able to successfully transfer genes for glycerol 3-phosphate acyltransferases into tobacco. Transfer of the gene from *Arabidopsis* (which gives rise to unsaturated moieties at the *sn*-1 position) conferred chilling resistance whereas that from squash (which has a strong substrate selectivity for palmitoyl-ACP) made the transgenic tobacco more chilling sensitive (Murata et al., 1992). This significant experiment showed that it was possible to engineer new plants which were capable of withstanding agriculturally-relevant environmental stresses. *Arabidopsis* plants have also been transformed with an *E. coli* gene which codes for a glycerol 3-phosphate acyltransferase which does not discriminate between 16:0- and 18:1-ACP substrates. Thus, transformed plants contained significant quantities of dipalmitoyl-phosphatidylglycerol (>50% total). Such plants suffered wilting and necrosis after exposure to low temperatures (Wolter et al., 1992) in support of Murata's hypothesis.

Other workers have also examined the significance of molecular species of phosphatidylglycerol in conferring chilling resistance or sensitivity. Although there is substantial agreement with Murata's hypothesis, there are a significant number of exceptions (see Norman et al., 1984; Roughan 1985; Bishop 1986). In particular, a recent *Arabidopsis* mutant has been described which, despite substantial amounts of dipalmitoyl-phosphatidylglycerol, does not show classic signs of chilling sensitivity (Wu and Browse, 1995). The mutant *Arabidopsis* plants showed failure to grow under long-term exposure to 2 °C. This sensitivity to chilling stress was thought to reflect features other than the possible phase separation of phosphatidylglycerols as proposed by Murata (1983). Some comments on these aspects of chilling sensitivity have been made previously (Harwood, 1989) and the reader is also referred to the discussion by Wu and Browse (1995).

The only other chloroplast lipid that could possibly give rise to chilling sensitivity is sulphoquinovosyldiacylglycerol. This sulfolipid also has a relatively high percentage of saturated acyl chains. Murata and Hoshi (1984) concluded that saturated molecular species enrichment was not correlated with chilling sensitivity. Kenrick and Bishop (1986) concurred and thought that the fatty acid content of sulfolipid was related to plant family rather than to chilling sensitivity. In only one case, *Carica papaya*, was the content of saturated fatty acids high enough to suggest that this lipid could undergo phase separations above 0 °C.

B. Acyl Chain-Length Control

Because longer-chain acids in a homologous series have higher transition temperatures than their shorter-chain equivalents, then control of acyl chain length provides another broad strategy for membrane fluidity control in response to temperature stress. However, because most organisms have a mixture of saturated and unsaturated fatty acids then many effects of chain-length will be masked by larger changes in unsaturation. Indeed, for many instances in higher plants or eukaryotic algae low temperature stress actually increases the average chain length which is hardly the expected adaptive response. However, this increase in chain length (see Harwood and Jones, 1989) is caused by the fact that the unsaturated fatty acids usually have longer chain lengths than the major saturated fatty acid in most plants and algae which is palmitate.

C. Low Temperature Stress Often Increases Unsaturation

The most commonly observed change in membrane lipids following low temperature stress is an increase in fatty acid unsaturation. There are good reasons for

this. First, desaturation of an existing (more saturated) fatty acid can be a very rapid response. Second, desaturation usually takes place using complex lipid substrates (Harwood, 1988) and, therefore, modifies membrane components with very little disruption of overall membrane structure. Thirdly, insertion of a single *cis*-double bond can make a very large change to the physical properties of a given lipid. The change in Tc is especially large for the first double bond but is also significant for the second. Therefore, most experiments on low temperature adaptation have concentrated on the desaturation of saturated or monounsaturated moieties.

A key aspect of the increase in unsaturation is the ability of a given plant or alga to raise the relative activity of its desaturase enzyme(s). Three main mechanisms have been proposed (Table 6). Availability of substrate, especially oxygen, was first proposed nearly twenty years ago (see Harwood et al., 1994). There is no doubt that, if oxygen is limited sufficiently, then it can change the pattern of membrane fatty acids. Thus, when Bligny et al. (1985) reduced oxygen from the normal 250μM to 10μM, then the average percentage of oleate and linoleate in their sycamore cell cultures changed from 3% and 49% to 45% and 22%, respectively. This pointed to a limitation in Δ12-desaturase activity at the low oxygen value. Since oxygen is more soluble at low temperatures, then chilling stress could increase unsaturation by raising substrate (oxygen) availability. However, Km values for purified enzymes (McKeon and Stumpf, 1982), detailed studies with non-plant organisms (e.g. Jones et al., 1993) as well as indirect measurements in plants (see Browse and Slack, 1991) suggest that oxygen solubility in vivo will not vary enough for its availability to be a major mechanism for controlling aerobic desaturation. Incidentally, one surprising result of the experiments of Bligny et al. (1985) was that large changes in membrane lipid unsaturation did not, necessarily, lead to expected changes in Arrhenius plots of

---

Table 6. Possible mechanisms to increase fatty acid desaturase activity

1. Substrate supply (e.g. oxygen) is increased.

2. New desaturase protein is produced (most probably through increased gene expression).

3. Changes in the membrane environment result in activation of pre-existing enzyme (e.g. by altering its conformation or by influencing the interaction between sub-units).

functional aspects. However, it is clear from data using other organisms that the correlation of physiological function with unsaturation and, hence, fluidity can be absolute in certain systems (Avery et al., 1995—see also Chapter 12.)

A second method for increasing unsaturation is by raising the amount of desaturase protein. This can be done in various ways. For example, there are at least three proteins needed for aerobic desaturases, the oxygen-binding component (cyanide-sensitive protein), a cytochrome and an oxido-reductase. Any one of these could be increased, although it would be expected that, of the three, the cyanide-sensitive protein would be subject to regulation and, indeed, this has been shown in *Tetrahymena* (Thompson and Nozawa, 1984). Raising protein amounts can be caused by increased transcription or translation, through decreased degradation or by a combination of these. Changes in gene expression have been detected following chilling stress in some organisms (see Chapter 13 and Jones et al., 1993) and interestingly oxygen appears to be necessary for this process in *Acanthamoeba* (S. Avery, A. J. Rutter, D. Lloyd and J. L. Harwood, unpublished). However, to date no higher plant system has been studied in detail and such experiments are urgently needed.

A third method of increasing the activity of a membrane-bound desaturase is through alterations in membrane 'fluidity'. It can be argued that chilling stress would instantly increase membrane order thus activating the desaturase and, hence, providing a self-regulatory mechanism for temperature compensation. Changes in membrane order could either work through conformational alteration of the desaturase proteins or by affecting their interaction. Moreover, supply of reduced equivalents could also be affected. Catalytic hydrogenation of membranes containing the $\Delta 12$-desaturase of *Candida lipolytica* have shown this to be a plausible mechanism (Horvath et al., 1991).

Although much of the above work has only been carried out in detail in non-plant organisms (but including Cyanobacteria; see Chapter 13) unsaturation certainly does play a role in chloroplast thylakoid function. The ability to manipulate unsaturation by genetic engineering (see Section VI.A below) demonstrated very clearly how phosphatidylglycerol unsaturation could be increased to render plants chilling-resistant.

D. Changes in Lipid Class Composition Caused by Temperature Stress

Alterations in the proportions of leaf lipid classes are often seen with temperature stress (and subsequent acclimation). This represents a slower method of adaptation compared to changes in unsaturation or molecular species (Harwood, 1991). Moreover, it must be borne in mind that in many experiments it is not possible to distinguish changes in the proportions of cellular membranes from alterations to the lipid composition of a given membrane. Therefore, it is important to analyze isolated membranes (or organelles) when looking for the latter effect and to look at subcellular morphology for the former.

Some pairs of lipids can be relatively easily interconverted. Among these are the galactosylglycerides, monogalactosyldiacylglycerol (MGDG) and digalactosyldiacylglycerol (DGDG). Conversion of, e.g. the distearoyl species of MGDG into DGDG would change the Tc from 82/69 to 41/47 °C for the I and II forms, respectively (Quinn and Williams, 1983). Thus, such a metabolic conversion should aid survival at lower growth temperatures. Indeed, low temperature growth has been reported to cause such a change in MGDG/DGDG proportions in some experiments with pea leaves but not in others (see Harwood et al., 1994). However, for the green alga *Dunaliella salina* a thorough analysis of the effect of temperature stress on metabolism included the

Table 7. An example of how low temperature stress in *Dunaliella salina* leads to alteration in the lipid class distribution of chloroplasts (see Lynch and Thompson, 1982)

Lipid Class	% Distribution			
	Whole Cells		Chloroplast fraction	
	30 °C	12 °C	30 °C	12 °C
Monogalactosyldiacylglycerol	67	55	66	58
Digalactosyldiacylglycerol	21	27	19	28
Sulphoquinovosyldiacylglycerol	12	18	15	14

isolation of subcellular organelles and membranes (Table 7). An increase in the ratio of DGDG/MGDG at low temperatures was clearly seen with both whole cells and the chloroplast fraction (Lynch and Thompson, 1982).

E. Low Temperatures May Increase the Ratio of Lipids to Proteins in Membranes

Another long term response to chilling temperatures may be an increase in the relative proportion of membrane lipids. Whether this reflects a true adaptive response or follows on from changes in carbon partitioning or difficulties in assembling protein complexes at low growth temperatures, remains to be established. Some data for experiments with isolated thylakoid membranes are shown in Table 8. The changed ratio was thought to reflect an increase in 'fluidity' (as measured by fluorescence spectroscopy) and, therefore, to be a useful adaptation (Millner et al., 1984). In fact, a general increase in membrane lipid to protein has been seen in a number of studies of low temperature effects—often in the absence of other expected (e.g. unsaturation) changes (Harwood, 1984, 1994).

V. Drought

The physiological responses of plants to drought stress are well documented (Levitt, 1980; Bradford and Hsiao, 1982). Among other responses, drought causes reduction in photosynthetic activity, closure of stomata, alterations of respiratory rates and alteration of osmotic potential. Several chloroplast enzymes have been reported to be inhibited and electron transport may be reduced (Boyer and Bowen, 1970; Mohanty and Boyer, 1976).

The actual response seen in plants to drought depends not only on the degree of water deficiency but also on the species or variety. Mild drought stress may reduce photosynthesis because of a limitation in carbon dioxide due to stromatal closure. However, high water deficits cause a direct inhibition of photosynthesis (Smirnoff, 1993). In the last few years there has been increasing evidence that activated forms of oxygen may have an important role in the effects seen during drought (Price et al., 1989; Leprince et al., 1990; Seel et al., 1991; Quartacci and Navari-Izzo, 1992). Thus, carbon dioxide limitation can give rise to over-excitation of the reaction centers. As a result, the photoprotective (antioxidant) mechanisms are exceeded and oxidative damage will occur (Smirnoff, 1993).

Quite a lot is known of the effects of drought on lipid metabolism in fully phototrophic plants (Liljenberg, 1992). In general, drought sensitive species (or cultivars) show larger changes in their lipid composition than do drought resistant plants. Furthermore, (as expected) greater water deficits cause larger changes in lipid metabolism than small deficiencies. Again, the usual observed effects are a decrease in the tissue content of phospho- and glycosylglycerides, although there are exceptions (see Liljenberg, 1992). At the same time, non-polar lipids such as triacylglycerols may often accumulate (Douglas and Paleg, 1981; Martin et al., 1986; Wilson et al., 1987; Navari-Izzo et al., 1990). This may be a mechanism for removing fatty acids which have been liberated during the hydrolysis of membrane lipids (Douglas and Paleg, 1981; Wilson et al., 1987; Navari-Izzo et al., 1990). In contrast to the above changes, fatty acid unsaturation is largely unaltered by drought stress (Liljenberg, 1992).

Increases in lipid hydrolysis leading first to a build-up in fatty acids and, subsequently, to triacylglycerol increases have been extensively recorded (e.g. Douglas and Paleg, 1981; Martin et

Table 8. Low temperature stress can increase the lipid to protein ratio of thylakoid membranes (from Harwood et al., 1994, with permission)

Plant	Growth temperature (°C)	Double bond index	Lipid/protein
Pea	6	5.11 ± 0.03	0.48 ± 0.05
	18	5.05 ± 0.02	0.42 ± 0.03
Lolium temulentum	6	5.14 ± 0.08	0.76 ± 0.10
	18	5.31 ± 0.06	0.66 ± 0.02
Spring oats	6	5.44 ± 0.02	0.75 ± 0.05
	18	5.42 ± 0.02	0.68 ± 0.09

al., 1986; Navari-Izzo et al., 1989). Undoubtedly connected with a rise in unesterified fatty acids, which form substrates for lipoxygenases (Vick and Zimmerman, 1987), is the fact that drought stress may induce lipid peroxidation (Price and Hendry, 1991). Since some plants seem able to mobilize defense strategies against radical-mediated lipid peroxidation (Smirnoff and Colombe, 1988) it is not surprising that drought tolerance has been related to the extent of peroxidation for a given species in the field (Seel et al., 1992). Increased production of the superoxide radical was observed in wheat chloroplasts prepared from drought-treated leaves (Price et al., 1989) and may take part in the formation of the highly damaging hydroxyl radical. Drought-related damage to plants, including the photosynthetic machinery, may be associated with the loss in specific activity of free radical processing enzymes, including superoxide dismutase, peroxidase and glutathione reductase (e.g. Leprince et al., 1990).

Further detailed aspects of drought-related changes in lipid biochemistry are reviewed by Liljenberg (1992) and the role of active oxygen in the overall response is summarized by Smirnoff (1993).

VI. Atmospheric Constituents

A. The 'Greenhouse Effect' and Raised Carbon Dioxide

The much published 'Greenhouse Effect' has been predicted to involve an approximate doubling of atmospheric CO_2 (650–700 ppm instead of 350 ppm) by the middle of the 21st century. In turn this is thought to bring about an increase in ambient temperature of about 4 °C. Thus, the 'Greenhouse Effect' involves a combination of the effects of these two major environmental factors as well as possible changes to other minor gases. We have studied the possible effects of raised CO_2 and temperature (in the predicted range) on plant lipid metabolism and have reported both quantitative and qualitative changes for a major cereal crop, wheat (Williams et al., 1994, 1995). Surprisingly, there are substantial changes in leaf metabolism caused by CO_2 alone. Of course, growth under raised CO_2 increases the availability of carbon and there is increased growth but, in addition, there are large changes in leaf morphology (Robertson et al., 1995). In several studies of lipid metabolism in wheat grown under elevated CO_2 and/or temperature, we have found substantial alterations in the partitioning of fatty acids between the 'eukaryotic' and 'prokaryotic' pathways (see Browse and Somerville, 1991). Since this modified metabolism leads to changes in both the total synthesis of chloroplast lipids and in their fatty acid quality (see Harwood 1995) it would be very interesting to measure the possible consequences of this on photosynthetic function. Indeed, significant actions of the 'Greenhouse Effect' on chloroplast development have already been noted (e.g. Robertson and Leech, 1995) although there have been no specific attempts to correlate these with membrane lipids.

B. Ozone Pollution

Among several atmospheric pollutants known to alter plant lipid metabolism (Heath, 1984) there has been increasing interest in ozone. Although there is a popular awareness of ozone layer depletion in the stratosphere, it is the local increase in ozone at ground levels which affects plants particularly. This increase is caused by photolysis of nitrogen dioxide (particularly from automobile exhaust gases) and occurs in a cyclic manner during the day. Ozone levels in excess of 50 ppb are thought to be damaging for some plants and episodal increases well in excess of this are detected regularly in Europe. The concentrations of ozone in particular world sites such as urban areas of Japan have been reported as high as 500 ppb (Sakaki et al., 1983).

There is no doubt that short-term exposure of plants to relatively high amounts (150–600 ppb) of ozone causes large decreases in the level of chloroplast galactolipids and in their fatty acid unsaturation (Fong and Heath 1981; Sakaki et al., 1985, 1994; Nouchi and Toyama 1988). These changes in thylakoid lipids have been rationalized as depicted in Table 9. An initial rise in non-esterified fatty acids (caused, presumably, by acyl hydrolase activation) leads to an increase in galactolipid:galactolipid galactosyltransferase (GGGT) activity, an inhibition of UDP-galactose:diacylglycerol (DAG) galactosyltransferase and an availability of substrates for acyl-CoA synthase. The DAG formed by GGGT can then be acylated by DAG acyltransferase to form triacylglycerol. These facts account for the loss of chloroplast thylakoid lipids, the rises in non-esterified fatty acids and triacylglycerols and the functional impairment of chloroplasts caused by ozone exposure (Wellburn, 1988).

Table 9. Sequence of events to explain ozone-induced changes in leaf lipids

1.   Acyl lipids $\xrightarrow{\text{acyl hydrolase}}$ NEFAs

2.   NEFAs $\xrightarrow{\text{lipoxygenase etc.}}$ oxidized products

3. $2 \times$ MGDG $\xrightarrow{\text{(NEFA)}\rightarrow\text{GGGT}}$ DAG + DGDG

4.   MGDG + UDP − gal $\xrightarrow{\text{(NEFA)}\rightarrow\text{gal. trans.}}$ DGDG + UDP

5.   NEFA + CoA + ATP $\xrightarrow{\text{Acyl-CoA synth}}$ Acyl − CoA + AMP + PPi

6.   DAG + Acyl − CoA $\xrightarrow{\text{DAG acyltrans.}}$ TAG

Ozone damage involves activation of acyl hydrolase activity at an early stage. The other reactions follow sequentially (see text for more details), with NEFAs activating reaction 3 and inhibiting reaction 4.
Abbreviations: GGGT – galactolipid:galactolipid galactosyltransferase; NEFAs – non-esterified (free) fatty acids.

In a more recent study using moderate (65 ppb) levels of ozone Sandelius and colleagues have shown different responses for 18:3 versus 16:3 plants, in monocotyledons versus dicotyledons and in young versus mature leaves. In sensitive plants there seemed to be a general decrease in chloroplast galactolipids without the change in MGDG/DGDG ratio reported for high ozone exposure (Sandelius et al., 1995). There was a clear shift in carbon flux from chloroplast to cytosolic membrane lipids. MGDG 18:3 seemed to be particularly affected (see also Carlsson et al., 1994). Fatty acid changes have also been noted for MGDG in Norway spruce exposed to ozone and these may cause the lack of frost resistance in such species (Wolfenden and Wellburn 1991; Wellburn et al., 1994).

VII. Salt Stress and the Effects of Minerals

Since lipids are important in membrane permeability, then such compounds could be predicted to play a role in the phenomenon of salt tolerance in plants. In a number of experiments, a salt-induced decrease in chloroplast lipids has been seen (Harwood, 1984). In contrast, other experiments have shown a decrease in all the acyl lipids of salt-stressed plants (Harwood, 1983). However, in the case of *Plantago* spp. it seems unlikely that salt tolerance is due to changes in phospholipid contents (Erdei et al., 1980). In sunflower leaves (as well as roots), NaCl stress led to a decrease in fatty acid unsaturation and in [^{14}C]oleate desaturation (Ellouze et al., 1982). Decreases in unsaturation seemed to be particularly marked in chloroplast lipids (Zarrouk et al., 1995) and, indeed, marked inhibition of galactolipid synthesis is seen

after salt stress of both olive (Zarrouk et al., 1995) and rape (Najine et al., 1995).

Whereas NaCl treatment has been shown to decrease the total lipid content (especially chloroplast) of leaves, $CaSO_4$ treatment caused an increase (Bettaieb et al., 1980). The effects of calcium have been studied most intensively by the use of plants which differ in their requirements for calcium. Such plants can be categorized as calcifuge (plants thriving on soils poor in $Ca(CO_3)_2$ e.g. *Lupinus luteus*) or calciole (plants thriving in soils rich in $Ca(CO_3)_2$ e.g. *Vicia faba*). Most experiments have related to root lipid metabolism (Harwood, 1989) but studies with calmodulin or calcium antagonists has revealed that these reagents can stimulate digalactosyldiacylglycerol breakdown in potato leaves (Piazza and Moreau, 1987).

For general discussions of salt effects on plant lipid metabolism the reader is referred to Harwood (1984, 1989) and Kuiper (1980, 1985).

VIII. Xenobiotics which Affect Chloroplast Lipids

A. General Remarks

Xenobiotics including non-herbicidal pesticides (e.g. insecticides, fungicides) may cause stress to plants through interactions with lipids in the cuticular surface layer or cellular membranes. Some have been reported to affect particular lipid metabolism pathways. In cases of chemical spillages, hydrocarbons are a good example which may affect plant health. Most noticeable within xenobiotics are, however, herbicides which act (primarily) on fatty acid synthesis.

B. Substituted Pyridazinones

The action of pyridazinones on lipid metabolism has been extensively reviewed (Harwood 1991a). Because there have been few recent experiments only a summary will be provided here.

Basically, substituted pyradizones can have inhibitory effects on fatty acid desaturation, carotenoid biosynthesis and electron transport. The relative effect of a given compound on these three parameters depends on the pyridazinone and also on the target system. Some herbicides have a predominant effect on fatty acid unsaturation with San 9785 being the best example. Sandoz 9785 inhibits ω3-desaturation on monogalactosyldiacylglycerol in higher plants. Although this inhibition may not in itself be toxic, it can render the treated plants susceptible to other environmental challenges, particularly chilling injury (Harwood, 1991a).

When treating susceptible plants with the substituted pyridazinones, San 9785 and San 6703, changes in chloroplast fatty acid compositions were correlated with ultrastructural alterations. It may be that these morphological alterations have consequences for further stress injury, although there has been little study of this aspect. For details of these structural changes and of individual experiments where substituted pyridazinones were tested on lipid metabolism see Harwood (1991a, b).

C. The Graminicides

The aryloxyphenoxypropionates (FOPS) and cyclohexanediones (DIMS) (Fig. 1) are herbicidal to grasses by virtue of their inhibition of acetyl-CoA carboxylase (Harwood, 1991a, b). The action of these herbicides has also been reviewed specifically (Walker et al., 1989; Gronwald, 1994). It is still a puzzle as to why the acetyl-CoA carboxylase from grasses is sensitive but not that from other monocotyledons. Moreover, grasses contain two isoforms of acetyl-CoA carboxylase both of which are multifunctional proteins over 200 KDa in molecular mass. The major isoform is located in the mesophyll chloroplasts and is herbicide-sensitive (Herbert et al., 1994; Egli et al., 1993). The minor isoform is probably located in the cytosol of epithelial cells where it would be used to supply malonyl-CoA for fatty acid elongation. It is herbicide-resistant yet, like the chloroplast isoform, shows Ter-Ter reaction kinetics (Herbert et al., 1996). By contrast, acetyl-CoA carboxylase from a resistant

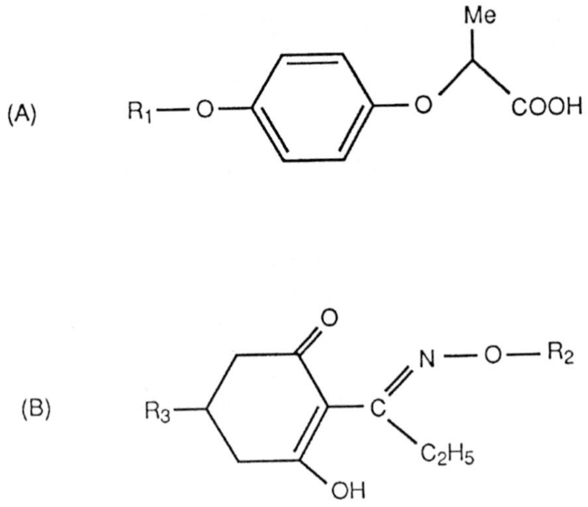

Fig. 1. Structure of the aryloxyphenoxypropionate (FOPS) and cyclohexanedione (DIMS) classes of Graminicides.
R_1 usually has halogen(s) attached to benzene, pyridine or quinoxalinyl moieties. R_2 such as ethyl. R_3 such as ethylthiopropyl or mesityl. Examples of aryloxyphenoxypropionates (FOPS) would be fluazifop, haloxyfop or quizalofop. Examples of cyclohexanediones (DIMS) would be Sethoxydim or Tralkoxydim.

grass, *Poa annua*, shows Ping-Pong kinetics, perhaps reflecting the insensitivity of this enzyme (Herbert et al., 1995). However, those insensitive acetyl-CoA carboxylases from graminaceae which have been examined show co-operativity in herbicide binding in contrast to sensitive enzymes. This property may be of significance for determining herbicide sensitivity (see Herbert et al., 1995).

Dicotyledons have been shown to have a different enzyme form for their acetyl-CoA carboxylase—in this case a multienzyme complex (see Alban et al., 1994). Therefore, it is not surprising that dicotyledons are insensitive to the FOPS and DIMS.

The FOPS and DIMS cause necrosis of target tissues, presumably because of the lack of normal membrane synthesis. Resistance may develop following exposure of grasses to these herbicides provided that they survive under stress. There have been a number of cases where grass weeds have become resistant and the usual reason is that an insensitive acetyl-CoA carboxylase is produced (Tardif and Powles, 1993). For example, resistant lines of an Italian ryegrass were developed by repeated stress of plants to field concentrations of diclofop-methyl (Gronwald et al., 1992). Since the graminicides have been shown to act on the carboxyltransferase partial reaction, one presumes that only a minor mutation of the part of the gene coding for that portion of acetyl-CoA carboxylase is necessary to confer resistance.

In contrast to acquired resistance, a few grasses are inheritantly resistant to particular graminicides or classes. This resistance may be due to metabolism of the herbicide (such as wheat and diclofop) or to the possession of a naturally insensitive acetyl-CoA carboxylase (e.g. in *Poa annua* or *Festuca rubra*). For a fuller discussion see Tardif and Powles (1993).

The action of the FOPS and the DIMS appears to be mainly effective on the rapidly-dividing meristematic tissue. In such cells, inhibition would prevent membrane biogenesis and, hence, it is unsurprising that necrosis rapidly develops. Death of the plant ensues within a week or so of herbicide application. For further details see references in Walker et al. (1989), Harwood (1991a) and Gronwald (1994).

D. Other Herbicides Which May Affect Chloroplast Lipids

Several other herbicides have been found to inhibit acyl lipid synthesis. These include chloroacetamides, triazines and diuron. Whether these inhibitory actions form any basis for their mode of action or are merely secondary consequences has not been determined. For detailed references the reader is referred to Harwood (1991a,b).

Acknowledgments

The author wishes to acknowledge support from AgrEvo, Rhone-Poulenc, Zeneca, the Natural Environment Research Council and the Biotechnology and Biological Sciences Research Council which has partly funded his research on the effects of stress on plant lipids.

References

Avery S, Lloyd D and Harwood JL (1995) Temperature-dependent changes in plasma membrane lipid order and the phagocytotic activity of the amoeba *Acanthamoeba castellanii* are closely correlated. Biochem J 312: 811–816

Bettaieb L, Gharsalli M and Cherif A (1980) Effect of sodium chloride and calcium sulphate on the lipid composition of sunflower lipids. In: Mazliak P, Benveniste P, Costes C and Douce R (eds) Biogenesis and Function of Plant Lipids, pp 243–247. Elsevier, Amsterdam

Bishop DG (1986) Chilling sensitivity in higher plants: The role of phosphatidylglycerol. Plant Cell Environ 9: 613–616

Bligny R, Rebeille F and Douce R (1985) O_2-triggered changes of membrane fatty acid composition have no effect on Arrhenius discontinuities of respiration in sycamore cells. J Biol Chem 260: 9166–9170

Bolton P and Harwood JL (1978) Lipid metabolism in green leaves of developing monocotyledons. Planta 139: 267–272

Bolton P, Wharfe J and Harwood JL (1978) The lipid composition of a mutant lacking chlorophyll *b*. Biochem J 174: 67–72

Boyer JS and Bowen BI (1970) Inhibition of oxygen evolution in chloroplasts isolated from leaves with low water potential. Plant Physiol 45: 612–615

Bradford KJ and Hsiao TC (1982) Physiological responses to water stress. In: Encyclopedia of Plant Physiology. New Series, Vol 12B Lang OL, Nobel PS, Osmund CB and Ziegler H (eds) pp 263–324. Springer-Verlag, Berlin

Browse J and Somerville C (1991) Glycerolipid synthesis: Biochemistry and regulation. Ann Rev Plant Physiol Plant Mol Biol 42: 467–506

Browse J, Miquel M, McConn M and Wu J (1994) *Arabidopsis* mutants and genetic approaches to the control of lipid composition. In: Cossins AR (ed) Temperature Adaptation of Biological Membranes, pp 141–154. Portland Press, London

Carlsson AS, Hellgren LI, Sellden G and Sandelius AS (1994) Effects of moderately enhanced levels of ozone on the acyl lipid composition of leaves of garden pea. Physiol Plant 91: 754–762

Douglas TJ and Paleg LG (1981) Lipid composition of *Zea mays* seedlings and water stress-induced changes. J Exp Bot 32: 499–508

Egli MA, Gengenbach BG, Gronwald JW, Somers DA and Wyse DL (1993) Characterisation of maize acetyl-CoA carboxylase. Plant Physiol 101: 499–506

Ellouze M, Gharsalli M and Cherif A (1982) The effect of sodium chloride on the biosynthesis of unsaturated fatty acids of sunflower plants. In: Wintermans JFGM and Kuiper PJC (eds) Biochemistry and Metabolism of Plant Lipids, pp 419–422. Elsevier, Amsterdam

Erdei L, Stuiver CEE and Kuiper PJC (1980) The effect of salinity on lipid composition and on activity of Ca^{2+} and Mg^{2+}-stimulated ATPases in salt-sensitive and salt-tolerant *Plantago* species. Physiol Plant Pathol 49: 315–319

Foley AA, Rowlands CC, Evans JC and Harwood JL (1988) Electron paramagnetic studies on copper phaeophytin in the presence and absence of phosphoglycerides. J Inorganic Biochem 32: 125–133

Fong F and Heath RL (1981) Lipid content in the primary leaf of bean after ozone fumigation. Z Pflanzenphysiol 104: 109–115

Frentzen M, Nishida I and Murata N (1987) Properties of the plastidial acyl-(acyl-carrier protein): Glycerol 3-phosphate acyltransferase from the chilling-sensitive plant squash. Plant Cell Physiol 28: 1195–1201

Gemmrich AR (1982) Effect of light on lipid metabolism of tissue cultures. In: Wintermans JFGM and Kuiper PJC (eds) Biochemistry and Metabolism of Plant Lipids, pp 213–216. Elsevier, Amsterdam

Gronwald J (1994) Herbicides inhibiting acetyl-CoA carboxylase. Biochem Soc Trans 22: 616–621

Gronwald JW, Eberlein CV, Betts KJ, Baerg RJ, Ehlke NJ and

Wyse DL (1992) Mechanism of diclofop resistance in an Italian ryegrass biotype. Pest Biochem Physiol 44: 126–139

Harwood JL (1983) Adaptive changes in the lipids of higher plant membranes. Biochem Soc Trans 11: 343–346

Harwood JL (1984) Effects of the environment on the acyl lipids of algae and higher plants. In: Siegenthaler P-A and Eichenberger W (eds) Structure, Function and Metabolism of Plant Lipids pp 543–550, Elsevier, Amsterdam

Harwood JL (1988) Fatty acid metabolism. Ann Rev Plant Physiol Plant Mol Biol 39: 101–138

Harwood JL (1989) Lipid metabolism. CRC Crit Revs Plant Sci 8: 1–43

Harwood JL (1991) Strategies for coping with low environmental temperatures. Trends Biochem Sci 16: 126–127

Harwood JL (1991a) Herbicides affecting chloroplast lipid synthesis. In: Baker NR and Percival MP (eds) Herbicides pp 209–246. Elsevier, Amsterdam

Harwood JL (1991b) Lipid synthesis. In: Kirkwood RC (ed) Target sites for herbicide action pp 57–94. Plenum Press, New York

Harwood JL (1994) Environmental factors affecting lipid metabolism. Prog Lipid Res 33: 193–202

Harwood JL (1995) Recent environmental concerns and lipid metabolism. In: Kader J-C and Mazliak P. (eds.) Plant Lipid Metabolism pp 561–568. Kluwer, Dordrecht

Harwood JL and James AT (1974) Metabolism of trans-3-hexadecenoic acid in broad bean. Europ J Biochem 50: 325–334

Harwood JL and Jones AL (1989) Lipid metabolism in algae. Adv Bot Res 16: 1–53

Harwood JL, Jones AL, Perry HJ, Rutter AJ, Smith KL and Williams M (1994) Changes in plant lipids during temperature adaptation. In: Cossins AR (ed) Temperature Adaptation of Biological Membranes, pp 107–118. Portland Press, London

Hawke JC, Rumsby MG and Leech RM (1974) Lipid biosynthesis in green leaves of developing maize. Plant Physiol 53: 555–561

Heinrich R and Rapoport TA (1974) A linear steady-state treatment of enzymatic chains. Europ J Biochem 42: 97–105

Herbert D, Alban C, Cole DJ, Pallett KE and Harwood JL (1994) Characterisation of two forms of acetyl-CoA carboxylase from maize leaves. Biochem Soc Trans 22: 261

Herbert D, Harwood JL, Cole DJ and Pallett KE (1995) Characteristics of aryloxyphenoxypropionate herbicide interactions with acetyl-CoA carboxylases of different graminicide sensitivities. Brighton Crop Protection Conference – Weeds 1995: 387–392

Herbert D, Price LJ, Alban C, Dehaye L, Job D, Cole DJ, Pallett KE and Harwood JL (1996) Kinetic studies on two isoforms of acetyl-CoA carboxylase from maize leaves. Biochem J 318: 997–1006

Horvath I, Torok Z, Vigh L and Kates M (1991) Lipid hydrogenation induces elevated 18:1-CoA desaturase activity in Candida lipolytica microsomes. Biochim Biophys Acta 1085: 126–130.

Huner NPA, Kool M, Williams JP, Maissan E, Low P, Roberts D and Thompson JE (1987) A low temperature induced decrease in 3-trans-hexadecenoic acid content and its influence on LHCII organisation. Plant Physiol 84: 12–18

Jones AL, Lloyd D and Harwood JL (1993) Rapid induction of microsomal $\Delta 12(\omega 6)$-desaturase activity in chilled Acanth-

amoeba castellanii. Biochem J 296: 183–188

Kacser HB and Burns JA (1973) The control of flux. Sym Soc Exp Biol 27: 65–104

Kenrick JR and Bishop DG (1986) The fatty acid composition of phosphatidylglycerol and sulphoquinovosyldiacylglycerol of higher plants in relation to chilling sensitivity. Plant Physiol 81: 946–949

Krol K, Huner NPA, Williams JP and Maissan E (1988) Chloroplast biogenesis at cold hardening temperatures. Kinetics of trans-3-hexadecenoic acid accumulation and the assembly of LHCll. Photosynth Res 15: 115–132

Krupa Z, Huner NPA, Williams JP, Maissan E and James DR (1987) Development of cold hardening temperatures. The structure and composition of purified rye light harvesting complex ll. Plant Physiol 84: 19–24

Krupa Z, Williams JP, Khan MU and Huner WPA (1992) The role of acyl lipids in reconstitution of lipid-depleted light-harvesting complex II from cold-hardened and non-hardened rape. Plant Physiol 100: 931–938

Kuiper PJC (1980) Lipid metabolism as a factor in environmental adaptation. In: Mazliak P, Benveniste P, Costes C and Douce R (eds) Biogenesis and Function of Plant Lipids, pp 169–176. Elsevier, Amsterdam

Kuiper PJC (1985) Environmental changes and lipid metabolism in higher plants. Physiol Plant 64: 118–122

Leprince O, Deltour R, Thorpe PC, Atherton NM and Hendry GAF (1990) The role of free radicals and radical processing systems in loss of desiccation tolerance in germinating maize. New Phytol 116: 573–580

Levitt J (1980) Water, radiation, salt and other stress. Responses of plants to environmental stresses, second edition, Vol II, Academic Press, New York

Liljenberg C (1992) The effects of water deficit stress on plant membrane lipids. Prog Lipid Res 31: 335–343

Lynch DV and Thompson GA (1982) Low temperature induced alterations in the chloroplast and microsomal membranes of Dunaliella salina. Plant Physiol 69: 1369–1375

Lynch D and Thompson GA (1984) Microsomal phospholipid molecular species alterations during low temperature acclimation in Dunaliella. Plant Physiol 74: 193–197

Lynch D, Norman HA and Thompson GA (1984) Changes in membrane lipid molecular species during acclimation to low temperature by Dunaliella salina. In: Siegenthaler P-A and Eichenberger WA (eds) Structure, Function and Metabolism of Plant Lipids, pp 567–570. Elsevier, Amsterdam

McCourt P, Browse J, Watson J, Arntzen CJ and Somerville CR (1985) Analysis of photosynthetic antenna function in a mutant of Arabidopsis thaliana lacking trans-hexadecenoic acid. Plant Physiol 78: 853–858

McKeon T and Stumpf PK (1982) Purification and characterisation of the stearoyl-acyl carrier protein desaturase and the acyl-acyl carrier protein thioesterase from maturing seeds of safflower. J Biol Chem 257: 12141–12147

Martin BA, Schoper JB and Rinne RW (1986) Changes in soybean glycerolipids in response to water stress. Plant Physiol 81: 798–801

Mohanty P and Boyer JS (1976) Chloroplast response to low leaf potentials. Plant Physiol 57: 704–709

Murata N (1983) Molecular species composition of phosphatidylglycerols from chilling-sensitive and chilling-resistant plants. Plant Cell Physiol 24: 81–86

Murata N and Hoshi H (1984) Sulphoquinovosyl diacylglycerols in chilling-sensitive and chilling-resistant plants. Plant Cell Physiol 25: 1241–1245

Murata N, Ishizaki-Nishizawa O, Higashi S, Hayashi H, Tasaka Y and Nishida I (1992) Genetically engineered alteration in the chilling sensitivity of plants. Nature 356: 710–712

Najine F, Marzouk B and Cherif A (1995) Sodium chloride effect on the evolution of fatty acid composition in developing rape seedlings. In: Kader J-C and Mazliak P (eds) Plant Lipid Metabolism, pp 435–437. Kluwer, Dordrecht

Navari-Izzo F, Vangioni N and Quartacci MF (1990) Lipids of soybean and sunflower seedlings grown under drought conditions. Phytochemistry 29: 2119–2123

Norman HA, McMillan C and Thompson GA (1984) Phosphatidylglycerol molecular species in chilling-sensitive and chilling-resistant populations of *Avicennia germinans*. Plant Cell Physiol 25: 1437–1444

Nouchi I and Toyama S (1988) Effects of ozone and peroxyacetyl nitrate on polar lipids and fatty acids in leaves of morning glory and kidney bean. Plant Physiol 87: 638–646

Page RA, Okada S and Harwood JL (1994) Acetyl-CoA carboxylase exerts strong flux control over lipid synthesis in plants. Biochim Biophys Acta 1210: 369–372

Picaud A, Creach A and Tremolieres A (1990) Light-stimulated fatty acid synthesis in *Chlamydomonas* whole cells. In: Quinn PJ and Harwood JL (eds) Plant Lipid Biochemistry, Structure and Utilization, pp 393–395. Portland Press, London

Post-Beittenmiller D, Jaworski JG and Ohlrogge JB (1991) In vivo pools of free and acylated acyl carrier proteins in spinach. J Biol Chem 266: 1858–1865.

Post-Beittenmiller D, Roughan PG and Ohlrogge JB (1992) Regulation of plant fatty acid biosynthesis. Analysis of acyl-CoA and acyl-ACP substrate pools in spinach and pea chloroplasts. Plant Physiol 100: 923–930

Price AH and Hendry GAF (1991) Iron-catalysed oxygen radical formation and its possible contribution to drought damage in nine native grasses and three cereals. Plant Cell Environ 14: 477–484

Price AH, Atherton N and Hendry GAF (1989) Plants under drought stress generate active oxygen. Free Rad Res Commun 8: 61–66

Quartacci MF and Navari-Izzo F (1992) Water stress and free radical mediated changes in sunflower seedlings. J Plant Physiol 139: 621–623

Quinn PJ and Williams WP (1983) The structural role of lipids in photosynthetic membranes. Biochim Biophys Acta 737: 223–266

Robertson EJ and Leech RM (1995) Significant changes in cell and chloroplast development in young wheat leaves grown in elevated CO_2. Plant Physiol 107: 63–71

Robertson EJ, Williams M, Harwood JL, Lindsay JG, Leaver CJ and Leech RM (1995) Mitochondria increase three-fold and mitochondrial proteins and lipids change dramatically in post-meristematic cells in young wheat leaves grown in elevated CO_2. Plant Physiol 108: 469–474

Roughan PG (1985) Phosphatidylglycerol and chilling sensitivity in plants. Plant Physiol 77: 740–746

Sakaki T, Kondo N and Sugahara K (1983) Breakdown of photosynthetic pigments and lipids in spinach leaves with ozone fumigation: Role of active oxygens. Physiol Plant 59: 28–34

Sakaki T, Ohnishi J, Kondo N and Yamada M (1985) Polar and neutral lipid changes in spinach leaves with ozone fumigation. Plant Cell Physiol 26: 253–262

Sakaki T, Tanaka K and Yamada M (1994) General metabolic changes in leaf lipids in response to ozone. Plant Cell Physiol 35: 53–62

Sanchez J (1994) Lipid photosynthesis in olive fruit. Prog Lipid Res 33: 97–104

Sanchez J (1995) Olive oil biogenesis. Contribution of fruit photosynthesis. In: Kader J-C and Mazliak P (eds) Plant Lipid Metabolism, pp 564–566. Kluwer, Dordrecht

Sandelius AS, Carlsson AS, Pleijel H, Hellgren LI, Wallin G and Sellden G (1995) The leaf acyl lipid composition of plants exposed to moderately enhanced levels of ozone: Species, age and dose dependence. In: Kader J-C and Mazliak P (eds) Plant Lipid Metabolism, pp 459–461. Kluwer, Dordrecht

Seel WE, Hendry GAF and Lee JA (1992) The combined effects of desiccation and irradiance on mosses from xeric and hydric habitats. J Exptl Bot 43: 1023–1030

Seel WS, Hendry G, Atherton N and Lee J (1991) Radical formation and accumulation in vivo, in desiccation tolerant and intolerant mosses. Free Rad Res Commun 15: 133–141

Selstam E (1980) Lipids, pigments, light-harvesting chlorophyll protein complex and structure of a virescent mutant of maize. In: Mazliak P, Benveniste P, Costes C, Douce R (eds) Biogenesis and Function of Plant Lipids, pp 379–383. Elsevier, Amsterdam

Smirnoff N (1993) The role of active oxygen in the response of plants to water deficit and desiccation (Tansley Review No 52). New Phytol 125: 27–58

Smirnoff H and Columbe SV (1988) Drought influences the activity of enzymes of the chloroplast hydrogen peroxide scavenging system. J. Exptl Bot 39: 1097–1108

Tardif FJ and Powles SB (1993) Target site-based resistance to herbicides inhibiting acetyl-CoA carboxylase. Proc Brighton Crop Protec Conf – Weeds, 533–540

Telfer A, Hodges M and Barber J (1983) Analysis of chlorophyll fluorescence induction curves in the presence of dichlorophenyldimethylurea as a function of magnesium concentration and NADPH-activated light-harvesting chlorophyll *a/b* protein phosphorylation. Biochim Biophys Acta 724: 167–175

Tevini M (1977) Light, function and lipids during plastid development. In: Tevini M and Lichtenthaler HK (eds) Lipids and Lipid Polymers in Higher Plants, pp 121–145. Springer-Verlag, Berlin

Tremolieres A, Guillot-Salomon TD, Dubacq JP, Jaques R, Mazliak P and Signol M (1979) The effects of monochromatic light on α-linolenic and *trans*-3-hexadecenoic acids biosynthesis and its correlation to the development of the plastid lamellar system. Physiol Plant 45: 429–436

Tremolieres A, Garnier J and Dubertret G (1992) Lipid protein interactions in relation to light energy distribution in photosynthetic membrane of eukaryotic organisms. In: Cherif A, Miled-Daoud DB, Marzouk B, Smaoui A and Zarrouk M (eds) Metabolism, Structure and Utilization of Plant Lipids, pp 289–292. Centre National Pedagogique, Tunis

Tuquet C, Guillot-Salomon TD, de Lubac MF and Signol M (1977) Granum formation and the presence of phosphatidylglycerol containing *trans*-3-hexadecenoic acid. Plant Sci Lett 8: 59–64

Vick BA and Zimmerman DC (1987) Oxidative systems for modification of fatty acids: The lipoxygenase pathway. In:

Stumpf PF and Conn EE (eds) The Biochemistry of Plants, Vol 9, pp 53–90. Academic Press, New York

Walker KA, Ridley SM, Lewis T and Harwood JL (1989) Action of aryloxy-phenoxy carboxylic acids on lipid metabolism. Rev Weed Sci 4: 71–84

Wellburn AR (1988) Air Pollution and Acid Rain. Longman Scientific, Harlow

Wellburn AR, Robinson DC, Thomson A and Leith ID (1994) Influence of episodes of summer ozone on Δ^5 and Δ^9 fatty acids in autumnal lipids of Norway spruce. New Phytol 127: 355–361

Williams M, Shewry PR and Harwood JL (1994) The influence of the 'greenhouse effect' on wheat grain lipids. J Exptl Bot 45: 1379–1385

Williams M, Shewry PR, Lawlor DW and Harwood JL (1995) The effects of elevated temperature and atmospheric carbon dioxide concentrations on the quality of grain lipids in wheat grown at two levels of nitrogen. Plant Cell Environ 18: 999–1009

Wilson RF, Burke JJ and Quisenberry JE (1987) Plant morphological and biochemical responses to field water deficits. Plant Physiol 84: 251–254

Wolfenden J and Wellburn AR (1991) Effects of summer ozone on membrane lipid composition during subsequent frost hardening in Norway spruce. New Phytol 118: 323–329

Wolter FP, Schmidt R and Heinz E (1992) Chilling sensitivity of *Arabidopsis thaliana* with genetically engineered membrane lipids. EMBO J 11: 4685–4692

Wu J and Browse J (1995) Elevated levels of high-melting-point phosphatidylglycerols do not induce chilling sensitivity in an *Arabidopsis* mutant. The Plant Cell 7: 17–27

Zarrouk M, Seqqat-Dakhma W and Cherif A (1995) Salt stress effect on polar lipid metabolism in olive leaves. In: Kader J-C and Mazliak P (eds) Plant Lipid Metabolism, pp 429–431. Kluwer, Dordrecht

Index

Advances in Photosynthesis

Series editor: Govindjee, University of Illinois, Urbana, Illinois, U.S.A.

KLUWER ACADEMIC PUBLISHERS – DORDRECHT / BOSTON / LONDON

Advances in Photosynthesis

Series editor: Govindjee, University of Illinois, Urbana, Illinois, U.S.A.

1. D.A. Bryant (ed.): The Molecular Biology of Cyanobacteria. 1994
   ISBN Hb: 0-7923-2179-6; Pb: 0-7923-3273-9
2. R.E. Blankenship, M.T. Madigan and C.E. Bauer (eds.): Anoxygenic Photosynthetic Bacteria. 1995
   ISBN Hb: 0-7923-3681-X; Pb: 0-7923-3682-8
3. J. Amesz and A.J. Hoff (eds.): Biophysical Techniques in Photosynthesis. 1996
   ISBN 0-7923-3642-9
4. D.R. Ort and C.F. Yocum (eds.) (Associate ed.): Oxygenic Photosynthesis: The Light Reactions. 1996
   ISBN Hb: 0-7923-3683-6; Pb: 0-7923-3684-4
5. N.R. Baker (ed.): Photosynthesis and the Environment. 1996
   ISBN 0-7923-4316-8
6. R.A. Sadava and P.J. Mullet (eds.) ... Photosynthesis: Structure,
   Function and Genetics. 1998
7. J.-D. Rochaix, M. Goldschmidt-Clermont and S. Merchant (eds.): The Molecular
   Biology of Chloroplasts and Mitochondria in Chlamydomonas. 1998
   ISBN 0-7923-3744-1

KLUWER ACADEMIC PUBLISHERS – DORDRECHT / BOSTON / LONDON